Signals and Systems with MATLAB®

Won Y. Yang · Tae G. Chang · Ik H. Song ·
Yong S. Cho · Jun Heo · Won G. Jeon ·
Jeong W. Lee · Jae K. Kim

Signals and Systems with MATLAB®

Springer

Limits of Liability and Disclaimer of Warranty of Software

The reader is expressly warned to consider and adopt all safety precautions that might be indicated by the activities herein and to avoid all potentials hazards. By following the instructions contained herein, the reader willingly assumes all risks in connection with such instructions.

The authors and publisher of this book have used their best efforts and knowledge in preparing this book as well as developing the computer programs in it. However, they make no warranty of any kind, expressed or implied, with regard to the programs or the documentation contained in this book. Accordingly, they shall not be liable for any incidental or consequential damages in connection with, or arising out of, the readers' use of, or reliance upon, the materials in this book.

MATLAB® and Simulink® are registered trademarks of The MathWorks, Inc.

ISBN 978-3-540-92953-6 e-ISBN 978-3-540-92954-3
DOI 10.1007/978-3-540-92954-3
Springer Dordrecht Heidelberg London New York

Library of Congress Control Number: 2009920196

Cover design: WMXDesign GmbH, Heidelberg

Springer is a part of Springer Science+Business Media (www.springer.com)

Yang et. al.: Signals and Systems with MATLAB®

First Indian Reprint 2011
Second Indian Reprint 2014

ISBN 978-81-322-0395-7

This edition is published by Springer (India) Private Limited, (a part of Springer Science+Business Media), Registered Office: 7th Floor, Vijaya Building, 17 Barakhamba Road, New Delhi – 110 001, India.

Printed in India by Rakmo Press Pvt. Ltd., Okhla.

To our parents and families
who love and support us
and
to our teachers and students
who enriched our knowledge

Preface

This book is primarily intended for junior-level students who take the courses on 'signals and systems'. It may be useful as a reference text for practicing engineers and scientists who want to acquire some of the concepts required for signal processing. The readers are assumed to know the basics about linear algebra, calculus (on complex numbers, differentiation, and integration), differential equations, Laplace transform, and MATLAB®. Some knowledge about circuit systems will be helpful.

Knowledge in signals and systems is crucial to students majoring in Electrical Engineering. The main objective of this book is to make the readers prepared for studying advanced subjects on signal processing, communication, and control by covering from the basic concepts of signals and systems to manual-like introductions of how to use the MATLAB® and Simulink® tools for signal analysis and filter design. The features of this book can be summarized as follows:

1. It not only introduces the four Fourier analysis tools, CTFS (continuous-time Fourier series), CTFT (continuous-time Fourier transform), DFT (discrete-time Fourier transform), and DTFS (discrete-time Fourier series), but also illuminates the relationship among them so that the readers can realize why only the DFT of the four tools is used for practical spectral analysis and why/how it differs from the other ones, and further, think about how to reduce the difference to get better information about the spectral characteristics of signals from the DFT analysis.
2. Continuous-time and discrete-time signals/systems are presented in parallel to save the time/space for explaining the two similar ones and increase the understanding as far as there is no concern over causing confusion.
3. It covers most of the theoretical foundations and mathematical derivations that will be used in higher-level related subjects such as signal processing, communication, and control, minimizing the mathematical difficulty and computational burden.
4. Most examples/problems are titled to illustrate key concepts, stimulate interest, or bring out connections with any application so that the readers can appreciate what the examples/problems should be studied for.
5. MATLAB® is integrated extensively into the text with a dual purpose. One is to let the readers know the existence and feel the power of such software tools as help them in computing and plotting. The other is to help them to

realize the physical meaning, interpretation, and/or application of such concepts as convolution, correlation, time/frequency response, Fourier analyses, and their results, etc.

6. The MATLAB® commands and Simulink® blocksets for signal processing application are summarized in the appendices in the expectation of being used like a manual. The authors made no assumption that the readers are proficient in MATLAB® . However, they do not hide their expectation that the readers will get interested in using the MATLAB® and Simulink® for signal analysis and filter design by trying to understand the MATLAB® programs attached to some conceptually or practically important examples/problems and be able to modify them for solving their own problems.

The contents of this book are derived from the works of many (known or unknown) great scientists, scholars, and researchers, all of whom are deeply appreciated. We would like to thank the reviewers for their valuable comments and suggestions, which contribute to enriching this book.

We also thank the people of the School of Electronic & Electrical Engineering, Chung-Ang University for giving us an academic environment. Without affections and supports of our families and friends, this book could not be written. Special thanks should be given to Senior Researcher Yong-Suk Park of KETI (Korea Electronics Technology Institute) for his invaluable help in correction. We gratefully acknowledge the editorial and production staff of Springer-Verlag, Inc. including Dr. Christoph Baumann and Ms. Divya Sreenivasan, Integra.

Any questions, comments, and suggestions regarding this book are welcome. They should be sent to wyyang53@hanmail.net.

Seoul, Korea

Won Y. Yang
Tae G. Chang
Ik H. Song
Yong S. Cho
Jun Heo
Won G. Jeon
Jeong W. Lee
Jae K. Kim

Contents

Chapter 1
Signals and Systems

Contents

In this chapter we introduce the mathematical descriptions of signals and systems. We also discuss the basic concepts on signal and system analysis such as linearity, time-invariance, causality, stability, impulse response, and system function (transfer function).

W.Y. Yang et al., *Signals and Systems with MATLAB®*,
DOI 10.1007/978-3-540-92954-3_1,

1.1 Signals

1.1.1 Various Types of Signal

A *signal*, conveying information generally about the state or behavior of a physical system, is represented mathematically as a function of one or more independent variables. For example, a speech signal may be represented as an amplitude function of time and a picture as a brightness function of two spatial variables. Depending on whether the independent variables and the values of a signal are continuous or discrete, the signal can be classified as follows (see Fig. 1.1 for examples):

- Continuous-time signal $x(t)$: defined at a continuum of times.
- Discrete-time signal (sequence) $x[n] = x(nT)$: defined at discrete times.
- Continuous-amplitude(value) signal x_c: continuous in value (amplitude).
- Discrete-amplitude(value) signal x_d: discrete in value (amplitude).

Here, the bracket [] indicates that the independent variable n takes only integer values. A continuous-time continuous-amplitude signal is called an *analog signal* while a discrete-time discrete-amplitude signal is called a *digital signal*. The ADC (analog-to-digital converter) converting an analog signal to a digital one usually performs the operations of sampling-and-hold, quantization, and encoding. However, throughout this book, we ignore the quantization effect and use "discrete-time signal/system" and "digital signal/system" interchangeably.

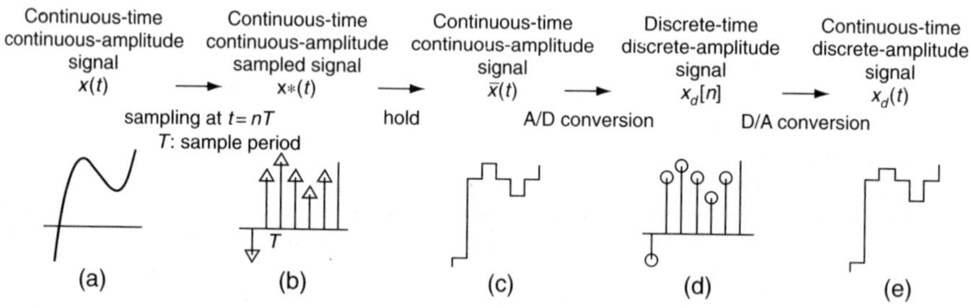

Fig. 1.1 Various types of signal

1.1.2 Continuous/Discrete-Time Signals

In this section, we introduce several elementary signals which not only occur frequently in nature, but also serve as building blocks for constructing many other signals. (See Figs. 1.2 and 1.3.)

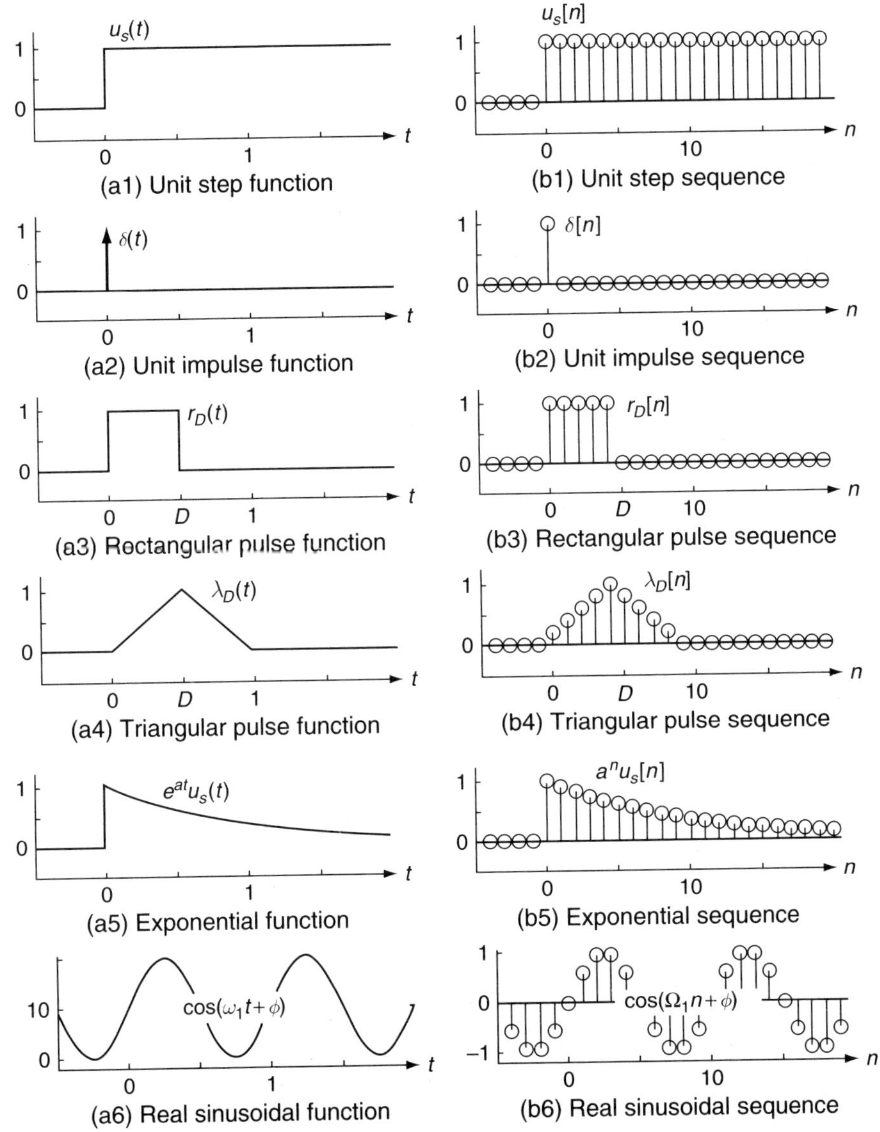

Fig. 1.2 Some continuous–time and discrete–time signals

1.1.2.1a Unit step function

$$u_s(t) = \begin{cases} 1 & \text{for } t \geq 0 \\ 0 & \text{for } t < 0 \end{cases} \tag{1.1.1a}$$

1.1.2.1b Unit step sequence

$$u_s[n] = \begin{cases} 1 & \text{for } n \geq 0 \\ 0 & \text{for } n < 0 \end{cases} \tag{1.1.1b}$$

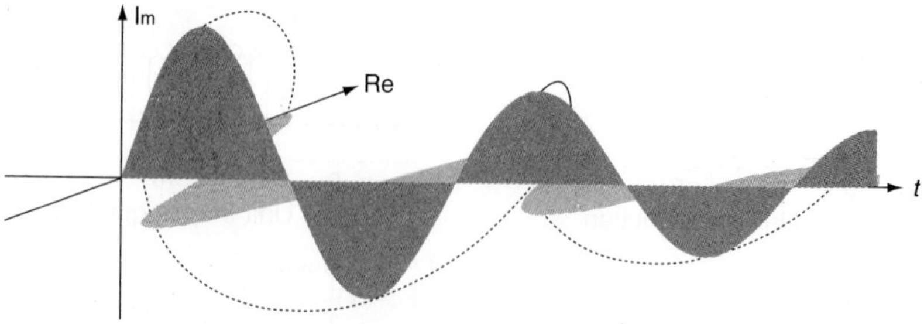

(a) Complex exponential function $x(t) = e^{s_1 t} = e^{\sigma_1 t} e^{j\omega_1 t}$

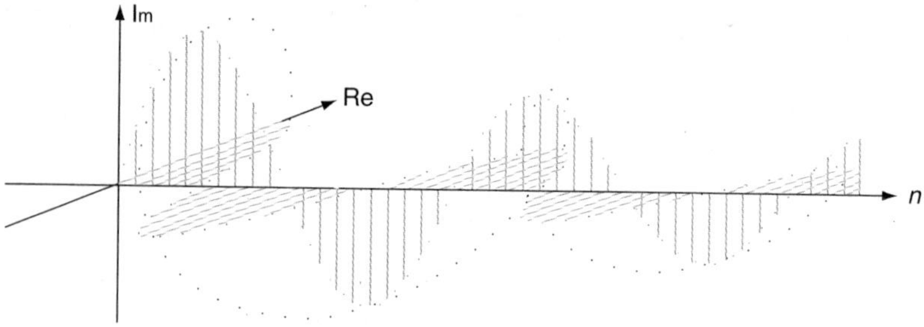

(b) Complex exponential sequence $x(n) = z_1^n = r_1^n e^{j\Omega_1 n}$

Fig. 1.3 Continuous–time/discrete–time complex exponential signals

(cf.) A delayed and scaled step function

$$Au_s(t - t_0) = \begin{cases} A & \text{for } t \geq t_0 \\ 0 & \text{for } t < t_0 \end{cases}$$

(cf.) A delayed and scaled step sequence

$$Au_s[n - n_0] = \begin{cases} A & \text{for } n \geq n_0 \\ 0 & \text{for } n < n_0 \end{cases}$$

1.1.2.2a Unit impulse function

$$\delta(t) = \frac{d}{dt}u_s(t) = \begin{cases} \infty & \text{for } t = 0 \\ 0 & \text{for } t \neq 0 \end{cases} \quad (1.1.2a)$$

(cf.) A delayed and scaled impulse function

$$A\delta(t - t_0) = \begin{cases} A\infty & \text{for } t = t_0 \\ 0 & \text{for } t \neq t_0 \end{cases}$$

1.1.2.2b Unit sample or impulse sequence

$$\delta[n] = \begin{cases} 1 & \text{for } n = 0 \\ 0 & \text{for } n \neq 0 \end{cases} \quad (1.1.2b)$$

(cf.) A delayed and scaled impulse sequence

$$A\,\delta[n - n_0] = \begin{cases} A & \text{for } n = n_0 \\ 0 & \text{for } n \neq n_0 \end{cases}$$

(cf.) Relationship between $\delta(t)$ and $u_s(t)$

$$\delta(t) = \frac{d}{dt}u_s(t) \tag{1.1.3a}$$

$$u_s(t) = \int_{-\infty}^{t} \delta(\tau)d\tau \tag{1.1.4a}$$

1.1.2.3a Rectangular pulse function

$$r_D(t) = u_s(t) - u_s(t-D) = \begin{cases} 1 & \text{for } 0 \le t < D \ (D: \text{duration}) \\ 0 & \text{elsewhere} \end{cases} \tag{1.1.5a}$$

1.1.2.4a Unit triangular pulse function

$$\lambda_D(t) = \begin{cases} 1 - |t \quad D|/D & \text{for } |t - D| \le D \\ 0 & \text{elsewhere} \end{cases} \tag{1.1.6a}$$

1.1.2.5a Real exponential function

$$x(t) = e^{at}u_s(t) = \begin{cases} e^{at} & \text{for } t \ge 0 \\ 0 & \text{for } t < 0 \end{cases} \tag{1.1.7a}$$

1.1.2.6a Real sinusoidal function

$$x(t) = \cos(\omega_1 t + \phi) = \mathrm{Re}\{e^{j(\omega_1 t + \phi)}\} = \frac{1}{2}\{e^{j\phi}e^{j\omega_1 t} + e^{-j\phi}e^{-j\omega_1 t}\} \tag{1.1.8a}$$

1.1.2.7a Complex exponential function

$$x(t) = e^{s_1 t} = e^{\sigma_1 t}e^{j\omega_1 t} \text{ with } s_1 = \sigma_1 + j\omega_1 \tag{1.1.9a}$$

Note that σ_1 determines the changing rate or the time constant and ω_1 the oscillation frequency.

1.1.2.8a Complex sinusoidal function

$$x(t) = e^{j\omega_1 t} = \cos(\omega_1 t) + j\sin(\omega_1 t) \tag{1.1.10a}$$

(cf) Relationship between $\delta[n]$ and $u_s[n]$

$$\delta[n] = u_s[n] - u_s[n-1] \tag{1.1.3b}$$

$$u_s[n] = \sum_{m=-\infty}^{n} \delta[m] \tag{1.1.4b}$$

1.1.2.3b Rectangular pulse sequence

$$r_D[n] = u_s[n] - u_s[n-D] = \begin{cases} 1 & \text{for } 0 \le n < D \ (D: \text{duration}) \\ 0 & \text{elsewhere} \end{cases} \tag{1.1.5b}$$

1.1.2.4b Unit triangular pulse sequence

$$\lambda_D[n] = \begin{cases} 1 - |n + 1 - D|/D & \text{for } |n + 1 - D| \le D - 1 \\ 0 & \text{elsewhere} \end{cases} \tag{1.1.6b}$$

1.1.2.5b Real exponential sequence

$$x[n] = a^n u_s[n] = \begin{cases} a^n & \text{for } n \ge 0 \\ 0 & \text{for } n < 0 \end{cases} \tag{1.1.7b}$$

1.1.2.6b Real sinusoidal sequence

$$x[n] = \cos(\Omega_1 n + \phi) = \mathrm{Re}\left\{e^{j(\Omega_1 n + \phi)}\right\} = \frac{1}{2}\left\{e^{j\phi}e^{j\Omega_1 n} + e^{-j\phi}e^{-j\Omega_1 n}\right\} \tag{1.1.8b}$$

1.1.2.7b Complex exponential function

$$x[n] = z_1^n = r_1^n e^{j\Omega_1 n} \text{ with } z_1 = r_1 e^{j\Omega_1} \tag{1.1.9b}$$

Note that r_1 determines the changing rate and Ω_1 the oscillation frequency.

1.1.2.8b Complex sinusoidal sequence

$$x[n] = e^{j\Omega_1 n} = \cos(\Omega_1 n) + j\sin(\Omega_1 n) \tag{1.1.10b}$$

1.1.3 Analog Frequency and Digital Frequency

A continuous-time signal $x(t)$ is *periodic* with *period P* if P is generally the smallest positive value such that $x(t+P) = x(t)$. Let us consider a continuous-time periodic signal described by

$$x(t) = e^{j\omega_1 t} \tag{1.1.11}$$

The *analog* or *continuous-time* (*angular*) *frequency*[1] of this signal is ω_1 [rad/s] and its period is

$$P = \frac{2\pi}{\omega_1} [\mathrm{s}] \tag{1.1.12}$$

where

$$e^{j\omega_1(t+P)} = e^{j\omega_1 t} \ \forall\, t \quad (\because\ \omega_1 P = 2\pi \Rightarrow e^{j\omega_1 P} = 1) \tag{1.1.13}$$

If we sample the signal $x(t) = e^{j\omega_1 t}$ periodically at $t = nT$, we get a discrete-time signal

$$x[n] = e^{j\omega_1 nT} = e^{j\Omega_1 n} \text{ with } \Omega_1 = \omega_1 T \tag{1.1.14}$$

Will this signal be periodic in n? You may well think that $x[n]$ is also periodic for any sampling interval T since it is obtained from the samples of a continuous-time periodic signal. However, the discrete-time signal $x[n]$ is periodic only when the sampling interval T is the continuous-time period P multiplied by a rational number, as can be seen from comparing Fig. 1.4(a) and (b). If we sample $x(t) = e^{j\omega_1 t}$ to get $x[n] = e^{j\omega_1 nT} = e^{j\Omega_1 n}$ with a sampling interval $T = mP/N$ [s/sample] where the two integers m and N are relatively prime (coprime), i.e., they have no common divisor except 1, the discrete-time signal $x[n]$ is also periodic with the *digital* or *discrete-time frequency*

$$\Omega_1 = \omega_1 T = \omega_1 \frac{mP}{N} = \frac{m}{N} 2\pi \text{ [rad/sample]} \tag{1.1.15}$$

The period of the discrete-time periodic signal $x[n]$ is

$$N = \frac{2m\pi}{\Omega_1} \text{ [sample]}, \tag{1.1.16}$$

where

$$e^{j\Omega_1(n+N)} = e^{j\Omega_1 n} e^{j2m\pi} = e^{j\Omega_1 n} \ \forall\, n \tag{1.1.17}$$

---

[1] Note that we call the angular or radian frequency measured in [rad/s] just the frequency without the modifier 'radian' or 'angular' as long as it can not be confused with the 'real' frequency measured in [Hz].

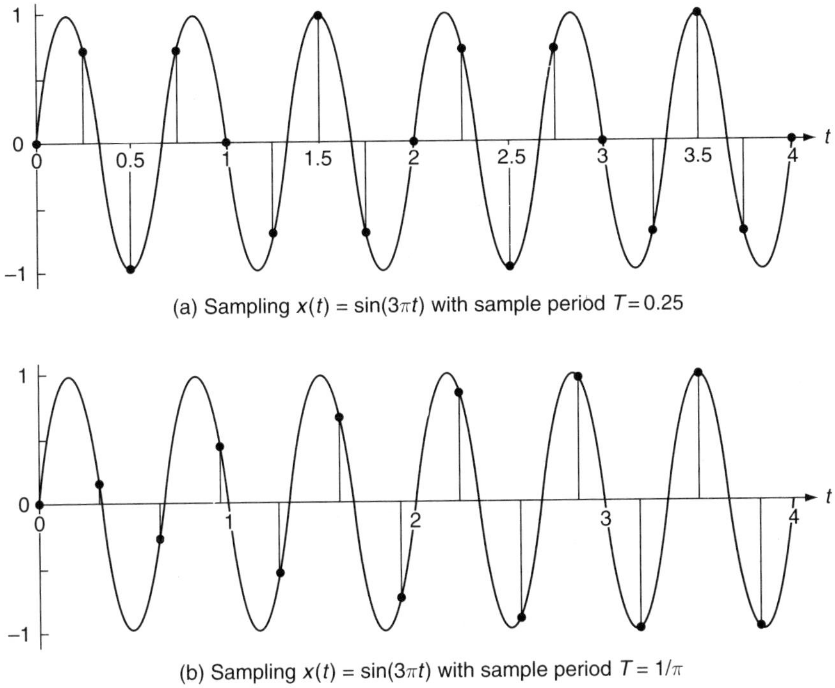

Fig. 1.4 Sampling a continuous–time periodic signal

This is the counterpart of Eq. (1.1.12) in the discrete-time case. There are several observations as summarized in the following remark:

Remark 1.1 Analog (Continuous-Time) Frequency and Digital (Discrete-Time) Frequency

(1) In order for a discrete-time signal to be periodic with period N (being an integer), the digital frequency Ω_1 must be π times a rational number.
(2) The period N of a discrete-time signal with digital frequency Ω_1 is the minimum positive integer to be multiplied by Ω_1 to make an integer times 2π like $2m\pi$ (m: an integer).
(3) In the case of a continuous-time periodic signal with analog frequency ω_1, it can be seen to oscillate with higher frequency as ω_1 increases. In the case of a discrete-time periodic signal with digital frequency Ω_1, it is seen to oscillate faster as Ω_1 increases from 0 to π (see Fig. 1.5(a)–(d)). However, it is seen to oscillate rather slower as Ω_1 increases from π to 2π (see Fig. 1.5(d)–(h)). Particularly with $\Omega_1 = 2\pi$ (Fig. 1.5(h)) or $2m\pi$, it is not distinguishable from a DC signal with $\Omega_1 = 0$. The discrete-time periodic signal is seen to oscillate faster as Ω_1 increases from 2π to 3π (Fig. 1.5(h) and (i)) and slower again as Ω_1 increases from 3π to 4π.

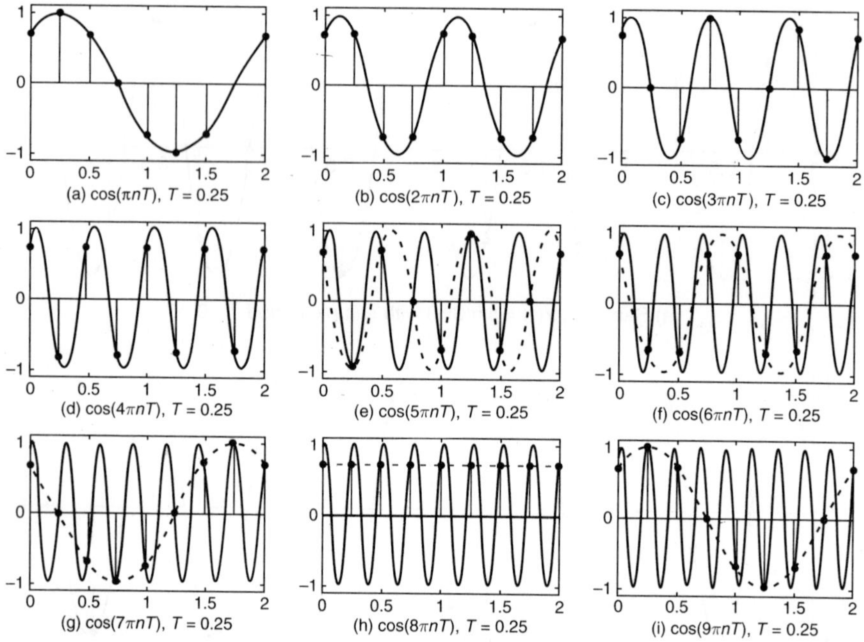

Fig. 1.5 Continuous–time/discrete–time periodic signals with increasing analog/digital frequency

This implies that the frequency characteristic of a discrete-time signal is periodic with period 2π in the digital frequency Ω. This is because $e^{j\Omega_1 n}$ is also periodic with period 2π in Ω_1, i.e., $e^{j(\Omega_1+2m\pi)n} = e^{j\Omega_1 n}e^{j2mn\pi} = e^{j\Omega_1 n}$ for any integer m.

(4) Note that if a discrete-time signal obtained from sampling a continuous-time periodic signal has the digital frequency higher than π [rad] (in its absolute value), it can be identical to a lower-frequency signal in discrete time. Such a phenomenon is called *aliasing*, which appears as the stroboscopic effect or the wagon-wheel effect that wagon wheels, helicopter rotors, or aircraft propellers in film seem to rotate more slowly than the true rotation, stand stationary, or even rotate in the opposite direction from the true rotation (the reverse rotation effect).[W-1]

1.1.4 Properties of the Unit Impulse Function and Unit Sample Sequence

In Sect. 1.1.2, the *unit impulse*, also called the *Dirac delta*, function is defined by Eq. (1.1.2a) as

$$\delta(t) = \frac{d}{dt}u_s(t) = \begin{cases} \infty & \text{for } t = 0 \\ 0 & \text{for } t \neq 0 \end{cases} \tag{1.1.18}$$

Several useful properties of the unit impulse function are summarized in the following remark:

Remark 1.2a Properties of the Unit Impulse Function $\delta(t)$

(1) The unit impulse function $\delta(t)$ has unity area around $t = 0$, which means

$$\int_{-\infty}^{+\infty} \delta(\tau)d\tau = \int_{0^-}^{0^+} \delta(\tau)d\tau = 1 \tag{1.1.19}$$

$$\left(\because \int_{-\infty}^{0^+} \delta(\tau)d\tau - \int_{-\infty}^{0^-} \delta(\tau)d\tau \overset{(1.1.4a)}{=} u_s(0^+) - u_s(0^-) = 1 - 0 = 1\right)$$

(2) The unit impulse function $\delta(t)$ is symmetric about $t = 0$, which is described by

$$\delta(t) = \delta(-t) \tag{1.1.20}$$

(3) The convolution of a time function $x(t)$ and the unit impulse function $\delta(t)$ makes the function itself:

$$x(t) * \delta(t) \underset{\text{definition of convolution integral}}{\overset{(A.17)}{=}} \int_{-\infty}^{+\infty} x(\tau)\delta(t-\tau)d\tau = x(t) \tag{1.1.21}$$

$$\left(\begin{array}{l}\because \int_{-\infty}^{+\infty} x(\tau)\delta(t-\tau)d\tau \overset{\delta(t-\tau)\neq 0 \text{ for only } \tau=t}{=} \int_{-\infty}^{+\infty} x(t)\delta(t-\tau)d\tau \overset{x(t) \text{ independent of } \tau}{=} x(t)\int_{-\infty}^{+\infty} \delta(t-\tau)d\tau \\ \overset{t-\tau\to t'}{=} x(t)\int_{t+\infty}^{t-\infty} \delta(t')(-dt') = x(t)\int_{t-\infty}^{t+\infty} \delta(t')dt' = x(t)\int_{-\infty}^{+\infty} \delta(\tau)d\tau \overset{(1.1.19)}{=} x(t)\end{array}\right)$$

What about the convolution of a time function $x(t)$ and a delayed unit impulse function $\delta(t - t_1)$? It becomes the delayed time function $x(t - t_1)$, that is,

$$x(t) * \delta(t - t_1) = \int_{-\infty}^{+\infty} x(\tau)\delta(t - \tau - t_1)d\tau = x(t - t_1) \tag{1.1.22}$$

What about the convolution of a delayed time function $x(t - t_2)$ and a delayed unit impulse function $\delta(t - t_1)$? It becomes another delayed time function $x(t - t_1 - t_2)$, that is,

$$x(t - t_2) * \delta(t - t_1) = \int_{-\infty}^{+\infty} x(\tau - t_2)\delta(t - \tau - t_1)d\tau = x(t - t_1 - t_2) \tag{1.1.23}$$

If $x(t) * y(t) = z(t)$, we have

$$x(t - t_1) * y(t - t_2) = z(t - t_1 - t_2) \tag{1.1.24}$$

However, with t replaced with $t - t_1$ on both sides, it does not hold, i.e.,

$$x(t - t_1) * y(t - t_1) \neq z(t - t_1), \text{ but } x(t - t_1) * y(t - t_1) = z(t - 2t_1)$$

(4) The unit impulse function $\delta(t)$ has the *sampling* or *sifting property* that

$$\int_{-\infty}^{+\infty} x(t)\delta(t-t_1)dt \overset{\delta(t-t_1)\neq 0 \text{ for only } t=t_1}{=} x(t_1)\int_{-\infty}^{+\infty} \delta(t-t_1)dt \overset{(1.1.19)}{=} x(t_1) \tag{1.1.25}$$

This property enables us to sample or sift the sample value of a continuous-time signal $x(t)$ at $t = t_1$. It can also be used to model a discrete-time signal obtained from sampling a continuous-time signal.

In Sect. 1.1.2, the *unit-sample*, also called the *Kronecker delta*, sequence is defined by Eq. (1.1.2b) as

$$\delta[n] = \begin{cases} 1 & \text{for } n = 0 \\ 0 & \text{for } n \neq 0 \end{cases} \tag{1.1.26}$$

This is the discrete-time counterpart of the unit impulse function $\delta(t)$ and thus is also called the discrete-time impulse. Several useful properties of the unit-sample sequence are summarized in the following remark:

Remark 1.2b Properties of the Unit-Sample Sequence $\delta[n]$

(1) Like Eq. (1.1.20) for the unit impulse $\delta(t)$, the unit-sample sequence $\delta[n]$ is also symmetric about $n = 0$, which is described by

$$\delta[n] = \delta[-n] \tag{1.1.27}$$

(2) Like Eq. (1.1.21) for the unit impulse $\delta(t)$, the convolution of a time sequence $x[n]$ and the unit-sample sequence $\delta[n]$ makes the sequence itself:

$$x[n] * \delta[n] \overset{\text{definition of convolution sum}}{=} \sum_{m=-\infty}^{\infty} x[m]\delta[n-m] = x[n] \tag{1.1.28}$$

$$\left(\begin{array}{l} \because \sum_{m=-\infty}^{\infty} x[m]\delta[n-m] \overset{\delta[n-m]\neq 0 \text{ for only } m=n}{=} \sum_{m=-\infty}^{\infty} x[n]\delta[n-m] \\ \overset{x[n] \text{ independent of } m}{=} x[n]\sum_{m=-\infty}^{\infty} \delta[n-m] \\ \overset{\delta[n-m]\neq 0 \text{ for only } m=n}{=} x[n]\delta[n-n] \overset{(1.1.26)}{=} x[n] \end{array}\right)$$

(3) Like Eqs. (1.1.22) and (1.1.23) for the unit impulse $\delta(t)$, the convolution of a time sequence $x[n]$ and a delayed unit-sample sequence $\delta[n-n_1]$ makes the delayed sequence $x[n-n_1]$:

$$x[n]*\delta[n-n_1]=\sum_{m=-\infty}^{\infty}x[m]\delta[n-m-n_1]=x[n-n_1] \tag{1.1.29}$$

$$x[n-n_2]*\delta[n-n_1]=\sum_{m=-\infty}^{\infty}x[m-n_2]\delta[n-m-n_1]=x[n-n_1-n_2] \tag{1.1.30}$$

Also like Eq. (1.1.24), we have

$$x[n]*y[n]=z[n]\Rightarrow x[n-n_1]*y[n-n_2]=z[n-n_1-n_2] \tag{1.1.31}$$

(4) Like (1.1.25), the unit-sample sequence $\delta[n]$ has the *sampling* or *sifting property* that

$$\sum_{n=-\infty}^{\infty}x[n]\delta[n-n_1]=\sum_{n=-\infty}^{\infty}x[n_1]\delta[n-n_1]=x[n_1] \tag{1.1.32}$$

1.1.5 Several Models for the Unit Impulse Function

As depicted in Fig. 1.6(a)–(d), the unit impulse function can be modeled by the limit of various functions as follows:

$$-\ \delta(t)=\lim_{D\to 0^+}\frac{1}{D}\frac{\sin(\pi t/D)}{\pi t/D}=\lim_{D\to 0^+}\frac{1}{D}\mathrm{sinc}(t/D)\overset{\pi/D\to w}{=}\lim_{w\to\infty}\frac{w}{\pi}\frac{\sin(wt)}{wt} \tag{1.1.33a}$$

$$-\ \delta(t)=\lim_{D\to 0^+}\frac{1}{D}r_D\left(t+\frac{D}{2}\right) \tag{1.1.33b}$$

$$-\ \delta(t)=\lim_{D\to 0^+}\frac{1}{D}\lambda_D\,(t+D) \tag{1.1.33c}$$

$$-\ \delta(t)=\lim_{D\to 0^+}\frac{1}{2D}e^{-|t|/D} \tag{1.1.33d}$$

Note that scaling up/down the impulse function horizontally is equivalent to scaling it up/down vertically, that is,

$$\delta(at)=\frac{1}{|a|}\delta(t) \tag{1.1.34}$$

It is easy to show this fact with any one of Eqs. (1.1.33a–d). Especially for Eq. (1.1.33a), that is the sinc function model of $\delta(t)$, we can prove the validity of Eq. (1.1.34) as follows:

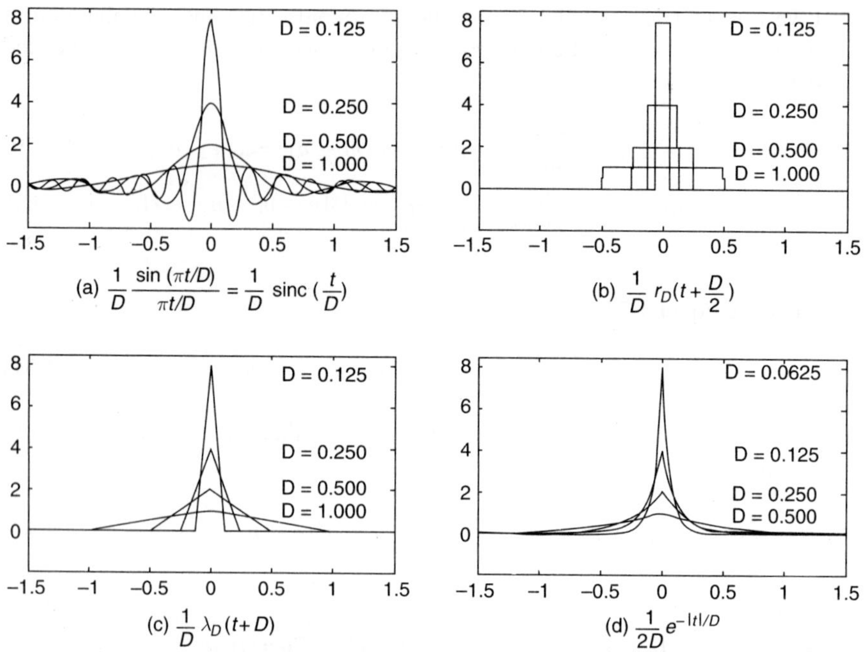

Fig. 1.6 Various models of the unit impulse function $\delta(t)$

$$
\begin{aligned}
\delta(at) &\overset{(1.1.33a)}{=} \lim_{D\to 0^+}\frac{1}{D}\frac{\sin(\pi at/D)}{\pi at/D} = \lim_{D/|a|\to 0^+}\frac{1}{|a|(D/|a|)}\frac{\sin(\pi t/(D/a))}{\pi t/(D/a)} \\
&\overset{D/|a|\to D'}{=} \frac{1}{|a|}\lim_{D'\to 0^+}\frac{1}{D'}\frac{\sin(\pi t/D')}{\pi t/D'} \overset{(1.1.33a)}{=} \frac{1}{|a|}\delta(t)
\end{aligned}
$$

On the other hand, the unit-sample or impulse sequence $\delta[n]$ can be written as

$$
\delta[n] = \frac{\sin(\pi n)}{\pi n} = \mathrm{sinc}(n) \tag{1.1.35}
$$

where the *sinc* function is defined as

$$
\mathrm{sinc}(x) = \frac{\sin(\pi x)}{\pi x} \tag{1.1.36}
$$

1.2 Systems

A *system* is a plant or a process that produces a response called an *output* in response to an excitation called an *input*. If a system's input and output signals are scalars, the system is called a *single-input single-output* (SISO) system. If a system's input and output signals are vectors, the system is called a *multiple-input*

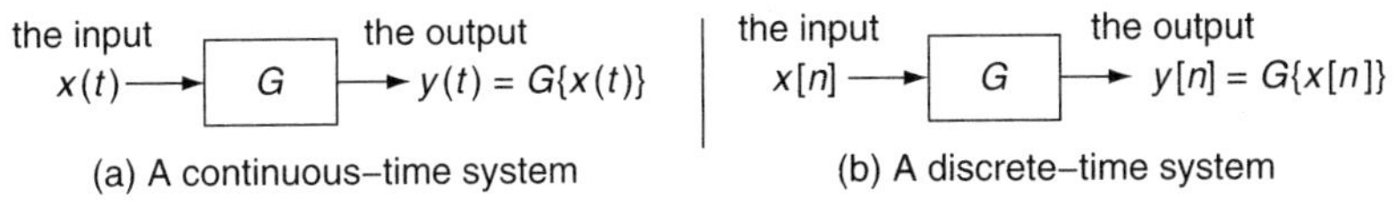

Fig. 1.7 A description of continuous–time and discrete–time systems

multiple-output (MIMO) system. A *single-input multiple-output* (SIMO) system and a *multiple-input single-output* (MISO) system can also be defined in a similar way. For example, a spring-damper-mass system is a mechanical system whose output to an input force is the displacement and velocity of the mass. Another example is an electric circuit whose inputs are voltage/current sources and whose outputs are voltages/currents/charges in the circuit. A mathematical operation or a computer program transforming input argument(s) (together with the initial conditions) into output argument(s) as a model of a plant or a process may also be called a system.

A system is called a continuous-time/discrete-time system if its input and output are both continuous-time/discrete-time signals. Continuous-time/discrete-time systems with the input and output are often described by the following equations and the block diagrams depicted in Fig. 1.7(a)/(b).

Continuous-time system	Discrete-time system
$x(t)\overset{G\{\}}{\rightarrow} y(t);\ y(t) = G\{x(t)\}$	$x[n]\overset{G\{\}}{\rightarrow} y[n];\ y[n] = G\{x[n]\}$

1.2.1 Linear System and Superposition Principle

A system is said to be *linear* if the superposition principle holds in the sense that it satisfies the following properties:

- *Additivity*: The output of the system excited by more than one independent input is the algebraic sum of its outputs to each of the inputs applied individually.
- *Homogeneity*: The output of the system to a single independent input is proportional to the input.

This *superposition principle* can be expressed as follows:

If the output to $x_i(t)$ is $y_i(t) = G\{x_i(t)\}$, the output to $\sum_i a_i x_i(t)$ is $\sum_i a_i G\{x_i(t)\}$, that is,	If the output to $x_i[n]$ is $y_i[n] = G\{x_i[n]\}$, the output to $\sum_i a_i x_i[n]$ is $\sum_i a_i G\{x_i[n]\}$, that is,

$$G\left\{\sum_i a_i x_i(t)\right\} = \sum_i a_i G\{x_i(t)\} = \sum_i a_i y_i \quad (1.2.1a)$$

$$G\left\{\sum_i a_i x_i[n]\right\} = \sum_i a_i G\{x_i[n]\} = \sum_i a_i y_i[n] \quad (1.2.1b)$$

(Ex) A continuous-time linear system

$$y(t) = 2x(t)$$

(Ex) A discrete-time linear system

$$y[n] = 2x[n]$$

(Ex) A continuous-time nonlinear system

$$y(t) = x(t) + 1$$

(Ex) A discrete-time nonlinear system

$$y[n] = x^2[n]$$

Remark 1.3 Linearity and Incremental Linearity

(1) Linear systems possess a property that zero input yields zero output.
(2) Suppose we have a system which is essentially linear and contains some memory (energy storage) elements. If the system has nonzero initial condition, it is not linear any longer, but just *incrementally linear*, since it violates the zero-input/zero-output condition and responds linearly to changes in the input. However, if the initial condition is regarded as a kind of input usually represented by impulse functions, then the system may be considered to be linear.

1.2.2 Time/Shift-Invariant System

A system is said to be *time/shift-invariant* if a delay/shift in the input causes only the same amount of delay/shift in the output without causing any change of the charactersitic (shape) of the output. Time/shift-invariance can be expressed as follows:

If the output to $x(t)$ is $y(t)$, the output to $x(t-t_1)$ is $y(t-t_1)$, i.e.,

$$G\{x(t-t_1)\} = y(t-t_1) \quad (1.2.2a)$$

If the output to $x[n]$ is $y[n]$, the output to $x[n-n_1]$ is $y[n-n_1]$, i.e.,

$$G\{x[n-n_1]\} = y[n-n_1] \quad (1.2.2b)$$

(Ex) A continuous-time time-invariant system

$$y(t) = \sin(x(t))$$

(Ex) A discrete-time time-invariant system

$$y[n] = \frac{1}{3}(x[n-1] + x[n] + x[n+1])$$

(Ex) A continuous-time time-varying system

$$y'(t) = (\sin(t) - 1)y(t) + x(t)$$

(Ex) A discrete-time time-varying system

$$y[n] = \frac{1}{n}y[n-1] + x[n]$$

1.2.3 *Input-Output Relationship of Linear Time-Invariant (LTI) System*

Let us consider the output of a continuous-time linear time-invariant (LTI) system G to an input $x(t)$. As depicted in Fig. 1.8, a continuous-time signal $x(t)$ of any arbitrary shape can be approximated by a linear combination of many scaled and shifted rectangular pulses as

$$\hat{x}(t) = \sum_{m=-\infty}^{\infty} x(mT)\frac{1}{T}r_T\left(t + \frac{T}{2} - mT\right)T \overset{T\to d\tau, mT\to\tau}{\underset{(1.1.33b)}{\rightarrow}}$$

$$x(t) = \lim_{T\to 0}\hat{x}(t) = \int_{-\infty}^{\infty} x(\tau)\delta(t-\tau)d\tau = x(t) * \delta(t) \qquad (1.2.3)$$

Based on the linearity and time-invariance of the system, we can apply the superposition principle to write the output $\hat{y}(t)$ to $\hat{x}(t)$ and its limit as $T \to 0$:

$$\hat{y}(t) = G\{\hat{x}(t)\} = \sum_{m=-\infty}^{\infty} x(mT)\hat{g}_T(t - mT)T$$

$$y(t) = G\{x(t)\} = \int_{-\infty}^{\infty} x(\tau)g(t-\tau)d\tau = x(t) * g(t) \qquad (1.2.4)$$

Here we have used the fact that the limit of the unit-area rectangular pulse response as $T \to 0$ is the *impulse response* $g(t)$, which is the output of a system to a unit impulse input:

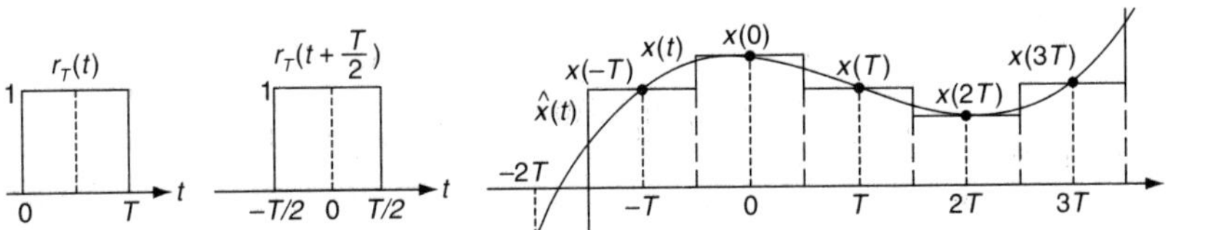

(a) Rectangular pulse (b) Rectangular pulse shifted by $-T/2$ (c) A continuous–time signal approximated by a linear combination of scaled/shifted rectangular pulses

Fig. 1.8 The approximation of a continuous-time signal using rectangular pulses

$$\hat{g}_T(t) = G\left\{\frac{1}{T}r_T\left(t+\frac{T}{2}\right)\right\}$$: the response of the LTI system to a unit-area rectangular pulse input

$$\overset{T\to 0}{\to} \lim_{T\to 0} g_T(t) = \lim_{T\to 0} G\left\{\frac{1}{T}r_T\left(t+\frac{T}{2}\right)\right\}$$

$$= G\left\{\lim_{T\to 0}\frac{1}{T}r_T\left(t+\frac{T}{2}\right)\right\} \overset{(1.1.33b)}{=} G\{\delta(t)\} = g(t) \quad (1.2.5)$$

To summarize, we have an important and fundamental input-output relationship (1.2.4) of a continuous-time LTI system (described by a convolution integral in the time domain) and its Laplace transform (described by a multiplication in the s-domain)

$$y(t) = x(t) * g(t) \underset{\text{Table B.7(4)}}{\overset{\text{Laplace transform}}{\rightarrow}} Y(s) = X(s)G(s) \quad (1.2.6)$$

where the *convolution integral*, also called the *continuous-time convolution*, is defined as

$$x(t) * g(t) = \int_{-\infty}^{\infty} x(\tau)g(t-\tau)d\tau = \int_{-\infty}^{\infty} g(\tau)x(t-\tau)d\tau = g(t) * x(t) \quad (1.2.7)$$

(cf.) This implies that the output $y(t)$ of an LTI system to an input can be expressed as the *convolution* (*integral*) of the input $x(t)$ and the impulse response $g(t)$.

Now we consider the output of a discrete-time linear time-invariant (LTI) system G to an input $x[n]$. We use Eq. (1.1.28) to express the discrete-time signal $x[n]$ of any arbitrary shape as

$$x[n] \overset{(1.1.28)}{=} x[n] * \delta[n] \overset{\text{definition of convolution sum}}{=} \sum_{m=-\infty}^{\infty} x[m]\delta[n-m] \quad (1.2.8)$$

Based on the linearity and time-invariance of the system, we can apply the superposition principle to write the output to $x[n]$:

$$y[n] = G\{x[n]\} \overset{(1.2.8)}{=} G\left\{\sum_{m=-\infty}^{\infty} x[m]\delta[n-m]\right\}$$

$$\overset{\text{linearity}}{=} \sum_{m=-\infty}^{\infty} x[m]G\{\delta[n-m]\}$$

$$\overset{\text{time-invariance}}{=} \sum_{m=-\infty}^{\infty} x[m]g[n-m] = x[n] * g[n] \quad (1.2.9)$$

Here we have used the definition of the *impulse response* or *unit-sample response* of a discrete-time system together with the linearity and time-invariance of the system as

$$G\{\delta[n]\} = g[n] \overset{\text{time-invariance}}{\rightarrow} G\{\delta[n-m]\} = g[n-m]$$

$$G\{x[m]\delta[n-m]\} \overset{\text{linearity}}{=} x[m]G\{\delta[n-m]\} \overset{\text{time-invariance}}{=} x[m]g[n-m]$$

To summarize, we have an important and fundamental input-output relationship (1.2.9) of a discrete-time LTI system (described by a convolution sum in the time domain) and its z-transform (described by a multiplication in the z-domain)

$$y[n] = x[n] * g[n] \underset{\text{Table B.7(4)}}{\overset{z\text{-transform}}{\rightarrow}} Y[z] = X[z]G[z] \tag{1.2.10}$$

where the *convolution sum*, also called the *discrete-time convolution*, is defined as

$$x[n] * g[n] = \sum\nolimits_{m=-\infty}^{\infty} x[m]g[n-m] = \sum\nolimits_{m=-\infty}^{\infty} g[m]x[n-m] = g[n] * x[n] \tag{1.2.11}$$

(cf.) If you do not know about the z-transform, just think of it as the discrete-time counterpart of the Laplace transform and skip the part involved with it. You will meet with the z-transform in Chap. 4.

Figure 1.9 shows the abstract models describing the input-output relationships of continuous-time and discrete-time systems.

1.2.4 Impulse Response and System (Transfer) Function

The *impulse response* of a continuous-time/discrete-time linear time-invariant (LTI) system G is defined to be the output to a unit impulse input $x(t) = \delta(t)/x[n] = \delta[n]$:

$$g(t) = G\{\delta(t)\} \tag{1.2.12a}$$

$$g[n] = G\{\delta[n]\} \tag{1.2.12b}$$

As derived in the previous section, the input-output relationship of the system can be written as

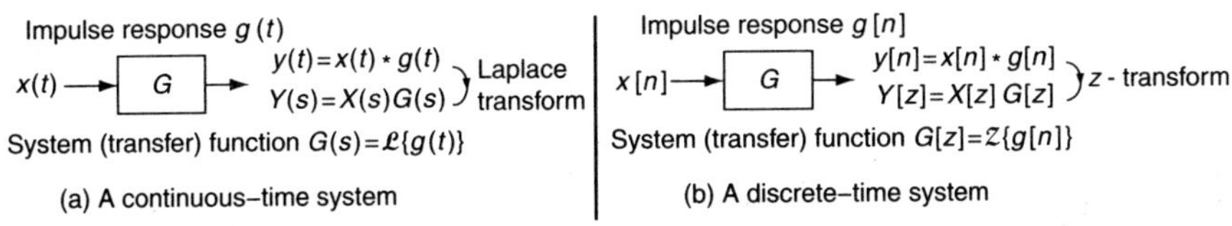

Fig. 1.9 The input-output relationships of continuous-time/discrete-time linear time-invariant (LTI) systems

$$y(t) \overset{(1.2.7)}{=} x(t) * g(t) \overset{\mathcal{L}}{\leftrightarrow} Y(s) \overset{(1.2.6)}{=} X(s)G(s) \quad \Bigg| \quad y[n] \overset{(1.2.11)}{=} x[n] * g[n] \overset{\mathcal{Z}}{\leftrightarrow} Y[z] \overset{(1.2.10)}{=} X[z]G[z]$$

where $x(t)/x[n]$, $y(t)/y[n]$, and $g(t)/g[n]$ are the input, output, and impulse response of the system. Here, the transform of the impulse response, $G(s)/G[z]$, is called the *system* or *transfer function* of the system. We can also rewrite these equations as

$$G(s) = \frac{Y(s)}{X(s)} = \frac{\mathcal{L}\{y(t)\}}{\mathcal{L}\{x(t)\}} = \mathcal{L}\{g(t)\} \quad (1.2.13a) \quad \Bigg| \quad G[z] = \frac{Y[z]}{X[z]} = \frac{\mathcal{Z}\{y[n]\}}{\mathcal{Z}\{x[n]\}} = \mathcal{Z}\{g[n]\} \quad (1.2.13b)$$

This implies another definition or interpretation of the system or transfer function as the ratio of the transformed output to the transformed input of a system with no initial condition.

1.2.5 Step Response, Pulse Response, and Impulse Response

Let us consider a continuous-time LTI system with the impulse response and transfer function given by

$$g(t) = e^{-at}u_s(t) \quad \text{and} \quad G(s) \overset{(1.2.13a)}{=} \mathcal{L}\{g(t)\} = \mathcal{L}\{e^{-at}u_s(t)\} \overset{\text{Table A.1(5)}}{=} \frac{1}{s+a}, \tag{1.2.14}$$

respectively. We can use Eq. (1.2.6) to get the *step response*, that is the output to the unit step input $x(t) = u_s(t)$ with $X(s) \overset{\text{Table A.1(3)}}{=} 1/s$, as

$$Y_s(s) = G(s)X(s) = \frac{1}{s+a}\frac{1}{s} = \frac{1}{a}\left(\frac{1}{s} - \frac{1}{s+a}\right);$$
$$y_s(t) = \mathcal{L}^{-1}\{Y_s(s)\} \overset{\text{Table A.1(3),(5)}}{=} \frac{1}{a}(1 - e^{-at})u_s(t) \tag{1.2.15}$$

Now, let a unity-area rectangular pulse input of duration (pulsewidth) T and height $1/T$

$$x(t) = \frac{1}{T}r_T(t) = \frac{1}{T}(u_s(t) - u_s(t-T)); X(s) = \mathcal{L}\{x(t)\} = \frac{1}{T}\mathcal{L}\{u_s(t) - u_s(t-T)\}$$
$$\overset{\text{Table A.1(3), A.2(2)}}{=} \frac{1}{T}\left(\frac{1}{s} - e^{-Ts}\frac{1}{s}\right) \tag{1.2.16}$$

be applied to the system. Then the output $g_T(t)$, called the *pulse response*, is obtained as

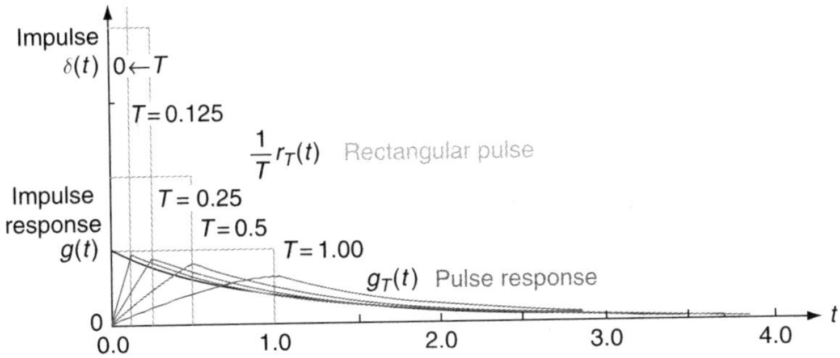

Fig. 1.10 The pulse response and the impulse response

$$Y_T(s) = G(s)X(s) = \frac{1}{T}\left(\frac{1}{s(s+a)} - e^{-Ts}\frac{1}{s(s+a)}\right)$$

$$= \frac{1}{aT}\left(\frac{1}{s} - \frac{1}{s+a} - e^{-Ts}\left(\frac{1}{s} - \frac{1}{s+a}\right)\right); \quad g_T(t) = \mathcal{L}^{-1}\{Y_T(s)\}$$

$$\overset{\text{Table A.1(3),(5),A.2(2)}}{=} \frac{1}{aT}\left((1-e^{-at})u_s(t) - (1-e^{-a(t-T)})u_s(t-T)\right) \tag{1.2.17}$$

If we let $T \rightarrow 0$ so that the rectangular pulse input becomes an impulse $\delta(t)$ (of instantaneous duration and infinite height), how can the output be expressed? Taking the limit of the output equation (1.2.17) with $T \rightarrow 0$, we can get the impulse response $g(t)$ (see Fig. 1.10):

$$g_T(t) \overset{T\rightarrow 0}{\rightarrow} \frac{1}{aT}\left((1-e^{-at})u_s(t) - (1-e^{-a(t-T)})u_s(t)\right) = \frac{1}{aT}(e^{aT}-1)e^{-at}u_s(t)$$

$$\underset{aT\rightarrow 0}{\overset{\text{(D.25)}}{\cong}} \frac{1}{aT}(1+aT-1)e^{-at}u_s(t) = e^{-at}u_s(t) \overset{(1.2.14)}{\equiv} g(t) \tag{1.2.18}$$

This implies that as the input gets close to an impulse, the output becomes close to the impulse response, which is quite natural for any linear time-invariant system.

On the other hand, Fig. 1.11 shows the validity of Eq. (1.2.4) insisting that the linear combination of scaled/shifted pulse responses gets closer to the true output as $T \rightarrow 0$.

1.2.6 Sinusoidal Steady-State Response and Frequency Response

Let us consider the *sinusoidal steady-state response*, which is defined to be the everlasting output of a continuous-time system with system function $G(s)$ to a sinusoidal input, say, $x(t) = A\cos(\omega t + \phi)$. The expression for the sinusoidal steady-state

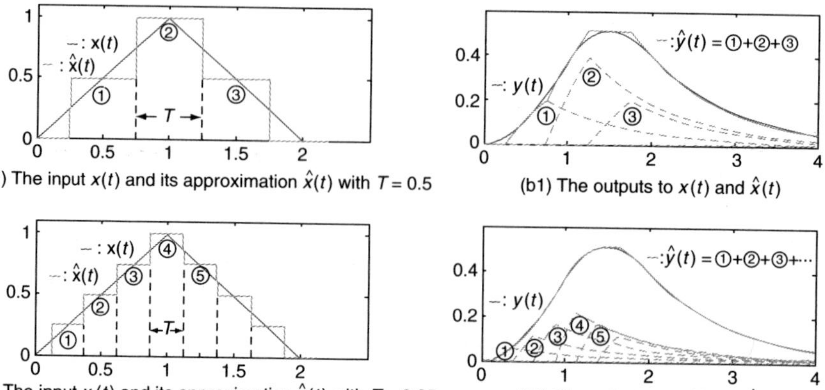

(a1) The input $x(t)$ and its approximation $\hat{x}(t)$ with $T = 0.5$

(b1) The outputs to $x(t)$ and $\hat{x}(t)$

(a2) The input $x(t)$ and its approximation $\hat{x}(t)$ with $T = 0.25$

(b2) The outputs to $x(t)$ and $\hat{x}(t)$

Fig. 1.11 The input–output relationship of a linear time-invariant (LTI) system – convolution

response can be obtained from the time-domain input-output relationship (1.2.4). That is, noting that the sinusoidal input can be written as the sum of two complex conjugate exponential functions

$$x(t) = A\cos(\omega t + \phi) \overset{(D.21)}{=} \frac{A}{2}(e^{j(\omega t+\phi)} + e^{-j(\omega t+\phi)}) = \frac{A}{2}(x_1(t) + x_2(t)), \quad (1.2.19)$$

we substitute $x_1(t) = e^{j(\omega t+\phi)}$ for $x(t)$ into Eq. (1.2.4) to obtain a partial steady-state response as

$$\begin{aligned} y_1(t) = G\{x_1(t)\} &\overset{(1.2.4)}{=} \int_{-\infty}^{\infty} x_1(\tau)g(t-\tau)d\tau = \int_{-\infty}^{\infty} e^{j(\omega\tau+\phi)}g(t-\tau)d\tau \\ &= e^{j(\omega t+\phi)}\int_{-\infty}^{\infty} e^{-j\omega(t-\tau)}g(t-\tau)d\tau = e^{j(\omega t+\phi)}\int_{-\infty}^{\infty} e^{-j\omega t}g(t)dt \\ &= e^{j(\omega t+\phi)}G(j\omega) \end{aligned} \quad (1.2.20)$$

with

$$G(j\omega) = \int_{-\infty}^{\infty} e^{-j\omega t}g(t)dt \underset{\text{causal system}}{\overset{g(t)=0 \text{ for } t<0}{=}} \int_{0}^{\infty} g(t)e^{-j\omega t}dt \overset{(A.1)}{=} G(s)|_{s=j\omega} \quad (1.2.21)$$

Here we have used the definition (A.1) of the Laplace transform under the assumption that the impulse response $g(t)$ is zero $\forall t < 0$ so that the system is *causal* (see Sect. 1.2.9). In fact, every physical system satisfies the assumption of causality that its output does not precede the input. Here, $G(j\omega)$ obtained by substituting $s = j\omega$ (ω: the analog frequency of the input signal) into the system function $G(s)$ is called the *frequency response*.

The total sinusoidal steady-state response to the sinusoidal input (1.2.19) can be expressed as the sum of two complex conjugate terms:

$$y(t) = \frac{A}{2}\left(y_1(t) + y_2(t)\right) = \frac{A}{2}\left\{e^{j(\omega t+\phi)}G(j\omega) + e^{-j(\omega t+\phi)}G(-j\omega)\right\}$$

$$= \frac{A}{2}\left\{e^{j(\omega t+\phi)}|G(j\omega)|e^{j\theta(\omega)} + e^{-j(\omega t+\phi)}|G(-j\omega)|e^{-j\theta(\omega)}\right.$$

$$= \frac{A}{2}|G(j\omega)|\left\{e^{j(\omega t+\phi+\theta(\omega))} + e^{-j(\omega t+\phi+\theta(\omega))}\right\}$$

$$\overset{(D.21)}{=} A|G(j\omega)|\cos(\omega t + \phi + \theta(\omega)) \tag{1.2.22}$$

where $|G(j\omega)|$ and $\theta(\omega) = \angle G(j\omega)$ are the magnitude and phase of the *frequency response* $G(j\omega)$, respectively. Comparing this steady-state response with the sinusoidal input (1.2.19), we see that its amplitude is $|G(j\omega)|$ times the amplitude A of the input and its phase is $\theta(\omega)$ plus the phase ϕ of the input at the frequency ω of the input signal.

(cf.) The system function $G(s)$ (Eq. (1.2.13a)) and frequency response $G(j\omega)$ (Eq. (1.2.21)) of a system are the Laplace transform and Fourier transform of the impulse response $g(t)$ of the system, respectively.

Likewise, the *sinusoidal steady-state response* of a discrete-time system to a sinusoidal input, say, $x[n] = A\cos(\Omega n + \phi)$ turns out to be

$$y[n] = A|G[e^{j\Omega}]|\cos(\Omega n + \phi + \theta(\Omega)) \tag{1.2.23}$$

where

$$G[e^{j\Omega}] = \sum_{n=-\infty}^{\infty} g[n]e^{-j\Omega n} \underset{\text{causal system}}{\overset{g[n]=0 \text{ for } n<0}{=}} \sum_{n=0}^{\infty} g[n]e^{-j\Omega n} \overset{\text{Remark 4.5}}{=} G[z]|_{z=e^{j\Omega}} \tag{1.2.24}$$

Here we have used the definition (4.1) of the z-transform. Note that $G[e^{j\Omega}]$ obtained by substituting $z = e^{j\Omega}$ (Ω: the digital frequency of the input signal) into the system function $G[z]$ is called the *frequency response*.

Remark 1.4 Frequency Response and Sinusoidal Steady-State Response

(1) The frequency response $G(j\omega)$ of a continuous-time system is obtained by substituting $s = j\omega$ (ω: the analog frequency of the input signal) into the system function $G(s)$. Likewise, the frequency response $G[e^{j\Omega}]$ of a discrete-time system is obtained by substituting $z = e^{j\Omega}$ (Ω: the digital frequency of the input signal) into the system function $G[z]$.

(2) The steady-state response of a system to a sinusoidal input is also a sinusoidal signal of the same frequency. Its amplitude is the amplitude of the input times the magnitude of the frequency response at the frequency of the input. Its phase is the phase of the input plus the phase of the frequency response at the frequency of the input (see Fig. 1.12).

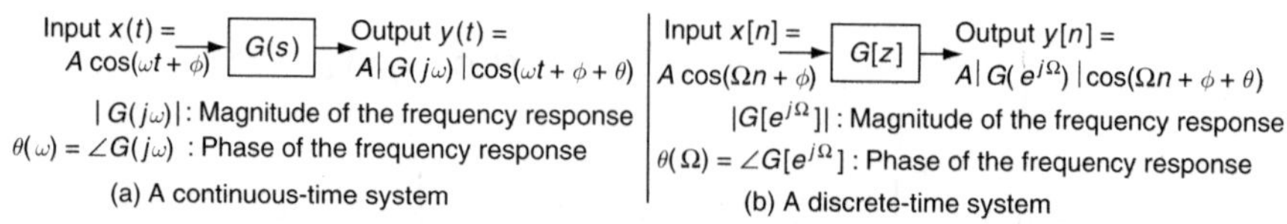

Fig. 1.12 The sinusoidal steady–state response of continuous-time/discrete-time linear time-invariant systems

1.2.7 Continuous/Discrete-Time Convolution

In Sect. 1.2.3, the output of an LTI system was found to be the convolution of the input and the impulse response. In this section, we take a look at the process of computing the convolution to comprehend its physical meaning and to be able to program the convolution process.

The continuous-time/discrete-time convolution $y(t)/y[n]$ of two functions/ sequences $x(\tau)/x[m]$ and $g(\tau)/g[m]$ can be obtained by time-reversing one of them, say, $g(\tau)/g[m]$ and time-shifting (sliding) it by t/n to $g(t-\tau)/g[n-m]$, multiplying it with the other, say, $x(\tau)/x[m]$, and then integrating/summing the multiplication, say, $x(\tau)g(t-\tau)/x[m]g[n-m]$. Let us take a look at an example.

Example 1.1 Continuous-Time/Discrete-Time Convolution of Two Rectangular Pulses

(a) Continuous-Time Convolution (Integral) of Two Rectangular Pulse Functions $r_{D_1}(t)$ and $r_{D_2}(t)$ Referring to Fig. 1.13(a1–a8), you can realize that the convolution of the two rectangular pulse functions $r_{D_1}(t)$ (of duration D_1) and $r_{D_2}(t)$ (of duration $D_2 < D_1$) is

$$r_{D1}(t)*r_{D_2}(t) = \begin{cases} t & \text{for } 0 \le t < D_2 \\ D_2 & \text{for } D_2 \le t < D_1 \\ -t + D & \text{for } D_1 \le t < D = D_1 + D_2 \\ 0 & \text{elsewhere} \end{cases} \tag{E1.1.1a}$$

The procedure of computing this convolution is as follows:

- (a1) and (a2) show $r_{D_1}(\tau)$ and $r_{D_2}(\tau)$, respectively.
- (a3) shows the time-reversed version of $r_{D_2}(\tau)$, that is $r_{D_2}(-\tau)$. Since there is no overlap between $r_{D_1}(\tau)$ and $r_{D_2}(-\tau)$, the value of the convolution $r_{D_1}(t)*r_{D_2}(t)$ at $t=0$ is zero.

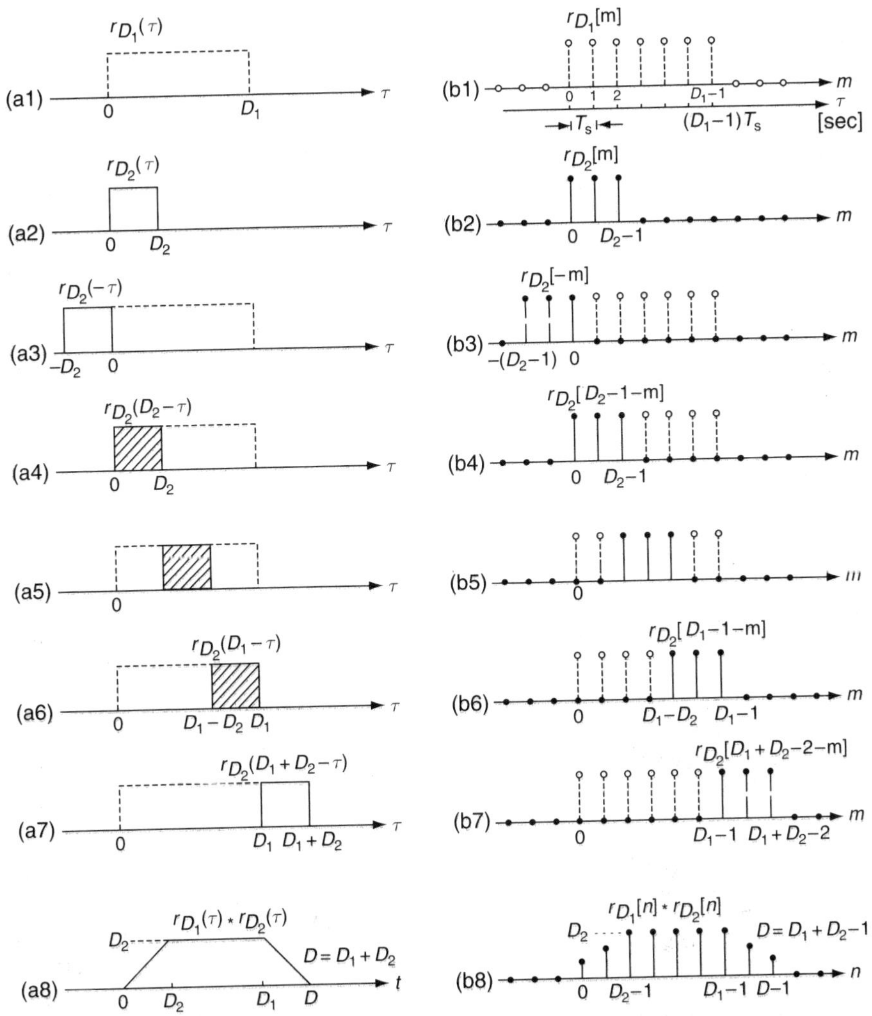

Fig. 1.13 Continuous–time/discrete–time convolution of two rectangular pulses

- (a4) shows the D_2-delayed version of $r_{D_2}(-\tau)$, that is $r_{D_2}(D_2-\tau)$. Since this overlaps with $r_{D_1}(\tau)$ for $0 \le \tau < D_2$ and the multiplication of them is 1 over the overlapping interval, the integration (area) is D_2, which will be the value of the convolution at $t = D_2$. In the meantime (from $t = 0$ to D_2), it gradually increases from $t = 0$ to D_2 in proportion to the lag time t.
- As can be seen from (a4)–(a6), the length of the overlapping interval between $r_{D_2}(t-\tau)$ and $r_{D_1}(\tau)$ and the integration of the multiplication is kept constant as D_2 till $r_{D_2}(t-\tau)$ is slided by D_1 to touch the right end of $r_{D_1}(\tau)$. Thus the value of the convolution is D_2 all over $D_2 \le t < D_1$.

- While the sliding $r_{D_2}(t-\tau)$ passes by the right end of $r_{D_1}(\tau)$, the length of the overlapping interval with $r_{D_1}(\tau)$ and the integration of the multiplication decreases gradually from D_2 to 0 till it passes through the right end of $r_{D_1}(\tau)$ at $t = D_1 + D_2$ as shown in (a7).
- After the left end of $r_{D_2}(t-\tau)$ passes by the right end of $r_{D_1}(\tau)$ at $t = D_1 + D_2$, there is no overlap and thus the value of the convolution is zero all over $t \geq D_1 + D_2$.
- The convolution obtained above is plotted against the time lag t in (a8).

(b) Discrete-Time Convolution (Sum) of Two Rectangular Pulse Sequences $r_{D_1}[n]$ and $r_{D_2}[n]$ Referring to Fig. 1.13(b1–b8), you can realize that the convolution of the two rectangular pulse sequences $r_{D_1}[n]$ (of duration D_1) and $r_{D_2}[n]$ (of duration $D_2 < D_1$) is as follows:

$$r_{D1}[n]*r_{D_2}[n] = \begin{cases} n+1 & \text{for } 0 \leq n < D_2 \\ D_2 & \text{for } D_2 \leq n < D_1 \\ -n+D & \text{for } D_1 \leq n < D = D_1 + D_2 - 1 \\ 0 & \text{elsewhere} \end{cases} \tag{E1.1.1b}$$

The procedure of computing this discrete-time convolution is similar to that taken to get the continuous-time convolution in (a). The difference is as follows:

- The value of $r_{D_1}[n] * r_{D_2}[n]$ is not zero at $n = 0$ while that of $r_{D_1}(t) * r_{D_2}(t)$ is zero at $t = 0$.
- The duration of $r_{D_1}[n] * r_{D_2}[n]$ is from $n = 0$ to $D_1 + D_2 - 2$ while that of $r_{D_1}(t) * r_{D_2}(t)$ is from $t = 0$ to $D_1 + D_2$.

(cf.) You had better try to understand the convolution process rather graphically than computationally.

(cf.) Visit the web site <http://www.jhu.edu/~signals/> to appreciate the joy of convolution.

Remark 1.5 Continuous-Time/Discrete-Time Convolution of Two Rectangular Pulses

(1) If the lengths of the two rectangular pulses are D_1 and D_2 (with $D_1 > D_2$), respectively, the continuous-time and discrete-time convolutions have the duration of $D_1 + D_2$ and $D_1 + D_2 - 1$, respectively and commonly the height of D_2.
(2) If the lengths of the two rectangular pulses are commonly D, the continuous-time and discrete-time convolutions are triangular pulses whose durations are $2D$ and $2D - 1$, respectively and whose heights are commonly D:

$$r_D(t)*r_D(t) = D\lambda_D(t) \tag{1.2.25a}$$

$$r_D[n]*r_D[n] = D\lambda_D[n] \tag{1.2.25b}$$

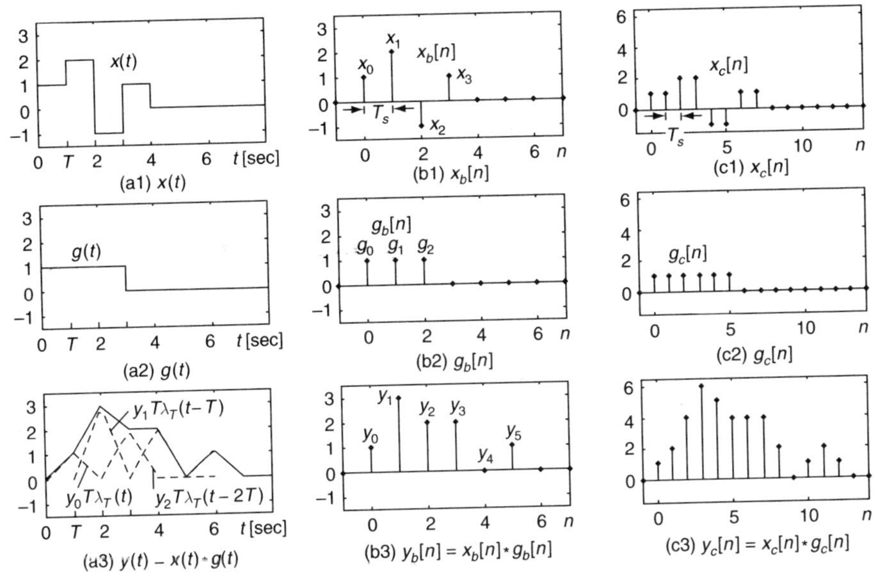

Fig. 1.14 Approximation of a continuous–time convolution by a discrete–time convolution

The above example implies that the continuous-time convolution $x(t) * g(t)$ can be approximated by the limit of a discrete-time convolution $x(nT_s) * g(nT_s)T_s$ as $T_s \to 0$ where T_s is the sampling period. Let us take a look at another example.

Example 1.2 Approximation of a Continuous-Time Convolution by a Discrete-Time One

Figure 1.14(a1), (a2), and (a3) show two continuous-time signals, $x(t)$, $g(t)$, and their convolution $y(t) = x(t) * g(t)$, respectively where $y(t)$ is actually a discrete-time version $y(nT_s)$, which is obtained as follows:

- We sample $x(t)$ and $g(t)$ every $T_s = 0.01$ [s] to make two sequences $x(nT_s)$ and $g(nT_s)$.
- We use the MATLAB function `conv()` to compute their discrete-time convolution and multiply it by the sampling interval T_s to get $y(nT_s) = x(nT_s) * g(nT_s)T_s$ as we do in the numerical integration.

Figure 1.14(b1)/(c1), (b2)/(c2), and (b3)/(c3) show two discrete-time signals $x[n]$, $g[n]$, and their discrete-time convolution $y[n] = x[n] * g[n]$ (computed using `conv()`), respectively where $x[n]$ and $g[n]$ are obtained by sampling $x(t)$ and $g(t)$ with sampling period $T_s = 1/T_s = 0.5$.

Interested readers are welcome to run the following MATLAB program "sig01f14.m".

```
%sig01f14.m: Fig.1.14
clear, clf
x=[1 2 -1 1]; g=[1 1 1];
t0_x=0;t0_g=0; %t0_x=1; t0_g=-2; % the initial time indices of x and g
% Continuous-Time Convolution
T=1; Ts=0.01; dt=Ts; M=T/Ts; %Interval T, Sampling period Ts, and # of Ts per T
xt= ones(M,1)*x; xt=xt(:).'; Nxt=length(xt);
gt= ones(M,1)*g; gt=gt(:).'; Ngt=length(gt);
yt= conv(xt,gt)*Ts; Nyt=length(yt); % the length of convolution Nxt+Ngt-1
NytM=Nyt+M; % plotting interval extended by one interval T
t0_y=t0_x+t0_g; tt= min([t0_x t0_g t0_y])+[0:NytM-1]*Ts;
xt_= [zeros(1,fix((t0_x-tt(1))/Ts)) xt]; xt_=[xt_ zeros(1,NytM-length(xt_))];
gt_= [zeros(1,fix((t0_g-tt(1))/Ts)) gt]; gt_=[gt_ zeros(1,NytM-length(gt_))];
yt_= [zeros(1,fix((t0_y-tt(1))/Ts)) yt]; yt_= [yt_ zeros(1,NytM-length(yt_))];
ymax= max([xt gt yt])+0.5; ymin= min([xt gt yt])-0.5;
subplot(331), plot(tt,xt_), axis([tt([1 end]) ymin ymax])
subplot(334), plot(tt,gt_), axis([tt([1 end]) ymin ymax])
subplot(337), plot(tt,yt_), axis([tt([1 end]) ymin ymax])
% Discrete-Time Convolution
Ts=1; n0_x = t0_x/Ts;  n0_g = t0_g/Ts;
xn=xt(1:Ts/dt:4*M), gn=gt(1:Ts/dt:3*M)
Nx=length(xn); Ng=length(gn); % the length of convolution Nx+Ng-1
yn=conv(xn,gn); Ny=length(yn);
xn_extended=[zeros(1,Ng-1) xn]; gn_reversed= fliplr(gn);
gr_extended= [gn_reversed zeros(1,Ny-Ng)];
for n=1:Ny % Instead of conv(), we slide g[-n], multiply x[n], and take sum.
   yn1(n)=xn_extended*gr_extended.';
   gr_extended= [0 gr_extended(1:Ny-1)];
end
Ny1=Ny+3; % plotting interval extended by two samples
n0_y=n0_x+n0_g; nn= min([n0_x n0_g n0_y])+[-1:Ny1-2];
xn_= [zeros(1,n0_x-nn(1)) xn]; xn_= [xn_ zeros(1,Ny1-length(xn_))];
gn_= [zeros(1,n0_g-nn(1)) gn]; gn_= [gn_ zeros(1,Ny1-length(gn_))];
yn_= [zeros(1,n0_y-nn(1)) yn]; yn_= [yn_ zeros(1,Ny1-length(yn_))];
ymax= max([xn gn yn])+0.5; ymin= min([xn gn yn])-0.5;
subplot(332), stem(nn,xn_,'.'), axis([nn([1 end]) ymin ymax])
subplot(335), stem(nn,gn_,'.'), axis([nn([1 end]) ymin ymax])
subplot(338), stem(nn,yn_,'.'), axis([nn([1 end]) ymin ymax])
```

The systems, called *tapped delay lines*, to be discussed in the following example are essentially continuous-time systems. However, they behave like discrete-time systems in terms of the input-output relationship and thus are expected to be a link between continuous-time systems and discrete-time ones.

Example 1.3 Tapped Delay Lines

(a) Tapped Delay Lines

Consider the continuous-time LTI system depicted in Fig. 1.15(a) where its impulse response and system function are

$$g(t) = \sum_{n=-\infty}^{\infty} g_n \delta(t - nT); \; G(s) = \mathcal{L}\{g(t)\} = \sum_{n=-\infty}^{\infty} g_n e^{-snT} \quad \text{(E1.3.1)}$$

Let the input and its Laplace transform be

$$x(t) = \sum_{m=-\infty}^{\infty} x_m \delta(t - mT); \; X(s) = \mathcal{L}\{x(t)\} = \sum_{m=-\infty}^{\infty} x_m e^{-smT} \tag{E1.3.2}$$

Then we claim that the output and its Laplace transform will be

$$y(t) = \sum_{k=-\infty}^{\infty} y_k \delta(t - kT); \; Y(s) = \mathcal{L}\{y(t)\} = \sum_{k=-\infty}^{\infty} y_k e^{-skT} \tag{E1.3.3}$$

with

$$y_k = x_k * g_k = \sum_{m=-\infty}^{\infty} x_m g_{k-m}; \; Y[z] = \mathcal{Z}\{y_k\} = X[z]G[z] \tag{E1.3.4}$$

Proof

$$\begin{aligned} y(t) = x(t)*g(t) &= \sum_{m=-\infty}^{\infty} x_m \delta(t - mT) * \sum_{n=-\infty}^{\infty} g_n \delta(t - nT) \\ &= \sum_{m=-\infty}^{\infty} \sum_{n=-\infty}^{\infty} x_m g_n \delta(t - mT) * \delta(t - nT) \\ &\overset{(1.1.23)}{=} \sum_{m=-\infty}^{\infty} \sum_{n=-\infty}^{\infty} x_m g_n \delta(t - mT - nT) \\ &\underset{n \to k-m}{\overset{m+n \to k}{=}} \sum_{k=-\infty}^{\infty} \sum_{m=-\infty}^{\infty} x_m g_{k-m} \delta(t - kT) \\ &= \sum_{k=-\infty}^{\infty} y_k \delta(t - kT) \text{ with } y_k = \sum_{m=-\infty}^{\infty} x_m g_{k-m} \end{aligned}$$

(b) Tapped Delay Lines with a Zero-Order Hold at its Output Consider the continuous-time LTI system with a zero-order hold (z.o.h.) device at its output as depicted in Fig. 1.15(b) where its impulse response is

$$\bar{g}(t) = \sum_{n=-\infty}^{\infty} g_n r_T(t - nT) = \sum_{n=-\infty}^{\infty} g_n (u_s(t - nT) - u_s(t - nT - T)) \tag{E1.3.5}$$

Let the input be

$$\bar{x}(t) = \sum_{m=-\infty}^{\infty} x_m r_T(t - mT) = \sum_{m=-\infty}^{\infty} x_m (u_s(t - mT) - u_s(t - mT - T)) \tag{E1.3.6}$$

Then we claim that the output will be

$$y(t) = \sum_{k=-\infty}^{\infty} y_k T \lambda_T(t - kT) \tag{E1.3.7}$$

with

$$y_k = x_k * g_k = \sum_{m=-\infty}^{\infty} x_m g_{k-m} \tag{E1.3.8}$$

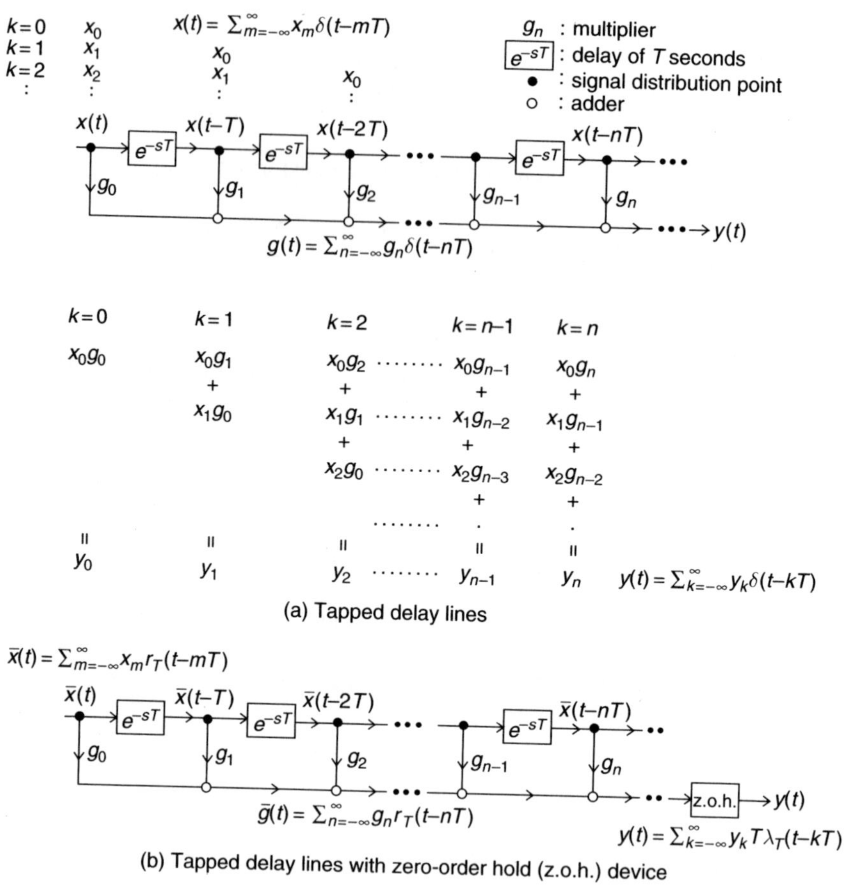

Fig. 1.15 Systems for Example 1.3(a) and (b)

Proof

$$y(t) = \bar{x}(t)*\bar{g}(t) = \sum_{m=-\infty}^{\infty} x_m r_T(t-mT)*\sum_{n=-\infty}^{\infty} g_n r_T(t-nT)$$

$$= \sum_{m=-\infty}^{\infty}\sum_{n=-\infty}^{\infty} x_m g_n r_T(t-mT)*r_T(t-nT)$$

$$\overset{(1.2.25a)}{\underset{(1.1.24)}{=}} \sum_{m=-\infty}^{\infty}\sum_{n=-\infty}^{\infty} x_m g_n T\lambda_T(t-mT-nT)$$

$$\overset{m+n\to k}{\underset{n\to k-m}{=}} \sum_{k=-\infty}^{\infty}\sum_{m=-\infty}^{\infty} x_m g_{k-m} T\lambda_T(t-kT)$$

$$= \sum_{k=-\infty}^{\infty} y_k T\lambda_T(t-kT) \text{ with } y_k = \sum_{m=-\infty}^{\infty} x_m g_{k-m}$$

(cf.) Fig. 1.14(a1–a3) shows the validity of Eq. (E1.3.7).

1.2.8 Bounded-Input Bounded-Output (BIBO) Stability

A system is said to be (BIBO) *stable* if every *b*ounded *i*nput produces a *b*ounded *o*utput. To find the stability condition for LTI systems with impulse response $g(t)/g[n]$, suppose we have a bounded input $x(t)/x[n]$ such that

$$|x(t)| < B < \infty \text{ for all } t \quad (1.2.26a) \qquad |x[n]| < B < \infty \text{ for all } n \quad (1.2.26b)$$

Then we can use the input-output relationship (1.2.6)/(1.2.9) of the continuous-time/discrete-time LTI system to express the magnitude of the output as

$$\begin{aligned} |y(t)| &\overset{(1.2.6)}{=} \left|\int_{-\infty}^{\infty} x(\tau)g(t-\tau)d\tau\right| \\ &\le \int_{-\infty}^{\infty} |x(\tau)||g(t-\tau)|d\tau \\ &\overset{(1.2.26a)}{\le} B\int_{-\infty}^{\infty} |g(t-\tau)|d\tau \\ &= B\int_{-\infty}^{\infty} |g(\tau)|d\tau \end{aligned}$$

$$\begin{aligned} |y[n]| &\overset{(1.2.9)}{=} \left|\sum_{m=-\infty}^{\infty} x[m]g[n-m]\right| \\ &\le \sum_{m=-\infty}^{\infty} |x[m]||g[n-m]| \\ &\overset{(1.2.26b)}{\le} B\sum_{m=-\infty}^{\infty} |g[n-m]| \\ &= B\sum_{m=-\infty}^{\infty} |g[m]| \end{aligned}$$

which means that the output is also bounded. This implies that the system is (BIBO) *stable* if the impulse response $g(t)/g[n]$ is absolutely integrable/summable, i.e.,

$$\int_{-\infty}^{\infty} |g(\tau)|d\tau < \infty \quad (1.2.27a) \qquad \sum_{m=-\infty}^{\infty} |g[m]| < \infty \quad (1.2.27b)$$

In fact, this condition is not only sufficient but also necessary for stability of continuous-time/discrete-time systems since if it is not satisfied, we can find a bounded input which makes the output unbounded (see Problem 1.5). The following remark restates the stability condition in terms of the pole locations of the system functions where the poles are the values of s or z at which the denominator of $G(s)$ or $G[z]$ becomes zero.

Remark 1.6 Stability of LTI Systems with System Function $G(s)/G[z]$

A continuous-time/discrete-time linear time-invariant (LTI) system having the system function $G(s) = \mathcal{L}\{g(t)\}$ / $G[z] = \mathcal{Z}\{g[n]\}$ is *stable* if and only if all the poles of $G(s)$ / $G[z]$ are strictly within the left-half s -plane/the unit circle in the z -plane (see Remarks 2.5, 3.2, 4.5, and Theorem A.1 in Appendix A).

1.2.9 Causality

A signal $x(t)/x[n]$ is said to be *causal* if it is zero for all $t < 0$ / $n < 0$. A system is said to be *causal* or *non-anticipatory* if the response (output) of the system does not precede the excitation (input), i.e., the output depends only on the present and past values of the input. In other words, the output of a causal system to an input appears only while or after the input is applied to the system. The necessary and sufficient condition for the system causality is that the impulse response is causal, i.e.,

$$g(t) = 0 \text{ for all } t < 0 \qquad (1.2.28a) \quad | \quad g[n] = 0 \text{ for all } n < 0 \qquad (1.2.28b)$$

since the impulse response $g(t)$ / $g[n]$ means the output measured at time t [s]/ n [samples] after the impulse signal is applied as an input.

We can use Eqs. (1.2.6)/(1.2.10) (with Eqs. (1.2.7)/(1.2.11)) to write the time-domain input-output relationships of continuous-time/discrete-time causal LTI systems as

$$y(t) = \int_{-\infty}^{\infty} x(\tau)g(t-\tau)d\tau \underset{\text{causal}}{\overset{g(t-\tau)=0 \text{ for } t-\tau<0}{=}} \int_{-\infty}^{t} g(\tau)x(t-\tau)d\tau$$

$$= \int_{-\infty}^{t_0} x(\tau)g(t-\tau)d\tau + \int_{t_0}^{t} x(\tau)g(t-\tau)d\tau;$$

$$y(t) = y(t_0) + \int_{t_0}^{t} x(\tau)g(t-\tau)d\tau \qquad (1.2.29a)$$

$$y[n] = \sum_{m=-\infty}^{\infty} x[m]g[n-m] \underset{\text{causal}}{\overset{g[n-m]=0 \text{ for } n-m<0}{=}} \sum_{m=-\infty}^{n} x[m]g[n-m]$$

$$= \sum_{m=-\infty}^{n_0} x[m]g[n-m] + \sum_{m=n_0+1}^{n} x[m]g[n-m];$$

$$y[n] = y[n_0] + \sum_{m=n_0+1}^{n} x[m]g[n-m] \qquad (1.2.29b)$$

In fact, all physical systems such as analog filters are causal in the sense that they do not react to any input before it is applied to them. As a contraposition, a *non-causal* or *anticipatory* system is not physically realizable. Causal filters are typically employed in applications where the inputs are (on-line) processed as they arrive. An example of non-causal system is a digital filter for image processing, which collects the input data $x[m, n]$'s in one frame for some period and processes them at a time where m and n are spatial indices.

1.2.10 Invertibility

A system is said to be *invertible* if distinct inputs produce distinct outputs. If a system G is invertible, then it is possible to design its inverse system H which, connected in cascade with G at the output port, receives the output of G to restore the input (applied to G) as its output.

Especially for an LTI system G with the impulse response $g(t)/g[n]$, the impulse response $h(t)/h[n]$ of its inverse system H must satisfy the following condition: (see Problem 1.6)

$$g(t)*h(t) = \delta(t);\ G(s)H(s) = 1 \tag{1.2.30a}$$

$$g[n]*h[n] = \delta[n];\ G[z]H[z] = 1 \tag{1.2.30b}$$

(cf.) An example of non-invertible system is a gambler (with a poker face) whose face (output) does not vary with his card (input) so that other gamblers cannot make the inverse system reading his card from his face.

1.3 Systems Described by Differential/Difference Equations

1.3.1 Differential/Difference Equation and System Function

The time-domain input-output relationships of continuous-time/discrete-time systems are often described by linear constant-coefficient differential/difference equations that are set up based on underlying physical laws or design specifications to make it perform desired operations:

$$\sum_{i=0}^{N} a_i \frac{d^i y(t)}{dt^i} = \sum_{j=0}^{M} b_j \frac{d^j x(t)}{dt^j}$$

with the initial conditions

$$y(t_0),\ y'(t_0),\ \cdots,\ y^{(N-1)}(t_0)$$

$$\sum_{i=0}^{N} a_i\, y[n-i] = \sum_{j=0}^{M} b_j x[n-j]$$

with the initial conditions

$$y[n_0],\ y[n_0-1],\ \cdots,\ y[n_0-N+1]$$

With zero initial conditions, this can be transformed to make the system or transfer function as

$$\sum_{i=0}^{N} a_i s^i Y(s) = \sum_{j=0}^{M} b_j s^j X(s)$$

$$A(s)Y(s) = B(s)X(s);$$

$$G(s) = \frac{Y(s)}{X(s)} = \frac{B(s)}{A(s)} \tag{1.3.1a}$$

where

$$A(s) = \sum_{i=0}^{N} a_i s^i,\ B(s) = \sum_{j=0}^{M} b_j s^j$$

$$\sum_{i=0}^{N} a_i z^{-i} Y[z] = \sum_{j=0}^{M} b_j z^{-j} X[z]$$

$$A[z]Y[z] = B[z]X[z];$$

$$G[z] = \frac{Y[z]}{X[z]} = \frac{B[z]}{A[z]} \tag{1.3.1b}$$

where

$$A[z] = \sum_{i=0}^{N} a_i z^{-i},\ B[z] = \sum_{j=0}^{M} b_j z^{-j}$$

Note the following:

- The *poles/zeros* of the system function $G(s)$ or $G[z]$ are the values of s or z at which its denominator/numerator becomes zero.

- The degree N of the denominator $A(s)/A[z]$ of the system function $G(s)/G[z]$ is called the *order* of the system. If $N \neq 0$, the system is said to be *recursive* in the sense that its output depends on not only the input but also the previous output; otherwise, it is said to be *non-recursive* or *memoryless* and its output depends only on the input.
- Especially, discrete-time recursive and non-recursive systems are called *IIR* (*i*nfinite-duration *i*mpulse *r*esponse) and *FIR* (*f*inite-duration *i*mpulse *r*esponse) systems, respectively since the duration of the impulse response of recursive/non-recursive system is infinite/finite, respectively.

1.3.2 Block Diagrams and Signal Flow Graphs

Systems are often described by graphical means such as block diagrams or signal flow graphs. As an example, let us consider an *RC* circuit or its equivalent depicted in Fig. 1.16(a). We can apply Kirchhoff's current law to write the node equation, take its Laplace transform with zero initial conditions, and find the system function as

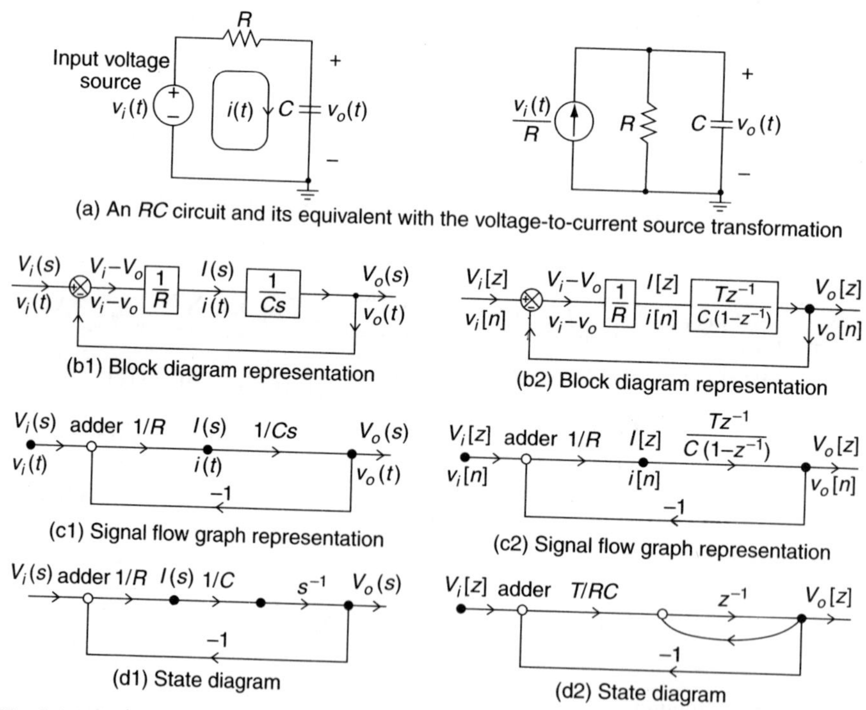

Fig. 1.16 Block diagram, signal flow graph, and state diagram representations of a system

$$C\frac{dv_o(t)}{dt} + \frac{v_o(t)}{R} = \frac{v_i(t)}{R} \tag{1.3.2a}$$

$$\underset{\text{B.7(5)}}{\overset{\text{Laplace transform}}{\rightarrow}} CsV_o(s) + \frac{1}{R}V_o(s) = \frac{1}{R}V_i(s) \rightarrow G(s) = \frac{V_o(s)}{V_i(s)} = \frac{1}{RCs+1} \tag{1.3.3a}$$

We may replace the derivative $dv_o(t)/dt$ with its difference approximation

$$\frac{dv_o(t)}{dt} \cong \frac{v_o((n+1)T) - v_o(nT)}{T} = \frac{v_o[n+1] - v_o[n]}{T}$$

with sampling interval T to discretize the differential equation into a difference equation, take its z -transform with zero initial conditions, and find the system function as

$$C\frac{v_o[n+1] - v_o[n]}{T} + \frac{v_o[n]}{R} = \frac{v_i[n]}{R} \tag{1.3.2b}$$

$$\underset{\text{B.7(2)}}{\overset{z\text{-transform}}{\rightarrow}} C\frac{zV_o[z] - V_o[z]}{T} + \frac{1}{R}V_o[z] = \frac{1}{R}V_i[z]$$

$$\rightarrow G[z] = \frac{V_o[z]}{V_i[z]} = \frac{1}{RC(z-1)/T + 1} \tag{1.3.3b}$$

Fig. 1.16(b1)/(b2) show the block diagram representing the continuous-time/discrete-time system whose input-output relationship is described by Eqs. (1.3.2a)/(1.3.2b) in the time domain and (1.3.3a)/(1.3.3b) in the s- / z -domain. Figure 1.16(c1)/(c2) and (d1)/(d2) show their *signal flow graph* representations where a branch from node j to node i denotes the causal relationship that the signal j is multiplied by the branch gain and contributes to the signal i. Especially, Fig. 1.16(d1) is called a *continuous-time state diagram* since all branch gains are constants or s^{-1} (denoting an integrator). Likewise, Fig. 1.16(d2) is called a *discrete-time state diagram* since all branch gains are constants or z^{-1} (denoting a delay T).

Since signal flow graphs are simpler to deal with than block diagrams, we will rather use signal flow graphs than block diagrams. A signal flow graph was originally introduced by S.J. Mason as a graphical means of describing the cause-effect relationship among the variables in a set of linear algebraic equations. It consists of nodes connected by line segments called *branch*es. Each *node* represents a signal (variable), whose value is the sum of signals coming along all the branches from other nodes and being multiplied by the branch gain. Every branch has the gain and direction that determine or are determined by the cause-effect relationship among the variables denoted by its nodes. Note that a branch with no gain indicated is supposed to have unity gain.

For example, we consider the following set of linear equations:

$$y_2 = a_{12}y_1 + a_{32}y_3, \qquad y_3 = a_{23}y_2 + a_{43}y_4,$$
$$y_4 = a_{34}y_3 + a_{44}y_4 + a_{54}y_5, \quad y_5 = a_{25}y_2 + a_{35}y_3 + a_{45}y_4$$

which describes the cause-effect relationship among the variables y_1, y_2, y_3, y_4, and y_5 with the causes/effects in the right/left-hand side, respectively. Figure 1.17 shows a signal flow graph representing the relationships described by this set of equations.

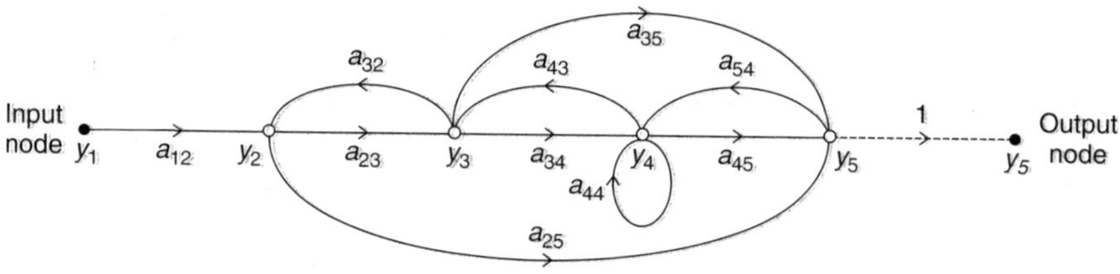

Fig. 1.17 A signal flow graph

1.3.3 General Gain Formula – Mason's Formula

In this section we introduce *Mason's gain formula*, which is applied to signal flow graphs to yield the overall gain from an input node to an output node. To understand how to use the formula, we need to know the following terms:

- Input Node (Source) : A node having only outgoing branches.
- Output Node (Sink) : A node having only incoming branches.

(cf.) Note that, in contrast with the input node, the output node may not be clearly seen. Suppose we don't have the dotted branch in the signal flow graph depicted in Fig. 1.17. In that case, if we regard y_5 as an output, we may draw a branch with unity gain from the node for y_5 so that the node appears to be an output node.

- Path: A continuous connection of branches traversed in the same direction.
- Forward Path: A path connecting an input node to an output node along which no node is encountered more than once.
- Loop: A closed path that originates and terminates at the same node and encounters no other node more than once.
- Path Gain: The product of all the branch gains along a path.
- Loop Gain: The path gain of a loop.
- Non-touching: Having no nodes in common.

The gain formula states that the overall gain is

$$G = \frac{y_{out}}{y_{in}} = \frac{1}{\Delta} \sum_{k=1}^{N} M_k \Delta_k \tag{1.3.4}$$

$$\text{with } \Delta = 1 - \sum_m P_{m1} + \sum_m P_{m2} - \sum_m P_{m3} + \cdots \tag{1.3.5}$$

where

N: Total number of forward paths from node y_{in} to node y_{out}
M_k: Path gain of the k th forward path
P_{mr}: Gain product of the m th combination of r nontouching loops
Δ_k : Δ (Eq. (1.3.5)) for the part of the signal flow graph not touching the k th forward path

It may seem to be formidable to use at first glance, but is not so complicated in practice since most systems have not so many non-touching loops. For example, we can apply Mason's gain formula to the signal flow graph depicted in Fig. 1.18 as follows:

$$G = \frac{y_5}{y_1} = \frac{1}{\Delta} \sum_{k=1}^{N} M_k \Delta_k$$

with

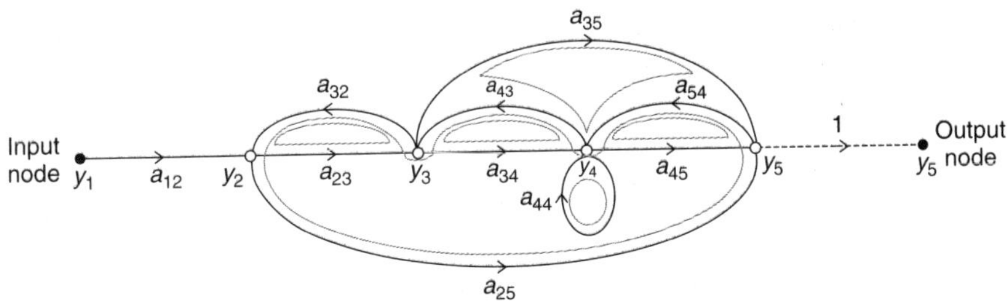

Fig. 1.18 A signal flow graph with its loops denoted by closed curves

$N = 3$ (the total number of forward paths from node y_1 to node y_5)
$M_1 = a_{12}a_{23}a_{34}a_{45}$, $\Delta_1 = 1$ for the forward path $y_1 - y_2 - y_3 - y_4 - y_5$
$M_2 = a_{12}a_{23}a_{35}$, $\Delta_2 = 1 - a_{44}$ for the forward path $y_1 - y_2 - y_3 - y_5$
$M_3 = a_{12}a_{25}$, $\Delta_3 = 1 - (a_{34}a_{43} + a_{44})$ for the forward path $y_1 - y_2 - y_5$

$$\Delta = 1 - (a_{23}a_{32} + a_{34}a_{43} + a_{44} + a_{45}a_{54} + a_{35}a_{54}a_{43} + a_{25}a_{54}a_{43}a_{32}) + (a_{23}a_{32}a_{44} + a_{23}a_{32}a_{45}a_{54})$$

1.3.4 State Diagrams

Now we introduce a special class of signal flow graphs, called the *state diagram* or *state transition signal flow graph*, in which every branch has a gain of constant or s^{-1} (integrator)/z^{-1} (delay). This is very useful for the purposes of system analysis, design, realization, and implementation. Systems represented by the state diagram need the following three basic software operations or hardware elements for implementation:

- addition(adder)	- addition
- multiplication(amplifier)	- multiplication
- integration(integrator) s^{-1}	- delay (z^{-1})/advance (z)

It is good enough to take a look at the following examples.

Example 1.4a Differential Equation and Continuous-Time State Diagram

Figure 1.19(a) and (b) show the controllable and observable form of state diagrams, respectively, both of which represent the following differential equation or its Laplace transform (see Problem 1.9):

$$y''(t) + a_1 y'(t) + a_0 y(t) = b_1 u'(t) + b_0 u(t) \text{ with zero initial conditions} \tag{E1.4a.1}$$

$$(s^2 + a_1 s + a_0)Y(s) = (b_1 s + b_0)U(s) \tag{E1.4a.2}$$

Example 1.4b Difference Equation and Discrete-Time State Diagram

Figure 1.20(a)/(b)/(c)/(d) show the direct I/transposed direct I/direct II/transposed direct II form of state diagrams, respectively, all of which represent the following difference equation or its z -transform (see Problems 1.8 and/or 8.4):

$$y[n+2] + a_1 y[n+1] + a_0 y[n] = b_1 u[n+1] + b_0 u[n] \text{ with zero initial conditions} \tag{E1.4b.1}$$

$$(z^2 + a_1 z + a_0)Y[z] = (b_1 z + b_0)U[z] \tag{E1.4b.2}$$

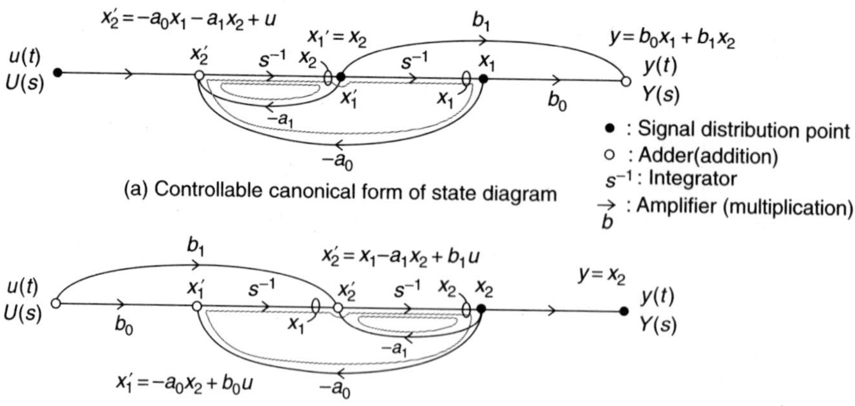

(a) Controllable canonical form of state diagram

(b) Observable canonical form of state diagram

Fig. 1.19 State diagrams for a given differential equation $y''(t)+a_1 y'(t)+a_0 y(t) = b_1 u'(t)+b_0 u(t)$

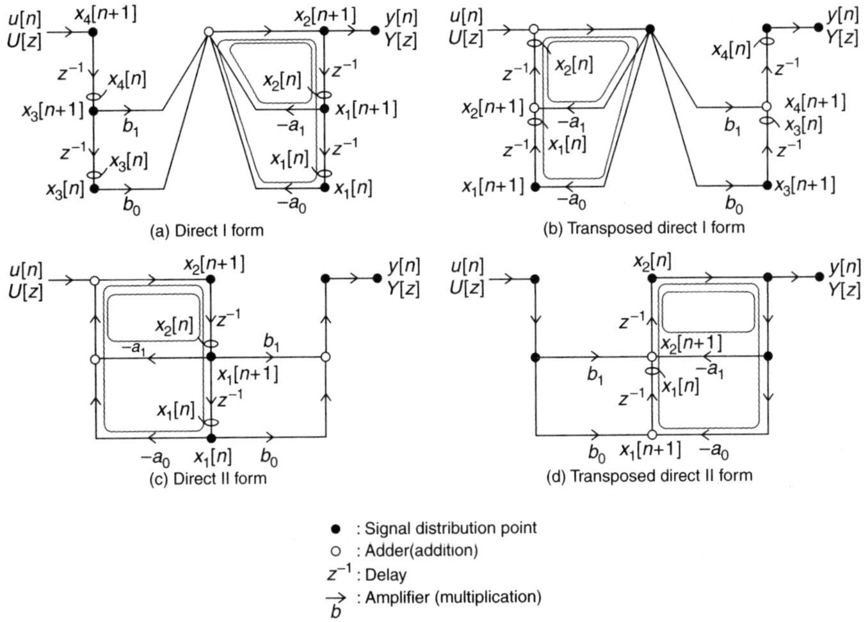

Fig. 1.20 State diagrams for a given difference equation $y[n + 2] + a_1 y[n + 1] + a_0 y[n] = b_1 u[n + 1] + b_0$

For example, we can write the state and output equations for the state diagram of Fig. 1.20(c) as

$$\begin{bmatrix} x_1[n+1] \\ x_2[n+1] \end{bmatrix} = \begin{bmatrix} 0 & 1 \\ -a_0 & -a_1 \end{bmatrix} \begin{bmatrix} x_1[n] \\ x_2[n] \end{bmatrix} + \begin{bmatrix} 0 \\ 1 \end{bmatrix} u[n] \tag{E1.4b.3}$$

$$y[n] = b_1 x_2[n] + b_0 x_1[n] \tag{E1.4b.4}$$

We can apply Eq. (8.3.2b) with Eqs. (1.4b.3,4) or Mason's gain formula for Fig. 1.20(c) to obtain the transfer function $G[z] = Y[z]/U[z]$ or equivalently, the input-output relationship (E1.4b.2).

The digital filter or controller may be realized either by using a general-purpose computer or a special digital hardware designed to perform the required computations. In the first case, the filter structure represented by the state diagram may be thought of as specifying a computational algorithm where the number of z or z^{-1} is proportional to the required memory size. In the latter case, it may be regarded as specifying a hardware configuration where z^{-1} denotes a delay element like a flip-flop. Note that, as shown in the above examples, the structure to solve or implement a given differential/difference equation is not unique and it can be rearranged or modified in a variety of ways without changing the overall input-output relationship or system function.

1.4 Deconvolution and Correlation

In this section we will introduce the two terms related with the convolution. One is the *deconvolution*, which is the inverse process of convolution. The other is the correlation, which is procedurally similar to, but totally different from the convolution in the physical meaning and usage.

1.4.1 Discrete-Time Deconvolution

In Sect. 1.2.3, the input-output relationship of a discrete-time LTI (linear time-invariant) system was derived as Eq. (1.2.9):

$$y[n] \overset{(1.2.9)}{=} g[n]*x[n] = \sum\nolimits_{m=-\infty}^{\infty} g[n-m]x[m] \tag{1.4.1}$$

where $x[n]$, $y[n]$, and $g[n]$ are the input, output, and impulse response of the system. Thus, for an FIR ($f$inite-duration $i$mpulse $r$esponse) system with

$$g[n] = 0 \text{ for } |n| \geq N, \tag{1.4.2}$$

we can write its output to the input $\{x[m] \text{ for } m = 0 : N-1\}$ applied from $m = 0$ to $N-1$ as

$$y[n] = \sum\nolimits_{m=0}^{N-1} g[n-m]x[m] \tag{1.4.3}$$

or in matrix-vector form as

$$\begin{bmatrix} y[0] \\ y[1] \\ \bullet \\ y[N-1] \end{bmatrix} = \begin{bmatrix} g[0] & g[-1] & \bullet & g[-N+1] \\ g[1] & g[0] & \bullet & g[-N+2] \\ \bullet & \bullet & \bullet & \bullet \\ g[N-1] & g[N-2] & \bullet & g[0] \end{bmatrix} \begin{bmatrix} x[0] \\ x[1] \\ \bullet \\ x[N-1] \end{bmatrix};$$

$$\mathbf{y} = G\mathbf{x} \text{ with } G: \ transmission\ matrix \tag{1.4.4}$$

One might ask a question, "Can we or how can we find the input sequence $x[n]$ for a system characterized by its impulse response $g[n]$ to produce a certain output sequence $y[n]$?". For the FIR system with the input-output relationship described by Eq. (1.4.4), we can find the input sequence as

$$\mathbf{x} = G^{-1}\mathbf{y}, \tag{1.4.5}$$

if the transmission matrix G is nonsingular. Furthermore, if the system is causal, i.e., $g[n] = 0$ for $n < 0$, then the transmission matrix becomes lower-triangular as

$$G = \begin{bmatrix} g[0] & 0 & \cdot & 0 \\ g[1] & g[0] & \cdot & 0 \\ \cdot & \cdot & \cdot & \cdot \\ g[N-1] & g[N-2] & \cdot & g[0] \end{bmatrix} \tag{1.4.6}$$

so that if $g[0] \neq 0$, the input sequence can be determined by forward substitution:

$$x[n] = \frac{y[n] - \sum_{m=0}^{n-1} g[n-m]x[m]}{g[0]} \tag{1.4.7}$$

starting from $x[0] = y[0]/g[0]$ (see Problem 1.11).

Note that the problem of determining the impulse response $g[n]$ for given input and output sequences can be dealt with in the same way.

1.4.2 Continuous/Discrete-Time Correlation

The cross-correlation between two signals measures how similar the two signals are over temporal or spatial difference. The auto-correlation of a signal measures how much the signal looks like itself over temporal or spatial difference. Correlation is widely used for various purposes including signal detection in communication systems.

Let $x(t)/x[n]$ and $y(t)/y[n]$ be two continuous-time/discrete-time transient or finite-duration signals. Then the cross-correlation between them is defined as follows:

$$\begin{aligned} \phi_{xy}(\tau) &= \int_{-\infty}^{\infty} x(t+\tau)y^*(t)dt \\ &= \int_{-\infty}^{\infty} x(t)y^*(t-\tau)dt \\ &= \int_{-\infty}^{\infty} x(-t+\tau)y^*(-t)dt \end{aligned} \tag{1.4.8a}$$

$$\begin{aligned} \phi_{xy}[m] &= \sum_{n=-\infty}^{\infty} x[n+m]y^*[n] \\ &= \sum_{n=-\infty}^{\infty} x[n]y^*[n-m] \\ &= \sum_{n=-\infty}^{\infty} x[-n+m]y^*[-n] \end{aligned} \tag{1.4.8b}$$

where τ/m, called a lag variable, represents the relative delay (shift) between the two signals. If the two signals happen to be identical, i.e., $x(t) = y(t)/x[n] = y[n]$, we have the auto-correlation as

$$\phi_{xx}(\tau) = \int_{-\infty}^{\infty} x(t+\tau)x^*(t)dt = \int_{-\infty}^{\infty} x(t)x^*(t-\tau)dt = \int_{-\infty}^{\infty} x(-t+\tau)x^*(-t)dt \quad (1.4.9a)$$

$$\phi_{xx}[m] = \sum\nolimits_{n=-\infty}^{\infty} x[n+m]x^*[n] = \sum\nolimits_{n=-\infty}^{\infty} x[n]x^*[n-m] = \sum\nolimits_{n=-\infty}^{\infty} x[-n+m]x^*[-n] \quad (1.4.9b)$$

There are some properties of correlation to note:

$$\phi_{xx}(\tau) = \phi_{xx}^*(-\tau) \quad (1.4.10a) \qquad \phi_{xx}[m] = \phi_{xx}^*[-m] \quad (1.4.10b)$$

$$\phi_{xy}(\tau) = \phi_{yx}^*(-\tau) \quad (1.4.11a) \qquad \phi_{xy}[m] = \phi_{yx}^*[-m] \quad (1.4.11b)$$

$$\phi_{xy}(\tau) = x(\tau)*y^*(-\tau) \quad (1.4.12a) \qquad \phi_{xy}[m] = x[m]*y^*[-m] \quad (1.4.12b)$$

If $x(t)/x[n]$ and $y(t)/y[n]$ are the input and output of a continuous-time/discrete-time LTI system so that $y(t) = g(t) * x(t)/y[n] = g[n] * x[n]$, then we have

$$\phi_{xy}(\tau) = g^*(-\tau)*\phi_{xx}(\tau) \quad (1.4.13a) \qquad \phi_{xy}[m] = g^*[-m]*\phi_{xx}[m] \quad (1.4.13b)$$

$$\phi_{yy}(\tau) = g^*(-\tau)*\phi_{yx}(\tau) \quad (1.4.14a) \qquad \phi_{yy}[m] = g^*[-m]*\phi_{yx}[m] \quad (1.4.14b)$$

(cf.) If the signals are random or periodic so that nonzero values appear all over the time interval, the correlation should be defined as a time-average version.

The correlation functions are very useful for extracting certain signals from noisy ones and determining the spectral density of random signals. Look at the following remark and example.

Remark 1.7 Properties of Autocorrelation

(1) In case two signals are periodic or random, their correlation should be defined as

$$\phi_{xy}(\tau) = \frac{1}{2T}\int_{-T}^{T} x(t+\tau)y^*(t)dt \quad (1.4.15a) \qquad \phi_{xy}[m] = \frac{1}{2N+1}\sum\nolimits_{n=-N}^{N} x[n+m]y^*[n] \quad (1.4.15b)$$

(2) The autocorrelation (1.4.9a)/(1.4.9b) is even and has its maximum at $\tau = 0/m = 0$. In the case of periodic or random signals, the maximum value is the mean squared value of the signal.

Remark 1.8 Convolution vs. Correlation and Matched Filter

(1) Equation (1.2.7)/(1.2.11) implies that the continuous-time/discrete-time convolution of two time functions/sequences can be obtained by time-reversing one of them and time-shifting (sliding) it, multiplying it with the other, and then integrating/summing the multiplication. The correlation differs from the convolution only in that the time-reversal is not performed.
(2) If we time-reverse one of two signals and then take the convolution of the time-reversed signal and the other one, it will virtually yield the correlation of the original two signals since time-reversing the time-reversed signal for computing the convolution yields the original signal as if it had not been time-reversed. This presents the idea of matched filter, which is to determine the correlation between the input signal $x(t)/x[n]$ and a particular signal $w(t)/w[n]$ based on the output of the system with the impulse response $g(t) = w(-t)/g[n] = w[-n]$ to the input $x(t)/x[n]$. This system having the time-reversed and possibly delayed version of a particular signal as its impulse response is called the *matched filter* for that signal. Matched filter is used to detect a signal, i.e., to determine whether or not the signal arrives and find when it arrives.

Example 1.5 Correlation and Matched Filter

Consider the two signal waveforms of duration $T = 2$ [s], which represent 0 and 1 and are depicted in Fig. 1.21(a1) and (b1), respectively. According to Remark 1.8(2), the impulse responses of their matched filters are

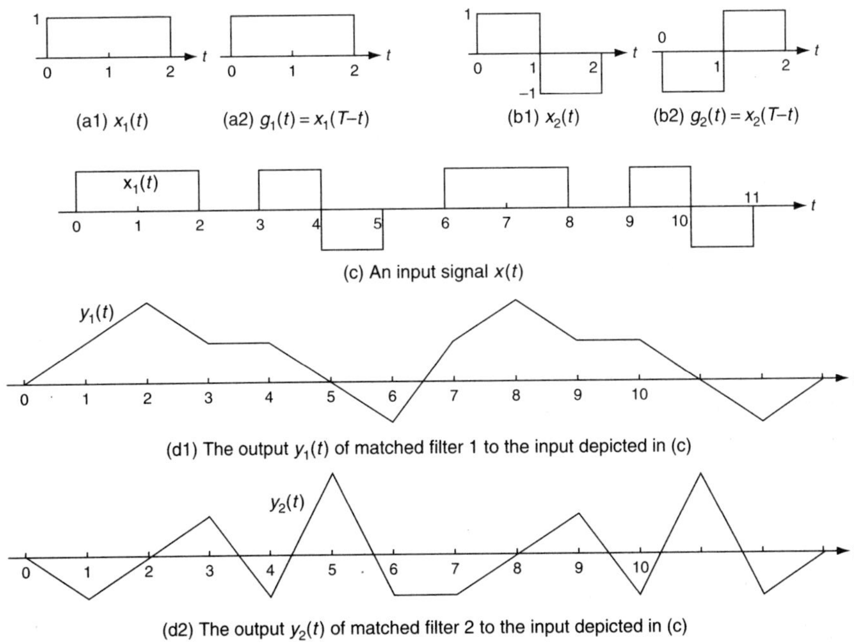

Fig. 1.21 Two matched filters for two signals and their outputs to an input signal

$$g_1(t) = x_1(-t) \quad \text{and} \quad g_2(t) = x_2(-t) \tag{E1.5.1}$$

We have, however, a problem that these filters are non-causal so that they are not physically realizable. A reasonable measure is to shift their impulse responses by $T = 2$ so that they become causal as depicted in Fig. 1.21(a2) and (b2):

$$g_1(t) = x_1(T-t) \quad \text{and} \quad g_2(t) = x_2(T-t) \tag{E1.5.2}$$

Suppose the input signal depicted in Fig. 1.21(c) is applied to the two matched filters with the impulse responses given by (E1.5.2). We can compose the following MATLAB program "sig01e05.m" to find their outputs and run it to get the results shown in Figs. 1.21(d1) and (d2). There are some points to think about:

- As long as the amplitude of the signals (expected to arrive) are the same, the output of each matched filter achieves its maximum 2 s after the corresponding signal arrives. Why?

```
%sig01e05.m
% Correlation/Convolution and Matched Filter
clear, clf
M=50; Ts=1/M;
x1=ones(M,1)*[1  1]; x1=x1(:).'; Nx=length(x1);
x2=ones(M,1)*[1 -1]; x2=x2(:).';
g1=fliplr(x1); g2=fliplr(x2);
x=[x1 zeros(1,M) x2 zeros(1,M) x1 zeros(1,M) x2]; % signal to transmit
length_x=length(x); Nbuffer= min(M*11,length_x); tt=[0:Nbuffer-1]*Ts;
% Noise_amp=0.3; x = x + Noise_amp*randn(1,length_x);
xbuffer=zeros(1,Nbuffer); ybuffer=zeros(2,Nbuffer);
for n=1:length_x
   xbuffer= [x(n) xbuffer(1:end-1)];
   y= [g1; g2]*xbuffer(1:Nx).'*Ts;  ybuffer= [ybuffer(:,2:end) y];
   subplot(312), plot(tt,ybuffer(1,:)), subplot(313), plot(tt,ybuffer(2,:))
   pause(0.01), if n<length_x, clf; end
end
y1=xcorr(x,x1)*Ts; y1=y1([end-Nbuffer+1:end]-Nx); %correlation delayed by Nx
y2=xcorr(x,x2)*Ts; y2=y2([end-Nbuffer+1:end]-Nx);
subplot(312), hold on,plot(tt,y1,'m') % only for cross-check
subplot(313), hold on, plot(tt,y2,'m')
```

- If we remove the zero period between the signals and generate a signal every 2 s in the input, could we still notice the signal arrival times from (local) maxima of the output? (See Problem 1.12(a).)

Remark 1.9 `xcorr()` – MATLAB function for Correlation

(1) The MATLAB function `xcorr(x,y)` returns the correlation of two sequences $x[n]$ and $y[n]$ defined by Eq. (1.4.16). Depending on the third input argument, it returns different versions of correlation as follows:

`xcorr(x,y,'coeff')` →

$$\rho_{xy}[m] = \frac{\sum_{n=-\infty}^{\infty} x[n+m]\, y^*[n]}{(\sum |x[n]|^2 \sum |y[n]|^2)^{1/2}} (-1 \le \rho \le +1)$$
(correlation coefficient) (1.4.16a)

`xcorr(x,y,'biased')` →

$$\varphi_{xy}[m] = \begin{cases} \frac{1}{N} \sum_{n=0}^{N-1-|m|} x[n+m]\, y^*[n] & \text{for } 0 \le m \le N-1 \\ \frac{1}{N} \sum_{n=0}^{N-1-|m|} x[n]\, y^*[n-m] & \text{for } -(N-1) \le m \le 0 \end{cases} \quad (1.4.16b)$$

`xcorr(x,y,'unbiased')` →

$$\varphi_{xy}[m] = \begin{cases} \frac{1}{N-|m|} \sum_{n=0}^{N-1-|m|} x[n+m]\, y^*[n] & \text{for } 0 \le m \le N-1 \\ \frac{1}{N-|m|} \sum_{n=0}^{N-1-|m|} x[n]\, y^*[n-m] & \text{for } -(N-1) \le m \le 0 \end{cases} \quad (1.4.16c)$$

(2) If the two sequences are of different lengths, `xcorr(x,y)` appends one of them by zeros to make them have equal lengths and returns their convolution.
(3) The durations of the convolution and correlation of two sequences $x[n]$ (of duration $[n_{0\_x} : n_{f\_x}]$) and $y[n]$ (of duration $[n_{0\_y} : n_{f\_y}]$) are $[n_{0\_x} + n_{0\_y} : n_{f\_x} + n_{f\_y}]$ and $[n_{0\_x} - n_{f\_y} : n_{f\_x} - n_{0\_y}]$, respectively.

Example 1.6 Correlation for Periodic Signals with Random Noise

We can use MATLAB function `randn(M,N)` to generate a zero-mean and unit-variance normal (Gaussian) noise sequence in an M × N matrix. Figure 1.22(a1) and (a2) illustrate a Gaussian 1 × 128 random sequence $w[n]$ with variance $\sigma^2 = 0.5^2$ (generated by `0.5*randn(1,N)`) and its autocorrelation $\phi_{ww}[m]$ (obtained by `xcorr(w,w)`), respectively. On the other hand, Fig. 1.22(b1) shows a noise-contaminated sinusoidal sequence $x[n] = \sin(\pi n/16) + w[n]$. Figure 1.22(b2), (b3), and (b4) show the various autocorrelations of $x[n]$ that are obtained by using `xcorr(w,w)`, `xcorr(w,w,'coef')/xcorr(w,w, 'biased')`, and `xcorr(w,w,'unbiased')`, respectively.

Note the following observations:

- All the autocorrelations in Fig. 1.22(a2), (b2)–(b4) are even and have their maxima at $m = 0$ as stated in Remark 1.7(2). Especially, the maximum of $\phi_{ww}[m]$ in Fig. 1.22(a2) is

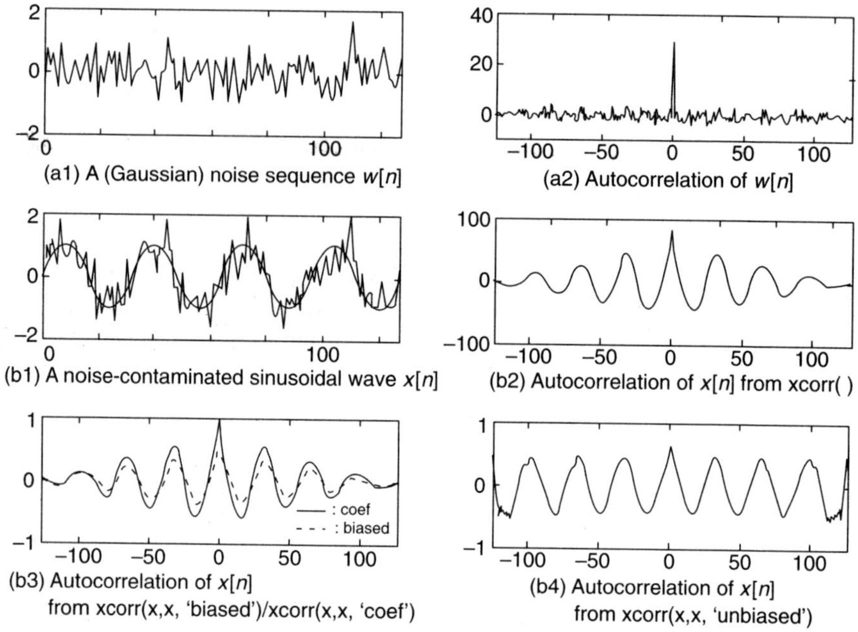

Fig. 1.22 Autocorrelation obtained using the MATLAB function `xcorr()`

$$\phi_{ww}[0] \simeq \text{Expectation}\left\{\sum_{n=0}^{N-1} x^2[n]\right\} \simeq \sum_{n=0}^{N-1} \sigma^2 = 128 \times 0.5^2 = 32 \tag{E1.6.1}$$

- Marvelously, the noise effect can hardly be found in the autocorrelation except for the spike at $m = 0$. Thanks to the noise reduction effect of autocorrelation, the periodicity of the signal can be observed more clearly from the autocorrelation than from the original noise-contaminated signal.
- Figure 1.22(b2) and (b3) show that the autocorrelation obtained from `xcorr(w,w,'biased')` is just a $1/N$-scaled version of that obtained from `xcorr(w,w)`. The normalized version obtained from `xcorr(w,w,'coef')` has similar shape. Any of these versions have some end effect, i.e., their magnitudes decrease as the time lag m increases not because the correlation or similarity gets loose, but because the number of samples in the summation decreases. In contrast, the autocorrelation obtained from `xcorr(w,w,'unbiased')` is relatively free from the end effect as shown in Fig. 1.22(b4).

```
%sig01e06.m : Autocorrelation of a noise-contaminated periodic signal
N=128; W=pi/16; nn=0:N-1;
s= sin(W*nn); w= 0.5*randn(1,N); x=s+w;
subplot(321), plot(nn,w)
[phi_w,mlag]=xcorr(w,w); % Autocorrelation of the noise
subplot(322), plot(mlag,phi_w)
```

```
subplot(323), plot(nn,x, nn,s,'k')
[phi_x,mlag]=xcorr(x,x); % Autocorrelation of the corrupted signal
subplot(324), plot(mlag,phi_x)
[phi_x_coef,mlag]=xcorr(x,x,'coef'); % Correlation coefficient
subplot(325), plot(mlag,phi_x_coef)
[phi_x_biased,mlag]=xcorr(x,x,'biased'); % Biased autocorrelation
hold on, plot(mlag,phi_x_biased,'r:')
[phi_x_unbiased,mlag]=xcorr(x,x,'unbiased'); % Unbiased autocorrelation
subplot(326), plot(mlag,phi_x_unbiased),axis([mlag([1 end]) -1 1])
```

1.5 Summary

In this chapter we have introduced some basic continuous-time/discrete-time signals and defined a number of important concepts related to systems such as linearity, time-invariance, causality, stability, impulse response, and system (transfer) function. We have also derived the convolution form of input-output relationship of LTI systems. We have also indicated that a continuous-time/discrete-time system can be described analytically by differential/difference equations and pictorially by signal flow graphs. Finally, we introduced the concept of correlation.

Problems

1.1 Representation of Unit Impulse Function by Sinc Function

(a) Referring to the web site <http://mathworld.wolfram.com/SincFunction.html> or [K-1] (Sect. 15.4, Example 2) or Eq. (E2.3.3) of this book, show that the sinc function representation (1.1.33a) of the unit impulse function $\delta(t)$ satisfies the property (1.1.19) regardless of D or w:

$$\int_{-\infty}^{\infty} \frac{w}{\pi} \frac{\sin(wt)}{wt} dt \overset{wt \to x}{=} \frac{1}{\pi} \int_{-\infty}^{\infty} \frac{\sin(x)}{x} dx = 1 \tag{P1.1.1}$$

(b) Plot the sinc function

$$\frac{1}{D}\text{sinc}\left(\frac{t}{D}\right) = \frac{\sin(\pi t/D)}{\pi t} \tag{P1.1.2}$$

against time t for $D = 1, 0.5, 0.25$, and 0.125 and check the following:

- It is an even function since both of the numerator and denominator functions are odd ones.
- Its amplitude decreases as $|t|$ increases since the magnitude of the denominator increases while the amplitude of the numerator is constant.
- It becomes zero at $t = mD(m \neq 0)$ and $1/D$ at $t = 0$ since $\lim_{x \to 0} \text{sinc}(x) = 1$.

1.2 Convolution
Consider the two continuous-time signals $x(t) = r_3(t) = u_s(t) - u_s(t-3)$ and $g(t) = e^{-0.5t}u_s(t)$.

(a) Show that the (continuous-time) convolution of these two signals is

$$x(t)*g(t) = \begin{cases} 0 & \text{for } t < 0 \\ 2(1-e^{-0.5t}) & \text{for } 0 \le t < 3 \\ 2(e^{-0.5(t-3)} - e^{-0.5t}) = 2(e^{1.5}-1)e^{-0.5t} & \text{for } t \ge 3 \end{cases} \quad \text{(P1.2.1)}$$

(b) As in Example 1.2, we can use the MATLAB function 'conv ()' to compute this continuous-time convolution approximately. Compose a program which samples the two signals at $t = nT_s(T_s = 0.1[\text{s}])$, use 'conv ()' to compute the convolution, and plot it together with the true convolution (P1.2.1) for $t = [0, 10]$s. Run it to see if the approximate convolution is close to the true one. If we decrease the sampling period, say, to $T_s = 0.01$ s, is it helpful?

1.3 Necessary Condition on BIBO (Bounded Input Bounded Output) Stability
In Sect. 1.2.8 it was shown that the absolute integrability/summability (1.2.27a)/(1.2.27b) of impulse response guarantees the BIBO stability and thus are sufficient conditions. Here, you can show that if the conditions do not hold, there is a bounded input yielding unbounded output, which implies that they are also necessary conditions for BIBO stability. To this end, suppose that the impulse response of a system does not satisfy the condition (1.2.27a)/(1.2.27b):

$$\int_{-\infty}^{\infty} |g(\tau)| d\tau = \infty \quad \text{(P1.3.1a)}$$

$$\sum\nolimits_{m=-\infty}^{\infty} |g[m]| = \infty \quad \text{(P1.3.1b)}$$

(a) Consider the following signal:

$$x(t) = \begin{cases} 0 & \text{for } t \text{ such that } g(-t) = 0 \\ \frac{g^*(-t)}{|g(-t)|} & \text{for } t \text{ such that } g(-t) \ne 0 \end{cases} \quad \text{(P1.3.2a)}$$

$$x[n] = \begin{cases} 0 & \text{for } n \text{ such that } g[-n] = 0 \\ \frac{g^*[-n]}{|g[-n]|} & \text{for } n \text{ such that } g[-n] \ne 0 \end{cases} \quad \text{(P1.3.2b)}$$

Are these signals bounded? If they are bounded, what is their upperbound?

(b) Show that the output $y(t)/y[n]$ to the (bounded) input $x(n)/x[n]$ at $t = 0/n = 0$ is unbounded, implying that the system is not BIBO stable.

1.4 Stability of Continuous-Time Systems
Remark 1.6 states that a continuous-time LTI system is stable if and only if its system function $G(s) = \mathcal{L}\{g(t)\}$ has all the poles strictly within the left-half s

-plane or equivalently, the real parts of all the poles are negative. Referring to the remark, consider the following two systems:

$$G_a(s) = \frac{2}{(s+1)(s+2)} \tag{P1.4.1a}$$

$$G_b(s) = \frac{6}{(s-1)(s+2)} \tag{P1.4.1b}$$

(a) Find the impulse responses $g_a(t)$ and $g_b(t)$ by taking the inverse Laplace transform of each system function. Check if each system satisfies the stability condition (1.2.27a) or not.
(b) Find the poles of each system function and determine whether or not each system satisfies the stability condition stated in Remark 1.6. Does the result agree to that obtained in (a)?
(c) Find the step responses of the two systems, i.e., their outputs to a unit-step input $x(t) = u_s(t)$ whose Laplace transform is $X(s) = 1/s$. Check if each step response converges or not.

(cf.) You might get $y_a(t)$ by typing the following commands into the MATLAB command window:

```
>> syms s; Gas=2/(s+1)/(s+2); Xs=1/s;
   Yas=Gas*Xs, yat=ilaplace(Yas)
```

1.5 Stability of Discrete-Time Systems
Remark 1.6 states that a discrete-time LTI System is stable if and only if its system function $G[z] = \mathcal{Z}\{g[n]\}$ has all the poles strictly within the unit circle (see Remark 4.5). Referring to the remark, consider the following two systems:

$$G_a[z] = \frac{3z}{(z-0.5)(z+0.5)} \tag{P1.5.1a}$$

$$G_b[z] = \frac{z}{(z-1.5)(z-0.5)} \tag{P1.5.1b}$$

(a) Find the impulse responses $g_a[n]$ and $g_b[n]$ by taking the inverse z -transform of each system function. Determine whether each system satisfies the stability condition (1.2.27b) or not.
(b) Find the poles of each system function and determine whether or not each system satisfies the stability condition stated in Remark 1.6. Does the result agree to that obtained in (a)?
(c) Find the step responses of the two systems to a unit-step input $x[n] = u_s[n]$ whose z -transform is $X[z] = z/(z-1)$. Determine whether each step response converges or not.

(cf.) You might get $y_a[n]$ by typing the following commands into the MATLAB command window:

```
>> syms z; Gaz=3*z/(z-0.5)/(z+0.5); Xz=z/(z-1);
   Yaz=Gaz*Xz, yan=iztrans(Yaz)
```

1.6 Inverse System
Consider a model for echo generation, which is depicted in Fig. P1.6(a).

(a) Noting that the time-domain input-output relationship is

$$y(t) = a\ y(t - T) + x(t) \tag{P1.6.1}$$

show that the impulse response is

$$g(t) = \sum_{k=0}^{\infty} a^k \delta(t - kT) \text{ with } 0 < a < 1 \tag{P1.6.2}$$

Also find the system function $G(s)$ by taking the Laplace transform of this impulse response or the time-domain input-output relationship. Referring to Eq. (1.2.27a) or Remark 1.6, determine the stability condition of the model system.

(b) Use Eq. (1.2.30a) to find the system function $H(s)$ of the inverse system. Show that the inverse system can be modeled as Fig. P1.6(b), whose time-domain input-output relationship is

$$\hat{x}(t) = y(t) - a\ y(t - T) \tag{P1.6.3}$$

1.7 Inverse System
Consider a discrete-time model for duo-binary signaling, which is depicted in Fig. P1.7(a).

(a) Noting that the time-domain input-output relationship is

$$y[n] = x[n] + x[n - 1] \tag{P1.7.1}$$

find the system function $G[z]$ by taking its z -transform.

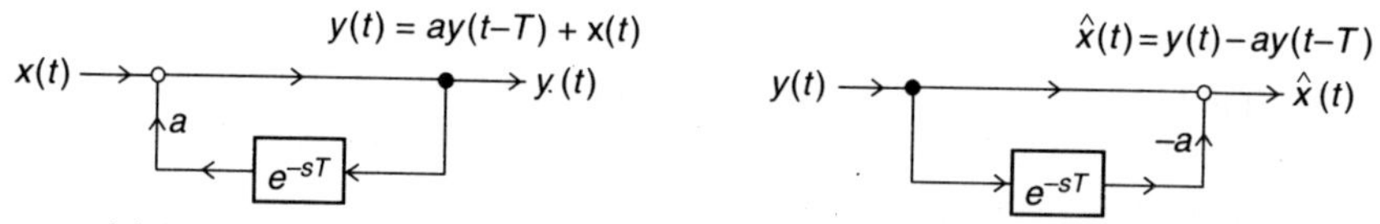

(a) A model for echoing

(b) A model for removing the echo

Fig. P1.6

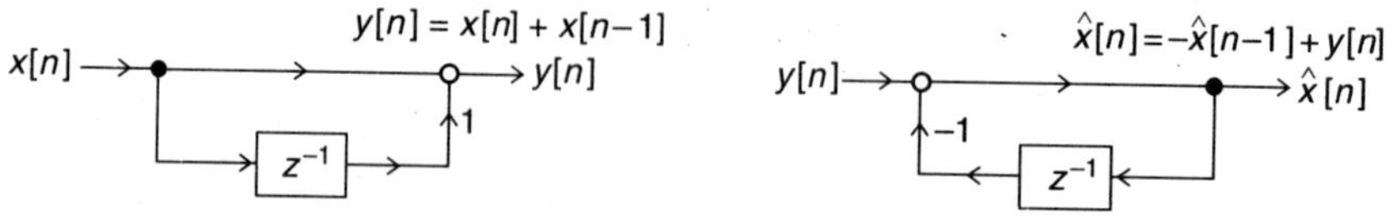

(a) A model for duo–binary signaling

(b) A model for detecting system

Fig. P1.7

(b) Use Eq. (1.2.30b) to find the system function $H[z]$ of the inverse system. Show that the inverse system can be modeled as Fig. P1.7(b), whose time-domain input-output relationship is

$$\hat{x}[n] = -\hat{x}[n-1] + y[n] \tag{P1.7.2}$$

1.8 Simulation of Continuous-Time/Discrete-Time System Using MATLAB and Simulink

(a) Find the system function of the system described by the block diagram in Fig. 1.16(b1) or the signal flow graph in Fig. 1.16(c1) or (d1) with $RC = 1$. Referring to the upper part of Fig. P1.8(a) and the parameter setting dialog boxes in Fig. P1.8(b1), (b2), and (b3), perform the Simulink simulation to plot the output signal $v_o(t)$ for $0 \le t < 4$ s.

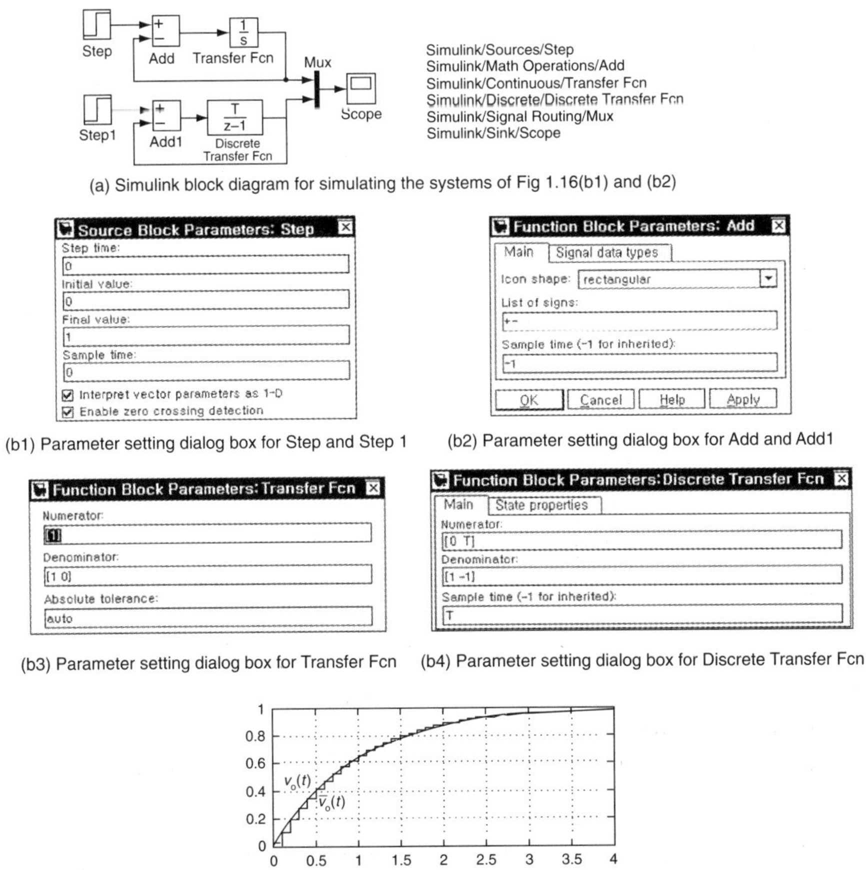

(a) Simulink block diagram for simulating the systems of Fig 1.16(b1) and (b2)

(b1) Parameter setting dialog box for Step and Step 1

(b2) Parameter setting dialog box for Add and Add1

(b3) Parameter setting dialog box for Transfer Fcn

(b4) Parameter setting dialog box for Discrete Transfer Fcn

(c) The outputs obtained from the Simulink simulation

Fig. P1.8

(b) Find the system function of the system described by the block diagram in Fig. 1.16(b2) or the signal flow graph in Fig. 1.16(c2) or (d2) with $RC = 1$. Referring to the lower part of Fig. P1.8(a) and the parameter setting dialog boxes in Fig. P1.8(b1), (b2), and (b4), perform the Simulink simulation with sampling interval $T = 0.1$ to plot $\bar{v}_o(t)$ for $0 \le t < 4$ s. Also referring to the following program "sig01p_08b.m", perform the MATLAB simulation to plot $v_o(t)$ for $0 \le t < 4$ s. Does decreasing the sampling interval to, say, $T = 0.01$ make the output close to that of the continuous-time system obtained in (a)?

(cf.) If you parameterize a constant such as T in the Simulink block diagram so that it can be easily changed, you can set it to a new value in the MATLAB command window.

```
% sig01p_08a.m
% Continuous-Time System Simulation clear, clf
R=1; C=1; RC=R*C; tf=4;
T=0.1; tt=[0:T:tf];
vi = ones(size(tt)); % step input
vo_t0=0;
options=odeset('RelTol',1e-4);
vo = ode45(dvodt,tt,vo_t0,options,RC);
plot(tt,vo), hold on
```

```
function dv=dvodt(t,vo,RC)
dv = (-vo+(t>=0))/RC;
```

```
% sig01p_08b.m
% Discrete-Time System Simulation
clear, clf
R=1; C=1; RC=R*C; tf=4;
T=0.1; tt=[0:T:tf];
vi = ones(size(tt)); % step input
TRC=T/RC; vo(1)=0;
for n=1:length(tt)-1
   vo(n+1)=vo(n)+(vi(n)-vo(n))*TRC;
end
stairs(tt,vo)
```

1.9 Continuous-Time State Diagram, State Equation, Transfer Function, and Input-Output Relationship

We can write the state equation for a continuous-time state diagram by taking the following procedure.

1. Assign a state variable $x_i(t)$ to the output of each integrator s^{-1}.
2. Express the input $x_i'(t)$ of each integrator s^{-1} in terms of the state variables and system input(s).
3. Express each output in terms of the state variables and input(s).

(a) Show that we can apply the above procedure to the state diagram of Fig. 1.19(a) and write the state equation together with the output equation as

$$\begin{bmatrix} x_1'(t) \\ x_2'(t) \end{bmatrix} = \begin{bmatrix} 0 & 1 \\ -a_0 & -a_1 \end{bmatrix} \begin{bmatrix} x_1(t) \\ x_2(t) \end{bmatrix} + \begin{bmatrix} 0 \\ 1 \end{bmatrix} u(t) \tag{P1.9.1}$$

$$y(t) = \begin{bmatrix} b_0 & b_1 \end{bmatrix} \begin{bmatrix} x_1(t) \\ x_2(t) \end{bmatrix} \tag{P1.9.2}$$

We can also substitute the output equation (P1.9.2) into the left-hand side of Eq. (E1.4a.1) and use the state equation (P1.9.1) to get the right-hand side of Eq. (E1.4a.1):

$$y''(t) + a_1 y'(t) + a_0 y(t) = b_1 u'(t) + b_0 u(t) \tag{P1.9.3}$$

(b) Show that we can apply the above procedure to the state diagram of Fig. 1.19(b) and write the state equation together with the output equation as

$$\begin{bmatrix} x_1'(t) \\ x_2'(t) \end{bmatrix} = \begin{bmatrix} 0 & -a_0 \\ 1 & -a_1 \end{bmatrix} \begin{bmatrix} x_1(t) \\ x_2(t) \end{bmatrix} + \begin{bmatrix} b_0 \\ b_1 \end{bmatrix} u(t) \tag{P1.9.4}$$

$$y(t) = \begin{bmatrix} 0 & 1 \end{bmatrix} \begin{bmatrix} x_1(t) \\ x_2(t) \end{bmatrix} \tag{P1.9.5}$$

We can also combine these state and output equations to get the input-output relationship (E1.4a.1).

(c) Apply Eq. (8.3.2a) with Eqs. (P1.9.1,2) or (P1.9.4,5) to find the transfer function $G(s) = Y(s)/U(s)$ or equivalently, the input-output relationship (E1.4a.2). Also apply Mason's gain formula for Fig. 1.19(a) or (b) to find $G(s)$.

(d) With $a_0 = 2$, $a_1 = 3$, $b_0 = 4$, and $b_1 = 3$, use the MATLAB function `ss2tf()` to find the transfer function $G(s)$ of the system described by Eqs. (P1.9.1,2) or (P1.9.4,5). Reversely, use the MATLAB function `tf2ss()` to find a state equation for the transfer function

$$G(s) = \frac{b_1 s + b_0}{s^2 + a_1 s + a_0} = \frac{3s + 4}{s^2 + 3s + 2} \tag{P1.9.6}$$

Which one does it yield, the equivalent of Eq. (P1.9.1,2) or (P1.9.4,5)?

(e) Referring to the following program "sig01p_09.m", simulate the systems described by Eqs. (P1.9.1,2) and (P1.9.4,5) with $a_0 = 2$, $a_1 = 3$, $b_0 = 4$, and $b_1 = 3$ to find their outputs ($y(t)$'s) to an input $u(t) = \sin(2t)$ for $0 \le t < 10\,\text{s}$ (with sampling interval $T = 0.01\,\text{s}$). Do the state equations conform to each other in terms of the input-output relationship?

```
a0=2; a1=3; b0=4; b1=3;
A1=[0 1;-a0 -a1]; B1=[0;1]; C1=[b0 b1]; [num1,den1]= ss2tf(A1,B1,C1,D1)
A2=[0 -a0;1 -a1]; B2=[b0;b1]; C2=[0 1]; [num2,den2]= ss2tf(A2,B2,C2,D2)
num=[b1 b0]; den=[1 a1 a0]; [A,B,C,D]=tf2ss(num,den)
```

```
%sig01p_09.m
% to simulate a continuous-time system described by state equations
clear, clf
a0=2; a1=3; b0=4; b1=3; w=2;

% Use ode45() to solve the continuous-time state equations
% dx_sig01p09= inline('A*x+B*sin(w*t)','t','x','A','B','w');
t0=0; tf=10; x0=[0; 0]; % initial time, final time, and initial state
A1=[0 1;-a0 -a1]; B1=[0;1]; C1=[b0 b1]; D1=0;
[tt1,xx1]= ode45(@dx_sig01p09,[t0 tf],x0,[],A1,B1,w); y1= xx1*C1.';
A2=[0-a0;1 -a1]; B2=[b0;b1]; C2=[0 1]; D2=0;
[tt2,xx2]= ode45(@dx_sig01p09,[t0 tf],x0,[],A2,B2,w); y2= xx2*C2.';
subplot(211), plot(tt1,y1, tt2,y2,'r')

% Use lsim(A,B,C,D,u,t,x0) to simulate continuous-time linear systems
T=0.01; tt=[t0:T:tf];
[num1,den1]= ss2tf(A1,B1,C1,D1)
[y1,x1]= lsim(A1,B1,C1,D1,sin(w*tt),tt,x0);
[num2,den2]=ss2tf(A2,B2,C2,D2)
[y2,x2]= lsim(A2,B2,C2,D2,sin(w*tt),tt,x0);
[y3,x3]= lsim(num1,den1,sin(w*tt),tt);
subplot(212), plot(tt,y1, tt,y2,'r', tt,y3,'m')
```

```
function dx=dx_sig01p09(t,x,A,B,w)
dx= A*x + B*sin(w*t);
```

(f) Referring to Fig. P1.9, perform the Simulink simulation for the systems described by Eqs. (P1.9.1,2), (P1.9.4,5), and (P1.9.6) to find their outputs ($y(t)$'s) to an input $u(t) = \sin(2t)$ for $0 \le t < 10\,\mathrm{s}$. Do the simulation results agree with each other and also with those obtained in (e)?

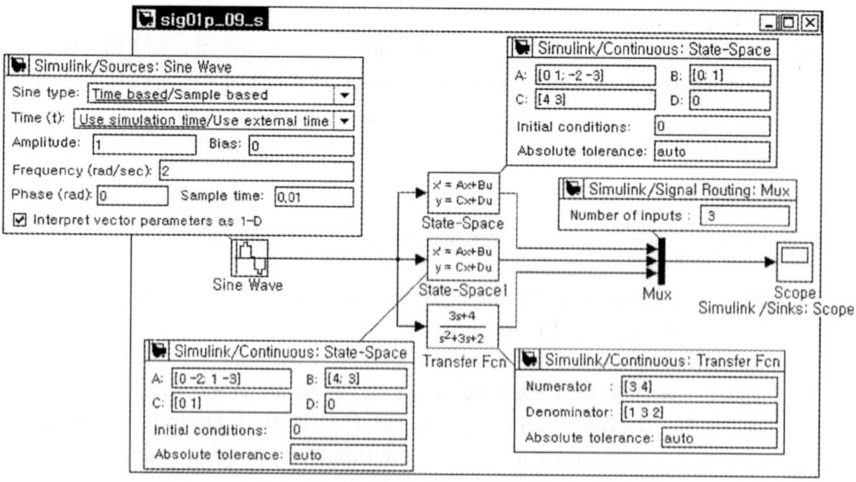

Fig. P1.9 Simulink block diagram for Problem 1.9(e)

1.10 Discrete-Time State Diagram, State Equation, Transfer Function, and Input-Output Relationship

We can write the state equation for a discrete-time state diagram by taking the following procedure.

1. Assign a state variable $x_i[n]$ to the output of each delay element z^{-1}.
2. Express the input $x_i[n+1]$ of each delay element z^{-1} in terms of the state variables and system input(s).
3. Express each output in terms of the state variables and input(s).

(a) Show that we can apply the above procedure to the state diagram of Fig. 1.20(c) and write the state equation together with the output equation as

$$\begin{bmatrix} x_1[n+1] \\ x_2[n+1] \end{bmatrix} = \begin{bmatrix} 0 & 1 \\ -a_0 & -a_1 \end{bmatrix} \begin{bmatrix} x_1[n] \\ x_2[n] \end{bmatrix} + \begin{bmatrix} b_0 \\ b_1 \end{bmatrix} u[n] \quad \text{(P1.10.1)}$$

$$y[n] = \begin{bmatrix} b_0 & b_1 \end{bmatrix} \begin{bmatrix} x_1[n] \\ x_2[n] \end{bmatrix} \quad \text{(P1.10.2)}$$

We can also substitute the output equation (P1.10.2) into the left-hand side of Eq. (E1.4b.1) and use the state equation (P1.10.1) to get the right-hand side of Eq. (E1.4b.1):

$$y[n+2] + a_1 y[n+1] + a_0 y[n] = b_1 u[n+1] + b_0 u[n] \quad \text{(P1.10.3)}$$

(b) Show that we can apply the above procedure to the state diagram of Fig. 1.20(b) and write the state equation together with the output equation as

$$\begin{bmatrix} x_1[n+1] \\ x_2[n+1] \end{bmatrix} = \begin{bmatrix} 0 & -a_0 \\ 1 & -a_1 \end{bmatrix} \begin{bmatrix} x_1[n] \\ x_2[n] \end{bmatrix} + \begin{bmatrix} b_0 \\ b_1 \end{bmatrix} u[n] \quad \text{(P1.10.4)}$$

$$y[n] = \begin{bmatrix} 0 & 1 \end{bmatrix} \begin{bmatrix} x_1[n] \\ x_2[n] \end{bmatrix} \quad \text{(P1.10.5)}$$

We can also combine these state and output equations to get the input-output relationship (E1.4b.1).

(c) Apply Eq. (8.3.2b) with Eqs. (P1.10.1,2) or (P1.10.4,5) to find the transfer function $G[z] = Y[z]/U[z]$ or equivalently, the input-output relationship (E1.4b.2). Also apply Mason's gain formula for Fig. 1.20(a), (b), (c), or (d) to find $G[z]$.

(d) Referring to the following program "sig01p_10.m" or Fig. P1.10, simulate the systems described by Eqs. (P1.10.1,2) and (P1.10.4,5) with $a_0 = 1/8$, $a_1 = 3/4$, $b_0 = 2$, and $b_1 = 1$ to find their outputs ($y(t)$'s) to an input $u(t) = \sin(2t)$ for $0 \le t < 10$ s (with sampling interval $T = 0.01$ s).

Do the state equations conform to each other in terms of the input-output relationship?

```
%sig01p_10.m
% to simulate a discrete-time system described by state equations
a0=1/8; a1=3/4; b0=2; b1=1; w=2;

% Use difference equation to solve the discrete-time state equations
t0=0; tf=10; x0=[0; 0]; % initial time, final time, and initial state
T=0.01; tt=[t0:T:tf];
A1=[0 1;-a0 -a1]; B1=[0;1]; C1=[b0 b1]; D1=0; x1(1,:)=[0 0];
A2=[0 -a0;1 -a1]; B2=[b0;b1]; C2=[0 1]; D2=0; x2(1,:)=[0 0];
for n=1:length(tt)
    t=tt(n);
    x1(n+1,:)= x1(n,:)*A1.'+sin(w*t)*B1.'; y1(n)= x1(n,:)*C1.';
    x2(n+1,:)= x2(n,:)*A2.'+sin(w*t)*B2.'; y2(n)= x2(n,:)*C2.';
end
subplot(211), plot(tt,y1, tt,y2,'r')

% Use dlsim(A,B,C,D,u,x0) to simulate discrete-time linear systems
[y1,x1]= dlsim(A1,B1,C1,D1,sin(w*tt),x0);
[y2,x2]= dlsim(A2,B2,C2,D2,sin(w*tt),x0);
[num2,den2]= ss2tf(A2,B2,C2,D2)
[y3,x3]= dlsim(num2,den2,sin(w*tt));
subplot(212), plot(tt,y1, tt,y2,'r', tt,y3,'m')
```

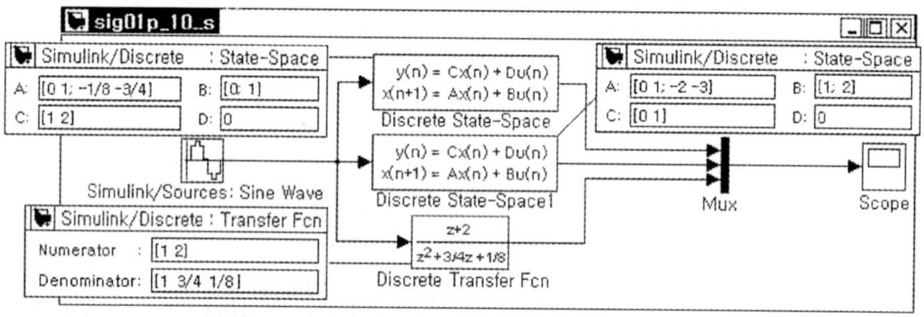

Fig. P1.10 Simulink block diagram for Problem 1.10(d)

1.11 Deconvolution

(a) We can make two discrete-time sequences as

$$\begin{array}{rl} n & = 0\ 1 \cdots 9\ 10\ 11 \cdots 19\ 20\ 21 \cdots 29\ 30\ 31 \cdots 39 \\ x[n] & = 1\ 1 \cdots 1\ 9\ 9 \cdots 9\ -6\ -6 \cdots -6\ 2\ 2 \cdots 2 \\ g[n] & = 1\ 1 \cdots 1\ 1\ 1 \cdots 1\ 1\ 1 \cdots 1\ 0\ 0 \cdots 0 \end{array}$$

Compute their convolution $y[n] = x[n] * g[n]$ and plot it.

(b) Referring to the following MATLAB program "sig01p_11.m", compute the estimate $\hat{x}[n]$ of $x[n]$ from $y[n]$ and $g[n]$ two times, once by using

Eq. (1.4.5) and once by using Eq. (1.4.7). Compare the two results in terms of how close they are to the original input sequence $x[n]$.

```
%sig01p_11.m
clear, clf
x=[ones(1,30) zeros(1,10)]; Nx=length(x);
g=ones(10,1)*[1 9 -6 2]; g=g(:).'; Ng=length(g);
n=0:Nx+Ng-2;
y=conv(x,g);
subplot(313), stem(n,y,'.')
% Deconvolution
for m=1:Ng
   for n=1:m, G(m,n)=g(m-n+1); end  % Eq.(1.4.5)
end
x0=(G^-1*y(1:Ng)')';
x1(1)=y(1)/g(1);
for n=2:Nx
   x1(n)=y(n);
   for m=1:n-1, x1(n)=x1(n)-g(n-m+1)*x1(m); end  % Eq.(1.4.7)
   x1(n)=x1(n)/g(1);
end
err0=norm(x0-x)
err1=norm(x1-x)
```

(cf.) Another way to compute deconvolution is to use the MATLAB command `x=deconv(y,g)`.

1.12 Correlation and Matched Filter for Signal Detection
Consider Example 1.5 and the related program "sig01e05.m" again.

(a) To see the transmitted signal x, insert the following statement and run the program to check if you can get the results as depicted in Fig. 1.21(c), (d1), and (d2).

```
subplot(311), plot(tt,xbuffer)
```

(b) Remove the zero periods between the signals in the input. Can you still notice the corresponding signal arrival times from (local) maxima of the output of each matched filter? To simulate this situation, modify the statement for generating the transmitted signal as follows:

```
x= [x1 x2 x1 x2 x2];
```

(c) If we sample the matched filter outputs every 2 s, can we use the sampled output to detect the signal successfully based on which filter output is the maximum? To simulate this situation, insert the following statements into the last part of the for loop:

```
if mod(n,Nx)==0
  fprintf('At t=%5.2f, Matched filter output=%5.2f %5.2f\n', n*Ts,y);
end
```

(d) If the input is contaminated with a Gaussian noise with zero mean and variance of $\sigma^2 = 0.3^2$, will the detection performance be harmed severely? To simulate this situation, modify one of the statements in the for loop as follows:

```
xbuffer= [x(n)+0.3*randn xbuffer(1:end-1)];
```

(cf.) The system with impulse response $g_i(t) = x_i(T - t)$ is called the 'matched filter' for the signal $x_i(t)$ because it is tuned to $x_i(t)$ so that it can produce the maximum output to the input $x_i(t)$.

1.13 Walsh Functions [W-3] and Matched Filter

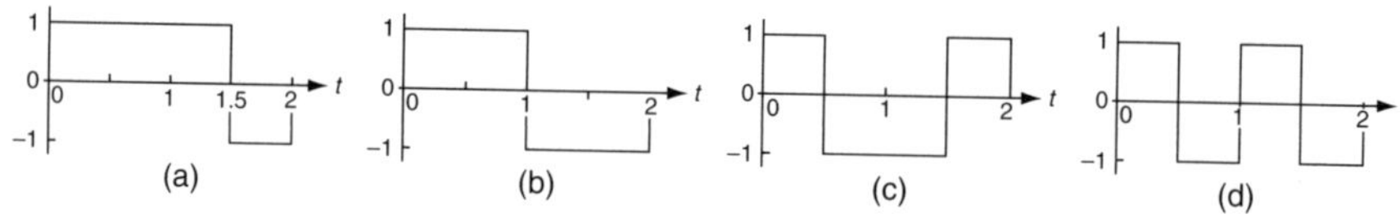

Fig. P1.13

With the program "sig01e05.m" modified in Problem 1.12, answer the following questions.

(a) If we use the signal waveform of Fig. P1.13(a) in place of that of Fig. 1.21(a1), how are the matched filter outputs sampled every 2 s changed? Can you explain why the output of a matched filter is not close to zero 2 s after another signal arrives in contrast with the case where the signal waveforms of Fig. 1.21(a1) and (b1) are used?

(cf.) This gives us a chance to think about how we can design the signal waveforms in such a way the difference between the (sampled) output of a matched filter to the input for which it is tuned and the (sampled) output to other input signals can be maximized, for the purpose of minimizing the suffer from possible noise/distortion/interference in a communication system. It is desirable to have the signal waveforms orthogonal to each other so that the integrations of their products are zero.

(b) Consider the three signal waveforms $x_1(t)$, $x_2(t)$, and $x_3(t)$ of duration $T = 2$ [s], known as Walsh functions, which are depicted in Fig. P1.13(b), (c), and (d) and represent *a, b*, and *c*, respectively.

- Are the three signals orthonormal in the following sense?

$$\int_0^T x_i(t)x_j(t)dt = \delta[i-j] = \begin{cases} 1 & \text{for } i = j \\ 0 & \text{for } i \neq j \end{cases} \qquad \text{(P1.13.1)}$$

- Modify the program "sig01e05.m" to simulate the situation that the following input signal is applied to the three matched filters with the impulse response $g_i(t) = x_i(T-t)$:

$$x(t) = [x_1(t)\; x_2(t-T)\; x_3(t-2T)\; x_2(t-3T)\; x_1(t-4T)] \quad \text{(P1.13.2)}$$

Does the receiver detect the received signals successfully from the sampled outputs of the matched filters for all the zero-mean Gaussian noise of variance $\sigma^2 = 0.3^2$?

1.14 Correlation Function and Radar (*Ra*dio *D*etection *a*nd *R*anging) System (Problem 3.28(e) of [O-1])
A radar transmits a short pulse of radio signal (electromagnetic radiation) and measures the time it takes for the reflection to return from the target. The distance is one-half the product of the round trip time and the speed of the signal [W-1].

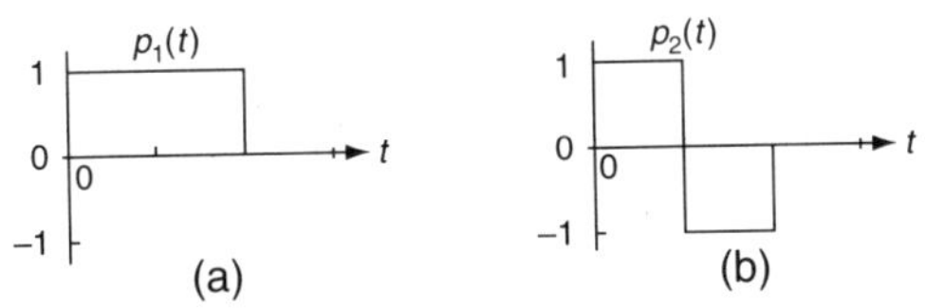

Fig. P1.14

(a) Let the transmitted pulse be $p(t)$ and the received pulse $a\,p(t-t_1)$ where t_1 is the round trip time. Show that their crosscorrelation $\phi_{xp}(t)$ is maximized at $t = t_1$:

$$\phi_{xp}(t_1) = \operatorname*{Max}_t \phi_{xp}(t) \quad \text{(P1.14.1)}$$

where

$$\phi_{xp}(t) = \int x(t+\tau)p(\tau)d\tau = a\int p(t-t_1+\tau)p(\tau)d\tau \quad \text{(P1.14.2)}$$

<*Hint*> To find the time at which the crosscorrelation is maximized, use the Cauchy-Schwartz inequality:

$$\int_a^b u(\tau)v(\tau)d\tau \le \left(\int_a^b u^2(\tau)d\tau\right)^{1/2}\left(\int_a^b v^2(\tau)d\tau\right)^{1/2} \quad \text{(P1.14.3)}$$

where the equality holds if and only if $u(\tau) = v(\tau)$.

(b) Since the radar looks for the peak of the correlation function to find the arrival time of returning signal, it would be good to use a pulse signal

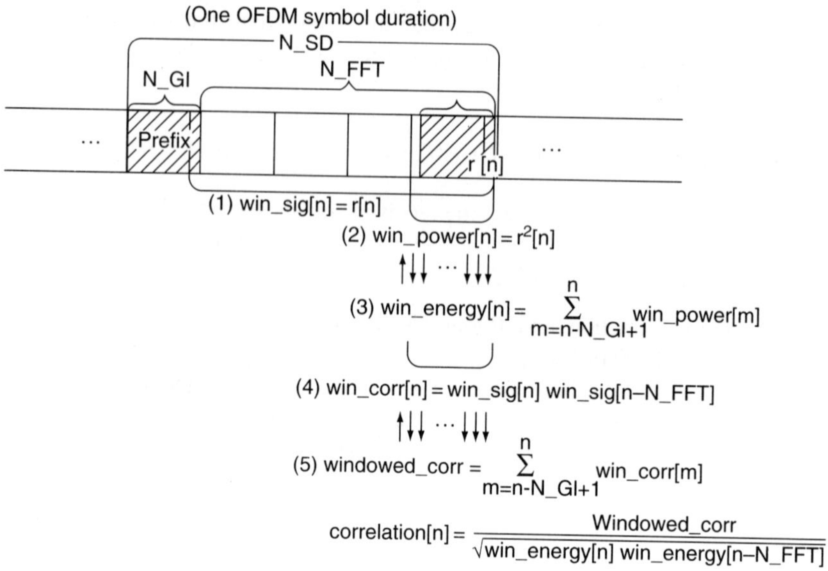

Fig. P1.15 Windows to detect an OFDM symbol using correlation

$p(t)$ that has a sharply peaked autocorrelation function against possible distortion in the returning signal. Which of the two pulses in Fig. P1.14 is better?

1.15 OFDM Symbol Timing Using Correlation

Suppose an OFDM (Orthogonal Frequency Division Multiplexing) communication system in which an OFDM symbol consists of N_FFT (Fast Fourier Transform) points and N_GI = 16 prefix points where the *cyclic prefix* is a repeat of the end of the symbol at the beginning. Let us consider a problem of detecting the start time of each OFDM symbol by using the correlation between the prefix and the end part of the symbol. Since we cannot store uncountably many samples of the received signal r[n], we use several windows (buffers) to store some duration of signal samples, powers, energies, and correlations as shown in Fig. P1.15 where the contents of each window is as follows:

(1) win_sig[n] = r[n] with the size of at least N_FFT + 1
(2) win_power[n] = $r[n]^2$ with the size of N_GI + 1
(3) win_energy[n] = $\sum_{m=n-N\_GI+1}^{n}$ win_power[m] with the size of N_FFT + 1
(4) win_corr[n] = r[n]r[n − N_FFT] with the size of N_GI + 1

```
%detect_OFDM_symbol_with_correlation.m
% Copyleft: Won Y. Yang, wyyang53@hanmail.net, CAU for academic use only
clear, clf
N_FFT=64; N_GI=16; % N_Prefix=16;
N_SD=N_FFT+N_GI; N_SD1=N_SD+1; % Symbol Duration
N_Null=N_SD; Nw=N_Null; % Null Duration
N_d=N_SD/4; % remaining period of the last symbol in the previous frame
N_OFDM=2; % One Null + N_OFDM symbols
symbols = []; Max_energy_ratio=0; Min_energy_ratio=1e10;
for i=1:N_OFDM
   symbol=2*rand(1,N_FFT)-1; symbol=[symbol(end-N_GI+1:end) symbol];
   symbols = [symbols symbol];
end
Nulls= zeros(1,N_Null);
received_signal = [rand(1,N_d) Nulls symbols];
length_received= length(received_signal);
noise = 0.1*(rand(size(received_signal))-0.5);
received_signal = received_signal + noise;
Nw1=Nw+1; N_GI1=N_GI+1; N_FFT1=N_FFT+1;
win_power= zeros(1,Nw1); win_corr= zeros(1,N_GI1);
win_sig= zeros(1,N_FFT1); win_energy= zeros(1,N_FFT1);
signal_buffer = zeros(1,length_received);
correlations = zeros(1,N_SD);
True_start_points= [N_d+N_SD:N_SD:length_received]
OFDM_start_points= [0]; windowed_corr=0;
nn = 1:length_received;
for n=1:length_received
   signal_buffer = [received_signal(n) signal_buffer(1:end-1)];
   win_power = [win_power(2:end) received_signal(n)^2]; % the power window
   win_sig = [win_sig(2:end) received_signal(n)]; % the signal window
   win_energy = [win_energy(2:end) win_energy(end)+win_power(end)];
   if n>N_GI, win_energy(end) = win_energy(end)-win_power(end-N_GI); end
   win_corr(1:end-1) = win_corr(2:end);
   if n>N_FFT
     win_corr(end) = win_sig(???)'*win_sig(1);
     windowed_corr = windowed_corr + win_corr(end);
   end
   if n>N_SD, windowed_corr= windowed_corr - win_corr(?); end
   % CP-based Symbol Timing
   subplot(311)
   stem(nn,signal_buffer,'.')
   axis([0 N_SD*4 -2 2]), hold on
    title('Received Signal and Estimated Starting Points of Symbols')
   if n>N_SD %+N_GI
     %normalized/windowed correlation across N_FFT samples for N_GI points
     normalized_corr = windowed_corr/sqrt(win_energy(???)*win_energy(?));
     correlations = [correlations normalized_corr];
     if normalized_corr>0.99&n-N_SD>OFDM_start_points(end)+N_FFT
       OFDM_start_points = [OFDM_start_points n-N_SD];
     end
     start_points = OFDM_start_points(2:end);
     subplot(312), stem(1:length(correlations),correlations,'.')
      axis([0 N_SD*4 -1 1.5]), hold on
     title('Correlation across NFFT samples')
   end
   if n<length_received,  clf;  end
end
Estimated_OFDM_start_points = start_points
```

At each iteration when a sampled signal arrives, we compute the normalized and windowed correlation

$$\text{correlation}[n] = \frac{\text{windowed\_corr}[n]}{\sqrt{\text{win\_energy}[n]\text{win\_energy}[n - \text{N\_FFT}]}} \overset{?}{>} \text{Threshold}(0.99) \tag{P1.15.1}$$

$$\text{with windowed\_corr}[n] = \sum_{m=n-\text{N\_GI}+1}^{n} \text{win\_corr}[m]$$

to determine whether the current sample is the end of an OFDM symbol or not. If the normalized correlation is found to exceed the threshold value, say, 0.99, we set the start time of the detected symbol to N_SD (one OFDM symbol duration) samples before the detection time. Complete the above program "detect_OFDM_symbol_with_correlation.m" so that it implements this OFDM symbol timing scheme. Also run it to check if it works.

Chapter 2
Continuous-Time Fourier Analysis

Contents

Since the great contribution and revolutionary discovery of Jean Baptiste Joseph Fourier saw the light in 1822 after passing through the long dark tunnel of J.L. Lagrange's stubborn objection and was supported by P.L. Dirichlet's rigorous mathematical proof in 1829, the Fourier series and transform techniques have played very significant role in so many disciplines within the fields of mathematics, science, and engineering.

Joseph Fourier (1768∼1830) was a French mathematician and physicist who initiated the investigation of Fourier series and its application. Born as a son of a tailor, he was orphaned at age 8. In 1784, at only 16 years of age, he became a mathematics teacher at the Ecole Royale Militaire of Auxerre, debarred from entering the army

W.Y. Yang et al., *Signals and Systems with MATLAB®*,
DOI 10.1007/978-3-540-92954-3_2,

on account of his obscurity and poverty. In 1795, Fourier took a faculty position at the Ecole Normale (Polytechnique) in Paris, which is an elite institution training high school teachers, university professors, and researchers. In 1798, he joined Napoleon's army in its expedition to Egypt as scientific advisor to help establish educational facilities there and carry out archaeological explorations.

Eluding the Nelson's British fleet, the Napoleon's Egyption expedition fleet of 300 ships, 30,000 infantry, 2,800 cavalry, and 1,000 cannons started Toulon on May 19,1798, sailing for Alexandria. The great expedition plan plotted by Napoleon attached a library with lots of books, many measurement instruments, various laboratory apparatuses, and about 500 civilians to the army; 150 of them were artists, scientists, scholars, engineers, and technicians. These human and physical resources formed the Institut d'Égypte in Cairo after Egypt was conquered by Napoleon. Napoleon Bonaparte (1769~1821) not yet 30 years old, a great hero in the human history, and Joseph Fourier, a great scholar of about the same age in their youth were on board the flagship L'Orient of the expedition fleet. What were they thinking of when walking around on the deck and looking up the stars twinkling in the sky above the Mediterranean Sea at several nights in May of 1798? One might have dreamed of Julius Caesar, who conquered Egypt about 1,800 years ago, falling in love with the Queen Cleopatra, or might have paid a tribute to the monumental achievements of the great king Alexander, who conquered one third of the earth, opening the road between the West and the Orient. The other might have refreshed his memory on what he wrote in his diary on his 21st birthday, *Yesterday was my 21st birthday, at that age Newton and Pascal had already acquired many claims to immortality*, arranging his ideas on Fourier series and heat diffusion or recollecting his political ideology which had swept him and made him get very close to guillotine in the vortex of French Revolution.

2.1 Continuous-Time Fourier Series (CTFS) of Periodic Signals

2.1.1 Definition and Convergence Conditions of CTFS Representation

Let a function $x(t)$ be *periodic* with period P in t, that is,

$$x(t) = x(t + P) \quad \forall\, t \tag{2.1.1}$$

where P [s] and $\omega_0 = 2\pi/P$ [rad/s] are referred to as the *fundamental period* and *fundamental (angular) frequency*, respectively, if P is the smallest positive real number to satisfy Eq. (2.1.1) for periodicity. Suppose $x(t)$ satisfies at least one of the following conditions A and B:

< Condition A >

(A1) The periodic function $x(t)$ is square-integrable over the period P, i.e.,

$$\int_P |x(t)|^2\, dt < \infty \tag{2.1.2a}$$

where $\int_P$ means the integration over any interval of length P. This implies that the signal described by $x(t)$ has finite power.

< Condition B : *Dirichlet condition* >

(B1) The periodic function $x(t)$ has only a finite number of extrema and discontinuities in any one period.
(B2) These extrema are finite.
(B3) The periodic function $x(t)$ is absolutely-integrable over the period P, i.e.,

$$\int_P |x(t)|\, dt < \infty \tag{2.1.2b}$$

Then the periodic function $x(t)$ can be represented by the following forms of *continuous-time Fourier series* (CTFS), each of which is called the *Fourier series representation*:
<*Trigonometric form*>

$$x(t) = a_0 + \sum_{k=1}^{\infty} a_k \cos k\omega_0 t + \sum_{k=1}^{\infty} b_k \sin k\omega_0 t \tag{2.1.3a}$$

$$\text{with } \omega_0 = \frac{2\pi}{P} \quad (P : \text{the period of } x(t))$$

where the Fourier coefficients a_0, a_k, and b_k are

$$\begin{aligned} a_0 &= \frac{1}{P}\int_P x(t)\, dt \text{ (the integral over one period } P) \\ a_k &= \frac{2}{P}\int_P x(t) \cos k\omega_0 t\, dt \\ b_k &= \frac{2}{P}\int_P x(t) \sin k\omega_0 t\, dt \end{aligned} \tag{2.1.3b}$$

<*Magnitude-and-Phase form*>

$$x(t) = d_0 + \sum_{k=1}^{\infty} d_k \cos(k\omega_0 t + \phi_k) \tag{2.1.4a}$$

where the Fourier coefficients are

$$d_0 = a_0, \quad d_k = \sqrt{a_k^2 + b_k^2}, \quad \phi_k = \tan^{-1}(-b_k/a_k) \tag{2.1.4b}$$

<Complex Exponential form>

$$x(t) = \frac{1}{P}\sum_{k=-\infty}^{\infty} c_k\, e^{jk\omega_0 t} \tag{2.1.5a}$$

where the Fourier coefficients are

$$c_k = \int_P x(t)\, e^{-jk\omega_0 t}\, dt \text{ (the integral over one period } P\text{)} \tag{2.1.5b}$$

Here, the k th frequency $k\omega_0$ ($|k| > 1$) with fundamental frequency $\omega_0 = 2\pi/P = 2\pi f_0$ [rad/s](P: period) is referred to as the k th *harmonic*. The above three forms of Fourier series representation are equivalent and their Fourier coefficients are related with each other as follows:

$$c_0 = \int_P x(t)\, dt = P d_0 = P a_0 \tag{2.1.6a}$$

$$\begin{aligned} c_k &= \int_P x(t)\, e^{-jk\omega_0 t}\, dt = \int_P x(t)\, (\cos k\omega_0 t - j \sin k\omega_0 t)\, dt \\ &= \frac{P}{2}(a_k - jb_k) = \frac{P}{2} d_k \angle \phi_k \end{aligned} \tag{2.1.6b}$$

$$\begin{aligned} c_{-k} &= \int_P x(t)\, e^{jk\omega_0 t}\, dt = \int_P x(t)\, (\cos k\omega_0 t + j \sin k\omega_0 t)\, dt \\ &= \frac{P}{2}(a_k + jb_k) = \frac{P}{2} d_k \angle - \phi_k = c_k^* \end{aligned} \tag{2.1.6c}$$

$$a_0 = \frac{c_0}{P}, \quad a_k = \frac{c_k + c_{-k}}{P} = \frac{2\mathrm{Re}\{c_k\}}{P}, \quad b_k = \frac{c_{-k} - c_k}{jP} = -\frac{2\mathrm{Im}\{c_k\}}{P} \tag{2.1.6d}$$

The plot of Fourier coefficients (2.1.4b) or (2.1.5b) against frequency $k\omega_0$ is referred to as the *spectrum*. It can be used to describe the spectral contents of a signal, i.e., depict what frequency components are contained in the signal and how they are distributed over the low/medium/high frequency range. We will mainly use Eqs. (2.1.5a) and (2.1.5b) for spectral analysis.

<Proof of the Complex Exponential Fourier Analysis Formula (2.1.5b)>

To show the validity of Eq. (2.1.5b), we substitute the Fourier synthesis formula (2.1.5a) with the (dummy) summation index k replaced by n into Eq. (2.1.5b) as

$$c_k \overset{?}{=} \int_P \frac{1}{P}\sum_{n=-\infty}^{\infty} c_n\, e^{jn\omega_0 t}\, e^{-jk\omega_0 t}\, dt = \sum_{n=-\infty}^{\infty} c_n\, \frac{1}{P}\int_P e^{j(n-k)\frac{2\pi}{P}t}\, dt \overset{\text{O.K.}}{=} c_k \tag{2.1.7}$$

This equality holds since

$$\frac{1}{P}\int_{-P/2}^{P/2} e^{j(n-k)\frac{2\pi}{P}t}\,dt$$

$$= \begin{cases} \frac{1}{P\cdot j(n-k)2\pi/P}\, e^{j(n-k)\frac{2\pi}{P}t}\Big|_{-P/2}^{P/2} = 0 & \text{for } n \neq k \\ \frac{1}{P}\int_{-P/2}^{P/2} dt = 1 & \text{for } n = k \end{cases} = \delta[n-k] \quad (2.1.8)$$

2.1.2 Examples of CTFS Representation

Example 2.1 Fourier Spectra of a Rectangular (Square) Wave and a Triangular Wave

(a) CTFS Spectrum of a Rectangular (Square) Wave (Fig. 2.1(a1))
Consider an even rectangular wave $x(t)$ with height A, duration D, and period P:

$$x(t) = A\,\tilde{r}_{D/P}(t) \quad (E2.1.1)$$

$$\text{where} \quad \tilde{r}_{D/P}(t) = \begin{cases} 1 & \text{for } |t - mP| \le D/2 (m: \text{ an integer}) \\ 0 & \text{elsewhere} \end{cases}$$

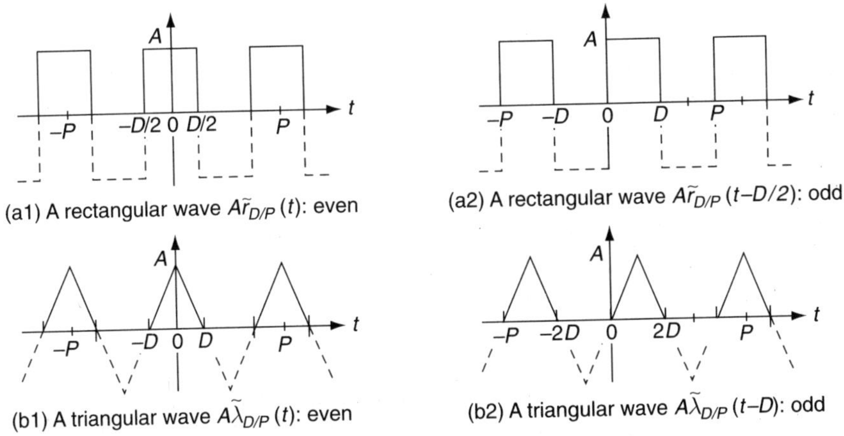

Fig. 2.1 Rectangular waves and triangular waves

We use Eq. (2.1.5b) to obtain the Fourier coefficients as

$$c_k = \int_{-P/2}^{P/2} A\,\tilde{r}_{D/P}(t)\,e^{-jk\omega_0 t}\,dt = A\int_{-D/2}^{D/2} e^{-jk\omega_0 t}\,dt = \frac{A}{-jk\omega_0}e^{-jk\omega_0 t}\Big|_{-D/2}^{D/2}$$

$$= A\frac{e^{jk\omega_0 D/2} - e^{-jk\omega_0 D/2}}{jk\omega_0} = AD\frac{\sin(k\omega_0 D/2)}{k\omega_0 D/2}$$

$$= AD\,\text{sinc}\left(k\frac{D}{P}\right) \text{ with } \omega_0 = \frac{2\pi}{P} \tag{E2.1.2}$$

Now we can use Eq. (2.1.5a) to write the Fourier series representation of the rectangular wave as

$$A\tilde{r}_{D/P}(t) \overset{(2.1.5a)}{=} \frac{1}{P}\sum_{k=-\infty}^{\infty} AD\,\text{sinc}\left(k\frac{D}{P}\right)e^{jk\omega_0 t}$$

$$= \frac{AD}{P} + \sum_{k=1}^{\infty} \frac{2AD}{P}\frac{\sin(k\pi D/P)}{k\pi D/P}\cos k\omega_0 t \tag{E2.1.3}$$

In the case of $D = 1$ and $P = 2D = 2$ as depicted in Fig. 2.2(a1), the (magnitude) spectrum is plotted in Fig. 2.2(b1).
(Q) What about the case of $P = D$, which corresponds to a constant DC (Direct Current) signal?

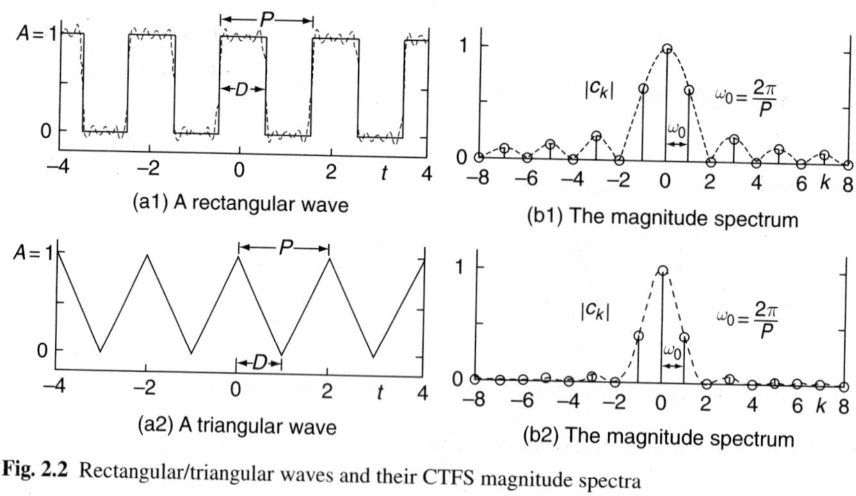

Fig. 2.2 Rectangular/triangular waves and their CTFS magnitude spectra

(b) CTFS Spectrum of a Triangular Wave (Fig. 2.1(b1))
Consider an even triangular wave $x(t)$ with maximum height A, duration $2D$, and period P:

$$x(t) = A\tilde{\lambda}_{D/P}(t) \tag{E2.1.4}$$

$$\text{where } \tilde{\lambda}_{D/P}(t) = \begin{cases} 1 - t/D & \text{for } |t - mP| \le D \ (m : \text{an integer}) \\ 0 & \text{elsewhere} \end{cases}$$

We use Eq. (2.1.5b) to obtain the Fourier coefficients as

$$\begin{aligned} c_k &= \int_{-P/2}^{P/2} A\left(1 - \frac{|t|}{D}\right) e^{-jk\omega_0 t}\, dt = \int_{-D}^{D} A\left(1 - \frac{|t|}{D}\right) \cos(k\omega_0 t)\, dt \\ &= 2\int_0^D A\left(1 - \frac{t}{D}\right) \cos(k\omega_0 t)\, dt = 2A\left(1 - \frac{t}{D}\right) \frac{1}{k\omega_0} \sin(k\omega_0 t)\Big|_0^D \\ &\quad - \int_0^D 2A\left(-\frac{1}{D}\right) \frac{1}{k\omega_0} \sin(k\omega_0 t)\, dt = -2A\frac{1}{(k\omega_0)^2 D} \cos(k\omega_0 t)\Big|_0^D \\ &= 2AD\frac{1 - \cos(k\omega_0 D)}{(k\omega_0 D)^2} = AD\frac{4\sin^2(k\omega_0 D/2)}{(k\omega_0 D/2)^2} \\ &= AD\,\mathrm{sin\,c}^2\left(k\frac{D}{P}\right) \quad \text{with} \quad \omega_0 = \frac{2\pi}{P} \end{aligned} \tag{E2.1.5}$$

Now we can use Eq. (2.1.5a) to write the Fourier series representation of the triangular wave as

$$A\tilde{\lambda}_{D/P}(t) \overset{(2.1.5a)}{=} \frac{1}{P} \sum_{k=-\infty}^{\infty} AD\,\mathrm{sinc}^2\left(k\frac{D}{P}\right) e^{jk\omega_0 t} \tag{E2.1.6}$$

In the case of $D = 1$ and $P = 2D = 2$ as depicted in Fig. 2.2(a2), the corresponding (magnitude) spectrum is plotted in Fig. 2.2(b2).

(c) MATLAB program to get the Fourier spectra
Once you have defined and saved a periodic function as an M-file, you can use the MATLAB routine "CTFS_exponential ()" to find its complex exponential Fourier series coefficients (c_k's). Interested readers are invited to run the following program "cir02e01.m" to get the Fourier coefficients and plot the spectra for a rectangular wave and a triangular wave as in Fig. 2.2.

```
%sig02e01.m : plot Fig. 2.2 (CTFS spectra of rectangular/triangular waves
clear, clf
global P D
N=8; k= -N:N; % the range of frequency indices
for i=1:2
   if i==1 % true Fourier series coefficients for a rectangular wave
     x = 'rectangular_wave'; P=2; D=1; c_true= D*sinc(k*D/P);
    else % true Fourier series coefficients for a triangular wave
     x = 'triangular_wave'; P=2; D=1; c_true= D*sinc(k*D/P).^2;
   end
    w0=2*pi/P; % fundamental frequency
    tt=[-400:400]*P/200; % time interval
   xt = feval(x,tt); % original signal
   [c,kk] = CTFS_exponential(x,P,N);
   [c; c_true] % to compare with true Fourier series coefficients
   discrepancy_between_numeric_and_analytic=norm(c-c_true)
   jkw0t= j*kk.'*w0*tt;
    xht = real(c/P*exp(jkw0t)); % Eq. (2.1.5a)
   subplot(219+i*2), plot(tt,xt,'k-', tt,xht,'b:')
   axis([tt(1) tt(end) -0.2 1.2]), title('Periodic function x(t)')
   c_mag = abs(c); c_phase = angle(c);
   subplot(220+i*2), stem(kk, c_mag), title('CTFS Spectrum |X(k)|')
end
```

```
function y=rectangular_wave(t)
global P D
tmp=min(abs(mod(t,P)),abs(mod(-t,P)));  y= (tmp<=D/2);
```

```
function y=triangular_wave(t)
global P D
tmp= min(abs(mod(t,P)),abs(mod(-t,P)));  y=(tmp<=D).*(1-tmp/D);
```

```
function [c,kk]=CTFS_exponential(x,P,N)
% Find the complex exponential Fourier coefficients c(k) for k=-N:N
% x: A periodic function with period P
% P: Period,  N: Maximum frequency index to specify the frequency range
w0=2*pi/P; % the fundamental frequency [rad/s]
xexp_jkw0t_= [x '(t).*exp(-j*k*w0*t)'];
xexp_jkw0t= inline(xexpjkw0t_,'t','k','w0');
kk=-N:N; tol=1e-6; % the frequency range tolerance on numerical error
for k=kk
   c(k+N+1)= quadl(xexp_jkw0t,-P/2,P/2,tol,[],k,w0); % Eq. (2.1.5b)
end
```

```
%sig02_01.m : plot Fig. 2.3 (CTFS reconstruction)
clear, clf
global P D
P=2; w0=2*pi/P; D=1; % period, fundamental frequency, and duration
tt=[-400:400]*P/400; % time interval of 4 periods
x = 'rectangular_wave';
xt = feval(x,tt); % original signal
plot(tt,xt,'k:'), hold on
Ns= [1 3 9 19];
for N=Ns
   k= -N:N; jkw0t= j*k.'*w0*tt; % the set of Fourier reconstruction terms
   c= D*sinc(k*D/P);
   xht = real(c/P*exp(jkw0t)); % Eq. (2.1.9)
   plot(tt,xht,'b'), hold on, pause
end
axis([tt(1) tt(end) -0.2 1.2])
```

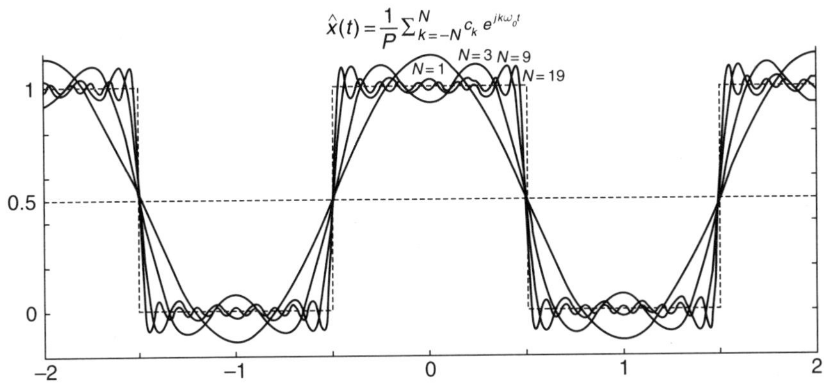

Fig. 2.3 Examples of the approximate Fourier reconstruction for a rectangular pulse

At this point, you may wonder how a rectangular wave with discontinuities can be represented by the sum of trigonometric or complex exponential functions that are continuous for all t. To satisfy your curiosity, let us consider the approximate Fourier series reconstruction formula.

$$\hat{x}_N(t) = \frac{1}{P}\sum_{k=-N}^{N} c_k\, e^{jk\omega_0 t} \tag{2.1.9}$$

This can be used to reconstruct the original time function $x(t)$ from its Fourier series coefficients. We can use the above MATLAB program 'sig02_01.m' to plot the Fourier series reconstructions of a rectangular wave with increasing number of terms $N = 1, 3, 9, 19, \ldots$ as in Fig. 2.3.

The following remark with Fig. 2.3 will satisfy your curiosity:

Remark 2.1 Convergence of *Fourier Series Reconstruction*

(1) The Fourier series convergence condition A stated in Sect. 2.1.1 guarantees that the Fourier coefficients are finite and the Fourier series reconstruction $\hat{x}_N(t)$ converges to the original time function $x(t)$ in the sense that

$$\int_P |\hat{x}_N(t) - x(t)|^2\, dt \to 0 \text{ as } N \to \infty$$

(2) The Fourier series convergence condition B stated in Sect. 2.1.1 guarantees the following:

- The Fourier coefficients are finite.
- The Fourier series reconstruction $\hat{x}_N(t)$ converges to the original time function $x(t)$ at every t except the discontinuities of $x(t)$ and to the average value of the limits from the left/right at each discontinuity.

(3) Figure 2.3 illustrates that $\hat{x}_N(t)$ has ripples around the discontinuities of $x(t)$, whose magnitude does not decrease as $N \to \infty$. This is called the *Gibbs phenomenon*.

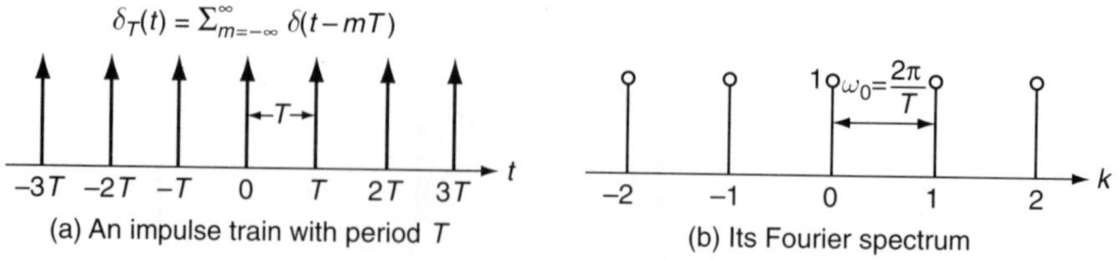

Fig. 2.4 An impulse train and its CTFT spectrum

(4) For practical purposes, we do not need to pay attention to the convergence condition because the "weird" signals that do not satisfy the condition are not important in the study of signals and systems.

Example 2.2 Fourier Spectrum of an Impulse Train

Consider an *impulse train* consisting of infinitely many shifted unit impulses that are equally spaced on the time axis:

$$\delta_T(t) = \sum\nolimits_{m=-\infty}^{\infty} \delta(t - mT) \tag{E2.1.1}$$

We can use Eq. (2.1.5b) with $P = T$ and $\omega_0 = 2\pi/T$ to obtain the Fourier coefficients as

$$c_k = \int_{-T/2}^{T/2} \delta_T(t)\, e^{-jk\omega_0 t}\, dt \overset{(E2.1.1)}{=} \int_{-T/2}^{T/2} \sum\nolimits_{m=-\infty}^{\infty} \delta(t - mT) e^{-jk\omega_0 t}\, dt$$

(since there is only one impulse $\delta(t)$ within the integration interval $[-T/2, T/2]$)

$$= \int_{-T/2}^{T/2} \delta(t)\, e^{-jk\omega_0 t}\, dt \underset{\text{with } t_1=0}{\overset{(1.1.25)}{=}} e^{-jk\omega_0 t}\Big|_{t=0} = 1\ \forall\, k$$

This means a flat spectrum that is uniformly distributed for every frequency index. Now we can use Eq. (2.1.5a) to write the Fourier series representation of the impulse train as

$$\delta_T(t) \underset{P=T}{\overset{(2.1.5a)}{=}} \frac{1}{P} \sum\nolimits_{k=-\infty}^{\infty} c_k\, e^{jk\omega_0 t} \overset{P=T}{=} \frac{1}{T} \sum\nolimits_{k=-\infty}^{\infty} e^{jk\omega_0 t} \quad \text{with} \quad \omega_0 = \frac{2\pi}{T} \tag{2.1.10}$$

Fig. 2.4(a) and (b) show an impulse train and its spectrum, respectively.

2.1.3 Physical Meaning of CTFS Coefficients – Spectrum

To understand the physical meaning of spectrum, let us see Fig. 2.5, which shows the major Fourier coefficients c_k of a zero-mean rectangular wave for $k = -3, -1,$ 1, and 3 (excluding the DC component c_0) and the corresponding time functions

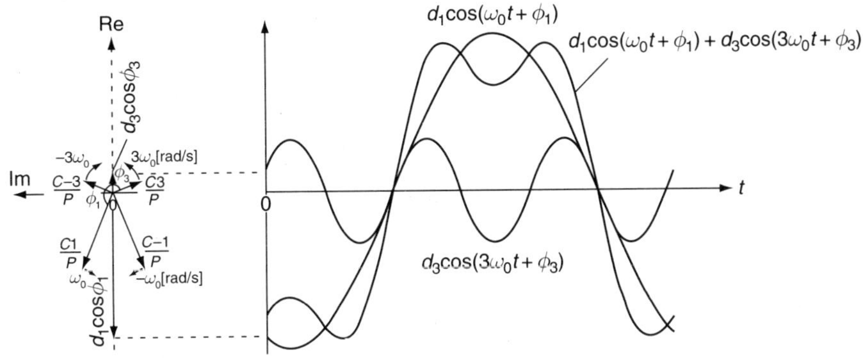

Fig. 2.5 Physical meaning of complex exponential Fourier series coefficients

$$\frac{1}{P}\left(c_{-1}\,e^{-j\omega_0 t}+c_1\,e^{j\omega_0 t}\right) \text{ and } \frac{1}{P}\left(c_{-3}\,e^{-j3\omega_0 t}+c_3\,e^{j3\omega_0 t}\right)$$

The observation of Fig. 2.5 gives us the following interpretations of Fourier spectrum:

Remark 2.2 Physical Meaning of Complex Exponential Fourier Series Coefficients

(1) While the trigonometric or magnitude-and-phase form of Fourier series has only nonnegative frequency components, the complex exponential Fourier series has positive/negative frequency ($\pm k\omega_0$) components that are conjugate-symmetric about $k = 0$, i.e.,

$$c_k \overset{(2.1.6b)}{=} \frac{P}{2}d_k e^{j\phi_k} \text{ and } c_{-k} \overset{(2.1.6c)}{=} \frac{P}{2}d_k e^{-j\phi_k} \rightarrow |c_{-k}| = |c_k| \text{ and } \phi_{-k} = -\phi_k$$

as shown in Sect. 2.1.1. This implies that the magnitude spectrum $|c_k|$ is (even) symmetric about the vertical axis $k = 0$ and the phase spectrum ϕ_k is odd symmetric about the origin.

(2) As illustrated above, the k th component appears as the sum of the positive and negative frequency components

$$\begin{aligned}\frac{c_k}{P}\,e^{jk\omega_0 t}+\frac{c_{-k}}{P}e^{-jk\omega_0 t} &\overset{(2.1.6)}{=} \frac{1}{2}d_k e^{j\phi_k}e^{jk\omega_0 t}+\frac{1}{2}d_k e^{-j\phi_k}e^{-jk\omega_0 t}\\ &= d_k\cos(k\omega_0 t+\phi_k)\end{aligned}$$

which denote two vectors (phasors) revolving in the opposite direction with positive (counter-clockwise) and negative (clockwise) angular velocities $\pm k\omega_0$ [rad/s] round the origin, respectively.

(3) Figure 2.6 also shows that the spectrum presents the descriptive information of a signal about its frequency components.

To get familiar with Fourier spectrum further, let us see and compare the spectra of the three signals, i.e., a rectangular wave, a triangular wave, and a constant

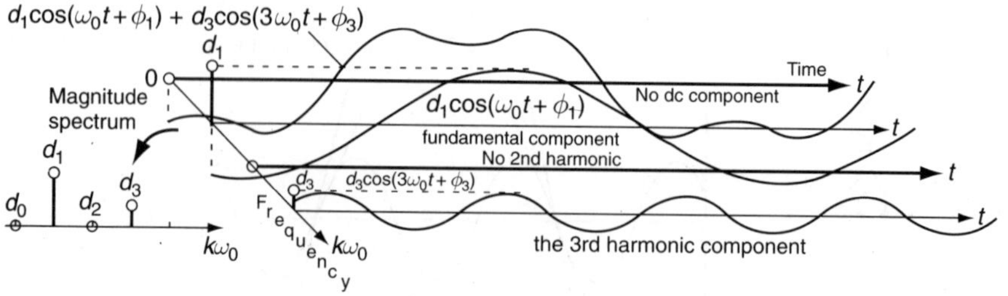

Fig. 2.6 Physical meaning of spectrum – time domain vs. frequency domain

(DC: Direct Current) signal depicted in Fig. 2.7. The observations are stated in the following remarks:

Remark 2.3 Effects of Smoothness and Period on Spectrum

(1) The smoother a time function is, the larger the relative magnitude of low frequency components to high frequency ones is. Compare Fig. 2.7(a1–b1), (a2–b2), and (a3–b3).

(cf.) The CTFS of the unit constant function can be obtained from Eq. (E2.1.3) with $A = 1$ and $P = D$.

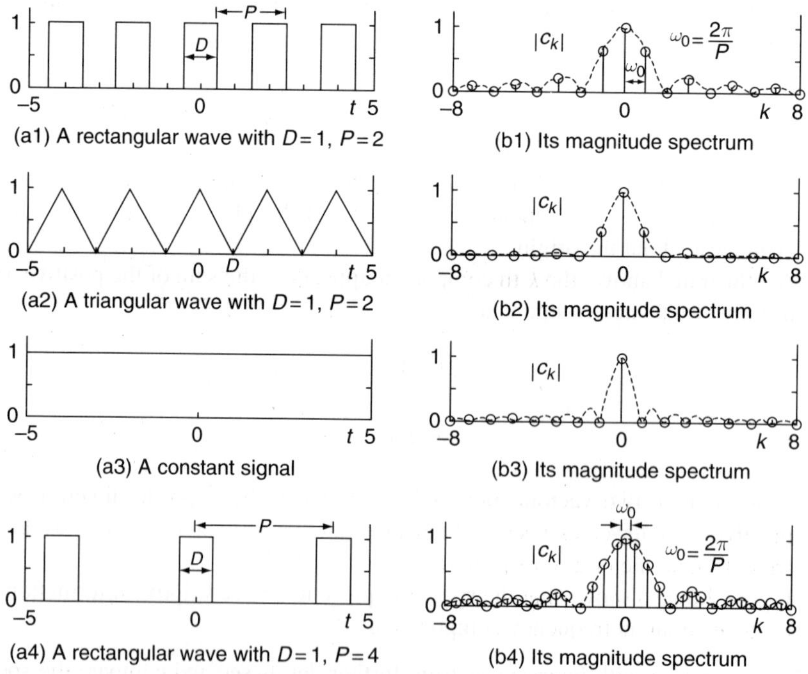

Fig. 2.7 The CTFS spectra of rectangular/triangular waves and a DC signal

(2) The longer the period P is, the lower the fundamental frequency $\omega_0 = 2\pi/P$ becomes and the denser the CTFS spectrum becomes. Compare Fig. 2.7(a4–b4) with (a1–b1).
   (cf.) This presages the continuous-time Fourier transform, which will be introduced in the next section.

Now, let us see how the horizontal/vertical translations of a time function $x(t)$ affect the Fourier coefficients.

<Effects of Vertical/Horizontal Translations of $x(t)$ on the Fourier coefficients>
Translating $x(t)$ by $\pm A$ (+: upward, −: downward) along the vertical axis causes only the change of Fourier coefficient $d_0 = a_0$ for $k = 0$ (DC component or average value) by $\pm A$. On the other side, translating $x(t)$ along the horizontal (time) axis by $\pm t_1$ (+: rightward, −: leftward) causes only the change of phases (ϕ_k's) by $\mp k\omega_0 t_1$, not affecting the magnitudes d_k of Eq. (2.1.4b) or $|c_k|$ of Eq. (2.1.5b):

$$
\begin{aligned}
c'_k &\overset{(2.1.5b)}{=} \int_P x(t-t_1)\, e^{-jk\omega_0 t}\, dt = \int_P x(t-t_1)\, e^{-jk\omega_0 (t-t_1+t_1)} dt \\
&= e^{-jk\omega_0 t_1} \int_P x(t-t_1)\, e^{-jk\omega_0 (t-t_1)} dt \overset{(2.1.5b)}{=} c_k e^{-jk\omega_0 t_1} = |c_k| \angle(\phi_k - k\omega_0 t_1)
\end{aligned} \tag{2.1.11}
$$

Note that $x(t - t_1)$ is obtained by translating $x(t)$ by t_1 in the positive (rightward) direction for $t_1 > 0$ and by $-t_1$ in the negative (leftward) direction for $t_1 < 0$ along the horizontal (time) axis. Eq. (2.1.11) implies that horizontal shift of $x(t)$ causes a change not in its magnitude spectrum but in its phase spectrum.

2.2 Continuous-Time Fourier Transform of Aperiodic Signals

In this section we will define the Fourier transform for aperiodic signals. Suppose we have an aperiodic signal $x(t)$ of finite duration $D > 0$ and its periodic extension $\tilde{x}_P(t)$ with period $P > D$ that is obtained by repeating $x(t)$ every P s.

Noting that, as we choose the period P to be longer, $\tilde{x}_P(t)$ appears to be identical to $x(t)$ over a longer interval, we can think of $x(t)$ as the limit of $\tilde{x}_P(t)$ as $P \to \infty$. Since $\tilde{x}_P(t)$ is periodic with period P, it can be represented by the Fourier series of the form

$$
\tilde{x}_P(t) = \frac{1}{P} \sum_{k=-\infty}^{\infty} c_k\, e^{jk\omega_0 t} = \frac{1}{2\pi} \sum_{k=-\infty}^{\infty} X(jk\omega_0)\, e^{jk\omega_0 t} \omega_0 \quad \text{with} \quad \omega_0 = \frac{2\pi}{P}
$$

where the CTFS coefficients are

$$
X(jk\omega_0) = c_k = \int_P \tilde{x}_P(t)\, e^{-jk\omega_0 t}\, dt = \int_{-\infty}^{\infty} x(t)\, e^{-jk\omega_0 t}\, dt
$$

Noting that $\tilde{x}_P(t) \to x(t)$ and $\omega_0 = 2\pi/P \to 0$ as $P \to \infty$, we let $\omega_0 = d\omega$ and $k\omega_0 = \omega$ and take the limits of the above two equations as $P \to \infty$ to write the *continuous-time Fourier transform* (CTFT) pair:

$$X(j\omega) = \mathcal{F}\{x(t)\} = \int_{-\infty}^{\infty} x(t)e^{-j\omega t}\,dt \text{ (Fourier transform/integral)} \tag{2.2.1a}$$

$$x(t) = \mathcal{F}^{-1}\{X(j\omega)\} = \frac{1}{2\pi}\int_{-\infty}^{\infty} X(j\omega)e^{j\omega t}\,d\omega \text{ (Inverse Fourier transform)} \tag{2.2.1b}$$

where $X(j\omega)$, called the spectrum of $x(t)$, has values at a continuum of frequencies and is often written as $X(\omega)$ with the constant j omitted. Like the CTFS of periodic signals, the CTFT of aperiodic signals provides us with the information concerning the frequency contents of signals, while the concept of frequency describes rather how rapidly a signal changes than how fast it oscillates.

Note that the sufficient condition for the convergence of CTFT is obtained by replacing the square-integrability condition (2.1.2a) of Condition A with

$$\int_{-\infty}^{\infty} |x(t)|^2\,dt < \infty \tag{2.2.2a}$$

or by replacing the absolute-integrability condition (2.1.2b) of Condition B with

$$\int_{-\infty}^{\infty} |x(t)|\,dt < \infty \tag{2.2.2b}$$

Remark 2.4 Physical Meaning of Fourier Transform – Signal Spectrum and Frequency Response

(1) If a time function $x(t)$ represents a physical signal, its Fourier transform $X(j\omega) = \mathcal{F}\{x(t)\}$ means the signal spectrum, which describes the frequency contents of the signal.
(2) In particular, if a time function $g(t)$ represents the impulse response of a continuous-time LTI (linear time-invariant) system, its Fourier transform $G(j\omega) = \mathcal{F}\{g(t)\}$ means the frequency response, which describes how the system responds to a sinusoidal input of (angular) frequency ω (refer to Sect. 1.2.6 for the definition of frequency response).

Remark 2.5 Frequency Response Existence Condition and Stability Condition of a System

For the impulse response $g(t)$ of a continuous-time LTI system, the absolute-integrability condition (2.2.2b) is identical with the stability condition (1.2.27a). This implies that a stable LTI system has a well-defined system function (frequency response) $G(j\omega) = \mathcal{F}\{g(t)\}$.

Remark 2.6 Fourier Transform and Laplace Transform

For any square-integrable or absolutely-integrable causal function $x(t)$ such that $x(t) = 0\ \forall\, t < 0$, the Fourier transform can be obtained by substituting $s = j\omega$ into the Laplace transform:

$$X(j\omega) \overset{(2.2.1a)}{=} \int_{-\infty}^{\infty} x(t)e^{-j\omega t}dt \underset{\text{causal signal}}{\overset{x(t)=0 \text{ for } t<0}{=}} \int_{0}^{\infty} x(t)e^{-j\omega t}dt \overset{(A.1)}{=} X(s)|_{s=j\omega} \quad (2.2.3)$$

This argues that for a physical system having causal impulse response $g(t)$, the Fourier transform $G(j\omega)$ of $g(t)$, that is the frequency response, can be obtained by substituting $s = j\omega$ into the system function $G(s)$, which is the Laplace transform of $g(t)$. (See Eq. (1.2.21).)

Example 2.3 CTFT Spectra of a Rectangular (Square) Pulse and a Triangular Pulse

(a) CTFT Spectrum of a Rectangular (Square) Pulse (Fig. 2.8(a))
Consider a single rectangular pulse with height 1 and duration D on the interval $[-D/2,\ D/2]$:

$$Ar_D(t) = A\left(u_s\left(t+\frac{D}{2}\right) - u_s\left(t-\frac{D}{2}\right)\right) = \begin{cases} A & \text{for } -D/2 \le |t| < D/2 \\ 0 & \text{elsewhere} \end{cases} \quad (E2.3.1)$$

We use Eq. (2.2.1a) to obtain the CTFT coefficients as

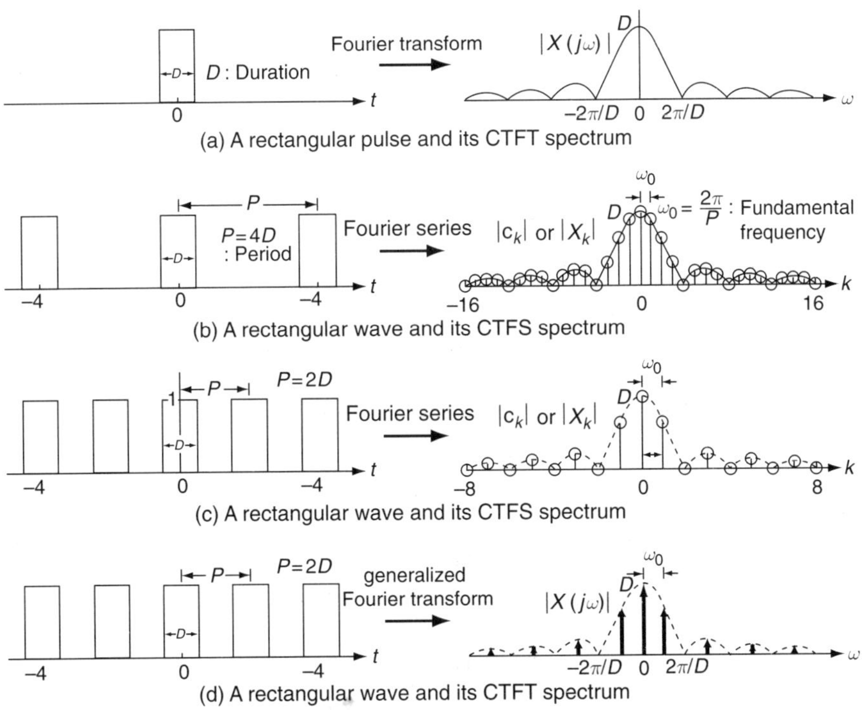

Fig. 2.8 The CTFT or CTFS spectra of rectangular pulses or waves

$$AR_D(j\omega) \overset{(2.2.1a)}{=} \int_{-\infty}^{\infty} Ar_D(t)e^{-j\omega t}\,dt = A\int_{-D/2}^{D/2} e^{-j\omega t}dt = \frac{A}{-j\omega}e^{-j\omega t}\Big|_{-D/2}^{D/2}$$
$$= A\frac{e^{j\omega D/2} - e^{-j\omega D/2}}{j\omega} = AD\frac{\sin(\omega D/2)}{\omega D/2}$$
$$= AD\mathrm{sinc}\left(\frac{\omega D}{2\pi}\right) \tag{E2.3.2}$$

This CTFT spectrum is depicted in Fig. 2.8(a). The first zero crossing $B = 2\pi/D$ [rad/s] of the magnitude spectrum is often used as a measure of the frequency spread of a signal and called the *zero-crossing (null-to-null) bandwidth* of the signal.

(cf.) As a by-product, we can apply the inverse CTFT formula (2.2.1b) for Eq. (E2.3.2) to get the integral of a sinc function:

$$r_D(t) \overset{(2.2.1b)}{=} \frac{1}{2\pi}\int_{-\infty}^{\infty} R_D(j\omega)e^{j\omega t}\,d\omega \overset{(E2.3.2)}{=} \frac{1}{2\pi}\int_{-\infty}^{\infty} D\frac{\sin(\omega D/2)}{\omega D/2}e^{j\omega t}\,d\omega$$
$$u_s\left(t+\frac{D}{2}\right) - u_s\left(t-\frac{D}{2}\right) \overset{\omega=2w}{=} \frac{1}{\pi}\int_{-\infty}^{\infty} \frac{\sin(wD)}{w}e^{j2wt}\,dw$$

Substituting $t = 0$ into this equation yields

$$\frac{1}{\pi}\int_{-\infty}^{\infty} \frac{\sin(wD)}{w}dw = u_s\left(\frac{D}{2}\right) - u_s\left(-\frac{D}{2}\right) = \mathrm{sign}(D) = \begin{cases} 1 & \text{for } D>0 \\ 0 & \text{for } D=0 \\ -1 & \text{for } D<0 \end{cases} \tag{E2.3.3}$$

(b) CTFT Spectrum of a Triangular Pulse
Like Eq. (E2.1.5), which is the CTFS coefficient of a triangular wave, we can find the CTFT spectrum of a single triangular pulse $x(t) = A\lambda_D(t)$ as

$$X(j\omega) = A\Lambda_D\,(j\omega) = AD\frac{\sin^2(\omega D/2)}{(\omega D/2)^2} = AD\,\mathrm{sinc}^2\left(\frac{\omega D}{2\pi}\right) \tag{E2.3.4}$$

Figure 2.8(a) shows a rectangular pulse and its CTFT spectrum, while Fig. 2.8(b) and (c) show two rectangular waves and their CTFS spectra. These figures present us an observation about the relationship between the CTFT of a single pulse $x(t)$ and the CTFS of its periodic repetition $\tilde{x}_P(t)$ with period P, which is summarized in the following remark.

Remark 2.7 Fourier Series and Fourier Transform

(1) We will mainly use the complex exponential Fourier coefficients, but seldom use the trigonometric or magnitude-and-phase form of Fourier series. Thus, from now on, we denote the complex exponential Fourier coefficients of $x(t)$

by X_k instead of c_k, which has been used so far to distinguish it from other Fourier coefficients a_k, b_k, or d_k.

(2) As can be seen from comparing Eqs. (E2.1.2) and (E2.3.2) or (E2.1.5) and (E2.3.4), the relationship between the CTFT $X(j\omega)$ of $x(t)$ and the CTFS coefficient X_k of $\tilde{x}_P(t)$ (the periodic extension of $x(t)$ with period P) is as follows:

$$X(j\omega)|_{\omega=k\omega_0=2\pi k/P} = X(jk\omega_0) = X_k \tag{2.2.4}$$

As the period P gets longer so that the fundamental frequency or frequency spacing $\omega_0=2\pi/P$ decreases, the Fourier coefficients X_k's become more closely spaced samples of the CTFT $X(j\omega)$, implying that the set of CTFS coefficients approaches the CTFT as $P \to \infty$ (see Fig. 2.8(c), (b), and (a)).

(3) Unlike the discrete frequency $k\omega_0$ of CTFS, the continuous frequency ω of CTFT describes how abruptly the signal changes rather than how often it oscillates.

(4) If the CTFT of a single pulse $x(t)$ and the CTFS of the periodic extension $\tilde{x}_P(t)$ were of different shape in spite of the same shape of $x(t)$ and $\tilde{x}_P(t)$ over one period P, it would be so confusing for one who wants the spectral information about a signal without knowing whether it is of finite duration or periodic. In this context, how lucky we are to have the same shape of spectrum (in the sense that CTFS are just samples of CTFT) whether we take the CTFT of $x(t)$ or the CTFS of $\tilde{x}_P(t)$! Furthermore, you will be happier to see that even the CTFT of $\tilde{x}_P(t)$ (Fig. 2.8(d)) is also of the same shape as the CTFS of $\tilde{x}_P(t)$, because one might observe one period of $\tilde{x}_P(t)$ and mistake it for $x(t)$ so that he or she would happen to apply the CTFT for periodic signals. Are you puzzled at the CTFT of a periodic signal? Then rush into the next section.

2.3 (Generalized) Fourier Transform of Periodic Signals

Since a periodic signal can satisfy neither the square-integrability condition (2.2.2a) nor the absolute-integrability condition (2.2.2b), the CTFTs of periodic signals are not only singular but also difficult to compute directly. For example, let us try to compute the CTFT of $x_k(t) = e^{jk\omega_0 t}$ by using Eq. (2.2.1a):

$$\mathcal{F}\{x_k(t)\} \overset{(2.2.1a)}{=} \int_{-\infty}^{\infty} e^{jk\omega_0 t}\, e^{-j\omega t}\, dt = \int_{-\infty}^{\infty} e^{-j(\omega-k\omega_0)t}\, dt$$

$$\overset{(D.33)}{=} \frac{1}{-j(\omega-k\omega_0)}\, e^{-j(\omega-k\omega_0)t}\Big|_{-T}^{T}\Big|_{T\to\infty}$$

$$= \frac{1}{j(\omega-k\omega_0)}\left(e^{j(\omega-k\omega_0)T} - e^{-j(\omega-k\omega_0)T}\right)\Big|_{T=\infty}$$

$$= \left(\frac{2\sin(\omega-k\omega_0)T}{(\omega-k\omega_0)}\right)\Big|_{T=\infty} =? \tag{2.3.1}$$

To get around this mathematical difficulty, let us find the inverse CTFT of

$$X_k(j\omega) = 2\pi\ \delta(\omega - k\omega_0)$$

by applying the inverse Fourier transform (2.2.1b):

$$\mathcal{F}^{-1}\{X_k(j\omega)\} \overset{(2.2.1b)}{=} \frac{1}{2\pi}\int_{-\infty}^{\infty} 2\pi\ \delta(\omega - k\omega_0)e^{j\omega t}\,d\omega \overset{(1.1.25)}{=} e^{jk\omega_0 t}$$

This implies a CTFT pair as

$$x_k(t) = e^{jk\omega_0 t} \overset{\mathcal{F}}{\leftrightarrow} X_k(j\omega) = 2\pi\ \delta(\omega - k\omega_0) \tag{2.3.2}$$

Based on this relation, we get the Fourier transform of a periodic function $x(t)$ from its Fourier series representation as follows:

$$x(t) \overset{(2.1.5a)}{=} \frac{1}{P}\sum_{k=-\infty}^{\infty} X_k\, e^{jk\omega_0 t} \quad \text{with} \quad \omega_0 = \frac{2\pi}{P}$$

$$\overset{\mathcal{F}}{\leftrightarrow} X(j\omega) \overset{(2.3.2)}{=} \frac{2\pi}{P}\sum_{k=-\infty}^{\infty} X_k\ \delta(\omega - k\omega_0) \tag{2.3.3}$$

This implies that the CTFT of a periodic signal consists of a train of impulses on the frequency axis having the same shape of envelope as the CTFS spectrum. Figure 2.8(d) is an illustration of Eq. (2.3.3) as stated in Remark 2.7(4).

Remark 2.8 Fourier Transform of a Periodic Signal
It would be cumbersome to directly find the CTFT of a periodic function. Thus we had better find the CTFS coefficients first and then use Eq. (2.3.2) as illustrated in Eq. (2.3.3).

2.4 Examples of the Continuous-Time Fourier Transform

Example 2.4 Fourier Transform of an Exponential Function

For an exponential function (Fig. 2.9(a)) with time constant $T > 0$

$$e_1(t) = \frac{1}{T}e^{-t/T}u_s(t) \quad \text{with} \quad T > 0, \tag{E2.4.1}$$

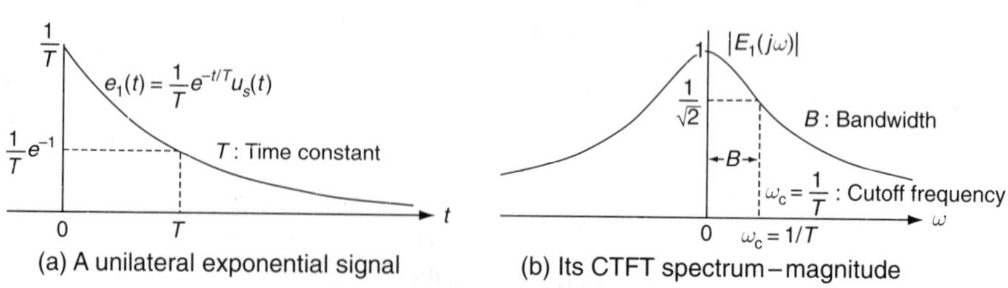

Fig. 2.9 A unilateral exponential signal and its CTFT spectrum

we have the Fourier transform

$$E_1(j\omega) = \mathcal{F}\{e_1(t)\} \overset{(2.2.1a)}{=} \int_{-\infty}^{\infty} \frac{1}{T} e^{-t/T} u_s(t)\, e^{-j\omega t}\, dt = \frac{1}{T}\int_0^{\infty} e^{-t/T} e^{-j\omega t}\, dt$$

$$= \frac{1}{T}\int_0^{\infty} e^{-(1/T+j\omega)t}\, dt \overset{(D.33)}{=} -\frac{1}{T(1/T+j\omega)} e^{-(1/T+j\omega)t}\Big|_0^{\infty}$$

$$= \frac{1}{1+j\omega T} = \frac{1}{\sqrt{1+(\omega T)^2}} \angle - \tan^{-1}(\omega T) \tag{E2.4.2}$$

whose magnitude is depicted in Fig. 2.9(b). From this magnitude spectrum, we see that $\omega_c = 1/T$ is the half-power frequency at which $|E_1(j\omega)|$ is $1/\sqrt{2}$ times the maximum magnitude 1:

$$|E_1(j\omega)| = \frac{1}{\sqrt{1+(\omega\, T)^2}} = \frac{1}{\sqrt{2}};\ 1+(\omega\, T)^2 = 2;\ \omega T = 1;\ \omega_c = \frac{1}{T} \tag{E2.4.3}$$

This example makes possible the following interpretations:

Remark 2.9 Signal Bandwidth and System Bandwidth – Uncertainty Principle

(1) In case the function $e_1(t)$ represents a physical signal itself, the above example illustrates the inverse relationship (a kind of duality) between the time and frequency domains that the bandwidth B [rad/s] of the signal is inversely proportional to the time-duration T [s], i.e., BT = constant(= 1). Note that the *bandwidth* of a signal, i.e., the width of the frequency band carrying the major portion of the signal energy, describes how rich frequency contents the signal contains. Such a relationship could also be observed in Example 2.3 and Fig. 2.8 where the time-duration of the rectangular pulse is D and the *signal bandwidth*, defined as the frequency range to the first zero-crossing of the magnitude spectrum, is $2\pi/D$. This observation, called the reciprocal *duration-bandwidth* relationship, is generalized into the *uncertainty principle* that the time-duration and bandwidth of a signal cannot be simultaneously made arbitrarily small, implying that a signal with short/long time-duration must have a wide/narrow bandwidth [S-1, Sect. 4.5.2]. This has got the name from the Heisenberg uncertainty principle in quantum mechanics that the product of the uncertainties in the measurements of the position and momentum of a particle cannot be made arbitrarily small.

(2) In case the function $e_1(t)$ represents the impulse response of a filter such as the RC circuit shown in Fig. 2.10(a), it has another interpretation that the bandwidth of the system behaving as a low-pass filter is inversely proportional to the time constant $T = RC$[s]. Note that the *system bandwidth* of a filter describes the width of frequency band of the signal to be relatively well passed and that it is usually defined as the frequency range to the 3dB-magnitude (half-power) frequency $B = 1/T$ [rad/s]. Also note that, in comparison with the bandwidth as a frequency-domain "capacity" concept, the *time constant* describes how fast the filter produces the response (output) to an applied input

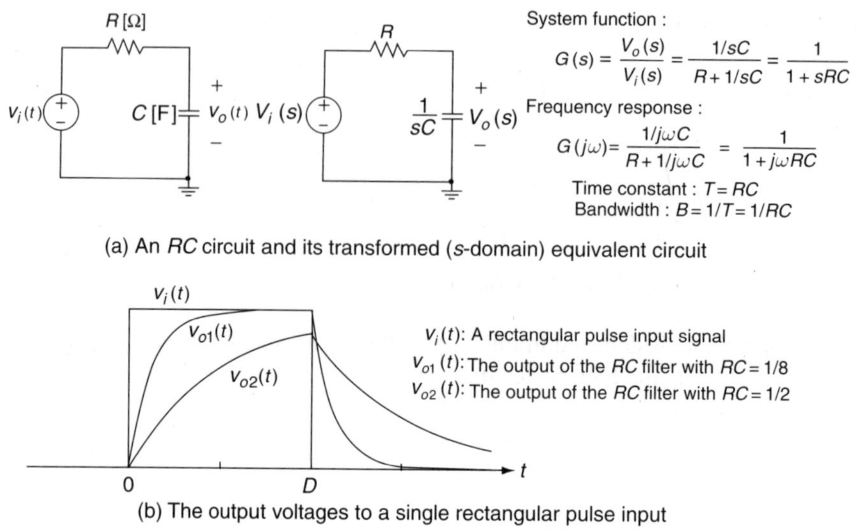

(a) An *RC* circuit and its transformed (*s*-domain) equivalent circuit

(b) The output voltages to a single rectangular pulse input

Fig. 2.10 An RC circuit and its outputs to rectangular pulse input signals

signal and that it is defined as the time taken for the output to reach $68.2\%(e^{-1})$ of the final (steady-state) value, or equivalently, the time measured until the (slowest) term of the transient response (converging towards zero) becomes as small as 32.8% of the initial value.

(3) Referring to Fig. 2.10(b), suppose a rectangular pulse of duration D is applied to the RC filter. Then, in order for the low-pass filter to have *high fidelity* of reproduction so that the output pulse will appear very much like the input pulses, the system bandwidth (the reciprocal of time constant RC) of the filter had better be greater than the signal bandwidth $2\pi/D$ of the input pulse.

Example 2.5 Fourier Transform of an Even-Symmetric Exponential Function

For an exponential function (Fig. 2.11(a)) with time constant $T > 0$

$$e_2(t) = \frac{1}{2T}e^{-|t|/T} \quad \text{with} \quad T > 0, \tag{E2.5.1}$$

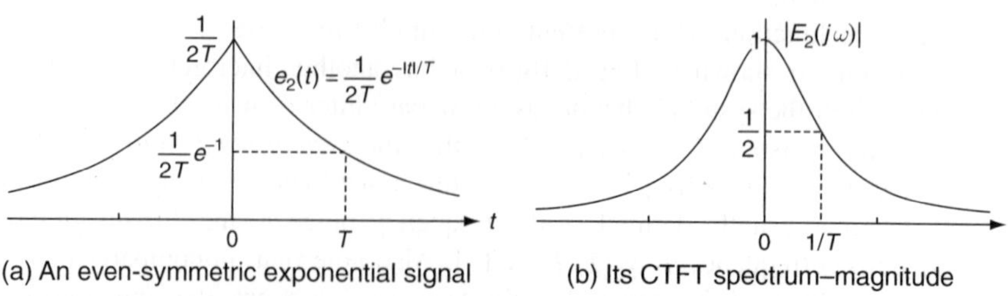

(a) An even-symmetric exponential signal (b) Its CTFT spectrum–magnitude

Fig. 2.11 An even-symmetric exponential signal and its CTFT spectrum we have the Fourier transform

$$E_2(j\omega) = \mathcal{F}\{e_2(t)\} \overset{(2.2.1a)}{=} \int_{-\infty}^{\infty} \frac{1}{2T} e^{-|t|/T} e^{-j\omega t}\, dt$$

$$= \frac{1}{2T}\left\{\int_{-\infty}^{0} e^{t/T} e^{-j\omega t}\, dt + \int_{0}^{\infty} e^{-t/T} e^{-j\omega t}\, dt\right\}$$

$$= \frac{1}{2T(1/T - j\omega)} e^{(1/T - j\omega)t}\Big|_{-\infty}^{0} - \frac{1}{2T(1/T + j\omega)} e^{-(1/T + j\omega)t}\Big|_{0}^{\infty}$$

$$= \frac{1/2}{1 - j\omega T} + \frac{1/2}{1 + j\omega T} = \frac{1}{1 + (\omega T)^2} \tag{E2.5.2}$$

whose magnitude is depicted in Fig. 2.11(b).

(Q) Why has the magnitude spectrum shorter bandwidth than that in Fig. 2.9(b)?
(A) It is because the signal in Fig. 2.11(a) is smoother than that in Fig. 2.9(a).

Example 2.6 Fourier Transform of the Unit Impulse (Dirac Delta) Function
We can obtain the Fourier transform of the unit impulse function $\delta(t)$ as

$$\mathcal{D}(j\omega) = \mathcal{F}\{\delta(t)\} \overset{(2.2.1a)}{=} \int_{-\infty}^{\infty} \delta(t) e^{-j\omega t}\, dt \overset{(1.1.25)}{=} 1\ \forall\, \omega \tag{E2.6.1}$$

This implies that an impulse signal has a flat or white spectrum, which is evenly distributed over all frequencies (see Fig. 2.12).

It is interesting to see that this can be derived by taking the limit of Eq. (E2.3.2) as $D \to 0$ in Example 2.3 (with $A = 1/D$) or Eq. (E2.5.2) as $T \to 0$ in Example 2.5:

$$\delta(t) \overset{(1.1.33b)}{=} \lim_{D\to 0} \frac{1}{D} r_D\left(t + \frac{D}{2}\right) \overset{(E2.3.1)}{=} \lim_{D\to 0} \frac{1}{D}\left(u_s\left(t + \frac{D}{2}\right) - u_s\left(t - \frac{D}{2}\right)\right)$$

$$\to \lim_{D\to 0} \frac{1}{D} R_D(j\omega) \overset{(E2.3.2)}{=} \lim_{D\to 0} \mathrm{sinc}\left(\frac{\omega D}{2\pi}\right) = 1$$

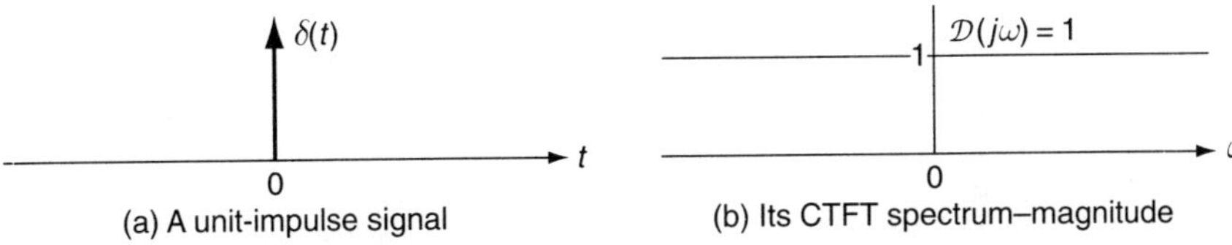

Fig. 2.12 A unit impulse signal and its CTFT spectrum

$$\delta(t) \overset{(1.1.33d)}{=} \lim_{T\to 0} e_2(t) \overset{(E2.5.1)}{=} \lim_{T\to 0} \frac{1}{2T} e^{-|t|/T}$$

$$\to \lim_{T\to 0} E_2(j\omega) \overset{(E2.5.2)}{=} \lim_{T\to 0} \frac{1}{1+(\omega T)^2} = 1$$

As a byproduct, we can apply the inverse Fourier transform (2.2.1b) to obtain an expression of the impulse function as

$$\delta(t) = \mathcal{F}^{-1}\{\mathcal{D}(j\omega)\} \overset{(2.2.1b)}{=} \frac{1}{2\pi}\int_{-\infty}^{\infty} \mathcal{D}(j\omega)e^{j\omega t}\, d\omega \overset{(E2.6.1)}{=} \frac{1}{2\pi}\int_{-\infty}^{\infty} 1e^{j\omega t}\, d\omega \tag{E2.6.2}$$

$$\overset{(D.33)}{=} \lim_{w\to\infty} \frac{1}{2\pi jt} e^{j\omega t}\Big|_{-w}^{w} = \lim_{w\to\infty} \frac{e^{j\omega t} - e^{-j\omega t}}{2\pi jt} \overset{(D.22)}{=} \lim_{w\to\infty} \frac{w}{\pi}\frac{\sin(wt)}{wt}$$

which is identical with Eq. (1.1.33a).

Remark 2.10 An Impulse Signal and Its (White/Flat) Spectrum

(1) Comparing Figs. 2.11 and 2.12, we see that short-duration signals contain stronger high-frequency components than long-duration ones do. This idea supports why a lightning stroke of very short duration produces an observable noise effect over all communication signals from the relatively low frequencies (550∼1600kHz) used in radio system to the relatively higher ones used in television system (60MHz for VHF ∼470MHz for UHF).
(2) We often use the impulse function as a typical input to determine the important characteristics (frequency response or system/transfer function) of linear time-invariant (LTI) systems. One practical reason is because the impulse function has uniform (flat) spectrum, i.e., contains equal power at all frequencies so that applying the impulse signal (or a good approximation to it) as the input to a system would be equivalent to simultaneously exciting the system with every frequency component of equal amplitude and phase.

Example 2.7 Fourier Transform of a Constant Function

We can obtain the Fourier transform of the unit constant function $c(t) = 1$ (Fig. 2.13(a)) as

$$C(j\omega) = \mathcal{F}\{1\} \overset{(2.2.1a)}{=} \int_{-\infty}^{\infty} 1e^{-j\omega t}\, dt \underset{\text{with } k=0}{\overset{(2.3.2)}{=}} 2\pi\,\delta(\omega) \tag{E2.7.1}$$

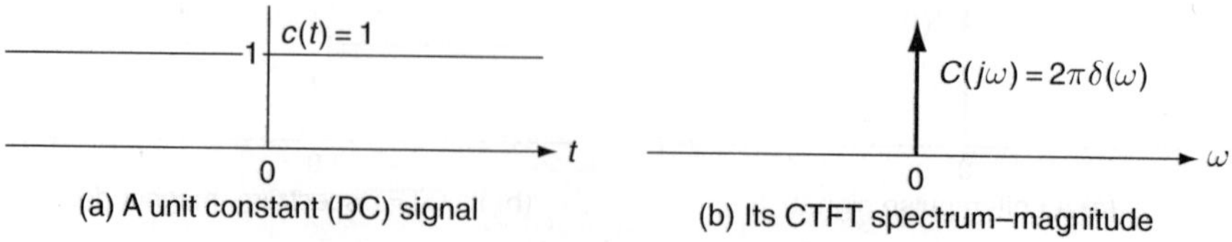

Fig. 2.13 A unit constant (DC) signal and its CTFT spectrum

This spectrum, that is depicted in Fig. 2.13(b), shows that a constant signal has only DC component with zero frequency, i.e., has all of its (infinite) energy at $\omega = 0$.

This can also be obtained by swapping t and ω in (E2.6.2) and verified by using the inverse Fourier transform (2.2.1b) to show that

$$\mathcal{F}^{-1}\{2\pi\delta(\omega)\} \overset{(2.2.1b)}{=} \frac{1}{2\pi}\int_{-\infty}^{\infty} 2\pi\,\delta(\omega)e^{j\omega t}\,d\omega \overset{(1.1.25)}{=} 1 \tag{E2.7.2}$$

Example 2.8 Fourier Transform of the Unit Step Function

To compute the Fourier transform of the unit step function $u_s(t)$, let us write its even-odd decomposition, i.e., decompose $u_s(t)$ into the sum of an even function and an odd function as

$$u_s(t) = u_e(t) + u_o(t) \tag{E2.8.1}$$

where

$$u_e(t) = \frac{1}{2}(u_s(t) + u_s(-t)) = \begin{cases} 1/2 & \text{for } t \neq 0 \\ 1 & \text{for } t = 0 \end{cases} \tag{E2.8.2}$$

$$u_o(t) = \frac{1}{2}(u_s(t) - u_s(-t)) = \frac{1}{2}\text{sign}(t) = \begin{cases} 1/2 & \text{for } t > 0 \\ 0 & \text{for } t \neq 0 \\ -1/2 & \text{for } t < 0 \end{cases} \tag{E2.8.3}$$

Then, noting that the even part $u_e(t)$ is a constant function of amplitude 1/2 except at $t = 0$, we can use Eq. (E2.7.1) to write its Fourier transform as

$$U_e(j\omega) = \mathcal{F}\{u_e(t)\} = \frac{1}{2}\mathcal{F}\{1\} \overset{(E2.7.1)}{=} \pi\,\delta(\omega) \tag{E2.8.4}$$

On the other side, the Fourier transform of the odd part can be computed as

$$U_o(j\omega) = \mathcal{F}\{u_o(t)\} \overset{(2.2.1a)}{=} \int_{-\infty}^{\infty} u_o(t)e^{-j\omega t}\,dt = \int_{-\infty}^{\infty} \overset{\text{odd}}{u_o(t)}(\overset{\text{even}}{\cos\omega t} - j\,\overset{\text{odd}}{\sin\omega t})\,dt$$

$$= -j\int_{-\infty}^{\infty} \overset{\text{odd}\times\text{odd}=\text{even}}{u_o(t)\sin\,\omega t}\,dt = -j2\int_{0}^{\infty} \overset{\text{even}}{u_o(t)\sin\,\omega t}\,dt$$

$$\overset{u_o(t)=1/2 \text{ for } t>0}{=} -j\int_0^{\infty}\sin\,\omega t\,dt = -j\int_0^{\infty}\sin\,\omega t\,e^{-st}\,dt|_{s=0}$$

$$\overset{(A.1)}{=} -j\,\mathcal{L}\{\sin\,\omega t\}|_{s=0} \overset{B.8(9)}{=} -j\left.\frac{\omega}{s^2+\omega^2}\right|_{s=0} = \frac{1}{j\omega} \tag{E2.8.5}$$

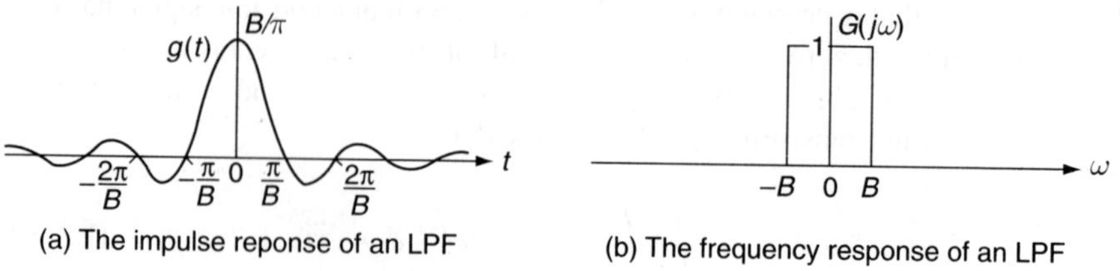

Fig. 2.14 The impulse response and frequency response of an ideal LPF

where we have used the Laplace transform. Now we add these two results to obtain the Fourier transform of the unit step function as

$$\mathcal{F}\{u_s(t)\} = \mathcal{F}\{u_e(t) + u_o(t)\} \overset{(2.5.1)}{=} U_e(j\omega) + U_o(j\omega) \overset{(E2.8.4),(E2.8.5)}{=} \pi\,\delta(\omega) + \frac{1}{j\omega} \tag{E2.8.6}$$

Example 2.9 Inverse Fourier Transform of an Ideal LPF Frequency Response

Let us consider the frequency response of an ideal lowpass filter (LPF) depicted in Fig. 2.14(b):

$$G(j\omega) = \begin{cases} 1 & \text{for } |\omega| \le B \\ 0 & \text{for } |\omega| > B \end{cases} \tag{E2.9.1}$$

Taking the inverse Fourier transform of this LPF frequency response yields the impulse response as

$$\begin{aligned} g(t) &= \mathcal{F}^{-1}\{G(j\omega)\} \overset{(2.2.1b)}{=} \frac{1}{2\pi}\int_{-\infty}^{\infty} G(j\omega)e^{j\omega t}\,d\omega = \frac{1}{2\pi}\int_{-B}^{B} 1e^{-j\omega t}\,d\omega \\ &= \frac{1}{2\pi jt}(e^{jBt} - e^{-jBt}) = \frac{\sin Bt}{\pi t} = \frac{B}{\pi}\text{sinc}\left(\frac{Bt}{\pi}\right) \end{aligned} \tag{E2.9.2}$$

which is depicted in Fig. 2.14(a).

(cf.) It may be interesting to see that a rectangular pulse and a sinc function constitute a Fourier transform pair, i.e., the Fourier transforms of rectangular pulse and sinc function turn out to be the spectra of sinc function and rectangular pulse function form, respectively (see Figs. 2.8(a) and 2.14). This is a direct consequence of the duality relationship between Fourier transform pairs, which will be explained in detail in Sect. 2.5.4.

Remark 2.11 Physical Realizability and Causality Condition

If a system has the frequency response that is strictly bandlimited like $G(j\omega)$ given by Eq. (E2.9.1) and depicted in Fig. 2.14(b), the system is not *physically realizable* because it violates the causality condition, i.e., $g(t) \neq 0$ for some $t < 0$ while every physical system must be causal (see Sect. 1.2.9).

Example 2.10 Fourier Transform of an Impulse Train

Let us consider an impulse train

$$\delta_T(t) = \sum\nolimits_{m=-\infty}^{\infty} \delta(t - mT) \tag{E2.10.1}$$

Since this is a periodic function, we first write its Fourier series representation from Eq. (2.1.10) as

$$\delta_T(t) \overset{(2.1.10)}{=} \frac{1}{T} \sum\nolimits_{k=-\infty}^{\infty} e^{jk\omega_s t} \quad \text{with} \quad \omega_s = \frac{2\pi}{T} \tag{E2.10.2}$$

Then we use Eq. (2.3.2) with $\omega_0 = \omega_s = 2\pi/T$ to obtain the Fourier transform as

$$\mathcal{D}_T(j\omega) \underset{(2.3.2)}{\overset{(E2.10.2)}{=}} \frac{2\pi}{T} \sum\nolimits_{k=-\infty}^{\infty} \delta(\omega - k\omega_s)$$

$$= \frac{2\pi}{T} \sum\nolimits_{k=-\infty}^{\infty} \delta(\omega + k\omega_s) \quad \text{with} \quad \omega_s = \frac{2\pi}{T} \tag{E2.10.3}$$

(cf.) Applying Eq. (2.2.1b) to take the inverse Fourier transform of Eq. (E2.10.3) will produce Eq. (2.10.2).

(cf.) Note that, as the period T (the interval between the impulses in time) increases, the fundamental frequency $\omega_s = 2\pi/T$ (the spacing between the impulses in frequency) decreases. This is also a consequence of the duality relationship between Fourier transform pairs.

Example 2.11 Fourier Transform of Cosine/Sine Functions

(a) The Fourier transform of $x(t) = \sin(\omega_1 t) = (e^{j\omega_1 t} - e^{-j\omega_1 t})/j2$ can be obtained as

$$X(j\omega) = \mathcal{F}\{\sin(\omega_1 t)\} = \frac{1}{j2}\mathcal{F}\{e^{j\omega_1 t} - e^{-j\omega_1 t}\} \overset{(2.3.2)}{=} j\pi(\delta(\omega + \omega_1) - \delta(\omega - \omega_1)) \tag{E2.11.1}$$

(b) The Fourier transform of $x(t) = \cos(\omega_1 t) = (e^{j\omega_1 t} + e^{-j\omega_1 t})/2$ can be obtained as

$$X(j\omega) = \mathcal{F}\{\cos(\omega_1 t)\} = \frac{1}{2}\mathcal{F}\{e^{j\omega_1 t} + e^{-j\omega_1 t}\} \overset{(2.3.2)}{=} \pi(\delta(\omega + \omega_1) + \delta(\omega - \omega_1)) \tag{E2.11.2}$$

2.5 Properties of the Continuous-Time Fourier Transform

In this section we are about to discuss some basic properties of continuous-time Fourier transform (CTFT), which will provide us with an insight into the Fourier transform and the capability of taking easy ways to get the Fourier transforms or inverse Fourier transforms.

(cf.) From now on, we will use $X(\omega)$ instead of $X(j\omega)$ to denote the Fourier transform of $x(t)$.

2.5.1 Linearity

With $\mathcal{F}\{x(t)\} = X(\omega)$ and $\mathcal{F}\{y(t)\} = Y(\omega)$, we have

$$a\ x(t) + b\ y(t) \overset{\mathcal{F}}{\leftrightarrow} a\ X(\omega) + b\ Y(\omega), \tag{2.5.1}$$

which implies that the Fourier transform of a linear combination of many functions is the same linear combination of the individual transforms.

2.5.2 (Conjugate) Symmetry

In general, Fourier transform has the time reversal property:

$$\begin{aligned}\mathcal{F}\{x(-t)\} &\overset{(2.2.1a)}{=} \int_{-\infty}^{\infty} x(-t)e^{-j\omega t}\,dt \overset{-t=\tau}{=} \int_{\infty}^{-\infty} x(\tau)e^{j\omega\tau}(-d\tau) \overset{\tau=t}{=} \int_{-\infty}^{\infty} x(t)e^{j\omega t}\,dt \\ &\underset{\omega\to-\omega}{\overset{(2.2.1a)}{=}} X(-\omega);\ x(-t) \overset{\mathcal{F}}{\leftrightarrow} X(-\omega)\end{aligned} \tag{2.5.2}$$

In case $x(t)$ is a real-valued function, we have

$$\begin{aligned}X(-\omega) &\underset{\omega\to-\omega}{\overset{(2.2.1a)}{=}} \int_{-\infty}^{\infty} x(t)e^{-j(-\omega)t}\,dt = \int_{-\infty}^{\infty} x(t)e^{-(-j)\omega t}dt \\ &\underset{j\to-j}{\overset{(2.2.1a)}{=}} X^*(\omega) \text{ (complex conjugate of } X\ (\omega))\end{aligned}$$

or equivalently,

$$\begin{aligned}\text{Re}\{X(-\omega)\} + j\text{Im}\{X(-\omega)\} &= \text{Re}\{X(\omega)\} - j\text{Im}\{X(\omega)\} \\ |X(-\omega)|\angle X(-\omega) &= |X(\omega)|\angle - X(\omega)\end{aligned} \tag{2.5.3}$$

This implies that the magnitude/phase of the CTFT of a real-valued function is an even/odd function of frequency ω. Thus, when obtaining the Fourier transform of a real-valued time function, we need to compute it only for $\omega \geq 0$ since we can use the conjugate symmetry to generate the values for $\omega < 0$ from those for $\omega > 0$. In other words, for a real-valued time function, its magnitude and phase spectra are symmetrical about the vertical axis and the origin, respectively.

For an even and real-valued function $x_e(t)$ such that $x_e(-t) = x_e(t)$, its Fourier transform is also an even and real-valued function in frequency ω:

$$\begin{aligned} X_e(-\omega) &\underset{\omega=-\omega}{\overset{(2.2.1a)}{=}} \int_{-\infty}^{\infty} x_e(t)\, e^{-j(-\omega)t}\, dt \overset{t=-\tau}{=} \int_{-\infty}^{\infty} x_e(-\tau)\, e^{-j\omega\tau} d\tau \\ &\underset{\text{even}}{\overset{x_e(-\tau)=x_e(\tau)}{=}} \int_{-\infty}^{\infty} x_e(\tau)\, e^{-j\omega\tau}\, d\tau \overset{(2.2.1a)}{=} X_e(\omega) \end{aligned} \tag{2.5.4a}$$

Also for an odd and real-valued function $x_o(t)$ such that $x_o(-t) = -x_o(t)$, its Fourier transform is an odd and imaginary-valued function in frequency ω:

$$\begin{aligned} X_o(-\omega) &\underset{\omega=-\omega}{\overset{(2.2.1a)}{=}} \int_{-\infty}^{\infty} x_o(t)\, e^{-j(-\omega)t}\, dt \overset{t=-\tau}{=} \int_{-\infty}^{\infty} x_o(-\tau)\, e^{-j\omega\tau} d\tau \\ &\underset{\text{odd}}{\overset{x_o(-\tau)=-x_o(\tau)}{=}} -\int_{-\infty}^{\infty} x_o(\tau)\, e^{-j\omega\tau}\, d\tau \overset{(2.2.1a)}{=} -X_o(\omega) \end{aligned} \tag{2.5.4b}$$

Note that any real-valued function $x(t)$ can be expressed as the sum of an even function and an odd one:

$$x(t) = x_e(t) + x_o(t)$$

where

$$x_e(t) = \frac{1}{2}(x(t) + x(-t)) \quad \text{and} \quad x_o(t) = \frac{1}{2}(x(t) - x(-t))$$

Thus we can get the relations

$$\begin{aligned} X(\omega) = \mathcal{F}\{x_e(t)\} + \mathcal{F}\{x_o(t)\} &= X_e(\omega) + X_o(\omega) \\ \text{Re}\{X(\omega)\} + j\text{Im}\{X(\omega)\} &= X_e(\omega) + X_o(\omega) \end{aligned}$$

which implies

$$\text{even and real-valued } x_e(t) \overset{\mathcal{F}}{\leftrightarrow} \text{Re}\{X(\omega)\} \text{ even and real-valued} \tag{2.5.5a}$$

$$\text{odd and real-valued } x_o(t) \overset{\mathcal{F}}{\leftrightarrow} j\text{Im}\{X(\omega)\} \text{ odd and imaginary-valued} \tag{2.5.5b}$$

2.5.3 Time/Frequency Shifting (Real/Complex Translation)

For a shifted time function $x_1(t) = x(t - t_1)$, we have its Fourier transform

$$\mathcal{F}\{x(t-t_1)\} \overset{(2.2.1a)}{=} \int_{-\infty}^{\infty} x(t-t_1)\, e^{-j\omega t}\, dt \overset{t-t_1\to t}{=} \int_{-\infty}^{\infty} x(t)\, e^{-j\omega(t+t_1)} dt$$

$$\overset{(2.2.1a)}{=} \mathcal{F}\{x(t)\}\, e^{-j\omega\, t_1};\ x(t-t_1) \overset{\mathcal{F}}{\leftrightarrow} X(\omega)\, e^{-j\omega t_1} \tag{2.5.6}$$

This implies that real translation (time shifting) of $x(t)$ by t_1 along the t -axis will affect the Fourier transform on its phase by $-\omega t_1$, but not on its magnitude. This is similar to Eq. (2.1.11), which is the time shifting property of CTFS.

In duality with the time shifting property (2.5.6), the complex translation (frequency shifting) property holds

$$x(t)\, e^{j\omega_1 t} \overset{\mathcal{F}}{\leftrightarrow} X(\,\omega - \omega_1\,) \tag{2.5.7}$$

2.5.4 Duality

As exhibited by some examples and properties, there exists a definite symmetry between Fourier transform pairs, stemming from a general property of *duality* between time and frequency. It can be proved by considering the following integral expression

$$f(u) = \int_{-\infty}^{\infty} g(v)\, e^{-juv} dv \overset{v\to -v}{=} \int_{-\infty}^{\infty} g(-v)\, e^{juv} dv \tag{2.5.8}$$

This, with $(\omega,\ t)$ or $(\pm t, \mp\omega)$ for $(u,\ v)$, yields the Fourier transform or inverse transform relation, respectively:

$$f(\omega) = \int_{-\infty}^{\infty} g(t)\, e^{-j\omega\, t}\, dt \overset{(2.2.1a)}{=} \mathcal{F}\{g(t)\} : \quad g(t) \overset{\mathcal{F}}{\leftrightarrow} f(\omega) \tag{2.5.9a}$$

$$f(\pm\, t) = \int_{-\infty}^{\infty} g(\mp\omega)\, e^{j\omega\, t}\, d\omega \overset{(2.2.1b)}{=} 2\pi \mathcal{F}^{-1}\{g(\mp\omega)\} : \quad f(\pm\, t) \overset{\mathcal{F}}{\leftrightarrow} 2\pi\, g(\mp\omega) \tag{2.5.9b}$$

It is implied that, if one Fourier transform relation holds, the substitution of $(\pm t, \mp\omega)$ for $(\omega,\ t)$ yields the other one, which is also valid. This property of duality can be used to recall or show Fourier series relations. We have the following examples:

(Ex 0) $$g(t) = x(t) \underset{(2.5.9a)}{\overset{\mathcal{F}}{\leftrightarrow}} f(\omega) = X(\omega)$$

$$\Leftrightarrow \frac{1}{2\pi} f(t) = \frac{1}{2\pi} X(t) \underset{(2.5.9b)}{\overset{\mathcal{F}}{\leftrightarrow}} g(\omega) = x(\omega)$$

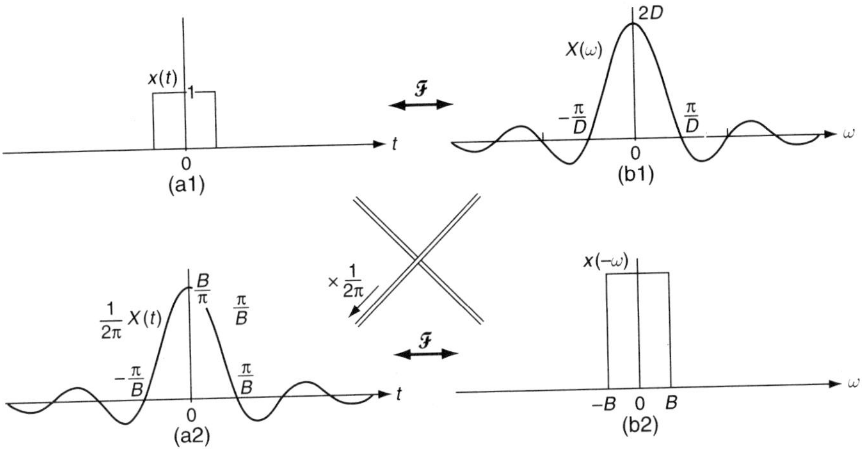

Fig. 2.15 Dual relationship between the Fourier transform pairs

(Ex 1) $x(t) = A\delta(t) \underset{(2.5.9a)}{\overset{\mathcal{F}}{\leftrightarrow}} X(\omega) \overset{(E2.6.1)}{=} A$

$$\Leftrightarrow X(t) = A \underset{(2.5.9b)}{\overset{\mathcal{F}}{\leftrightarrow}} 2\pi\, x(-\omega) = 2\pi A\, \delta(-\omega) \qquad \text{(See Example 2.7.)}$$

(Ex 2) $x(t) = \begin{cases} 1 & \text{for } |t| \le D \\ 0 & \text{for } |t| > D \end{cases} \underset{(2.5.9a)}{\overset{F}{\leftrightarrow}}$

$$X(\omega) \overset{(E2.3.1)}{=} 2D \,\text{sinc}\left(\frac{D\omega}{\pi}\right) \qquad \text{(See Example 2.9 and Fig.2.15)}$$

$$\Leftrightarrow \frac{1}{2\pi} X(t) = \frac{B}{\pi}\,\text{sinc}\left(\frac{Bt}{\pi}\right) \underset{(2.5.9b)}{\overset{\mathcal{F}}{\leftrightarrow}} x(-\omega) = \begin{cases} 1 & \text{for } |\omega| \le B \\ 0 & \text{for } |\omega| > B \end{cases}$$

(Ex 3) $x(t) = A\delta(t - t_1) \underset{(2.5.9a)}{\overset{\mathcal{F}}{\leftrightarrow}} X(\omega) \overset{(E2.6.1)\&(2.5.6)}{=} A\, e^{-j\omega\, t_1}$

$$\Leftrightarrow X(-t) = A\, e^{j\omega_1 t} \underset{(2.5.9b)}{\overset{\mathcal{F}}{\leftrightarrow}} 2\pi x(\omega) = 2\pi A\delta(\omega - \omega_1) \quad \text{(See Eq.(2.3.2))}$$

2.5.5 Real Convolution

For the (linear) convolution of two functions $x(t)$ and $g(t)$

$$y(t) = x(t) * g(t) = \int_{-\infty}^{\infty} x(\tau)g(t-\tau)\, d\tau = \int_{-\infty}^{\infty} g(\tau)x(t-\tau)\, d\tau = g(t) * x(t), \tag{2.5.10}$$

we can get its Fourier transform as

$$Y(\omega) = \mathcal{F}\{y(t)\} \underset{(2.5.10)}{\overset{(2.2.1a)}{=}} \int_{-\infty}^{\infty} \left\{ \int_{-\infty}^{\infty} x(\tau)g(t-\tau)\,d\tau \right\} e^{-j\omega t}dt$$

$$= \int_{-\infty}^{\infty} x(\tau)e^{-j\omega\tau} \left\{ \int_{-\infty}^{\infty} g(t-\tau)e^{-j\omega\,(t-\tau)}\,dt \right\} d\tau$$

$$\underset{(2.2.1a)}{\overset{t-\tau\to t}{=}} \int_{-\infty}^{\infty} x(\tau)e^{-j\omega\tau}\, G(\omega)\,d\tau = G(\omega)\int_{-\infty}^{\infty} x\,(\tau)e^{-j\omega\tau}d\tau \overset{(2.2.1a)}{=} G(\omega)X(\omega)$$

$$y(t) = x(t) * g(t) \overset{\mathcal{F}}{\leftrightarrow} Y(\omega) = X(\omega)\,G(\omega) \tag{2.5.11}$$

where $Y(\omega) = \mathcal{F}\{y(t)\}$, $X(\omega) = \mathcal{F}\{x(t)\}$, and $G(\omega) = \mathcal{F}\{g(t)\}$. This is the frequency-domain input-output relationship of a linear time-invariant (LTI) system with the input $x(t)$, the output $y(t)$, and the impulse response $g(t)$ where $G(\omega)$ is called the frequency response of the system.

On the other hand, if two functions $\tilde{x}_P(t)$ and $\tilde{g}_P(t)$ are periodic with common period P, their linear convolution does not converge so that we need another definition of convolution, which is the *periodic* or *circular convolution* with the integration performed over one period:

$$\begin{aligned}\tilde{y}_P(t) = \tilde{x}_P(t) \underset{P}{*} \tilde{g}_P(t) &= \int_P \tilde{x}_P(\tau)\tilde{g}_P(t-\tau)\,d\tau \\ &= \int_P \tilde{g}_P(\tau)\tilde{x}_P(t-\tau)\,d\tau = \tilde{g}_P(t) \underset{P}{*} \tilde{x}_P(t)\end{aligned} \tag{2.5.12}$$

where $\underset{P}{*}$ denotes the circular convolution with period P. Like the Fourier transform of a linear convolution, the Fourier series coefficient of the periodic (circular) convolution turns out to be the multiplication of the Fourier series coefficients of two periodic functions $\tilde{x}_P(t)$ and $\tilde{g}_P(t)$ (see Problem 2.8(a)):

$$\tilde{y}_P(t) = \tilde{x}_P(t) \underset{P}{*} \tilde{g}_P(t) \overset{\text{Fourier series}}{\leftrightarrow} Y_k = X_k\,G_k \tag{2.5.13}$$

where Y_k, X_k, and G_k are the Fourier coefficients of $\tilde{y}_P(t)$, $\tilde{x}_P(t)$, and $\tilde{g}_P(t)$, respectively.

2.5.6 Complex Convolution (Modulation/Windowing)

In duality with the convolution property (2.5.11) that convolution in the time domain corresponds to multiplication in the frequency domain, we may expect the *modulation property* that multiplication in the time domain corresponds to convolution in the frequency domain:

$$y(t) = x(t)m(t) \overset{\mathcal{F}}{\leftrightarrow} Y(\omega) = \frac{1}{2\pi} X(\omega) * M(\omega) \tag{2.5.14}$$

where $Y(\omega) = \mathcal{F}\{y(t)\}$, $X(\omega) = \mathcal{F}\{x(t)\}$, and $M(\omega) = \mathcal{F}\{m(t)\}$.

On the other hand, if two functions $\tilde{x}_P(t)$ and $\tilde{m}_P(t)$ are periodic with common period P, then their multiplication is also periodic and its Fourier series coefficient can be obtained from the convolution sum (see Problem 2.8(b)):

$$\tilde{y}_P(t) = \tilde{x}_P(t)\tilde{m}_P(t) \overset{\text{Fourier series}}{\leftrightarrow} Y_k = \frac{1}{P} \sum\nolimits_{n=-\infty}^{\infty} X_n \, M_{k-n} \tag{2.5.15}$$

where Y_k, X_k, and M_k are the Fourier coefficients of $\tilde{y}_P(t)$, $\tilde{x}_P(t)$, and $\tilde{m}_P(t)$, respectively.

Example 2.12 Sinusoidal Amplitude Modulation and Demodulation

(a) Sinusoidal Amplitude Modulation
Consider a sinusoidal amplitude-modulated (AM) signal

$$x_m(t) = x(t)m(t) = x(t)\cos(\omega_c t) \tag{E2.12.1}$$

Noting that the Fourier transform of the carrier signal $m(t) = \cos(\omega_c \, t)$ is

$$M(\omega) = \mathcal{F}\{m(t)\} = \mathcal{F}\{\cos(\omega_c t)\} \overset{\text{(E2.11.2)}}{=} \pi(\delta(\omega + \omega_c) + \delta(\omega - \omega_c)) \tag{E2.12.2}$$

we can use the modulation property (2.5.14) to get the Fourier transform of the AM signal as

$$\begin{aligned} X_m(\omega) = \mathcal{F}\{x_m(t)\} = \mathcal{F}\{x(t)\cos\omega_c \, t\} &\overset{\text{(2.5.14)}}{=} \frac{1}{2\pi} X(\omega) * M(\omega) \\ &\overset{\text{(E2.12.2)}}{=} \frac{1}{2\pi} X(\omega) * \pi(\delta(\omega + \omega_c) + \delta(\omega - \omega_c)) \\ &\overset{\text{(1.1.22)}}{=} \frac{1}{2}(X(\omega + \omega_c) + X(\omega - \omega_c)) \end{aligned} \tag{E2.12.3}$$

This implies that the spectrum of the AM signal $x(t)\cos(\omega_c t)$ consists of the sum of two shifted and scaled versions of $X(\omega) = \mathcal{F}\{x(t)\}$. Note that this result can also be obtained by applying the frequency shifting property (2.5.7) to the following expression:

$$x_m(t) = x(t)\cos(\omega_c t) \overset{\text{(D.21)}}{=} \frac{1}{2}(x(t)e^{j\omega_c t} + x(t)\, e^{-j\omega_c t}) \tag{E2.12.4}$$

(b) Sinusoidal Amplitude Demodulation
In an AM communication system, the receiver demodulates the modulated signal $x_m(t)$ by multiplying the carrier signal as is done at the transmitter:

$$x_d(t) = 2x_m(t)\cos(\omega_c t) \tag{E2.12.5}$$

We can use the modulation property (2.5.14) together with Eqs. (E2.12.3) and (E2.12.2) to express the Fourier transform of the demodulated signal in terms of the signal spectrum $X(\omega)$ as

$$\begin{aligned}
X_d(\omega) &= \mathcal{F}\{x_d(t)\} = \mathcal{F}\{x_m(t) 2\cos\omega_c\, t\} \overset{(2.5.14)}{=} \frac{1}{2\pi} X_m(\omega) * 2\,M(\omega) \\
&\underset{(E2.12.3)}{\overset{(E2.12.2)}{=}} \frac{1}{2\pi}\frac{1}{2}(X(\omega+\omega_c) + X(\omega-\omega_c)) * \ 2\pi(\delta(\omega+\omega_c) + \delta(\omega-\omega_c)) \\
&\overset{(1.1.22)}{=} \frac{1}{2}(X(\omega+2\omega_c) + X(\omega) + X(\omega) + X(\omega-2\omega_c)) \\
&= \frac{1}{2}X(\omega+2\omega_c) + X(\omega) + \frac{1}{2}X(\omega-2\omega_c)
\end{aligned} \tag{E2.12.6}$$

Example 2.13 Ideal (Impulse or Instant) Sampler and Finite Pulsewidth Sampler

(a) Ideal Sampling
We can describe the output of the *ideal* (or *impulse* or *instant*) *sampler* to a given input $x(t)$ as

$$x_*(t) = x(t)\,\delta_T(t) \left(\delta_T(t) = \sum\nolimits_{m=-\infty}^{\infty} \delta(t-mT) : \text{the impulse train}\right) \tag{E2.13.1}$$

This is illustrated in Fig. 2.16(a1). Here, the switching function has been modeled as an impulse train with period T and its Fourier transform is given by Eq. (E2.10.3) as

$$\mathcal{D}_T(\omega) \overset{(E2.10.3)}{=} \frac{2\pi}{T}\sum\nolimits_{k=-\infty}^{\infty} \delta(\omega + k\omega_s) \text{ with } \omega_s = \frac{2\pi}{T} \tag{E2.13.2}$$

which is shown in Fig. 2.16(c1). We can use the modulation property (2.5.14) together with Eq. (E2.13.2) to express the Fourier transform of the ideal sampler output in terms of the input spectrum $X(\omega)$ as

$$\begin{aligned}
X_*(\omega) &\overset{(2.5.14)}{=} \frac{1}{2\pi} X(\omega) * \mathcal{D}_T(\omega) \overset{(E2.13.2)}{=} \frac{1}{T}\sum\nolimits_{k=-\infty}^{\infty} X(\omega) * \delta(\omega + k\omega_s) \\
&\overset{(1.1.22)}{=} \frac{1}{T}\sum\nolimits_{k=-\infty}^{\infty} X(\omega + k\omega_s)
\end{aligned} \tag{E2.13.3}$$

which is depicted in Fig. 2.16(d1).

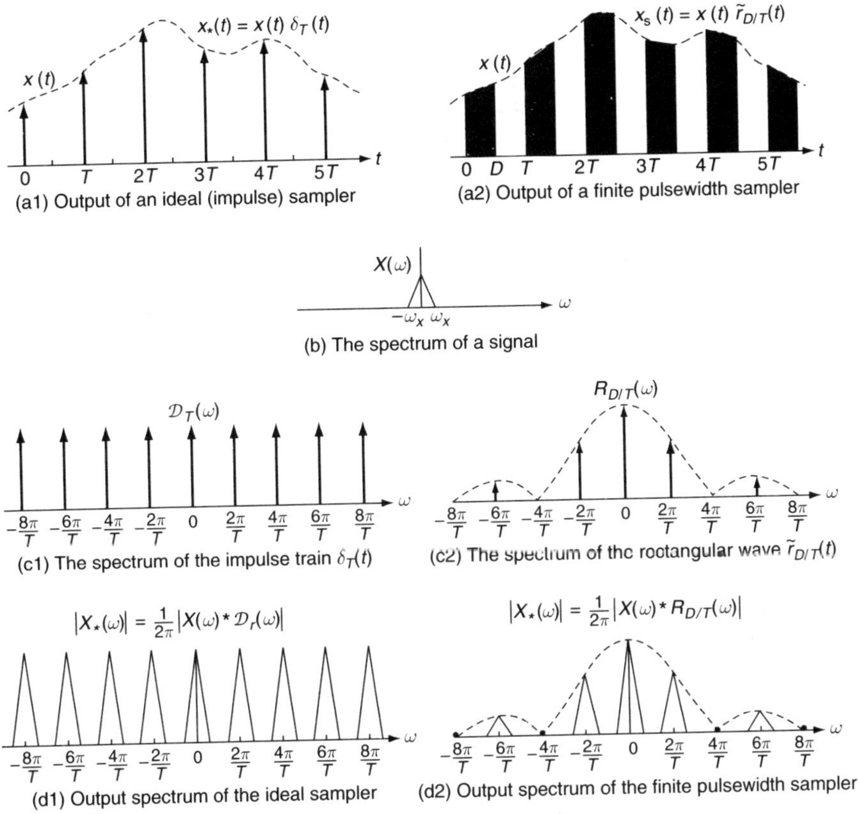

Fig. 2.16 Ideal sampler and finite pulsewidth sampler

(b) Finite Pulsewidth Sampling

We can describe the output of a *finite pulsewidth sampler* with period T to a given input $x(t)$ as

$$x_s(t) = x(t)\,\tilde{r}_{D/T}(t) \tag{E2.13.4}$$

$(\tilde{r}_{D/T}(t)$: a rectangular wave with duration D and period $T)$

This is illustrated in Fig. 2.16(a2). Here, the switching function has been modeled as a rectangular wave, $\tilde{r}_{D/P}(t)$, with duration D and period T. Noting that from Eq. (E2.1.3), the Fourier series representation of this rectangular wave is

$$\tilde{r}_{D/P}(t) \underset{A=1,P=T}{\overset{\text{(E2.1.3)}}{=}} \frac{1}{T}\sum_{k=-\infty}^{\infty} D\,\text{sinc}\left(k\frac{D}{T}\right) e^{jk\omega_s t} \text{ with } \omega_s = \frac{2\pi}{T}$$

(E2.13.5)

we can use Eq. (2.3.3) to write its Fourier transform as

$$R_{D/P}(\omega) \overset{(2.3.2)}{\underset{P=T}{=}} \frac{2\pi}{T}\sum\nolimits_{k=-\infty}^{\infty} c_k\,\delta(\omega - k\omega_s) \text{ with } c_k = D \operatorname{sinc}\left(k\frac{D}{T}\right) \tag{E2.13.6}$$

which is shown in Fig. 2.16(c2). Now we can apply the modulation property (2.5.14) together with Eq. (E2.13.6) to express the Fourier transform of the finite pulsewidth sampler output in terms of the input spectrum $X(\omega)$ as

$$\begin{aligned} X_s(\omega) &\overset{(2.5.14)}{=} \frac{1}{2\pi} X(\omega) * R_{D/P}(\omega) \overset{(E2.13.6)}{=} \frac{1}{T}\sum\nolimits_{k=-\infty}^{\infty} X(\omega) * c_k\,\delta(\omega - k\omega_s) \\ &\overset{(1.1.22)}{=} \frac{1}{T}\sum\nolimits_{k=-\infty}^{\infty} c_k\, X(\omega - k\omega_s) \end{aligned} \tag{E2.13.7}$$

which is depicted in Fig. 2.16(d2).

2.5.7 Time Differential/Integration – Frequency Multiplication/Division

By differentiating both sides of the inverse Fourier transform formula (2.2.1b) w.r.t. t, we obtain

$$\frac{dx(t)}{dt} \overset{(2.2.1b)}{=} \frac{1}{2\pi}\int_{-\infty}^{\infty} X(\omega)\left(\frac{d}{dt}e^{j\omega t}\right)d\omega = \frac{1}{2\pi}\int_{-\infty}^{\infty}(j\omega X(\omega))e^{j\omega t}\,d\omega \tag{2.5.16}$$

which yields the time-differentiation property of the Fourier transform as

$$\frac{dx(t)}{dt} \overset{F}{\longleftrightarrow} j\omega X(\omega) \tag{2.5.17}$$

This means that differentiation in the time domain results in multiplication by $j\omega$ in the frequency domain.

On the other hand, the time-integration property is obtained by expressing the integration of $x(t)$ in the form of the convolution of $x(t)$ and the unit step function $u_s(t)$ and then applying the convolution property (2.5.11) together with Eq. (E2.8.6) as

$$\begin{aligned} &\int_{-\infty}^{t} x(\tau)\,d\tau = \int_{-\infty}^{\infty} x(\tau)\,u_s(t-\tau)\,d\tau = x(t) * u_s(t) \underset{(2.5.11)}{\overset{\mathcal{F}}{\leftrightarrow}} \\ &\mathcal{F}\{x(t)\}\mathcal{F}\{u_s(t)\} \overset{(E2.8.6)}{=} \pi\, X(\omega)\delta(\omega) + \frac{1}{j\omega}X(\omega) = \pi\, X(0)\delta(\omega) + \frac{1}{j\omega}X(\omega) \end{aligned} \tag{2.5.18}$$

where the additional impulse term $\pi X(0)\delta(\omega)$ on the RHS reflects the DC value resulting from the integration. This equation is slightly above our intuition that integration/differentiation in the time domain results in division/multiplication by $j\omega$ in the frequency domain.

The differentiation/integration properties (2.5.17)/(2.5.18) imply that differentiating/integrating a signal increases the high/low frequency components, respectively because the magnitude spectrum is multiplied by $|j\omega| = \omega$ (proportional to frequency ω) or $|1/j\omega| = 1/\omega$ (inversely proportional to frequency ω). That is why a differentiating filter on the image frame is used to highlight the edge at which the brightness changes rapidly, while an integrating filter is used to remove impulse-like noises. Note also that a differentiator type filter tends to amplify high-frequency noise components and an integrator type filter would blur the image.

2.5.8 Frequency Differentiation – Time Multiplication

By differentiating both sides of the Fourier transform formula (2.2.1a) w.r.t. ω, we obtain

$$\frac{dX(\omega)}{d\omega} \overset{(2.2.1a)}{=} \int_{-\infty}^{\infty} x(t)\left(\frac{d}{d\omega}e^{-j\omega t}\right)dt = -j\int_{-\infty}^{\infty}(tx(t))e^{-j\omega t}dt \tag{2.5.19}$$

which yields the frequency-differentiation property of the Fourier transform as

$$t\,x(t) \overset{\mathcal{F}}{\leftrightarrow} j\frac{dX(\omega)}{d\omega} \tag{2.5.20}$$

This means that multiplication by t in the time domain results in differentiation w.r.t. ω and multiplication by j in the frequency domain.

2.5.9 Time and Frequency Scaling

The Fourier transform of $x(at)$ scaled along the time axis can be obtained as

$$F\{x(at)\} = \int_{-\infty}^{\infty} x(at)\,e^{-j\omega t}dt$$
$$\overset{t=\tau/a}{=} \frac{1}{|a|}\int_{-\infty}^{\infty} x(\tau)e^{-j\omega\tau/a}d\tau \overset{(2.2.1a)}{=} \frac{1}{|a|}X\left(\frac{\omega}{a}\right)$$

which yields the time and frequency scaling property of Fourier transform:

$$x(at) \overset{\mathcal{F}}{\leftrightarrow} \frac{1}{|a|}X\left(\frac{\omega}{a}\right) \tag{2.5.21}$$

This is another example of the dual (inverse) relationship between time and frequency. A common illustration of this property is the effect on frequency contents of playing back an audio tape at different speeds. If the playback speed is higher/slower than the recording speed, corresponding to compression($a>1$)/expansion($a<1$) in time, then the playback sounds get higher/lower, corresponding to expansion/compression in frequency.

2.5.10 Parseval's Relation (Rayleigh Theorem)

If $x(t)$ has finite energy and $\mathcal{F}\{x(t)\} = X(\omega)$, then we have

$$\int_{-\infty}^{\infty} |x(t)|^2\,dt = \frac{1}{2\pi}\int_{-\infty}^{\infty} |X(\omega)|^2\,d\omega \tag{2.5.22}$$

(Proof)

$$\begin{aligned}\int_{-\infty}^{\infty} |x(t)|^2\,dt &= \int_{-\infty}^{\infty} x(t)x^*(t)\,dt \\ &\overset{(2.2.1b)}{=} \int_{-\infty}^{\infty} x(t)\left\{\frac{1}{2\pi}\int_{-\infty}^{\infty} X^*(\omega)e^{-j\omega t}\,d\omega\right\}dt \\ &= \frac{1}{2\pi}\int_{-\infty}^{\infty} X^*(\omega)\left\{\int_{-\infty}^{\infty} x(t)\,e^{-j\omega t}\,dt\right\}d\omega \\ &\overset{(2.2.1a)}{=} \frac{1}{2\pi}\int_{-\infty}^{\infty} X^*(\omega)X(\omega)\,d\omega = \frac{1}{2\pi}\int_{-\infty}^{\infty} |X(\omega)|^2\,d\omega\end{aligned}$$

This implies that the total energy in the signal $x(t)$ can be determined either by integrating $|x(t)|^2$ over all time or by integrating $|X(\omega)|^2/2\pi$ over all frequencies. For this reason, $|X(\omega)|^2$ is called the *energy-density spectrum* of the signal $x(t)$.

On the other hand, if $\tilde{x}_P(t)$ is periodic with period P and its Fourier series coefficients are X_k's, then we have an analogous relation

$$\int_{-\infty}^{\infty} |\tilde{x}_P(t)|^2\,dt = \frac{1}{P}\sum_{k=-\infty}^{\infty} |X_k|^2 \tag{2.5.23}$$

where $|X_k|^2/P$ is called the *power-density spectrum* of the periodic signal $\tilde{x}_P(t)$.

2.6 Polar Representation and Graphical Plot of CTFT

Noting that a signal $x(t)$ can be completely recovered from its Fourier transform $X(\omega)$ via the inverse Fourier transform formula (2.3.1b), we may say that $X(\omega)$ contains all the information in $x(t)$. In this section we consider the polar or magnitude-phase representation of $X(\omega)$ to gain more insight to the (generally complex-valued) Fourier transform. We can write it as

$$X(\omega) = |X(\omega)|\angle X(\omega)$$

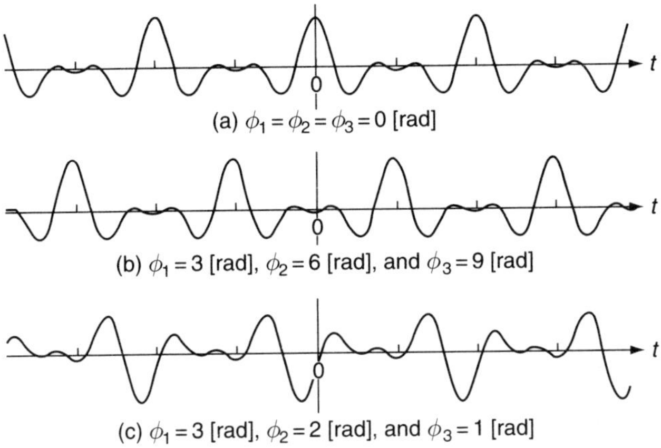

Fig. 2.17 Plots of $x(t) = 0.5\cos(2\pi t - \phi_1) + \cos(4\pi t - \phi_2) + (2/3)\cos(6\pi t - \phi_3)$ with different phases

where $|X(\omega)|$ and $\angle X(\omega)$ give us the information about the magnitudes and phases of the complex exponentials making up $x(t)$. Notice that if $x(t)$ is real, $|X(\omega)|$ is an even function of ω and $\angle X(\omega)$ is an odd function of ω and thus we need to plot the spectrum for $\omega \geq 0$ only (see Sect. 2.5.2).

The signals having the same magnitude spectrum may look very different depending on their phase spectra, which is illustrated in Fig. 2.17. Therefore, in some instances phase distortion may be serious.

2.6.1 Linear Phase

There is a particular type of phase characteristic, called *linear phase*, that the phase shift at frequency ω is a linear function of ω. Specifically, the Fourier transform of $x(t)$ changed in the phase by $-\alpha\omega$, by the time shifting property (2.5.6), simply results in a time-shifted version $x(t - \alpha)$:

$$X_1(\omega) = X(\omega)\angle - \alpha\omega = X(\omega)e^{-j\alpha\omega} \overset{\mathcal{F}}{\leftrightarrow} x_1(t) = x(t - \alpha) \tag{2.6.1}$$

For example, Fig. 2.17(b) illustrates how the linear phase shift affects the shape of $x(t)$.

2.6.2 Bode Plot

To show a Fourier transform $X(\omega)$, we often use a graphical representation consisting of the plots of the magnitude $|X(\omega)|$ and phase $\angle X(\omega)$ as functions of frequency ω. Although this is useful and will be used extensively in this book, we introduce another representation called the *Bode plot*, which is composed of two graphs, i.e., magnitude curve of log-magnitude $20\log_{10}|X(\omega)|$ [dB] and the phase curve of $\angle X(\omega)$ [degree] plotted against the log frequency $\log_{10}\omega$. Such a representation using the logarithm facilitates the graphical manipulations performed in analyzing LTI systems since the product and division factors in $X(\omega)$ become additions and subtractions, respectively. For example, let us consider a physical system whose system or transfer function is

$$G(s) = \frac{K(1+T_1 s)(1+T_2 s)}{s(1+T_a s)(1+2\zeta T_b s+(T_b s)^2)} \tag{2.6.2}$$

As explained in Sect. 1.2.6, its frequency response, that is the Fourier transform of the impulse response, is obtained by substituting $s = j\omega$ into $G(s)$:

$$G(j\omega) = G(s)|_{s=j\omega} = \frac{K(1+j\omega\, T_1)(1+j\omega\, T_2)}{j\omega(1+j\omega\, T_a)(1+j\omega 2\zeta T_b-(\omega T_b)^2)} \tag{2.6.3}$$

The magnitude of $G(j\omega)$ in decibels is obtained by taking the logarithm on the base 10 and then multiplying by 20 as follows:

$$\begin{aligned}|G(j\omega)| &= 20\log_{10}|G(j\omega)|[\text{dB}]\\ &= 20\log_{10}|K| + 20\log_{10}|1+j\omega\, T_1| + 20\log_{10}|1+j\omega\, T_2|\\ &\quad - 20\log_{10}|j\omega| - 20\log_{10}|1+j\omega\, T_a|\\ &\quad - 20\log_{10}|1+j\omega 2\zeta T_b-(\omega T_b)^2|\end{aligned} \tag{2.6.4a}$$

The phase of $G(j\omega)$ can be written as

$$\begin{aligned}\angle G(\omega) &= \angle K + \angle(1+j\omega\, T_1) + \angle(1+j\omega\, T_2) - \angle j\omega - \angle(1+j\omega\, T_a)\\ &\quad - \angle(1+j\omega 2\zeta T_b-(\omega T_b)^2)\end{aligned} \tag{2.6.4b}$$

The convenience of analyzing the effect of each factor on the frequency response explains why Bode plots are widely used in the analysis and design of linear time-invariant (LTI) systems.

(cf.) The MATLAB function "`bode(n,d,..)`" can be used to plot Bode plots.

2.7 Summary

In this chapter we have studied the CTFS (continuous-time Fourier series) and CTFT (continuous-time Fourier transform) as tools for analyzing the frequency characteristic of continuous-time signals and systems. One of the primary motivations for the use of Fourier analysis is the fact that we can use the Fourier series or transform to represent most signals in terms of complex exponential signals which are eigenfunctions of LTI systems (see Problem 2.14). The Fourier transform possesses a wide variety of important properties. For example, the convolution property allows us to describe an LTI system in terms of its frequency response and the modulation property provides the basis for the frequency-domain analysis of modulation, sampling, and windowing techniques.

Problems

2.1 Fourier Series Analysis and Synthesis of Several Periodic Functions

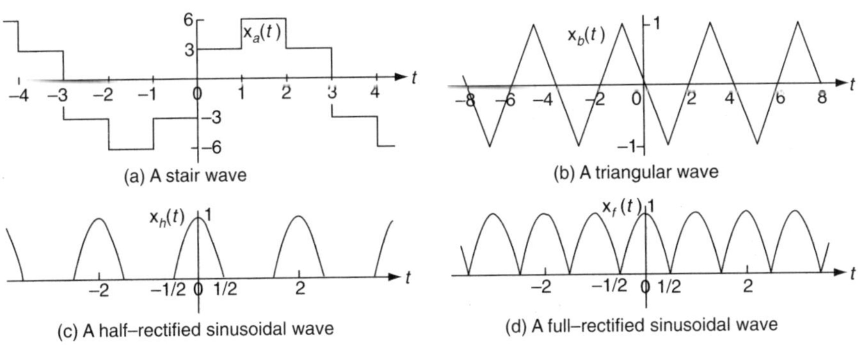

Fig. P2.1

(a) Noting that the stair wave in Fig. P2.1(a) can be regarded as the sum of three scaled/time-shifted rectangular (square) waves with common period 6[s] and different durations

$$x_a(t) = 6\tilde{r}_{3/6}(t-1.5) - 3 + 3\tilde{r}_{1/6}(t-1.5) - 3\tilde{r}_{1/6}(t-4.5), \quad \text{(P2.1.1)}$$

use Eq. (E2.1.3) together with (2.1.9) to find its complex exponential Fourier series coefficients. Then complete the following MATLAB program "cir02p_01a.m" and run it to plot the Fourier series representation (2.1.7) to see if it becomes close to $x_a(t)$ as the number of terms in the summation increases, say, from 5 to 20. Also, compare the Fourier coefficients with those obtained by using "CtFS_exponential()".

```
%cir02p_01a.m
clear, clf
P= 6; w0= 2*pi/P; % Period, Fundamental frequency
tt=[-400:400]*P/400; % time interval of 4 periods
x = 'x_a'; N=20; k= -N:N; % the range of frequency indices
c= 6*3*sinc(k*3/P).*exp(-j*1.5*k*w0)+3*sinc(k/P).*exp(-j*1.5*k*w0) ...
    - ??????????????????????????????;
c(N+1) = c(N+1) - ???;
% [c_n,k] = CtFS_exponential(x,P,N);
xt = feval(x,tt); % original signal
jkw0t= j*k.'*w0*tt; xht = real((c/P)*exp(jkw0t)); % Eq. (2.1.5a)
subplot(221), plot(tt,xt,'k-', tt,xht,'r:')
c_mag = abs(c); c_phase = angle(c);
subplot(222), stem(k, c_mag)
function y=x_a(t)
P=6; t= mod(t,P);
y= 3*(0<=t&t<1) +6*(1<=t&t<2) +3*(2<=t&t<3) ...
    -3*(3<=t&t<4) -6*(4<=t&t<5) -3*(5<=t&t<6);
```

(b) Use Eqs. (E.2.1.6) together with (2.1.9) to find the complex exponential Fourier series coefficients for the triangular wave in Fig. P2.1(b). Then compose a MATLAB program, say, "cir02p01b.m" to plot the Fourier series representation (2.1.7) and run it to see if the Fourier series representation becomes close to $x_b(t)$ as the number of terms in the summation increases, say, from 3 to 6. Also, compare the Fourier coefficients with those obtained by using "CtFS_exponential()".

(c) Consider a half-wave rectified cosine wave $x_h(t) = \max(\cos\omega_0 t, 0)$, which is depicted in Fig. P2.1(c). We can use Eq. (2.1.5b) to obtain the Fourier coefficients as

$$
\begin{aligned}
c_k &= \int_{-P/4}^{P/4} \cos\omega_0 t e^{-jk\omega_0 t} dt \\
&= \frac{1}{2}\int_{-P/4}^{P/4} (e^{j\omega_0 t} + e^{-j\omega_0 t})e^{-jk\omega_0 t}\, dt \ \text{ with } \omega_0 = \frac{2\pi}{P} = \pi \\
&\overset{k\neq 1\text{or}-1}{=} \frac{1}{2}\left(\frac{1}{-j(k-1)\omega_0}e^{-j(k-1)\omega_0 t} + \frac{1}{-j(k+1)\omega_0}e^{-j(k+1)\omega_0 t}\right)\Bigg|_{-1/2}^{1/2} \\
&= \frac{1}{(k-1)\pi}\sin(k-1)\frac{\pi}{2} + \frac{1}{(k+1)\pi}\sin(k+1)\frac{\pi}{2} \\
&\overset{k=2m(\text{even})}{=} (-1)^m\frac{1}{\pi}\left(\frac{1}{k+1} - \frac{1}{k-1}\right) = \frac{(-1)^{m+1}2}{(k^2-1)\pi}
\end{aligned} \tag{P2.1.2}
$$

$$
\begin{aligned}
c_k &\overset{k=1\text{or}-1}{=} \int_{-P/4}^{P/4} (e^{j\omega_0 t} + e^{-j\omega_0 t})e^{\pm j\omega_0 t} dt \\
&= \frac{1}{2}\int_{-P/4}^{P/4} (1 + e^{\pm j2\omega_0 t})dt = \frac{P}{4} = \frac{1}{2}
\end{aligned} \tag{P2.1.3}
$$

Thus compose a MATLAB program, say, "cir02p01c.m" to plot the Fourier series representation (2.1.7) and run it to see if the Fourier series representation becomes close to $x_h(t)$ as the number of terms in the summation increases, say, from 3 to 6. Also, compare the Fourier coefficients with those obtained by using "CtFS_exponential()".

(d) Consider a full-wave rectified cosine wave $x_f(t) = |\cos\omega_0 t|$, which is depicted in Fig. P2.1(d). Noting that it can be regarded as the sum of a half-wave rectified cosine wave $x_h(t)$ and its $P/2$-shifted version $x_h(t - P/2)$, compose a MATLAB program, say, "cir02p01d.m" to plot the Fourier series representation (2.1.7) and run it to see if the Fourier series representation becomes close to $x_f(t)$ as the number of terms in the summation increases, say, from 3 to 6. Also, compare the Fourier coefficients with those obtained by using "CtFS_exponential()".

(cf.) In fact, the fundamental frequency of a full-wave rectified cosine wave is $2\omega_0 = 4\pi/P$, which is two times that of the original cosine wave.

2.2 Fourier Analysis of *RC* Circuit Excited by a Square Wave Voltage Source
Figure P2.2(a) shows the PSpice schematic of an *RC* circuit excited by a rectangular (square) wave voltage source of height $\pm V_m = \pm\pi$, period $P = 2$[s], and duration (pulsewidth) $D = 1$[s], where the voltage across the capacitor is taken as the output. Figure P2.2(b) shows the input and output waveforms obtained from the PSpice simulation. Figure P2.2(c) shows the Fourier spectra of the input and output obtained by clicking the FFT button on the toolbar in the PSpice A/D (Probe) window. Figure P2.2(d) shows how to fill in the Simulation Settings dialog box to get the Fourier analysis results (for chosen variables) printed in the output file. Figure P2.2(e) shows the output file that can be viewed by clicking View/Output_File on the top menu bar and pulldown menu in the Probe window.

(a) Let us find the three leading frequency components of the input $v_i(t)$ and output $v_o(t)$. To this end, we first write the Fourier series representation of the rectangular wave input $v_i(t)$ by using Eq. (E2.1.3) as follows:

$$v_i(t) = 2\pi\,\tilde{r}_{1/2}(t-0.5) - \pi \overset{\text{(E2.1.3)}}{=} \sum_{k=1}^{\infty} 2\pi\frac{\sin(k\pi/2)}{k\pi/2}\cos k\pi(t-0.5)$$
$$= \sum_{k=\text{odd}}^{\infty}\frac{4}{k}\sin k\pi t = 4\sin\pi t + \frac{4}{3}\sin 3\pi t + \frac{4}{5}\sin 5\pi t + \cdots \qquad \text{(P2.2.1)}$$

Since the *RC* circuit has the system (transfer) function and frequency response

$$G(s) = \frac{1/sC}{R+1/sC} = \frac{1}{1+s\,RC};$$
$$G(j\omega) = \frac{1}{1+j\omega RC} = \frac{1}{\sqrt{1+(\omega RC)^2}}\angle -\tan^{-1}\omega RC, \qquad \text{(P2.2.2)}$$

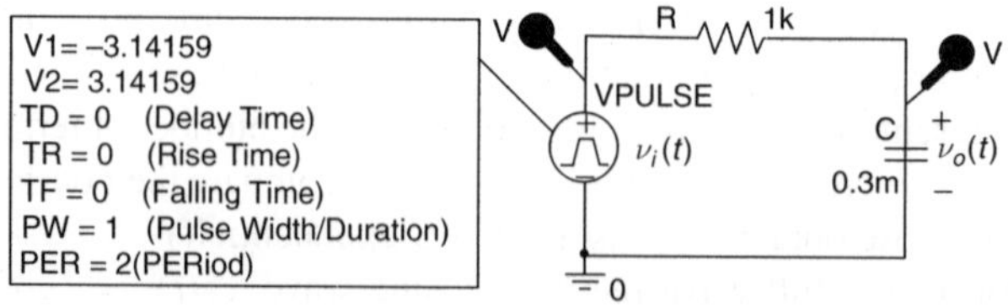

(a) PSpice schematic for the *RC* circuit excited by a square-wave source

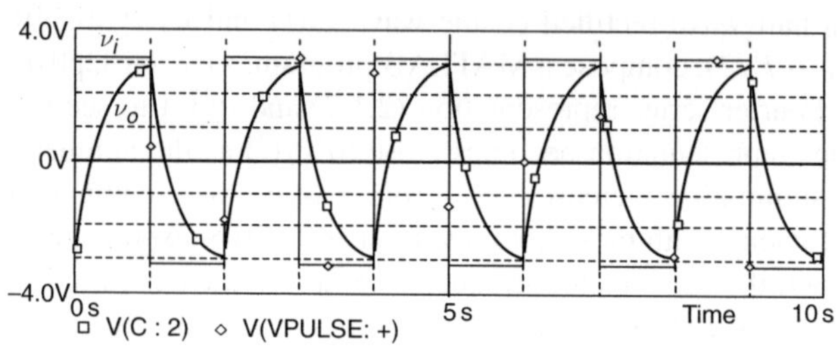

(b) The input and output voltage waveforms

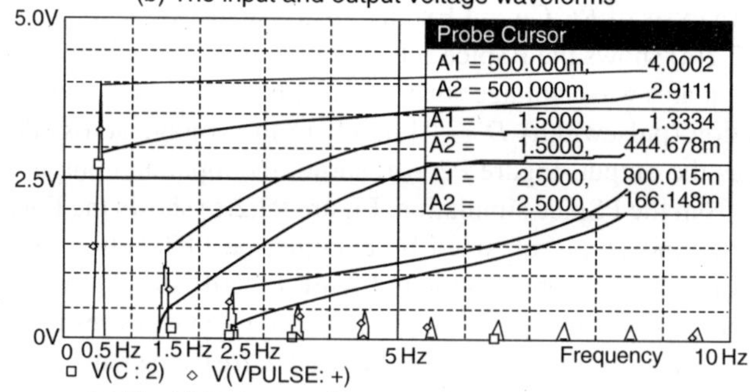

(c) The FFT spectra of the input and output voltages

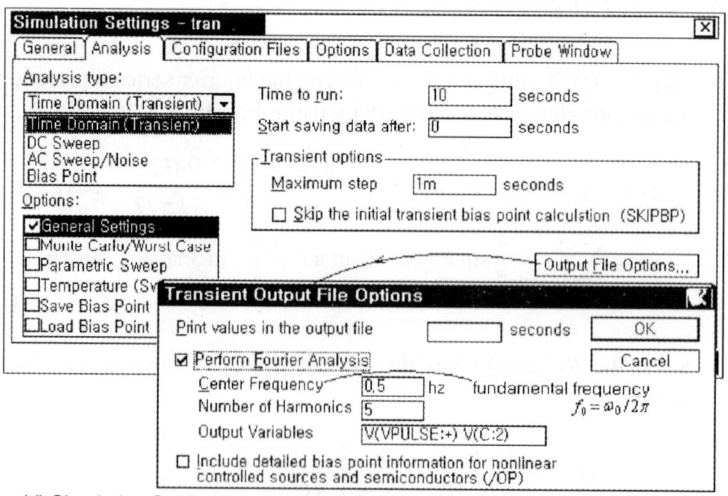

(d) Simulation Settings window and Transient Output File Options dialog box

Fig. P2.2 (continued)

```
** Profile: "SCHEMATIC1-tran" [C:\ORCAD\sig02p02-pspicefiles\tran.sim]
*Analysis directives:
.TRAN  0 10 0 1m
.FOUR 0.5 5 V([N00183]) V([N00070])

****     FOURIER ANALYSIS                    TEMPERATURE =   27.000 DEG C
FOURIER COMPONENTS OF TRANSIENT RESPONSE V(N00183)
 DC COMPONENT =    3.140322E-03(d'0)
 HARMONIC  FREQUENCY      FOURIER      NORMALIZED     PHASE          NORMALIZED
   NO(k)     (HZ)        COMPONENT     COMPONENT      (DEG)          PHASE (DEG)

     1    5.000E-01(f0)  4.000E+00(d'1) 1.000E+00  -1.800E-01(φ'1)  0.000E+00
     2    1.000E+00(2f0) 6.281E-03(d'2) 1.570E-03   8.964E+01(φ'2)  9.000E+01
     3    1.500E+00(3f0) 1.333E+00(d'3) 3.333E-01  -5.401E-01(φ'3) -7.176E-10
     4    2.000E+00(4f0) 6.281E-03(d'4) 1.570E-03   8.928E+01(φ'4)  9.000E+01
     5    2.500E+00(5f0) 8.000E-01(d'5) 2.000E-01  -9.002E-01(φ'5) -3.588E-09

     TOTAL HARMONIC DISTORTION =    3.887326E+01 PERCENT

FOURIER COMPONENTS OF TRANSIENT RESPONSE V(N00070)
 DC COMPONENT =    3.141583E-03(d'0)
 HARMONIC  FREQUENCY      FOURIER      NORMALIZED     PHASE          NORMALIZED
   NO(k)     (HZ)        COMPONENT     COMPONENT      (DEG)          PHASE (DEG)

     1    5.000E-01(f0)  2.911E+00(d'1) 1.000E+00  -4.348E+01(φ'1)  0.000E+00
     2    1.000E+00(2f0) 2.945E-03(d'2) 1.012E-03   2.759E+01(φ'2)  1.146E+02
     3    1.500E+00(3f0) 4.446E-01(d'3) 1.527E-01  -7.106E+01(φ'3)  5.939E+01
     4    2.000E+00(4f0) 1.611E-03(d'4) 5.534E-04   1.414E+01(φ'4)  1.881E+02
     5    2.500E+00(5f0) 1.661E-01(d'5) 5.705E-02  -7.892E+01(φ'5)  1.385E+02

     TOTAL HARMONIC DISTORTION =    1.630388E+01 PERCENT

         JOB CONCLUDED
```

(e) The PSpice output file with FFT analysis

Fig. P2.2 PSpice simulation and analysis for Problem 2.2

its output to the input $v_i(t)$ (described by Eq. (P2.2.1)) can be written as

$$v_o(t) = \sum_{k=2m+1}^{\infty} \frac{4}{k\sqrt{1+(k\omega_0 RC)^2}} \sin\left(k\,\omega_0 t - \tan^{-1} k\omega_0 RC\right)$$

$$\text{with } \omega_0 = \frac{2\pi}{P} = \pi \tag{P2.2.3}$$

Show that the three leading frequency components of the output phasor are

$$\mathbf{V}_o^{(1)} \overset{k=1}{=} G(j\omega_0)\mathbf{V}_i^{(1)} = \frac{4}{\sqrt{1+(\omega_0 RC)^2}} \angle -\tan^{-1}\omega_0 RC$$

$$\overset{RC=0.3}{=} \frac{4}{\sqrt{1+(0.3\pi)^2}} \angle -\tan^{-1} 0.3\pi = 2.91\angle -43.3^\circ \tag{P2.2.4}$$

$$\mathbf{V}_o^{(3)} \overset{k=3}{=} G(j3\omega_0)\mathbf{V}_i^{(3)} = \frac{4}{3\sqrt{1+(3\omega_0 RC)^2}} \angle -\tan^{-1} 3\omega_0 RC$$

$$\overset{RC=0.3}{=} \frac{4}{3\sqrt{1+(0.9\pi)^2}} \angle -\tan^{-1} 0.9\pi = 0.4446\angle -70.5^\circ \tag{P2.2.5}$$

$$\mathbf{V}_o^{(5)} \overset{k=5}{=} G(j5\omega_0)\mathbf{V}_i^{(5)} = \frac{4}{5\sqrt{1+(5\omega_0 RC)^2}}\angle - \tan^{-1} 5\omega_0 RC$$
$$\overset{RC=0.3}{=} \frac{4}{5\sqrt{1+(1.5\pi)^2}}\angle - \tan^{-1} 1.5\pi = 0.166\angle - 78^{\circ} \qquad \text{(P2.2.6)}$$

where the phases are taken with that of $\sin \omega t$ as a reference in the sine-and-phase form of Fourier series representation in order to match this Fourier analysis result with that seen in the PSpice simulation output file (Fig. P2.2(e)).

(cf.) Note that PSpice uses the sine-and-phase form of Fourier series representation for spectrum analysis as below:

$$x(t) = d'_0 + \sum_{k=1}^{\infty} d'_k \sin(k\omega_0 t + \phi'_k) \qquad \text{(P2.2.7a)}$$

where the Fourier coefficients are

$$d'_0 = a_0 = d_0, \quad d'_k = \sqrt{a_k^2 + b_k^2} = d_k, \quad \phi'_k = \tan^{-1}(a_k/b_k) = \phi_k + \pi/2 \qquad \text{(P2.2.7b)}$$

The magnitude ratios among the leading three frequency components of the input and output are

$$\text{input}: \quad |\mathbf{V}_i^{(1)}| : |\mathbf{V}_i^{(3)}| : |\mathbf{V}_i^{(5)}| = 4 : \frac{4}{3} : \frac{5}{3} = 15 : 5 : 3$$
$$\text{output}: |\mathbf{V}_o^{(1)}| : |\mathbf{V}_o^{(3)}| : |\mathbf{V}_o^{(5)}| = 2.91 : 0.4446 : 0.166 = 15 : 2.3 : 0.86$$

This implies that the relative magnitudes of high frequency components to low ones become smaller after the input signal passes through the filter. This is a piece of evidence that the *RC* circuit with the capacitor voltage taken as the output functions as a lowpass filter.

(b) Find the Fourier analysis results on $\mathbf{V}_i^{(k)}$'s and $\mathbf{V}_o^{(k)}$'s for $k = 0 : 5$ (and compare them with those obtained from PSpice simulation in (a)) by completing and running the following MATLAB program "sig02p_02.m", which uses the MATLAB function "Fourier_analysis()" declared as

```
function [yt,Y,X]= Fourier_analysis(n,d,x,P,N)
```

This function takes the numerator (n) and denominator (d) of the transfer function $G(s)$, the (periodic) input function (x) defined for at least one period $[-P/2, +P/2]$, the period (P), and the order (N) of Fourier analysis as the input arguments and produces the output (yt) for one period and the sine-and-phase form of Fourier coefficients Y and X of the output and input (for $k = 0, \ldots, N$).

```
%sig02p_02.m : to solve Problem 2.2
% Perform Fourier analysis to solve the RC circuit excited by a square wave
clear, clf
global P D Vm
P=2; w0=2*pi/P; D=1; Vm=pi; % Period, Frequency, Duration, Amplitude
N=5; kk=0:N; % Frequency indices to analyze using Fourier analysis
tt=[-300:300]*P/200; % Time interval of 3 periods
vi= 'f_sig02p02'; % Bipolar square wave input function defined in an M-file
RC=0.3; % The parameter values of the RC circuit
n=1; d=[?? 1]; % Numerator/Denominator of transfer function (P2.2.2)
% Fourier analysis of the input & output spectra of a system [n/d]
[Vo,Vi]= ???????_analysis(n,d,vi,P,N);
% Magnitude_and_phase_of_input_output_spectrum
disp(' frequency X_magnitude X_phase Y_magnitude Y_phase')
[kk; abs(Vi); angle(Vi)*180/pi; abs(Vo); angle(Vo)*180/pi].'
vit = feval(vi,tt); % the input signal for tt
vot= Vo(1); % DC component of the output signal
for k=1:N % Fourier series representation of the output signal
   % PSpice dialect of Eq.(2.1.4a)
   vot = vot + abs(Vo(k+1))*sin(k*w0*tt + angle(Vo(k+1))); % Eq.(P2.2.7a)
end
subplot(221), plot(tt,vit, tt,vot,'r') % plot input/output signal waveform
subplot(222), stem(kk,abs(Vi)) % input spectrum
hold on, stem(kk,abs(Vo),'r') % output spectrum
```

```
function y=f_sig02p02(t)
% defines a bipolar square wave with period P, duration D, and amplitude Vm
global P D Vm
t= mod(t,P); y= ((t<=D) - (t>D))*Vm;
```

```
function [Y,X]= Fourier_analysis(n,d,x,P,N)
%Input:  n= Numerator polynomial of system function G(s)
%        d= Denominator polynomial of system function G(s)
%        x= Input periodic function
%        P= Period of the input function
%        N= highest frequency index of Fourier analysis
%Output: Y= Fourier coefficients [Y0,Y1,Y2,...] of the output
%        X= Fourier coefficients [X0,X1,X2,...] of the input
% Copyleft: Won Y. Yang, wyyang53@hanmail.net, CAU for academic use only
if nargin<5, N=5; end
w0=2*pi/P; % Fundamental frequency
kk=0:N; % Frequency index vector
c= CtFS_???????????(x,P,N); % complex exponential Fourier coefficients
Xmag=[c(N+1) 2*abs(c(N+2:end))]/P; % d(k) for k=0:N by Eq.(P2.2.7b)
Xph=[0 angle(c(N+2:end))+pi/2]; % phi'(k) for k=0:N by Eq.(P2.2.7b)
X= Xmag.*exp(j*Xph); % Input spectrum
Gw= freqs(n,d,kk*w0); % Frequency response
Y= X.*Gw; % Output spectrum
```

2.3 CTFT (Continuous-Time Fourier Transform) of a Periodic Function

In Eq. (2.3.1), we gave up computing the Fourier transform of a periodic function $e^{jk\omega_0 t}$:

$$\mathcal{F}\{x_k(t)\} \overset{(2.2.1a)}{=} \int_{-\infty}^{\infty} e^{jk\omega_0 t} e^{-j\omega t} dt = \int_{-\infty}^{\infty} e^{-j(\omega-k\omega_0)t} dt$$

$$= \frac{1}{-j(\omega-k\omega_0)} e^{-j(\omega-k\omega_0)t}\Big|_{-T}^{T} = \frac{1}{-j(\omega-k\omega_0)} \left(e^{j(\omega-k\omega_0)T} - e^{-j(\omega-k\omega_0)T} \right)\Big|_{T=\infty}$$

$$= \left(\frac{2\sin(\omega-k\omega_0)T}{-(\omega-k\omega_0)} \right)\Big|_{T=\infty} = ? \qquad \text{(P2.3.1)}$$

Instead, we took a roundabout way of finding the inverse Fourier transform of $2\pi\delta(\omega - k\omega_0)$ in expectation of that it will be the solution and showed the validity of the inverse relationship. Now, referring to Eq. (E2.6.2) or using Eq. (1.1.33a), retry to finish the derivation of Eq. (2.3.2) from Eq. (P2.3.1).

2.4 Two-Dimensional Fourier Transform of a Plane Impulse Train
The two-dimensional (2-D) Fourier transform of a 2-D signal $f(x, y)$ on the x-y plane such as an image frame is defined as

$$F(u, v) = \mathcal{F}_2\{f(x, y)\} = \int_{-\infty}^{\infty}\int_{-\infty}^{\infty} f(x, y)e^{-j(ux+vy)}dx\,dy$$

$$= \int_{-\infty}^{\infty}\left(\int_{-\infty}^{\infty} f(x, y)e^{-jux}dx\right)e^{-jvy}\,dy \tag{P2.4.1}$$

where x and y are the spatial cordinates and u and v are the spatial angular frequencies [rad/m] representing how abruptly or smoothly $f(x, y)$ changes w.r.t. the spatial shift along the x and y -axes, respectively.

(a) Show that the 2-D Fourier transform of $f(x, y) = \delta(y - \alpha x)$ is

$$f(x, y) = \delta(y - \alpha x) \overset{\mathcal{F}_2}{\leftrightarrow} F(u, v) = 2\pi\,\delta(u + \alpha v) \tag{P2.4.2}$$

(b) Show that the 2-D Fourier transform of $f(x, y) = \sum_{n=-\infty}^{\infty}\delta(y - \alpha x - nd\sqrt{1+\alpha^2})$ (Fig. P2.4(a)) is

$$F(u, v) = \int_{-\infty}^{\infty}\int_{-\infty}^{\infty}\sum_{n=-\infty}^{\infty}\delta(y - \alpha x - nd\sqrt{1+\alpha^2})e^{-j(ux+vy)}dx\,dy$$

$$\overset{(D.33)}{=} \sum_{n=-\infty}^{\infty}\int_{-\infty}^{\infty} e^{-j(ux+v(\alpha x+nd\sqrt{1+\alpha^2}))}\,dx$$

$$= \sum_{n=-\infty}^{\infty}\int_{-\infty}^{\infty} e^{-j(u+\alpha v)x}\,dx e^{-jvnd\sqrt{1+\alpha^2}}$$

$$\underset{\text{with } \omega=u+\alpha v,\ t=x,\ k=0}{\overset{(2.3.2)}{=}} \sum_{n=-\infty}^{\infty} 2\pi\delta(u + \alpha v)e^{-jvnd\sqrt{1+\alpha^2}}$$

$$\underset{\text{with } \Omega=vd\sqrt{1+\alpha^2},\Omega_0=0}{\overset{(3.1.5)}{=}} 2\pi\delta(u + \alpha v)\sum_{i=-\infty}^{\infty} 2\pi\delta(vd\sqrt{1+\alpha^2} - 2\pi\,i)$$

$$\overset{(1.1.34)}{=} \frac{(2\pi)^2}{d\sqrt{1+\alpha^2}}\sum_{i=-\infty}^{\infty}\delta(u + \frac{2\pi i\alpha}{d\sqrt{1+\alpha^2}},\ v - \frac{2\pi i}{d\sqrt{1+\alpha^2}})$$

$$= \frac{(2\pi)^2}{d\sqrt{1+\alpha^2}}\sum_{i=-\infty}^{\infty}\delta(u + \frac{2\pi i}{d}\sin\theta,\ v - \frac{2\pi i}{d}\cos\theta) \tag{P2.4.3}$$

as depicted in Fig. P2.4(b).

2.5 ICTFT (Inverse Continuous-Time Fourier Transform) of $U_o(j\omega) = 1/j\omega$
Using Eq. (E2.3.3), show that the ICTFT of $U_o(j\omega) = 1/j\omega$ is the odd component of the unit step function:

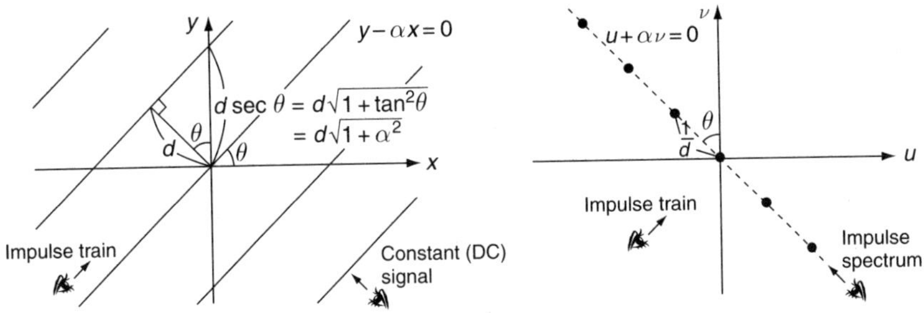

(a) A 2–D (plane) impulse wave in the spatial domain (b) The 2–D Fourier transform depicted in the spatial frequency domain

Fig. P2.4 An example of 2-D Fourier transform

$$\mathcal{F}^{-1}\{U_o(j\omega)\} \overset{(2.2.1\text{b})}{=} \frac{1}{2\pi}\int_{-\infty}^{\infty} \frac{1}{j\omega} e^{j\omega t}\, d\omega = \frac{1}{2}\text{sign}(t) = u_o(t) \qquad \text{(P2.5.1)}$$

which agrees with Eq. (E2.8.5).

2.6 Applications and Derivations of Several Fourier Transform Properties

(a) Applying the frequency shifting property (2.5.7) of Fourier transform to Eq. (E2.7.1), derive Eq. (2.3.2).
(b) Applying the duality relation (2.5.9) to the time shifting property (2.5.6) of Fourier transform, derive the frequency shifting property (2.5.7).
(c) From the time-differentiation property (2.5.7) of Fourier tansform that differentiation in the time domain corresponds to multiplication by $j\omega$ in the frequency domain, one might conjecture that integration in the time domain corresponds to division by $j\omega$ in the frequency domain. But, Eq. (2.5.18) shows that the Fourier transform of $\int_{-\infty}^{t} x(\tau)d\tau$ has an additional impulse term reflecting the DC or average value of $x(t)$. Can you apply the time-differentiation property to Eq. (2.5.18) to derive the original Fourier relation $\mathcal{F}\{x(t)\} = X(\omega)$?

$$\int_{-\infty}^{t} x(\tau)d\tau \overset{\mathcal{F}}{\leftrightarrow} \pi X(0)\delta(\omega) + \frac{1}{j\omega}X(\omega) \qquad (2.5.18)$$

(d) Applying the duality relation (2.5.9) to the convolution property (2.5.11) of Fourier transform, derive the modulation property (2.5.14).
(e) Apply the time-differentiation property to the time-integration property (2.5.18) to derive the original Fourier relation $\mathcal{F}\{x(\tau)\} = X(\omega)$.
(f) Applying the time-differentiation property (2.5.7) to the time-domain input-output relationship $y(t) = dx(t)/dt$ of a differentiator, find the frequency response of the differentiator.
(g) Applying the time-integration property (2.5.18) to the time-domain input-output relationship $y(t) = \int_{-\infty}^{t} x(\tau)d\tau$ of an integrator, find the frequency

response of the integrator on the assumption that the input has zero DC value.

(h) Find the time-domain and frequency-domain input-output relationships of a system whose impulse response is $g(t) = \delta(t - t_0)$.

2.7 Fourier Transform of a Triangular Pulse
Applying the convolution property (2.5.11), find the Fourier transform of a triangular pulse $D\lambda_D(t)$ (Eq. (1.2.25a)) obtained from the convolution of two rectangular pulses of height 1 and duration D.

2.8 Convolution/Modulation Properties of Fourier Series
Let $\tilde{x}_P(t)$ and $\tilde{y}_P(t)$ be periodic functions with common period P that have the following Fourier series representations, respectively:

$$\tilde{x}_P(t) = \frac{1}{P}\sum_{k=-\infty}^{\infty} X_k e^{jk\omega_0 t} \quad \text{and} \quad \tilde{y}_P(t) = \frac{1}{P}\sum_{k=-\infty}^{\infty} Y_k e^{jk\omega_0 t} \tag{P2.8.1}$$

(a) The periodic or circular convolution of two periodic functions is defined as

$$\begin{aligned}\tilde{z}_P(t) = \tilde{x}_P(t) * \tilde{y}_P(t) &\overset{(2.5.12)}{=} \int_P \tilde{x}_P(\tau)\tilde{y}_P(t-\tau)d\tau \\ &= \int_P \tilde{y}_P(\tau)\tilde{x}_P(t-\tau)d\tau = \tilde{y}_P(t) * \tilde{x}_P(t)\end{aligned} \tag{P2.8.2}$$

The product of the two Fourier series coefficients X_k and Y_k is the Fourier series coefficients of this convolution so that

$$\tilde{z}_P(t) = \frac{1}{P}\sum_{k=-\infty}^{\infty} Z_k e^{jk\omega_0 t} = \frac{1}{P}\sum_{k=-\infty}^{\infty} X_k Y_k e^{jk\omega_0 t} \tag{P2.8.3}$$

Applying the frequency shifting property (2.5.7) of Fourier transform to Eq. (E2.7.1), derive Eq. (2.3.2).

(b) Show that the Fourier series coefficients of $\tilde{w}_P(t) = \tilde{x}_P(t)\tilde{y}_P(t)$ are

$$\tilde{w}_P(t) = \tilde{x}_P(t)\tilde{y}_P(t) \underset{(2.5.15)}{\overset{\text{Fourier series}}{\leftrightarrow}} W_k = \frac{1}{P}\sum_{n=-\infty}^{\infty} X_n Y_{k-n} \tag{P2.8.4}$$

so that we have

$$\tilde{w}_P(t) = \frac{1}{P}\sum_{k=-\infty}^{\infty} W_k e^{jk\omega_0 t} \tag{P2.8.5}$$

2.9 Overall Input-Output Relationship of Cascaded System
Consider the systems of Fig. P2.9 consisting of two subsystems connected in cascade, i.e., in such a way that the output of the previous stage is applied to the input of the next stage. Note that the overall output $y(t)$ can be expressed as the convolution of the input $x(t)$ and the two subsystem's impulse responses

$g_1(t)$ and $g_2(t)$ as

$$y(t) = g_1(t) * g_2(t) * x(t) = g_2(t) * g_1(t) * x(t). \tag{P2.9.1}$$

Find the overall input-output relationship in the frequency domain.

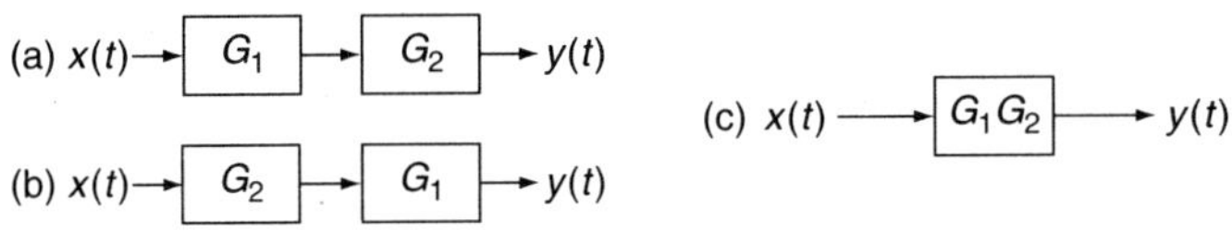

Fig. P2.9

2.10 Hilbert Transform and Analytic Signal

(a) Fourier Transform of Hilbert Transformed Signal
The *Hilbert transform* of a real-valued signal $x(t)$ is defined as the convolution of $x(t)$ and $1/\pi t$ or equivalently, the output of a *Hilbert transformer* (with the impulse response of $h(t) = 1/\pi t$) to the input $x(t)$:

$$\hat{x}(t) = h(t) * x(t) = \frac{1}{\pi t} * x(t) = \frac{1}{\pi} \int_{-\infty}^{\infty} x(\tau) \frac{1}{t - \tau} \, d\tau \tag{P2.10.1}$$

- First, let us find the Fourier transform of $h(t) = 1/\pi t$ by applying the duality (2.5.9) to Eq. (E2.8.5), which states that

$$\frac{1}{2}\mathrm{sign}(t) \overset{\mathcal{F}}{\leftrightarrow} \frac{1}{j\omega} \overset{\text{duality}}{\Leftrightarrow} \frac{1}{jt} \overset{\mathcal{F}}{\leftrightarrow} (\qquad\qquad\qquad\qquad) \tag{P2.10.2}$$

- We can multiply both sides by j/π to get the Fourier transform of $h(t) = 1/\pi t$ as

$$\frac{1}{\pi t} \overset{\mathcal{F}}{\leftrightarrow} (\qquad\qquad\qquad\qquad) \tag{P2.10.3}$$

- Now, we apply the convolution property to get the Fourier transform of $\hat{x}(t)$ as

$$\hat{x}(t) = \frac{1}{\pi t} * x(t) \underset{\text{convolution property}}{\overset{\mathcal{F}}{\leftrightarrow}}$$

$$\mathcal{F}\left(\frac{1}{\pi t}\right) \mathcal{F}(x(t)) = (\qquad\qquad\qquad\qquad) X(\omega) \tag{P2.10.4}$$

This implies that the Hilbert transform has the effect of shifting the phase of positive/negative frequency components of $x(t)$ by $-90^\circ/+90^\circ$, allowing the Hilbert transform a -90° *phase shifter*.

(b) Inverse Hilbert Transform
Note that multiplying $-H(\omega)$ by the Fourier transform of the Hilbert transformed signal $\hat{x}(t)$ brings the spectrum back into that of the original signal $x(t)$:

$$-H(\omega)H(\omega)\, X(\omega) = -(-j\ \text{sign}(\omega))(-j\ \text{sign}(\omega))\, X(\omega) = X(\omega) \tag{P2.10.5}$$

This implies the *inverse Hilbert transform* as

$$(-h(t)) * \hat{x}(t) = \left(-\frac{1}{\pi t}\right) * \frac{1}{\pi t} * x(t) = x(t) \tag{P2.10.6}$$

(c) Analytic Signal
The *analytic signal*, or analytic representation, of a real-valued signal $x(t)$ is defined as

$$x_a(t) = x(t) + j\hat{x}(t) \tag{P2.10.7}$$

where $\hat{x}(t)$ is the Hilbert transform of $x(t)$ and j is the imaginary unit. Show that the Fourier transform of the analytic signal is

$$X_a(\omega) = \mathcal{F}\{x(t)\} + j\mathcal{F}\{\hat{x}(t)\} \cong 2u_s(\omega)X(\omega) \tag{P2.10.8}$$

where $u_s(\omega)$ is the unit step function in frequency ω. This implies that $x_a(t)$ has only nonnegative frequency component of $x(t)$.

(d) Examples of Hilbert Transform and Analytic Signal
Show that the following Hilbert transform relations hold:

$$\cos(\omega_c\, t) \overset{\mathcal{H}}{\leftrightarrow} \sin(\omega_c\, t) \tag{P2.10.9}$$

$$\sin(\omega_c t) \overset{\mathcal{H}}{\leftrightarrow} -\cos(\omega_c t) \tag{P2.10.10}$$

In general, it would be difficult to get the Hilbert transform of a signal $x(t)$ directly from its definition (P2.10.1) and therefore, you had better take the inverse Fourier transform, $X(\omega)$, of $x(t)$, multiply $-j\ \text{sign}(\omega)$, and then take the inverse Fourier transform as

$$\hat{x}(t) = \mathcal{F}^{-1}\{-j\ \text{sign}(\omega)X(\omega)\} \tag{P2.10.11}$$

Now, let us consider a narrowband signal

$$s(t) = m(t)\cos(\omega_c t + \phi) \text{ with } m(t) : \text{a baseband message signal} \tag{P2.10.12}$$

whose spectrum is (narrowly) bandlimited around the carrier frequency ω_c. Show that the Hilbert transform of $s(t)$ is

$$s(t) = m(t)\cos(\omega_c t + \phi) \overset{\mathcal{H}}{\leftrightarrow} \hat{s}(t) = m(t)\sin(\omega_c t + \phi) \qquad \text{(P2.10.13)}$$

Also, show that the analytic signal for $s(t)$ can be expressed as

$$s_a(t) = s(t) + j\hat{s}(t) = m(t)e^{j(\omega_c t + \phi)} \qquad \text{(P2.10.14)}$$

This implies that we can multiply the analytic signal by $e^{-j\omega_c t}$ to obtain the envelope of the baseband message signal $m(t)$.

(e) Programs for Hilbert Transform
One may wonder how the convolution in the time domain results in just the phase shift. To satisfy your curiosity, try to understand the following program "sig02p_10e.m", which computes the output of a Hilbert transformer to a cosine wave $\cos\omega_c t$ with $\omega_c = \pi/16$ and sampling interval $T_s = 1/16$ [s]. For comparison, it uses the MATLAB built-in function "`hilbert( )`" to generate the analytic signal and take its imaginary part to get $\hat{x}(t)$.

(cf.) "`hilbert( )`" makes the Hilbert transform by using Eq. (P2.10.11).

```
%sig02p_10e.m
% To try the Hilbert transform in the time/frequency domain
clear, clf
Ts=1/16; t=[-149.5:149.5]*Ts; % Sample interval and Duration of h(t)
h= 1/pi./t; % h(t): Impulse response of a Hilbert transformer
Nh=length(t); Nh2=floor(Nh/2); % Sample duration of noncausal part
h_= fliplr(h); % h(-t): Time-reversed version of h(t)
wc= pi/16; % the frequency of an input signal
Nfft=512; Nfft2= Nfft/2; Nbuf=Nfft*2; % buffer size
tt= zeros(1,Nbuf); x_buf= zeros(1,Nbuf); xh_buf= zeros(1,Nbuf);
for n=-Nh:Nbuf-1 % To simulate the Hilbert transformer
  tn=n*Ts; tt=[tt(2:end) tn];
  x_buf = [x_buf(2:end) cos(wc*tn)];
  xh_buf = [xh_buf(2:end) h_*x_buf(end-Nh+1:end).'*Ts];
end
axis_limits= [tt([1 end]) -1.2 1.2];
subplot(321), plot(tt,x_buf), title('x(t)'), axis(axis_limits)
subplot(323), plot(t(1:Nh2),h(1:Nh2), t(Nh2+1:end),h(Nh2+1:end))
title('h(t)'), axis([t([1 end]) -11 11])
% To advance the delayed response of the causal Hilbert transformer
xh = xh_buf(Nh2+1:end); xh_1 = imag(hilbert(x_buf));
subplot(325), plot(tt(1:end-Nh2),xh,'k', tt,xh_1), axis(axis_limits)
subplot(326), plot(tt,xh_buf), axis(axis_limits)
ww= [-Nfft2:Nfft2]*(2*pi/Nfft);
Xw= fftshift(fft(x_buf,Nfft)); Xw= [Xw Xw(1)]; % X(w): spectrum of x(t)
Xhw_1= fftshift(fft(xh_1,Nfft)); Xhw_1= [Xhw_1 Xhw_1(1)]; % Xh(w)
norm(Xhw_1+j*sign(ww).*Xw)
```

In the program "sig02p_10e.m", identify the statements performing the following operations:

- The impulse response of the Hilbert transformer
- Generating an input signal to the Hilbert transformer

- Computing the output expected to be the Hilbert transformed signal

  Note the following:
- The output sequence has been advanced by Nh/2 for comparison with $\hat{x}(t)$ obtained using "`hilbert ( )`" because the real output of the causal Hilbert transformer is delayed by Nh/2, which is the length of the noncausal part of the Hilbert transformer impulse response h(n).

(f) Application of Hilbert Transform and Analytic Signal for Envelope Detection

To see an application of Hilbert transform and analytic signal, consider the following narrowband signal

$$s(t) = m(t)\cos(\omega_c t + \phi)$$
$$= \text{sinc}\left(\frac{B}{\pi}t\right)\cos(\omega_c t + \frac{\pi}{6}) \text{ with } B = 100,\ \omega_c = 400\pi \quad \text{(P2.10.15)}$$

```
%sig02p_10f.m
% To obtain the lowpass equivalent of Bandpass signal
Ts =0.001; fs =1/Ts; % Sampling Period/Frequency
t=[-511: 512]*Ts; % Duration of signal
fc=200; wc=2*pi*fc; B=100; % Center frequency & bandwidth of signal s(t)
m= sinc(B/pi*t);
s= m.*cos(wc*t+pi/6); % a narrowband (bandpass) signal
sa= hilbert(x);  % Analytic signal sa(t) = s(t) + j s^(t)
sl= sa.*exp(-j*wc*t);  % Lowpass Equivalent (Complex envelope) sl(t)
```

Referring to the above program "sig02p_10f.m", use the MATLAB function "`hilbert ( )`" to make the *analytic* or *pre-envelope* signal $s_a(t)$, multiply it by $e^{-j\omega_c t}$ to get the *lowpass equivalent* or *complex envelope* $s_l(t)$, take its absolute values $|s_l(t)|$ to obtain the envelope of the baseband message signal $m(t)$, and plot it to compare with $m(t) = |\text{sinc}\,(2Bt)\,|$.

(g) Hilbert Transform with Real-Valued Signals for Signal Detection

Consider the system of Fig. P2.10 where a narrowband signal $x(t)$ having *in-phase* component $x_c(t)$ and *quadrature* component $x_s(t)$ is applied as an input to the system:

$$x(t) = x_c(t)\cos\omega_c t - x_s(t)\sin\omega_c t \quad \text{(P2.10.16)}$$

Verify that the two outputs of the system are the same as $x_c(t)$ and $x_s(t)$, respectively by showing that their spectra are

$$X_{c\_d}(\omega) = X_c(\omega) \text{ and } X_{s\_d}(\omega) = X_s(\omega) \quad \text{(P2.10.17)}$$

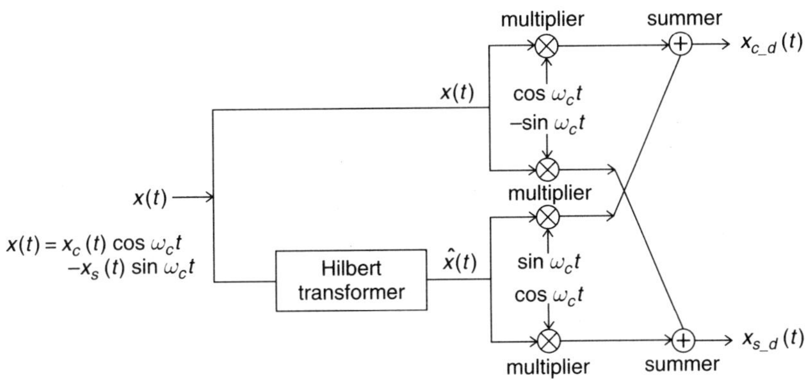

Fig. P2.10

Complete and run the following program "sig02p_10g.m" to check if these results are valid for

$$x_c(t) = \text{sinc}\,(100t)\,,\ x_s(t) = \text{sinc}\,(450t)\,,\ \text{and } \omega_c = 400\pi$$

Does the simulation result show that $x_{c\_d}(t) = x_c(t)$ and $x_{s\_d}(t) = x_s(t)$? If one of them does not hold, figure out the reason of the discrepancy.

```
%sig02p_10g.m
% To use the Hilbert transform for signal detection
clear, clf
Ts=0.0002; t=[-127:128]*Ts; % Sampling Period and Time interval
fc=200; wc=2*pi*fc; % Center frequency of signal x(t)
B1=100*pi; B2=450*pi; % Bandwidth of signal x(t)
xc= sinc(B1/pi*t); xs= sinc(B2/pi*t);
x= xc.*cos(wc*t) - xs.*sin(wc*t);
xh= imag(hilbert(x));  % imag(x(t)+j x^(t))=x^(t)
xc_d= x.*cos(wc*t) + xh.*?????????; % xc_detected
xs_d= -x.*sin(wc*t) + xh.*?????????; % xc_detected
subplot(221), plot(t,xc, t,xc_d,'r')
subplot(222), plot(t,xs, t,xs_d,'r')
norm(xc-xc_d)/norm(xc), norm(xs-xs_d)/norm(xs)
```

2.11 Spectrum Analysis of Amplitude Modulation (AM) Communication System
Consider the AM communication system of Fig. P2.11(a) where the message signal $x(t)$ is band-limited with maximum frequency ω_x and the spectrum $X(\omega)$ is depicted in Fig. P2.11(b). Assume that the frequency response of the (bandpass) channel is

$$H(\omega) = \begin{cases} 1 & \text{for } |\omega - \omega_c| < B_C/2 \\ 0 & \text{elsewhere} \end{cases} \text{ with } B_C \ge 2\omega_x \qquad \text{(P2.11.1)}$$

and that of the LPF (lowpass filter) at the receiver side is

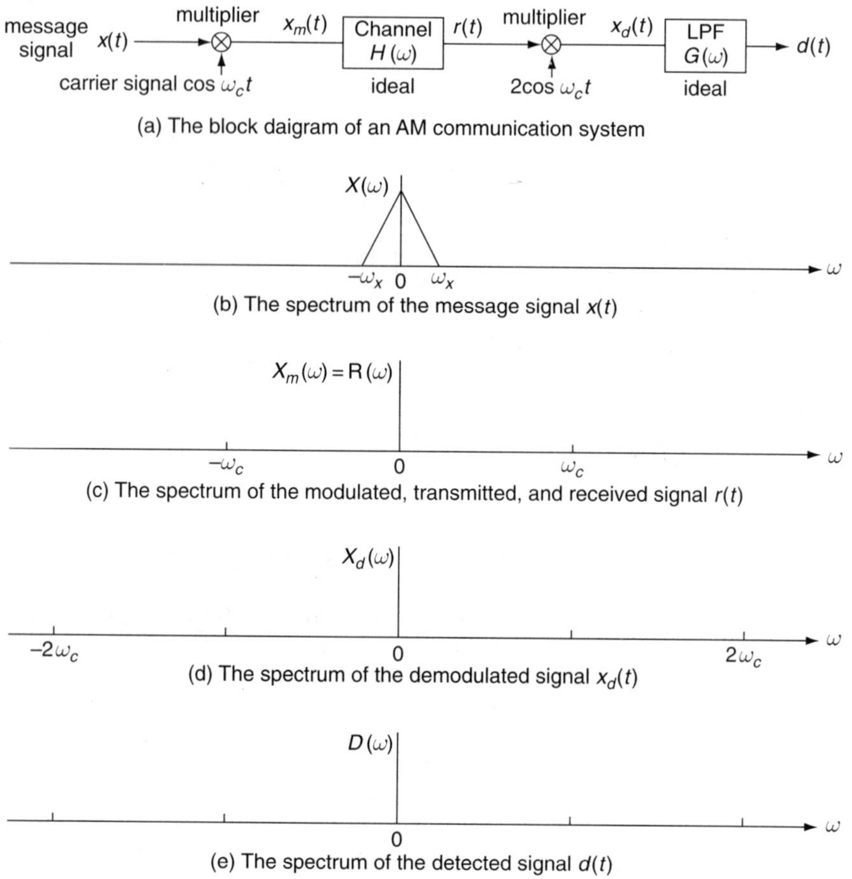

Fig. P2.11 An amplitude–modulation (AM) communication system

$$G(\omega) = \begin{cases} 1 & \text{for } |\omega| < B_R \\ 0 & \text{elsewhere} \end{cases} \text{ with } \omega_x < B_R < 2\omega_c - \omega_x \tag{P2.11.2}$$

(a) Express the spectrum, $X_d(\omega)$, of the demodulated signal $x_d(t)$ in terms of $X(\omega)$.

(b) Draw the spectra of the modulated and received signal $r(t)$, the demodulated signal $x_d(t)$, and the detected signal $d(t)$ on Fig. P2.11(c), (d), and (e).

(c) Complete and run the following program "sig02p_11.m" to simulate the AM communication system depicted in Fig. P2.11(a). Note that the spectrum of each signal is computed by using the MATLAB built-in function "`fft()`" instead of CTFT (Continuous-Time Fourier Transform) since DFT (Discrete Fourier Transform) is applied to any sequence (consisting of sampled values of an arbitrary continuous-time signal), while CTFT can be applied only to continuous-time signals expressed by a linear combination of basic functions. DFT and "`fft()`" will be introduced in Chap. 4.

```
%sig02p_11.m
% AM (amplitude-modulation) communication system
clear, clf
Ts=0.0001; t=[-127:128]*Ts; % Sampling Period and Time interval
fc=1000; wc=2*pi*fc; % Carrier frequency
B=1000; % (Signal) Bandwidth of x(t)
x= sinc(B/pi*t).^2; % a message signal x(t) having triangular spectrum
xm= x.*cos(wc*t);  % AM modulated signal
r= xm;  % received signal
xd= r.*(2*?????????); % demodulated signal
Br=wc/2; % (System) Bandwidth of an ideal LPF
g= Br/pi*sinc(Br/pi*t); % (truncated) Impulse Response of the ideal LPF
d= conv(xd,g)*Ts; % the output of the LPF to the demodulated signal
subplot(421), plot(t,x), title('A message signal x(t)')
Nfft=256; Nfft2=Nff/2; % FFT size and a half of Nfft
w= [-Nfft/2:Nfft/2]*(2*pi/Nfft/Ts); % Frequency vector
X= fftshift(fft(x)); X=[X X(1)]; % Spectrum of x(t)
subplot(422), plot(w,abs(X)), title('The spectrum X(w) of x(t)')
subplot(423), plot(t,xm), title('The modulated signal xm(t)')
Xm= fftshift(fft(xm)); Xm=[Xm Xm(1)]; % Spectrum of modulated signal
subplot(424), plot(w,abs(Xm)), title('The spectrum Xm(w) of xm(t)')
subplot(425), plot(t,xd), title('The demodulated signal xd(t)')
Xd= fftshift(fft(xd)); Xd=[Xd Xd(1)]; % Spectrum of demodulated signal
subplot(426), plot(w,abs(Xd)), title('The spectrum Xd(w) of xd(t)')
d= d(Nfft2+[0:Nfft-1]);
subplot(427), plot(t,d), title('The LPF output signal d(t)')
D= fftshift(fft(d)); D=[D D(1)]; % Spectrum of the LPF output
subplot(428), plot(w,abs(D)), title('The spectrum D(w) of d(t)')
```

2.12 Amplitude-Shift Keying (ASK) and Frequency-Shift Keying (FSK)

(a) Amplitude-Shift Keying (ASK) or On-Off Keying (OOK)
Find the spectrum $X_{ASK}(\omega)$ of a BASK (binary amplitude-shift keying) or OOK modulated signal $x_{ASK}(t) = r_D(t)\cos(\omega_1 t)$ (Fig. P2.12(a)) where $r_D(t)$ is a unit-height rectangular pulse with duration D.

(b) Frequency-Shift Keying (FSK)
Find the spectrum $X_{FSK}(\omega)$ of a BFSK (binary frequency-shift keying) modulated signal $x_{FSK}(t) = r_D(t)\cos(\omega_1 t) + r_D(t-D)\cos(\omega_2(t-D))$ (Fig. P2.12(b)).

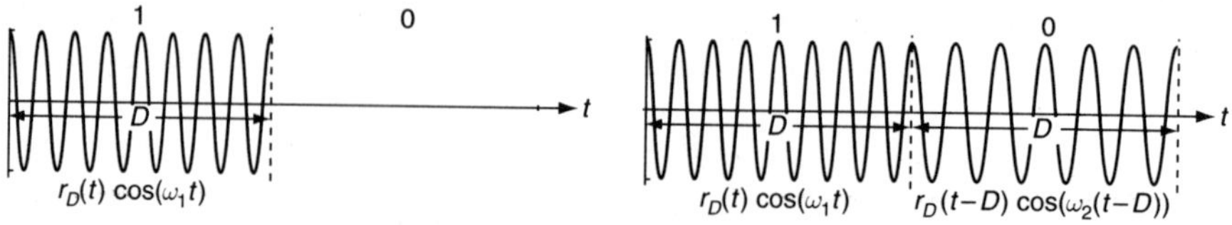

(a) A BASK or OOK signal

(b) A BFSK signal

Fig. P2.12

2.13 Autocorrelation/Crosscorrelation Theorem
Suppose we have $X(\omega) = \mathcal{F}\{x(t)\}$, $Y(\omega) = \mathcal{F}\{y(t)\}$, and $G(\omega) = \mathcal{F}\{g(t)\}$, where $x(t)$, $y(t)$, and $g(t)$ are the real-valued input, output, and impulse response of an LTI system G so that

$$y(t) = g(t) * x(t) \overset{\mathcal{F}}{\leftrightarrow} Y(\omega) = G(\omega) * X(\omega) \tag{P2.13.1}$$

(a) Prove the autocorrelation theorem:

$$\Phi_{xx}(\omega) = \mathcal{F}\{\phi_{xx}(t)\} \overset{(1.4.12a)}{=} \mathcal{F}\{x(t) * x(-t)\} = |X(\omega)|^2$$

(Energy Density Spectrum) (P2.13.2)

(b) Prove the crosscorrelation theorem:

$$\Phi_{xy}(\omega) = \mathcal{F}\{\phi_{xy}(t)\} \overset{(1.4.12a)}{=} \mathcal{F}\{x(t) * y(-t)\}$$
$$= X(\omega)Y^*(\omega) = G^*(\omega)\,|X(\omega)|^2 \tag{P2.13.3}$$

(c) Prove that

$$\Phi_{yy}(\omega) = \mathcal{F}\{\phi_{yy}(t)\} \overset{(1.4.12a)}{=} \mathcal{F}\{y(t) * y^*(-t)\} = Y(\omega)Y^*(\omega)$$
$$= |G(\omega)|^2|X(\omega)|^2 \tag{P2.13.4}$$

2.14 Power Theorem - Generalization of Parseval's Relation
Prove the power theorem that, with $X(\omega) = \mathcal{F}\{x(t)\}$ and $Y(\omega) = \mathcal{F}\{y(t)\}$, we have

$$\int_{-\infty}^{\infty} x(t)y^*(t)dt = \frac{1}{2\pi}\int_{-\infty}^{\infty} X(\omega)Y^*(\omega)\,d\omega \tag{P2.14.1}$$

2.15 Eigenfunction
If the output of a system to an input is just a (possibly complex-valued) constant times the input, then the input signal is called an *eigenfunction* of the system. Show that the output of an LTI (linear time-invariant) system with the (causal) impulse response $g(t)$ and system (transfer) function $G(s) = \mathcal{L}\{g(t)\}$ to a complex exponential input $e^{s_0 t}$ is,

$$G\{e^{s_0 t}\} = g(t) * e^{s_0 t} = G(s_0)e^{s_0 t} \text{ with } G(s_0) = G(s)|_{s=s_0} \qquad \text{(P2.15.1)}$$

where $G(s)$ is assumed not to have any pole at $s = s_0$ so that the amplitude factor $G(s_0)$, called the *eigenvalue*, is finite. This implies that a complex exponential function is an eigenfunction of LTI systems.

2.16 Fourier Transform of a Polygonal Signal (Problem 4.20 of [O-1])

(a) Show that the Fourier trnasform of $y(t) = (at + b)(u_s(t - t_1) - u_s(t - t_2))$ with $t_1 < t_2$ is

$$Y(\omega) = \frac{1}{\omega^2}(a+jb\omega)(e^{-j\omega t_2} - e^{-j\omega t_1}) + j\frac{a}{\omega}(t_1 e^{-j\omega t_2} - t_2 e^{-j\omega t_1}) \qquad \text{(P2.16.1)}$$

(b) We can describe a triangular pulse having three vertices $(t_{i-1}, 0)$, (t_i, x_i), and $(t_{i+1}, 0)$ as

$$\lambda_i(t) = \begin{cases} x_i + \frac{x_i}{t_i - t_{i-1}}(t - t_i) = -\frac{x_i t_{i-1}}{t_i - t_{i-1}} + \frac{x_i}{t_i - t_{i-1}}t & \text{for } t_{i-1} \le t < t_i \\ x_i + \frac{x_i}{t_i - t_{i+1}}(t - t_i) = -\frac{x_i t_{i+1}}{t_i - t_{i+1}} + \frac{x_i}{t_i - t_{i+1}}t & \text{for } t_i \le t < t_{i+1} \end{cases}$$

(P2.16.2)

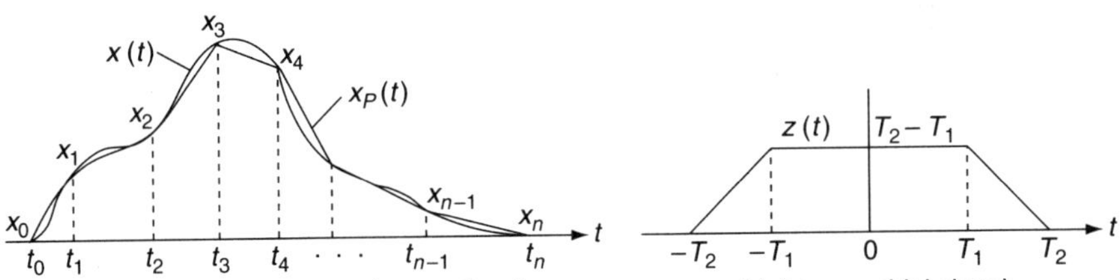

(a) A signal and its polygonal approximation (b) A trapezoidal signal

Fig. P2.16

Use Eq. (P2.16.1) to show that the Fourier transform of this pulse is

$$\Lambda_i(\omega) = \frac{1}{\omega^2}\left\{\frac{x_i}{t_i - t_{i-1}}(e^{-j\omega t_i} - e^{-j\omega t_{i-1}}) + \frac{x_i}{t_{i+1} - t_i}(e^{-j\omega t_i} - e^{-j\omega t_{i+1}})\right\}$$

(P2.16.3)

(c) Fig. P2.16(a) shows a finite-duration signal $x(t)$ and its piece-wise linear or polygonal approximation $x_P(t)$ where $x(t) = x_P(t) = 0\ \forall t \le t_0$ and $t \ge t_n$. Use Eq. (P2.16.3) to show that the Fourier transform of $x_P(t)$ is

$$X_P(\omega) = \mathcal{F}\{x_P(t)\} = \mathcal{F}\left\{\sum_{i=0}^{n} \lambda_i(t)\right\} = \frac{1}{\omega^2}\sum_{i=0}^{n} k_i e^{-j\omega t_i} \qquad \text{(P2.16.4)}$$

where

$$k_i = \frac{(t_{i-1} - t_i)x_{i+1} + (t_i - t_{i+1})x_{i-1} + (t_{i+1} - t_{i-1})x_i}{(t_i - t_{i-1})(t_{i+1} - t_i)} \quad \text{and}$$

$$x(t_{-1}) = x(t_{n+1}) = 0$$

Complete the following program "CTFT_poly ()" so that it implements this method of computing the Fourier transform.

```
function [X,w]=CTFT_poly(t,x,w)
% Computes the CTFT of x(t) for frequency w by using Eq.(P2.16.4).
N =length(x); n=1:N;
if nargin<3, w = [-100:100]*pi/100+1e-6; end
t=[2*t(1)-t(2) t 2*t(N)-t(N-1)]; x=[0 x 0];
X= 0;
for i=2:N+1
   ki=((t(i-1)-t(i))*x(i+1)+(t(i)-t(i+1))*x(i-1)+(t(i+1)-t(i-1))*x(i));
   k(i)= ki/(t(i)-t(i-1))/(t(i+1)-t(i));
   X= X + k(i)*exp(-j*w*t(i));
end
X = X./(w.^2+eps);
```

(d) For the trapezoidal signal $z(t)$ shown in Figure P2.16(b), find its Fourier transform $Z(\omega)$ in two ways:

(i) Regarding $z(t)$ as the convolution of two unit-height rectangular pulses each of length $D_1 = T_2 - T_1$ and $D_2 = T_2 + T_1$, you can use the convolution property (2.5.11) together with Eq. (E2.3.2).
(ii) Use Eq. (P2.16.4), which is the result of part (c).
(iii) Use the MATLAB routine "CTFT_poly ()" to get $Z(\omega) = \mathcal{F}\{z(t)\}$ with $T_1 = 1$ and $T_2 = 2$ and plot it to compare the result with that obtained (i) or (ii).
<Hint> You can complete and run the following program.

```
%sig02p_16.m
% Numerical approximation of CTFT for a trapezoidal signal
clear, clf
T1=1; T2=2; D1= T2-T1; D2= T1+T2;
t= [-T2 -T1 T1 T2];
x= [0 T2-T1 T2-T1 0];
w= [-200:200]/100*pi+1e-6; % frequency range on which to plot X(w)
X= D1*D2*sinc(w*D1/2/pi).*sinc(w*D2/2/pi);
X_poly= ?????????(?,?,?); % (P2.16.4)
Discrepancy=norm(X_poly-X)
plot(w,abs(X),'b', w,abs(X_poly),'r')
```

(e) Show that, if we choose the points t_i's sufficiently close together so that the piecewise linear approximation, $x_P(t)$, of $x(t)$ is accurate enough to satisfy the bounded error condition

$$|x(t) - x_P(t)| \le \varepsilon, \tag{P2.16.5}$$

then the Fourier transform (Eq. (P2.16.4)) of $x_P(t)$ is close to $X(\omega) = \mathcal{F}\{x(t)\}$ enough to satisfy

$$\int_{-\infty}^{\infty} |X(\omega) - X_P(\omega)|^2 d\omega \le 2\pi(t_n - t_0)\varepsilon^2 \tag{P2.16.6}$$

(cf.) This implies that even though a signal $x(t)$ cannot be expressed as a linear combination, a multiplication, or convolution of basic functions, we can find its approximate Fourier transform from the collection of sample values of $x(t)$ and that we can improve the accuracy by making the piecewise approximation $x_P(t)$ closer to $x(t)$.

(cf.) This result suggests us that it is possible to evaluate numerically the Fourier transform for a finite-duration signal whose values are measured experimentally at many (discrete) time instants, even though the closed-form analytic expression of the signal is not available or too complicated to deal with.

2.17 Impulse Response and Frequency Response of Ideal Bandpass Filter
Consider a sinc function multiplied by a cosine function, which is depicted in Fig. P2.17:

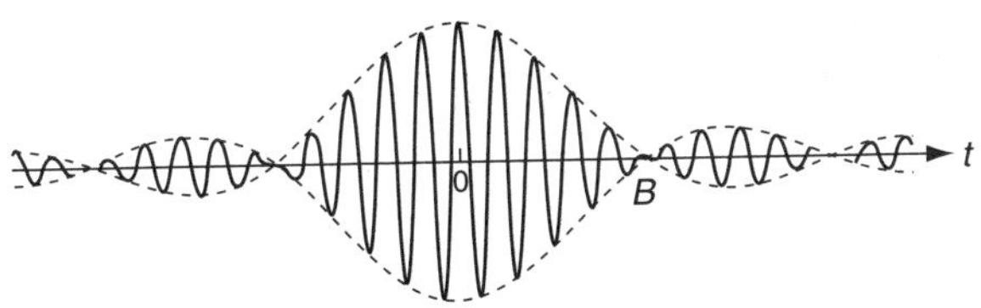

Fig. P2.17

$$g(t) = \frac{B}{\pi}\text{sinc}\left(\frac{B}{\pi}t\right)\cos\omega_p t \quad \text{with } \omega_p = 10B \tag{P2.17.1}$$

(a) Using Eqs. (E2.9.1&2) and the amplitude modulation (AM) property (E2.12.3), find the Fourier transform $G(\omega) = \mathcal{F}\{g(t)\}$ and sketch it.

(b) If Eq. (P2.17.1) describes the impulse response of a system, which type of filter is it?
(Lowpass, Bandpass, Highpass, Bandstop)

2.18 Design of Superheterodyne Receiver – Combination of Tunable/Selective Filters
We can build a tunable selective filter, called the *superheterodyne* or *supersonic heterodyne receiver* often abbreviated *superhet*), by combining a tunable, but poorly selective filter, a highly-selective, but untunable filter having

a fixed IF (intermediate frequency) passband, a mixer (multiplier), and a local oscillator as shown in Fig. P2.18.1.

Suppose the input signal $y(t)$ consists of many AM (amplitude modulated) signals which have been frequency-division-multiplexed (FDM) so that they each occupy different frequency bands. Let us consider one signal $y_1(t) = x_1(t)\cos\omega_c t$ with spectrum $Y_1(\omega)$ as depicted in Fig. P2.18.2(a). We want to use the system of Fig. P2.18.1 to demultiplex and demodulate for recovering the modulating signal $x_1(t)$, where the coarse tunable filter has the frequency response $H_c(\omega)$ shown in Fig. P2.18.2(b) and the fixed frequency selective filter is a bandpass filter (BPF) whose frequency response $H_f(\omega)$ is centered around fixed frequency ω_c as shown in Fig. P2.18.2(c).

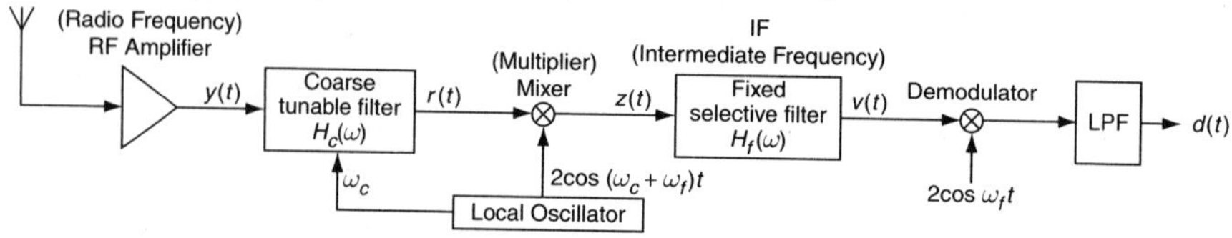

Fig. P2.18.1 Superheterodyne receiver – tunable and selective filter

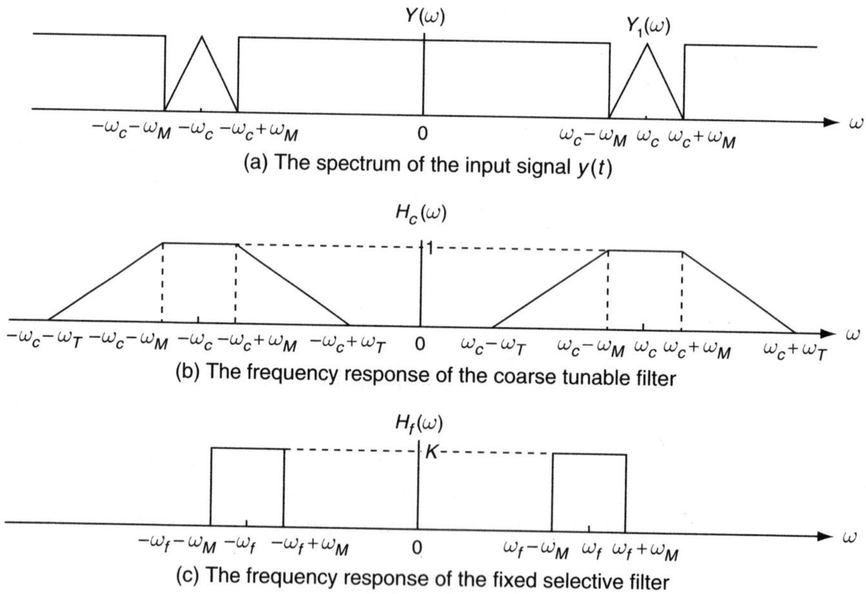

Fig. P2.18.2

(a) What constraint in terms of ω_c and ω_M must ω_T satisfy to guarantee that an undistorted spectrum of $x_1(t)$ is centered around ω_f so that the output of the fixed selective filter H_f can be $v(t) = x_1(t)\cos\omega_f t$?

(b) Determine and sketch the spectra $R(\omega)$, $Z(\omega)$, $V(\omega)$, and $D(\omega)$ of the output $r(t)$ of the coarse tunable filter H_c, the input $z(t)$ and output $v(t)$ of the fixed selective filter H_f, and the output $d(t)$ of the LPF.

(c) How could you determine the gain K of the fixed selective filter H_f required to have $v(t) = x_1(t)\cos\omega_f t$ at its output?

(d) Complete and run the following program "sig02p_18.m" to simulate the superheterodyne receiver twice, once with kT = 18 and once with kT = 19. Explain why the two simulation results are so different in terms of the error between $d(t)$ and $x_1(t)$. Note that the parameters kc, kf, kM, and kT correspond to ω_c, ω_f, ω_M, and ω_T, respectively.

```
%sig02p_18.m
% simulates a superheterodyne receiver combining a coarse tunable filter
%  and a fixed selective filter
% Copyleft: Won Y. Yang, wyyang53@hanmail.net, CAU for academic use only
clear, clf
N=128; kc=16; kf=12; kM=4; kT=12; Wcf=2*pi/N*(kc+kf); Wf=2*pi/N*kf;
%create a signal having the spectrum as Fig.P2.17.2(a)
n=[0:N-1]; kk=[-N/2:N/2]; % Time/Frequency ranges
y=make_signal(N,kc,kM);
Y=fftshift(fft(y)); Y_mag= abs([Y Y(1)]);
subplot(6,2,1), plot(n,y), title('y(t)')
subplot(6,2,2), plot(kk,Y_mag), title('Y(w)')
%tunable filter
r=??????????????(N,kc,kM,kT,?);
R=fftshift(fft(r)); R_mag= abs([R R(1)]);
subplot(6,2,3), plot(n,r), title('r(t) after tunable filter')
subplot(6,2,4), plot(kk,R_mag), title('R(w)')
%modulation
z=r.*(????????????);
Z=fftshift(fft(z)); Z_mag= abs([Z Z(1)]);
subplot(6,2,5), plot(n,z), title('z(t) after modulation')
subplot(6,2,6), plot(kk,Z_mag), title('Z(w)')
%selective filter
K=1; v=????????????????(N,kf,kM,K,?);
V=fftshift(fft(v)); V_mag= abs([V V(1)]);
subplot(6,2,7), plot(n,v), title('v(t) after selective filter')
subplot(6,2,8), plot(kk,V_mag), title('V(w)')
%demodulation
d0=v.*(???????????);
D0=fftshift(fft(d0)); D0_mag= abs([D0 D0(1)]);
subplot(6,2,9), plot(n,d0), title('d0(t) after demodulation')
subplot(6,2,10), plot(kk,D0_mag), title('D0(w)')
%tunable filter as LPF
d=tunable_filter(N,0,kM,kT,d0);
D=fftshift(fft(d)); D_mag= abs([D D(1)]);
x1=zeros(1,N);
for k=-kM:kM, x1(n+1)= x1(n+1)+(1-abs(k)/kM)*cos(2*pi*k*n/N); end
subplot(6,2,11), plot(n,d,'b', n,x1,'r')
title('detected/transmitted signal')
subplot(6,2,12), plot(kk,D_mag),
title('Spectrum D(w) of detected signal')
error_between_detected_transmitted_signals= norm(d-x1)
```

```
function x=make_signal(N,kc,kM)
% create a signal having the spectrum as Fig.P2.17.2(a)
n=1:N;
x=zeros(1,N);
for k=0:N/2-1
   tmp= 2*pi*k*(n-1)/N;
   if k<kc-kM, x(n)= x(n)+sin(tmp);
    elseif k>kc+kM, x(n)=x(n)+sin(tmp); % whatever, cos() or sin()
    else x(n)= x(n)+(1-abs(k-kc)/kM)*cos(tmp);
   end
end
```

```
function x_t=tunable_filter(N,kc,kM,kT,x)
% BPF with passband (kc-kM,kc+kM) and
%   stopband edge frequencies kc-kT and kc+kT
X=fft(x,N);
for k=1:N/2+1
   if k<=kc-kT|k>kc+kT+1, X(k)=0; X(N+2-k)=0;
    elseif k<=kc-kM+1
      X(k)=X(k)*(1-(kc-kM-k+1)/(kT-kM)); X(N+2-k)=conj(X(k));
    elseif k>kc+kM
      X(k)=X(k)*(1-(k-1-kc-kM)/(kT-kM)); X(N+2-k)=conj(X(k));
   end
end
x_t=real(ifft(X,N));
```

```
function x_t=selective_filter(N,kf,kM,K,x)
% passes only the freq (kf-kM,kf+kM) with the gain K
X=fft(x,N);
for k=1:N
   if (k>kf-kM&k<=kf+kM+1)|(k>N-kf-kM&k<=N-kf+kM+1), X(k)= K*X(k);
    else X(k)=0;
   end
end
x_t=real(ifft(X));
```

2.19 The Poisson Sum Formula (PSF) and Spectrum of Sampled Signal
Consider an LTI (linear time-invariant) system with impulse response $h(t)$ and frequency response $H(j\omega) = \mathcal{F}\{h(t)\}$. Note that the response (output) of the system to the impulse train can be expressed in terms of the impulse response as

$$\delta(t) \overset{G\{\}}{\rightarrow} h(t);\ \delta_T(t) = \sum_{n=-\infty}^{\infty} \delta(t-nT) \overset{G\{\}}{\rightarrow} \sum_{n=-\infty}^{\infty} h(t-nT) \quad \text{(P2.19.1)}$$

Likewise, the response (output) of the system to a single or multiple complex exponential signal can be written as

$$e^{jk\omega_0 t} \overset{G\{\}}{\rightarrow} H(jk\omega_0)e^{jk\omega_0 t};\ \frac{1}{T}\sum_{k=-\infty}^{\infty} e^{jk\omega_0 t} \overset{G\{\}}{\rightarrow} \frac{1}{T}\sum_{k=-\infty}^{\infty} H(jk\omega_0)e^{jk\omega_0 t} \quad \text{(P2.19.2)}$$

(a) Based on the above results (P2.19.1) & (P2.19.2) and Eq. (2.1.10), prove that, for any function $h(t)$ with its Fourier transform $H(j\omega) = \mathcal{F}\{h(t)\}$ well-defined, the following relation holds:

$$\sum_{n=-\infty}^{\infty} h(t-nT) = \frac{1}{T}\sum_{k=-\infty}^{\infty} H(jk\omega_0)e^{jk\omega_0 t} \text{ with } \omega_0 = \frac{2\pi}{T} \tag{P2.19.3}$$

which is called the *Poisson sum formula*.

(b) Using the Poisson sum formula, prove the following relation pertaining to the spectrum of a sampled signal (see Eq. (E2.13.3)):

$$\sum_{n=-\infty}^{\infty} x(nT)e^{-jn\omega T} = \frac{1}{T}\sum_{k=-\infty}^{\infty} X(j(\omega + k\omega_0)) \text{ with } X(j\omega) = \mathcal{F}\{x(t)\} \text{ and } \omega_0 = \frac{2\pi}{T} \tag{P2.19.4}$$

<Hint> Substitute $h(t) = x(t)e^{j\omega_1 t}$ and its Fourier transform $H(j\omega) = X(j(\omega + \omega_1))$ with $\omega = k\omega_0$ into Eq. (P2.19.3) and then substitute $t = 0$ and $\omega_1 = \omega$.

2.20 BPF (Bandpass Filter) Realization via Modulation-LPF-Demodulation
Consider the realization of BPF of Fig. P2.20(a), which consists of a modulator (multiplier), an LPF, and a demodulator (multiplier). Assuming that the spectrum of $x(t)$ is as depicted in Fig. P2.20(b), sketch the spectra of the signals $x_c(t)$, $x_s(t)$, $y_c(t)$, $y_s(t)$, and $y(t)$ to see to it that the composite system realizes a BPF and determine the passband of the realized BPF.

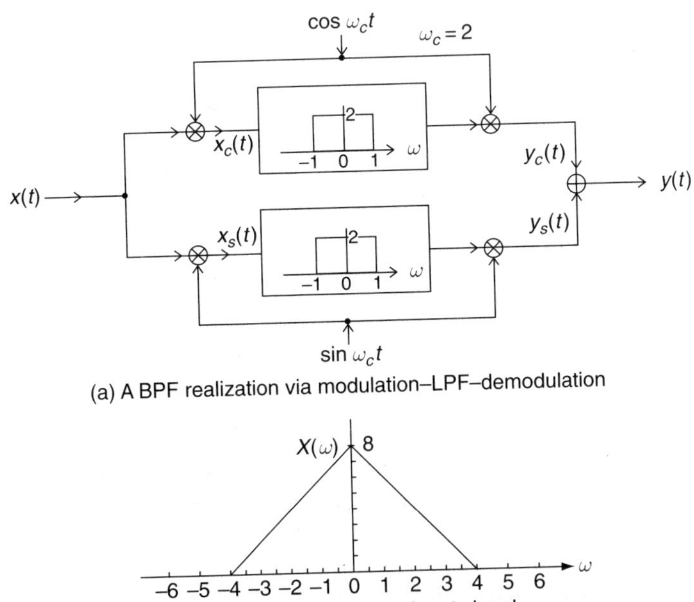

(a) A BPF realization via modulation–LPF–demodulation

(b) The spectrum of the input signal

Fig. P2.20

2.21 TDM (Time-Division Multiplexing)
As depicted in Fig. P2.21(a), *Time-Division multiplexing* (TDM) is to transmit two or more PAM (pulse amplitude modulation) signals not simultaneously as subchannels (separated in the frequency-domain), but physically by turns in one communication channel where each subchannel is assigned a timeslot of duration D every T s in such a way that the timeslots do not overlap. Sketch a TDM waveform of two signals $x_1(t)$ and $x_2(t)$ in Fig. P2.21(b).

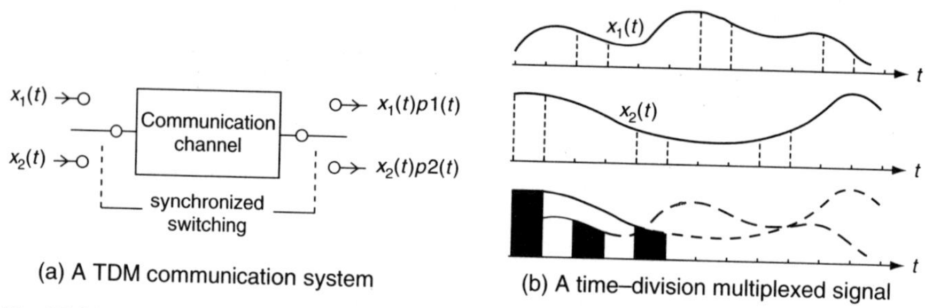

Fig. P2.21 TDM (Time-Division Multiplexing) communication

2.22 FDM (Frequency-Division Multiplexing)
As depicted in Fig. P2.22(a), *Frequency-Division multiplexing* (FDM) is to transmit two or more SAM (sinusoidal amplitude modulation) signals as *sub-carriers* (separated in the frequency-domain) in such a way that the frequency slots each carrying a signal (amplitude-modulated with different carrier frequency) do not overlap. Assuming that the channel is ideal so that $r(t) = s(t)$, sketch the spectrum of the FDM signal to be transmitted over the same communication channel where the spectra of two signals $x_1(t)$ and $x_2(t)$ to be frequency-multiplexed are shown in Fig. P2.22(b).

(cf.) Time/frequency-division multiplexing assigns different time/frequency intervals to each signal or subchannel.

2.23 Quadrature Multiplexing
Consider the quadrature multiplexing system depicted in Fig. P2.22, where the two signals are assumed to be bandlimited, that is,

$$X_1(\omega) = X_2(\omega) = 0 \text{ for } \omega > \omega_M \qquad \text{(P2.23.1)}$$

as illustrated in Fig. P2.23(b).

(a) Assuming that the channel is ideal so that $r(t) = s(t)$, express the spectra $S(\omega)$, $V_1(\omega)$, $V_2(\omega)$, $Y_1(\omega)$, $Y_2(\omega)$ of the signals $s(t)$, $v_1(t)$, $v_2(t)$, $y_1(t)$, and $y_2(t)$ in terms of $X_1(\omega)$ and $X_2(\omega)$.

(b) Complete and run the following MATLAB program "sig02p_23.m" to see if $y_1(t) = x_1(t)$ and $y_2(t) = x_2(t)$.

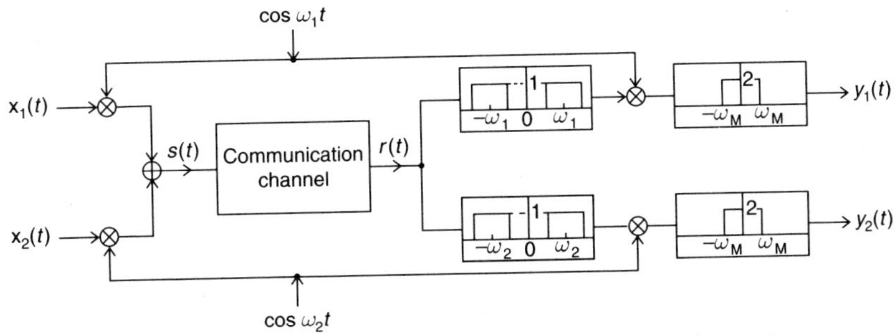

(a) A FDM (Frequency–Division Multiplexing) communication system

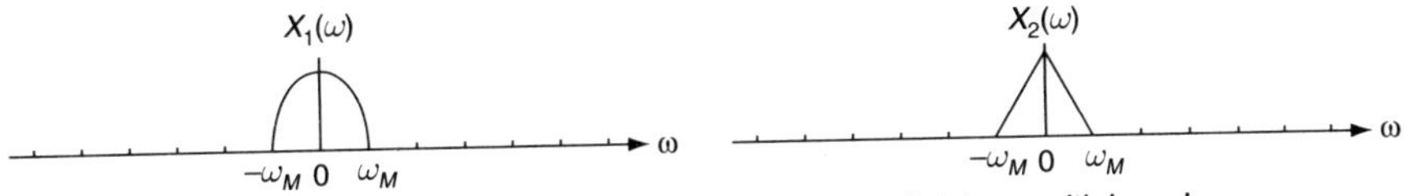

(b) The spectra of two signals to be frequency–division multiplexed

Fig. P2.22 FDM (Frequency Division Multiplexing) communication

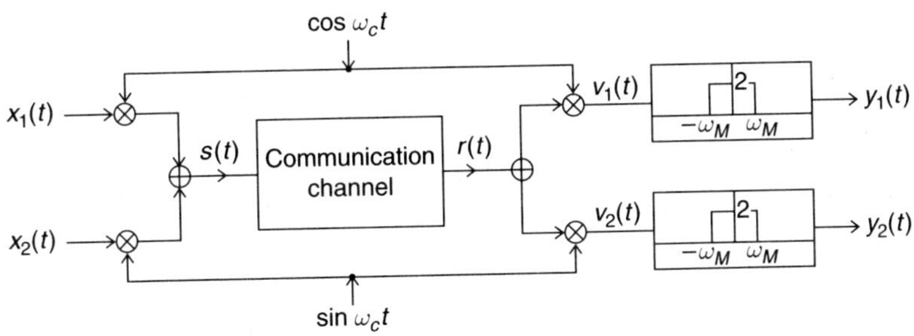

Fig. P2.23 Quadrature multiplexing

```
%sig02p_23.m
% Quadrature Multiplexing in Fig.P2.23
% Copyleft: Won Y. Yang, wyyang53@hanmail.net, CAU for academic use only
clear, clf
N=128; kc=16; kM=6; kx1=6; kx2=6; Wc=2*pi/N*kc;
n=[0:N-1]; kk=[-N/2:N/2];
x1= make_signal1(N,kx1); x2= make_signal2(N,kx2);
X1= fftshift(fft(x1)); X2= fftshift(fft(x2));
subplot(523), plot(kk,abs([X1 X1(1)])), title('X1(w)')
subplot(524), plot(kk,abs([X2 X2(1)])), title('X2(w)')
s= x1.*???(Wc*n) + x2.*sin(Wc*n);  S= fftshift(fft(s));
subplot(513), plot(kk,abs([S S(1)])),  title('S(w)')
v1= s.*cos(Wc*n); V1= fftshift(fft(v1));
v2= s.*???(Wc*n); V2= fftshift(fft(v2));
subplot(527), plot(kk,abs([V1 V1(1)])), title('V1(w)')
subplot(528), plot(kk,abs([V2 V2(1)])), title('V2(w)')
```

```
% selective filter(ideal LPF)
kf=0; K=2;
y1= selective_filter(N,kf,kM,K,v1); Y1= fftshift(fft(y1));
y2= ????????????????(N,kf,kM,K,v2); Y2= fftshift(fft(y2));
discrepancy1=norm(x1-y1), discrepancy2=norm(x2-y2)
subplot(529), plot(kk,abs([Y1 Y1(1)])), title('Y1(w)')
subplot(5,2,10), plot(kk,abs([Y2 Y2(1)])), title('Y2(w)')
```

```
function x=make_signal1(N,kx)
n=1:N; kk=1:kx-1; x= (1-kk/kx)*cos(2*pi*kk.'*n/N);
```

```
function x=make_signal2(N,kx)
n=1:N; kk=1:kx-1; x= kk/kx*cos(2*pi*kk.'*n/N);
```

2.24 Least Squares Error (LSE) and Fourier Series Representation

Consider a continuous-time function $x(t)$ on a time interval $[a, b]$ and a set of its sampled values, $\{x_1, x_2, \cdots, x_N\}$ (with $x_n = x(t_n)$). Let us find an approximate representation of $x(t)$ as a linear combination of some basis functions $\{\phi_k(t); k = 1 : K\}$:

$$\hat{x}(\mathbf{c}, t) = \sum\nolimits_{k=1}^{K} c_k\, \phi_k(t) \text{ for } a \le t \le b \qquad \text{(P2.24.1)}$$

where the coefficient vector $\mathbf{c} = [c_1\, c_2 \cdots c_K]^T$ is to be determined so that the following objective function can be minimized:

$$E^2(\mathbf{c}) = \int_a^b (x(t) - \hat{x}(\mathbf{c}, t))^2 dt \qquad \text{(P2.24.2)}$$

$$E^2(\mathbf{c}) = \sum\nolimits_{n=1}^{K} (x_n - \hat{x}(\mathbf{c}, t_n))^2 \qquad \text{(P2.24.3)}$$

The following notations will be used:

$$\mathbf{x} = [x_1\, x_2\, \cdots\, x_N]^T \text{ with } x_n = x(t_n) \qquad \text{(P2.24.4a)}$$

$$\mathbf{c} = [c_1\, c_2\, \cdots\, c_K]^T \qquad \text{(P2.24.4b)}$$

$$\varphi(t) = [\phi_1(t)\, \phi_2(t)\, \cdots\, \phi_K(t)]^T \qquad \text{(P2.24.4c)}$$

$$\varepsilon = [\varepsilon_1\, \varepsilon_2\, \cdots\, \varepsilon_N]^T \qquad \text{(P2.24.4d)}$$

$$\text{with } \varepsilon_n = x_n - \hat{x}(\mathbf{c},\ t_n) = x_n - \sum_{k=1}^{K} c_k\phi_k(t_n) = x_n - \varphi^T(t_n)\mathbf{c}$$

$$\Phi = \begin{bmatrix} \varphi^T(t_1) \\ \varphi^T(t_2) \\ \bullet \\ \bullet \\ \varphi^T(t_N) \end{bmatrix} = \begin{bmatrix} \phi_{11} & \phi_{21} & \bullet & \phi_{K1} \\ \phi_{12} & \phi_{22} & \bullet & \phi_{K2} \\ \bullet & \bullet & \bullet & \bullet \\ \bullet & \bullet & \bullet & \bullet \\ \phi_{1N} & \phi_{2N} & \bullet & \phi_{KN} \end{bmatrix} \text{ with } \phi_{kn} = \phi_k(t_n) \quad \text{(P2.24.4e)}$$

Based on these notations, the above objective function (P2.24.3) can be written as

$$\begin{aligned} E^2(\mathbf{c}) &= \varepsilon^{*T}\varepsilon = [\mathbf{x}^* - \Phi^*\mathbf{c}^*]^T[\mathbf{x} - \Phi\mathbf{c}] \\ &= \mathbf{x}^{*T}\mathbf{x} - \mathbf{x}^{*T}\Phi\mathbf{c} - \mathbf{c}^{*T}\Phi^{*T}\mathbf{x} + \mathbf{c}^{*T}\Phi^{*T}\Phi\mathbf{c} \\ &= \mathbf{x}^{*T}\left[I - \Phi[\Phi^{*T}\Phi]^{-1}\Phi^{*T}\right]\mathbf{x} \\ &\quad + [\Phi^T\Phi^*\mathbf{c}^* - \Phi^T\mathbf{x}^*]^T[\Phi^{*T}\Phi]^{-1}[\Phi^{*T}\Phi\mathbf{c} - \Phi^{*T}\mathbf{x}] \end{aligned} \quad \text{(P2.24.5)}$$

where only the second term of this equation depends on $\mathbf{c}$. Note that this objective function is minimized for

$$\Phi^{*T}\Phi\mathbf{c} - \Phi^{*T}\mathbf{x} = 0 \text{ (a normal equation); } \mathbf{c} = [\Phi^{*T}\Phi]^{-1}\Phi^{*T}\mathbf{x} \quad \text{(P2.24.6)}$$

(a) Let the K functions $\{\phi_k(t),\ k = 1 : K\}$ constitute an orthogonal set for the N discrete times $\{t_n,\ n = 1 :\ N\}$ in the sense that

$$\sum_{n=1}^{N} \phi_{kn}^*\phi_{mn} = 0\ \forall\ k \neq m \quad \text{(P2.24.7)}$$

so that $[\Phi^{*T}\Phi]$ is diagonal. Show that the coefficients can be determined as

$$c_k = \frac{\sum_{n=1}^{N} x_n\phi_{kn}^*}{\sum_{n=1}^{N} \phi_{kn}^*\phi_{kn}},\ \text{for } k = 1, 2, \cdots, K \quad \text{(P2.24.8)}$$

Likewise, if the K functions $\{\phi_k(t),\ k = 1 :\ K\}$ constitute an orthogonal set for the continuous time intrerval $[a, b]$ in the sense that

$$\int_a^b \phi_k^*(t)\phi_m(t)\,dt = 0\ \forall\ k \neq m, \quad \text{(P2.24.9)}$$

then the coefficients can be determined as

$$c_k = \frac{\int_a^b x(t)\phi_k^*(t)\,dt}{\int_a^b \phi_k^*(t)\phi_k(t)\,dt},\ \text{for } k = 1, 2, \cdots, K \quad \text{(P2.24.10)}$$

(b) Using Eq. (P2.24.1), find an approximate representation of a real-valued function $x(t)$ for $-P/2 \leq t \leq P/2$ in terms of the following basis functions

$$\phi_k(t) = \frac{1}{P} e^{jk\omega_0 t} \text{ with } \omega_0 = \frac{2\pi}{P} \text{ and } k = -K : K \tag{P2.24.11}$$

Chapter 3
Discrete-Time Fourier Analysis

Contents

In this chapter we study the discrete-time Fourier analysis techniques, i.e., the DTFT (Discrete-Time Fourier Transform), DFT (Discrete Fourier Transform), and DFS (Discrete Fourier Series) of a discrete-time sequence, which will be used to

W.Y. Yang et al., *Signals and Systems with MATLAB®*,
DOI 10.1007/978-3-540-92954-3_3, © Springer-Verlag Berlin Heidelberg 2009

describe and analyze the frequency characteristics of discrete-time signals and the frequency-domain behavior of discrete-time systems. We also deal with the fast Fourier transform (FFT), which is a very efficient algorithm for computing the DFT.

Each of the continuous-time and discrete-time Fourier techniques has its own application in the sense that they are used for analyzing continuous-time and discrete-time signals/systems, respectively. However, if you do not understand their organic inter-relationship beyond some similarity, you will miss the overall point of viewing the frequency characteristic of signals/systems and get only a confusing impression that there are too many Fourier techniques. Basically, we want to find, say, the CTFT of continuous-time signals since no inherently discrete-time signals exists in the physical world. However, the DFT, implemented by the FFT algorithm, of discrete-time signals obtained by sampling continuous-time signals is the only one practical Fourier analysis technique because of its outstanding computational convenience compared with the CTFT that is difficult to compute for general signals. That is why we need to know the relationship between the CTFT and DFT and ultimately, be able to get the "true" information about the frequency characteristic of a (continuous-time) signal from its FFT-based spectrum.

3.1 Discrete-Time Fourier Transform (DTFT)

3.1.1 Definition and Convergence Conditions of DTFT Representation

As the discrete-time counterpart of the CTFT (continuous-time Fourier transform)

$$X(j\omega) = \mathcal{F}\{x(t)\} \overset{(2.2.1a)}{=} \int_{-\infty}^{\infty} x(t)\, e^{-j\omega t}\, dt$$

we define the Fourier transform of a discrete-time sequence $x[n]$ as

$$X(j\Omega) = X[e^{j\Omega}] = \mathcal{F}\{x[n]\} = \sum\nolimits_{n=-\infty}^{\infty} x[n]\, e^{-j\Omega n} \tag{3.1.1}$$

which is called the DTFT (*discrete-time Fourier transform*). From now on, we will more often use $X(\Omega)$ than $X(j\Omega)$ or $X[e^{j\Omega}]$ for simplicity. Let us see the physical meaning of the DTFT.

Remark 3.1 Physical Meaning of the DTFT – Signal Spectrum and Frequency Response

(1) If a sequence $x[n]$ represents a physical signal, its DTFT $X(\Omega) = \mathcal{F}\{x[n]\}$ means the signal spectrum, which describes the frequency contents of the signal.
(2) In particular, if a time function $g[n]$ represents the impulse response of a discrete-time LTI (linear time-invariant) system, its DTFT $G(\Omega) = \mathcal{F}\{g[n]\}$ means the frequency response, which describes how the system responds to a

sinusoidal input sequence of digital (angular) frequency Ω (refer to Sect. 1.2.6 for the definition of frequency response).

One striking difference of the DTFT from the CTFT is its periodicity (with period 2π) in the (digital) frequency variable Ω, which results from the fact that it is a function of $e^{j\Omega}$ periodic with period 2π in Ω, i.e., $e^{j(\Omega+2\pi n)} = e^{j\Omega}$. Based on the periodicity of the DTFT, we are going to use the CTFS (for periodic functions) to derive the IDTFT (*inverse discrete-time Fourier transform*) formula. To this end, we can use Eq. (2.1.5a) with $P = 2\pi,\ \omega_0 = 2\pi/P = 1,\ t = \Omega$, and $k = -n$ to write the (continuous-frequency) Fourier series representation of $X(\Omega)$ as

$$X(\Omega) \overset{(2.1.5a)}{=} \frac{1}{2\pi} \sum\nolimits_{n=-\infty}^{\infty} x_n\, e^{-j\Omega n} \tag{3.1.2a}$$

where the Fourier coefficients are

$$x_n \underset{P=2\pi,\,\omega_0=1,\,t=\Omega,\,k=-n}{\overset{(2.1.5b)}{=}} \int_{2\pi} X(\Omega)\, e^{j\Omega n}\, d\Omega \text{ (the integral over one period of length } 2\pi) \tag{3.1.2b}$$

Noting that Eq. (3.1.2a) is the same as Eq. (3.1.1) multiplied by a scaling factor $1/2\pi$, we can multiply Eq. (3.1.2b) by the same scaling factor $1/2\pi$ to write the IDTFT formula as

$$x[n] = \mathcal{F}^{-1}\{X(\Omega)\} = \frac{1}{2\pi} \int_{2\pi} X(\Omega)\, e^{j\Omega n}\, d\Omega \tag{3.1.3}$$

We call Eqs. (3.1.1) and (3.1.3) the DTFT pair where Eq. (3.1.1) is the analysis equation and Eq. (3.1.3) is the synthesis equation.

Like the convergence conditions (2.2.2a) and (2.2.2b) for the CTFT, it can be stated that the DTFT will exist if the sequence $x[n]$ has finite energy, i.e.,

$$\sum\nolimits_{n=-\infty}^{\infty} |x[n]|^2 < \infty \tag{3.1.4a}$$

or if it is absolutely summable, i.e.,

$$\sum\nolimits_{n=-\infty}^{\infty} |x[n]| < \infty \tag{3.1.4b}$$

Remark 3.2 Frequency Response Existence Condition and Stability Condition of a System

Note that, for the impulse response $g[n]$ of a discrete-time LTI system, the absolute-summability condition (3.1.4b) is identical with the stability condition (1.2.27b). This implies that a stable LTI system has a well-defined frequency response $G(\Omega) = \mathcal{F}\{g[n]\}$.

3.1.2 Examples of DTFT Analysis

Example 3.1 DTFT of a Rectangular Pulse Sequence

For the rectangular pulse of duration $2M+1$ from $-M$ to M shown in Fig. 3.1(a1) and described by

$$r'_{2M+1}[n] = u_s[n+M] - u_s[n-M-1] \tag{E3.1.1}$$

we can apply Eq. (3.1.1) to get its DTFT as

$$\begin{aligned} R'_{2M+1}(\Omega) &\overset{(3.1.1)}{=} \sum_{n=-\infty}^{\infty} r'_{2M+1}[n]\, e^{-j\Omega n} \\ &\overset{(E3.1.1)}{=} \sum_{n=-M}^{M} e^{-j\Omega n} \overset{(D.23)}{=} e^{j\Omega M}\frac{1-e^{-j\Omega(2M+1)}}{1-e^{-j\Omega}} \text{(Dirichlet kernel)} \\ &= e^{j\Omega M}\frac{e^{-j\Omega(2M+1)/2}(e^{j\Omega(2M+1)/2}-e^{-j\Omega(2M+1)/2})}{e^{-j\Omega/2}(e^{j\Omega/2}-e^{-j\Omega/2})} \\ &\overset{(D.22)}{=} \frac{\sin(\Omega(2M+1)/2)}{\sin(\Omega/2)} \end{aligned} \tag{E3.1.2}$$

whose magnitude and phase are depicted in Fig. 3.1 (b1) and (c1), respectively. Likewise, for the rectangular pulse of duration $2M+1$ from 0 to $2M$ shown in Fig. 3.1(a2) and described by

$$r_{2M+1}[n] = u_s[n] - u_s[n-2M-1] = r'_{2M+1}[n-M] \tag{E3.1.3}$$

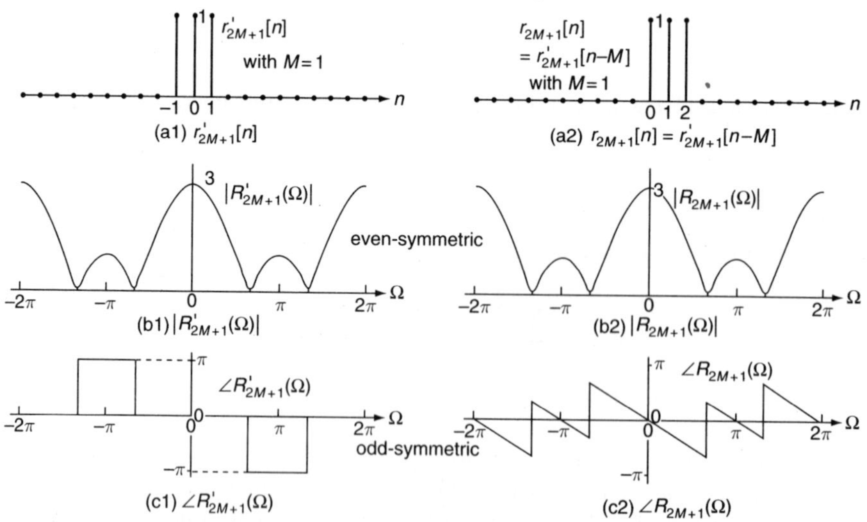

Fig. 3.1 Two rectangular pulse sequences and their magnitude & phase spectra

we can apply Eq. (3.1.1) to get its DTFT as

$$R_{2M+1}(\Omega) \overset{(3.1.1)}{=} \sum_{n=-\infty}^{\infty} r_{2M+1}[n]\, e^{-j\Omega n}$$

$$\overset{(E3.1.3)}{=} \sum_{n=0}^{2M} e^{-j\Omega n} \overset{(D.23)}{=} \frac{1 - e^{-j\Omega(2M+1)}}{1 - e^{-j\Omega}} = R'_{2M+1}(\Omega) e^{-j\Omega M}$$

$$\overset{(E3.1.2)}{=} \frac{\sin(\Omega(2M+1)/2)}{\sin(\Omega/2)} e^{-j\Omega M} = \frac{\sin(\Omega(2M+1)/2)}{\sin(\Omega/2)} \angle - M\Omega \tag{E3.1.4}$$

whose magnitude and phase are depicted in Figs. 3.1 (b2) and (c2), respectively. Note the following:

- In contrast with the continuous-time case (Fig. 2.8(a)), the DTFT of a rectangular pulse sequence is periodic, being no longer a pure sinc function, but an *aliased sinc* function. Still the magnitude spectra show that the rectangular pulses have low frequency components (around $\Omega = 0$) more than high frequency ones around $\Omega = \pm\pi$.
- The spectrum (E3.1.2) of an even-symmetric pulse sequence is real-valued and its phase is 0 or $\pm\pi$ depending on whether it is positive or negative. Especially for the frequency range such that $R'_{2M+1}(\Omega) < 0$, we set its phase to $+\pi$ or $-\pi$ so that the overall phase curve is odd-symmetric to comply with the conjugate symmetry (3.2.4) of DTFT (see Fig. 3.1(c1)).
- Due to the fact that the phase of $R_{2M+1}(\Omega)$ (Eq. (E3.1.4)) is proportional to the (digital) frequency Ω as $-M\Omega$, the (shifted) rectangular pulse sequence is said to have a *linear phase* despite the phase jumps (discontinuities) caused by the change of sign or the wrapping (modulo 2π) operation of angle (see the piecewise linear phase curve in Fig. 3.1(c2)).

Example 3.2 DTFT of an Exponential Sequence

For an exponential sequence $e_1[n] = a^n u_s[n]$ with $|a| < 1$, we can apply Eq. (3.1.1) to get its DTFT as

$$E_1(\Omega) = \sum_{n=-\infty}^{\infty} a^n u_s[n] e^{-j\Omega n} = \sum_{n=0}^{\infty} (ae^{-j\Omega})^n \overset{(D.23)}{=} \frac{1}{1 - a\, e^{-j\Omega}}$$

$$= \frac{1}{1 - a\, e^{-j\Omega}} \frac{1 - a\, e^{j\Omega}}{1 - a\, e^{j\Omega}} \overset{(D.20)}{=} \frac{1 - a\cos\Omega - j\, a\sin\Omega}{1 - 2a\cos\Omega + a^2} \tag{E3.2.1}$$

whose magnitude and phase are

$$|E_1(\Omega)| = \frac{1}{\sqrt{(1 - a\cos\Omega)^2 + (a\sin\Omega)^2}} \quad \text{and} \quad \angle E_1(\Omega) = \tan^{-1}\frac{-a\sin\Omega}{1 - a\cos\Omega} \tag{E3.2.2}$$

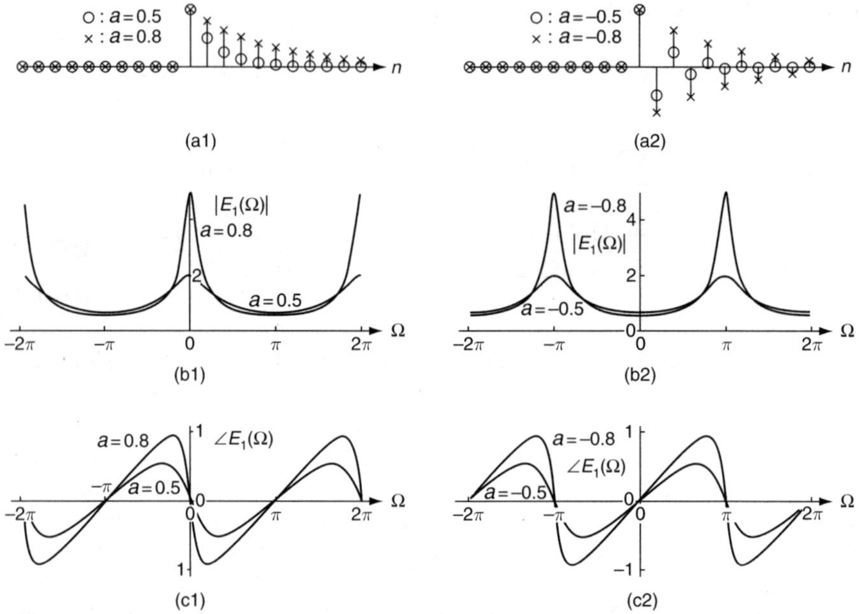

Fig. 3.2 Four exponential sequences and their magnitude & phase spectra

Fig. 3.2(a1–a2), (b1–b2), and (c2–c2) show the exponential sequences $e_1[n] = a^n u_s[n]$ with $a = \pm 0.5$ and ± 0.8 and their magnitude and phase spectra. Note the following:

- From Fig. 3.2(a1), we see that $e_1[n]$ with $a = 0.8$ is smoother than that with $a = 0.5$. This is reflected in Fig. 3.2(b1) showing that the magnitude spectrum $|E_1[\Omega]|$ with $a = 0.8$ is larger/smaller than that with $a = 0.5$ around $\Omega = 0$ (low frequency)/ $\Omega = \pm\pi$ (high frequency). Also from Fig. 3.2(a2), we see that $e_1[n]$ with $a = -0.8$ changes more rapidly than that with $a = -0.5$. This is reflected in Fig. 3.2(b2) showing that the magnitude spectrum $|E_1[\Omega]|$ with $a = -0.8$ is larger/smaller than that with $a = -0.5$ around $\Omega = \pm\pi$ (high frequency)/$\Omega = 0$ (low frequency).
- Comparing Fig. 3.2(a1) and (a2), we see that $e_1[n]$ with $a < 0$ changes much more rapidly than that with $a > 0$. This is reflected in that the magnitude spectrum $|E_1[\Omega]|$ with $a < 0$ (Fig. 3.2(b2)) is large around $\Omega = \pm\pi$ (high frequency) and that with $a > 0$ (Fig. 3.2(b1)) is large around $\Omega = 0$ (low frequency).

Example 3.3 DTFT of a Symmetric Exponential Sequence

For the exponential sequence $e_2[n] = a^{|n|}$ with $|a| < 1$, we can apply Eq. (3.1.1) to get its DTFT as

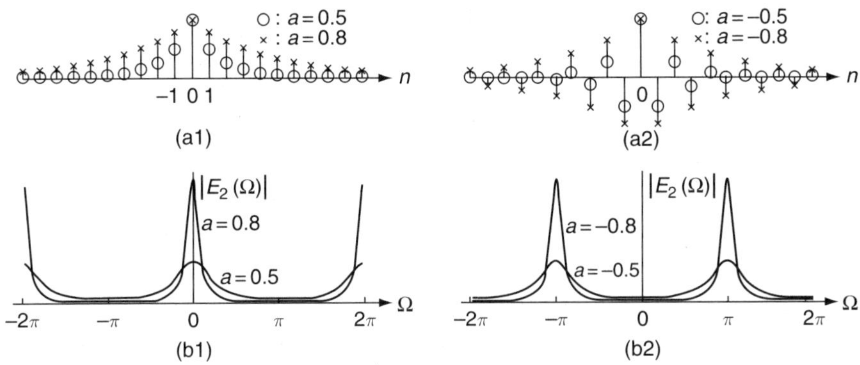

Fig. 3.3 Four exponential sequences and their magnitude spectra

$$E_2(\Omega) = \sum_{n=-\infty}^{\infty} a^{|n|} e^{-j\Omega n} = \sum_{n=-\infty}^{-1} a^{-n} e^{-j\Omega n}$$

$$+\sum_{n=0}^{\infty} a^n e^{-j\Omega n} = \sum_{n=1}^{\infty} (ae^{j\Omega})^n + \sum_{n=0}^{\infty} (ae^{-j\Omega})^n$$

$$\overset{(D.23)}{=} \frac{a\,e^{j\Omega}}{1-a\,e^{j\Omega}} + \frac{1}{1-a\,e^{-j\Omega}} \overset{(D.20)}{=} \frac{1-a^2}{1-2a\cos\Omega + a^2} : \text{real-valued} \tag{E3.3.1}$$

whose magnitudes with $a = \pm 0.5$ and ± 0.8 are depicted in Fig. 3.3.

Example 3.4 DTFT of the Unit Sample (Impulse) Sequence

For the impulse (unit sample) sequence $\delta[n]$, we can apply Eq. (3.1.1) to get its DTFT as

$$D(\Omega) = \sum_{n=-\infty}^{\infty} \delta[n]\, e^{-j\Omega n} = \delta[n]\, e^{-j\Omega n}\Big|_{n=0} = 1 \tag{E3.4.1}$$

As with the continuous-time case (Eq. (E2.6.1) in Example 2.6), this implies that a discrete-time impulse signal has a flat or white spectrum, which is evenly distributed over all digital frequencies.

(cf.) Very interestingly, applying the IDTFT formula (3.1.3) to (E3.4.1) yields a useful expression of the unit sample (impulse) sequence.

$$\delta[n] \overset{(3.1.3)}{=} \frac{1}{2\pi}\int_{2\pi} D(\Omega)\, e^{j\Omega n}\, d\Omega$$

$$\overset{(E3.4.1)}{=} \frac{1}{2\pi}\int_{-\pi}^{\pi} 1\, e^{j\Omega n} d\Omega \overset{(D.33)}{=} \frac{e^{j\pi n} - e^{-j\pi n}}{2\pi j\, n} \overset{(D.22)}{=} \frac{\sin n\pi}{n\,\pi} \tag{E3.4.2}$$

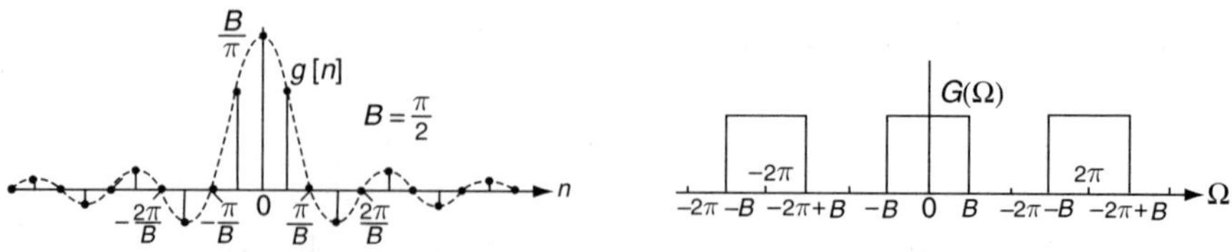

(a) The impulse response of an ideal LPF (b) The frequency response of an ideal LPF

Fig. 3.4 The impulse response and frequency response of an ideal LPF

Example 3.5 IDTFT of an Ideal Lowpass Filter Frequency Response

Let us consider the frequency response of an ideal lowpass filter (LPF) depicted in Fig. 3.4(b):

$$G(\Omega) = \begin{cases} 1 & \text{for } |\Omega - 2m\pi| \le B \le \pi \ (m: \text{ an integer}) \\ 0 & \text{elsewhere} \end{cases} \tag{E3.5.1}$$

We can take the IDTFT (inverse discrete-time Fourier transform) (3.1.3) to get the impulse response of the LPF, which is shown in Fig. 3.4(a), as follows:

$$g[n] = \mathcal{F}^{-1}\{G(\Omega)\} \overset{(3.1.3)}{=} \frac{1}{2\pi}\int_{-B}^{B} G(\Omega)\, e^{j\Omega n}\, d\Omega \underset{(D.33)}{\overset{(E3.5.1)}{=}} \frac{1}{2\pi j n}(e^{jBn} - e^{-jBn})$$

$$= \frac{\sin(Bn)}{\pi n} = \frac{B}{\pi}\text{sinc}\left(\frac{Bn}{\pi}\right) \tag{E3.5.2}$$

3.1.3 DTFT of Periodic Sequences

In analogy with the arguments in Sect. 2.3 for deriving Eq. (2.3.2) as the generalized CTFT of continuous-time periodic functions, we begin with applying the IDTFT formula (3.1.3) to an impulse train type spectrum

$$X(\Omega) = \sum_{i=-\infty}^{\infty} 2\pi\delta(\Omega - \Omega_0 - 2\pi\, i)$$

to get a complex sinusoidal sequence as

$$x[n] = \mathcal{F}^{-1}\{X(\Omega)\} \overset{(3.1.3)}{=} \frac{1}{2\pi}\int_{2\pi} X(\Omega)\, e^{j\Omega n} d\Omega$$

$$= \int_{2\pi} \delta(\Omega - \Omega_0)\, e^{j\Omega n}\, d\Omega \overset{(1.1.25)}{=} e^{j\Omega_0 n}$$

This implies the following DTFT relation:

$$e^{j\Omega_0 n} \overset{\mathcal{F}}{\leftrightarrow} X(\Omega) = \sum_{i=-\infty}^{\infty} 2\pi\delta(\Omega - \Omega_0 - 2\pi\, i) \tag{3.1.5}$$

Example 3.6 DTFT of a Constant Sequence

Let us consider a constant sequence $c[n] = 1$. Noting that this is a kind of periodic function obtained by substituting $\Omega_0 = 0$ into the LHS of DTFT relation (3.1.5), we can substitute $\Omega_0 = 0$ into the RHS of that DTFT relation to obtain the DTFT of $c[n] = 1$ as

$$c[n] = 1 \underset{(3.1.5)\text{ with }\Omega_0=0}{\overset{\mathcal{F}}{\leftrightarrow}} C(\Omega) = \sum_{i=-\infty}^{\infty} 2\pi\delta(\Omega - 2\pi\, i) \tag{E3.6.1}$$

Example 3.7 DTFT of a Sine/Cosine Sequence

(a) For $\sin(\Omega_0 n) \overset{(D.22)}{=} (e^{j\Omega_0 n} - e^{-j\Omega_0 n})/j2$, we use Eq. (3.1.5) to get its DTFT as

$$\sin(\Omega_0 n) \overset{\mathcal{F}}{\leftrightarrow} j\pi \sum_{i=-\infty}^{\infty} (\delta(\Omega + \Omega_0 - 2\pi\, i) - \delta(\Omega - \Omega_0 - 2\pi\, i)) \tag{E3.7.1}$$

(b) For $\cos(\Omega_0 n) \overset{(D.21)}{=} (e^{j\Omega_0 n} + e^{-j\Omega_0 n})/2$, we use Eq. (3.1.5) to get its DTFT as

$$\cos(\Omega_0 n) \overset{\mathcal{F}}{\leftrightarrow} \pi \sum_{i=-\infty}^{\infty} (\delta(\Omega + \Omega_0 - 2\pi i) + \delta(\Omega - \Omega_0 - 2\pi i)) \tag{E3.7.2}$$

Example 3.8 DTFT of the Unit Step Sequence

Similarly to Example 2.8 for deriving the CTFT of the unit step function, we first decompose the unit step sequence $u_s[n]$ into the sum of an even sequence and an odd sequence as

$$u_s[n] = u_e[n] + u_o[n] \tag{E3.8.1}$$

where

$$u_e[n] = \frac{1}{2}(u_s[n] + u_s[-n]) = \frac{1}{2}(1 + \delta[n]) \tag{E3.8.2}$$

$$u_o[n] = \frac{1}{2}(u_s[n] - u_s[-n]) = \frac{1}{2}\text{sign}(n) \tag{E3.8.3}$$

Then we can take the DTFT of the even and odd parts as

$$U_e(\Omega) = \mathcal{F}\{u_e[n]\} = \frac{1}{2}\mathcal{F}\{\delta[n]\} + \mathcal{F}\left\{\frac{1}{2}\right\}$$

$$\overset{(E3.4.1),(E3.6.1)}{=} \frac{1}{2} + \pi \sum_{i=-\infty}^{\infty} \delta(\Omega - 2\pi\, i) \tag{E3.8.4}$$

$$U_o(\Omega) = \mathcal{F}\{u_o[n]\} = \frac{1}{2}(\mathcal{F}\{u_s[n]\} - \mathcal{F}\{u_s[-n]\})$$

$$= \frac{1}{2}\left(\sum_{n=1}^{\infty} 1\, e^{-j\Omega n} - \sum_{n=-\infty}^{-1} 1\, e^{-j\Omega n}\right) = \frac{1}{2}\sum_{n=1}^{\infty} (e^{-j\Omega n} - e^{j\Omega n})$$

$$\overset{(D.23)}{=} \frac{1}{2}\left(\frac{e^{-j\Omega}}{1 - e^{-j\Omega}} - \frac{e^{j\Omega}}{1 - e^{j\Omega}}\right) \overset{(D.20)}{=} \frac{-j\sin\Omega}{2(1 - \cos\Omega)} \tag{E3.8.5}$$

Now we add these two results to obtain the Fourier transform of the unit step sequence as

$$\mathcal{F}\{u_s[n]\} = \frac{1}{2}\left(\frac{e^{-j\Omega}}{1-e^{-j\Omega}} + \frac{1}{1-e^{-j\Omega}}\right) + \frac{1}{2} + \pi \sum\nolimits_{i=-\infty}^{\infty} \delta(\Omega - 2\pi\, i)$$
$$= \frac{1}{1-e^{-j\Omega}} + \pi \sum\nolimits_{i=-\infty}^{\infty} \delta(\Omega - 2\pi\, i) \qquad \text{(E3.8.6)}$$

3.2 Properties of the Discrete-Time Fourier Transform

As with the continuous-time Fourier transform (CTFT), there are many properties of DTFT that provide us with further insight into the transform and can be used for reducing the computational complexity associated with it. Noting that there are striking similarities with the case of CTFT, we will simply state the properties unless their derivations and interpretations differ from those with the continuous-time case.

3.2.1 Periodicity

Since the DTFT $X(\Omega)$ defined by Eq. (3.1.1) is a function of $e^{j\Omega}$, it is always periodic with period 2π in Ω:

$$X(\Omega) = X[e^{j\Omega}] = X(\Omega + 2m\pi) \text{ for any integer } m \qquad (3.2.1)$$

(cf.) The periodicity lets us pay attention to the DTFT only for its one period, say, $-\pi \le \Omega \le \pi$.

3.2.2 Linearity

With $\mathcal{F}\{x[n]\} = X(\Omega)$ and $\mathcal{F}\{y[n]\} = Y(\Omega)$, we have

$$a\,x[n] + b\,y[n] \overset{\mathcal{F}}{\leftrightarrow} a\,X(\Omega) + b\,Y(\Omega), \qquad (3.2.2)$$

which implies that the DTFT of a linear combination of many sequences is the same linear combination of the individual DTFTs.

3.2.3 (Conjugate) Symmetry

In general, the DTFT has the time reversal property:

$$\mathcal{F}\{x[-n]\} \overset{(3.1.1)}{=} \sum\nolimits_{n=-\infty}^{\infty} x[-n]\,e^{-j\Omega n} \overset{-n=m}{=} \sum\nolimits_{m=-\infty}^{\infty} x[m]\,e^{-j(-\Omega)m} \overset{(3.1.1)}{=} X(-\Omega)$$

$$x[-n] \overset{\mathcal{F}}{\leftrightarrow} X(-\Omega) \qquad (3.2.3)$$

In case $x[n]$ is a real-valued sequence, we have

$$X(-\Omega) \overset{(3.1.1)}{=} \sum_{n=-\infty}^{\infty} x[n]\, e^{-j(-\Omega)n}$$
$$\overset{(2.2.1a)}{=} \sum_{n=-\infty}^{\infty} x[n]\, e^{-(-j)\Omega n} = X^*(\Omega) \text{ (complex conjugate of } X(\Omega))$$

or equivalently,

$$\mathrm{Re}\{X(-\Omega)\} + j\,\mathrm{Im}\{X(-\Omega)\} = \mathrm{Re}\{X(\Omega)\} - j\,\mathrm{Im}\{X(\Omega)\};$$
$$|X(-\Omega)|\angle X(-\Omega) = |X(\Omega)|\angle - X(\Omega) \tag{3.2.4}$$

This implies that the magnitude/phase of the DTFT of a real-valued sequence is an even/odd function of frequency Ω.

Also in analogy with Eq. (2.5.5) for the CTFT, we have

$$\text{even and real-valued } x_e[n] \overset{\mathcal{F}}{\leftrightarrow} \mathrm{Re}\{X(\Omega)\} \text{ even and real-valued} \tag{3.2.5a}$$

$$\text{odd and real-valued } x_o[n] \overset{\mathcal{F}}{\leftrightarrow} j\,\mathrm{Im}\{X(\Omega)\} \text{ odd and imaginary-valued} \tag{3.2.5b}$$

3.2.4 Time/Frequency Shifting (Real/Complex Translation)

The DTFT has the time-shifting and frequency-shifting properties as

$$x[n-n_1] \overset{\mathcal{F}}{\leftrightarrow} X(\Omega)\,e^{-j\Omega n_1} = X(\Omega)\angle - n_1\Omega \tag{3.2.6}$$

$$x[n]e^{j\Omega_1 n} \overset{\mathcal{F}}{\leftrightarrow} X(\Omega - \Omega_1) \tag{3.2.7}$$

3.2.5 Real Convolution

The DTFT has the convolution property

$$y[n] = x[n] * g[n] \overset{\mathcal{F}}{\leftrightarrow} Y(\Omega) = X(\Omega)\,G(\Omega) \tag{3.2.8}$$

which can be derived in the same way with Eq. (2.5.11). This is very useful for describing the input-output relationship of a discrete-time LTI system with the input $x[n]$, the output $y[n]$, and the impulse response $g[n]$ where $G(\Omega) = \mathcal{F}\{g[n]\}$ is called the frequency response of the system.

3.2.6 Complex Convolution (Modulation/Windowing)

In analogy with Eq. (2.5.14) for the CTFT, the DTFT also has the complex convolution (modulation) property as

$$y[n] = x[n]\,m[n] \overset{\mathcal{F}}{\leftrightarrow} Y(\Omega) = \frac{1}{2\pi} X(\Omega) \underset{2\pi}{*} M(\Omega) \tag{3.2.9}$$

where $\underset{2\pi}{*}$ denotes a circular or periodic convolution with period 2π.

Example 3.9 Effect of Rectangular Windowing on the DTFT of a Cosine Wave

From Eqs. (E3.7.2) and (E3.1.2), we can write the DTFTs of a cosine wave $x[n] = \cos(\Omega_0 n)$ and an even rectangular pulse sequence $r'_{2M+1}[n]$ of duration $2M+1$ as

$$X(\Omega) = \text{DTFT}\{\cos(\Omega_0 n)\}$$

$$\stackrel{\text{(E3.7.2)}}{=} \pi \sum_{i=-\infty}^{\infty} (\delta(\Omega + \Omega_0 - 2\pi\, i) + \delta(\Omega - \Omega_0 - 2\pi\, i)) \quad \text{(E3.9.1)}$$

$$R'_{2M+1}(\Omega) = \text{DTFT}\{r'_{2M+1}[n]\} \stackrel{\text{(E3.1.2)}}{=} \frac{\sin(\Omega(2M+1)/2)}{\sin(\Omega/2)} \quad \text{(E3.9.2)}$$

We can use the complex convolution property (3.2.9) to find the DTFT of a rectangular-windowed cosine wave $y[n] = \cos(\Omega_0 n) r'_{2M+1}[n]$ as

$$Y(\Omega) = \text{DTFT}\{y[n]\} = \text{DTFT}\{\cos(\Omega_0 n) r'_{2M+1}[n]\} \stackrel{(3.2.9)}{=} \frac{1}{2\pi} X(\Omega) \underset{2\pi}{*} R'_{2M+1}(\Omega)$$

$$\stackrel{\text{(E3.9.1)}}{=} \frac{1}{2\pi}\pi \sum_{i=-\infty}^{\infty} (\delta(\Omega + \Omega_0 - 2\pi\, i) + \delta(\Omega - \Omega_0 - 2\pi\, i)) \underset{2\pi}{*} R'_{2M+1}(\Omega)$$

$$\stackrel{\text{(D.37)}}{=} \frac{1}{2}(R'_{2M+1}(\Omega + \Omega_0 - 2\pi\, i) + R'_{2M+1}(\Omega - \Omega_0 - 2\pi\, i))$$

$$\stackrel{\text{(E3.9.2)}}{=} \frac{1}{2}\left(\frac{\sin((\Omega + \Omega_0 - 2\pi\, i)(2M+1)/2)}{\sin((\Omega + \Omega_0 - 2\pi\, i)/2)} + \frac{\sin((\Omega - \Omega_0 - 2\pi\, i)(2M+1)/2)}{\sin((\Omega - \Omega_0 - 2\pi\, i)/2)}\right) \quad \text{(E3.9.3)}$$

which is depicted together with $X(\Omega)$ (E3.9.1) and $R'_{2M+1}(\Omega)$ (E3.9.2) in Fig. 3.5. Compared with the spectrum $X(\Omega)$ of the cosine wave (Fig. 3.5(b1)), the spectrum $Y(\Omega)$ of the rectangular-windowed cosine wave (Fig. 3.5(b3)) has many side lobe ripples (with low amplitude) besides the two main peaks, which is interpreted as the *spectral leakage* due to the rectangular windowing.

Example 3.10 Impulse Response and Frequency Response of an FIR LPF (Lowpass Filter)

We can use Eq. (E3.5.2) to write the impulse response of an ideal LPF with bandwidth $B = \pi/4$ as

$$g[n] \stackrel{\text{(E3.5.2)}}{=} \left.\frac{\sin(Bn)}{\pi\, n}\right|_{B=\pi/4} = \frac{1}{4}\text{sinc}\left(\frac{n}{4}\right) \quad \text{(E3.10.1)}$$

which has an infinite duration so that it cannot be implemented by an FIR filter. Thus, to implement it with an FIR filter, we need to truncate the impulse response,

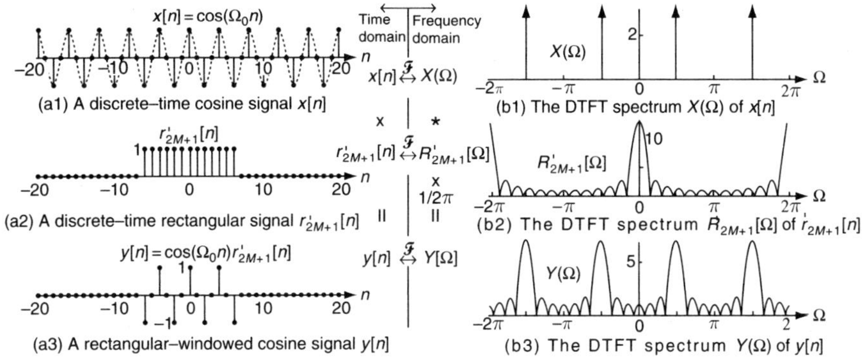

Fig. 3.5 Effect of rectangular windowing on the DTFT spectrum

say, by multiplying the rectangular window

$$w[n] = \begin{cases} 1 & \text{for } -3 \le n \le 3 \\ 0 & \text{elsewhere} \end{cases} \tag{E3.10.2}$$

whose DTFT is

$$W(\Omega) \overset{\text{(E3.1.2)}}{=} \frac{\sin(7\Omega/2)}{\sin(\Omega/2)} \tag{E3.10.3}$$

Then we obtain the windowed impulse response of an FIR LPF as

$$g_w[n] = g[n]\, w[n] \tag{E3.10.4}$$

whose DTFT is

$$G_w(\Omega) \underset{\text{complex convolution property}}{\overset{(3.2.9)}{=}} \frac{1}{2\pi} G(\Omega) \underset{2\pi}{*} W(\Omega) \tag{E3.10.5}$$

This is the frequency response of the FIR LPF. Figure 3.6 shows the impulse and frequency responses of the ideal and FIR LPFs together with the rectangular window sequence and its DTFT spectrum.

(cf.) Note that the frequency response of the FIR LPF (Fig. 3.6(b3)) has smooth transition in contrast with the sharp transition of the frequency response of the ideal LPF (Fig. 3.6(b1)).

We can run the following program "sig03e10.m" to plot Fig. 3.6 and compare the two DTFTs of the windowed impulse response $g_7[n]$, one obtained by using the DTFT formula (3.1.1) or the MATLAB routine "`DTFT()`" and one obtained by

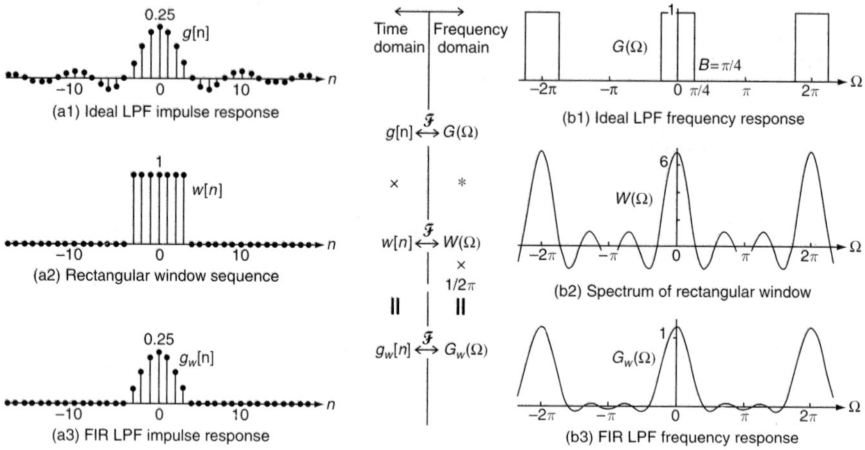

Fig. 3.6 Effect of windowing on the spectrum or frequency response

using the circular convolution (E3.10.5) or the MATLAB routine "`conv_circular()`".

```
%sig03e10.m
% Fig. 3.6: Windowing Effect on Impulse/Frequency Response of an ideal LPF
clear, clf
n= [-20: 20]; g= sinc(n/4)/4;
N=400; f=[-400:400]/N +1e-10; W=2*pi*f;
global P D
D=pi/2; P=2*pi; G=rectangular_wave(W);
subplot(321), stem(n,g,'.'), subplot(322), plot(f,G)
M=3; wdw= [zeros(1,20-M) ones(1,2*M+1) zeros(1,20-M)];
Wdw= DTFT(wdw,W,-20); % DTFT Wdw(W) of wdw[n] - Appendix E for DTFT
Wdw_t= sin(W*(2*M+1)/2)./sin(W/2); % Eq.(E3.1.2)
discrepancy_between_DTFT_and_E3_1_2 = norm(Wdw-Wdw_t)
subplot(323), stem(n,wdw,'.'), subplot(324), plot(f,real(Wdw))
gw= g.*wdw;
Gw= DTFT(gw,W,n(1)); % DTFT spectrum Gw(W) of gw[n]
Gw_1P= conv_circular(G,Wdw,N);
discrepancy_between_circonv_and_DTFT= norm(Gw(1:N)-Gw_1P)/norm(Gw_1P)
subplot(325), stem(n,gw,'.')
subplot(326), plot(f,real(Gw)), hold on
plot(f(1:N),real(Gw_1P),'r')
```

```
function z=conv_circular(x,y,N)
% Circular convolution z(n)= (1/N) sum_m=0^N-1 x(m)*y(n-m)
if nargin<3, N=min(length(x),length(y)); end
x=x(1:N); y=y(1:N); y_circulated= fliplr(y);
for n=1:N
   y_circulated= [y_circulated(N) y_circulated(1:N-1)];
   z(n)= x*y_circulated'/N;
end
```

3.2.7 Differencing and Summation in Time

As with the differentiation and integration property of the CTFT, the DTFT has the following differencing and summation properties:

$$x[n] - x[n-1] \overset{\mathcal{F}}{\leftrightarrow} (1 - e^{-j\Omega})\, X(\Omega) \tag{3.2.10}$$

$$\sum_{m=-\infty}^{n} x[m] = x[n] * u_s[n] \overset{\mathcal{F}}{\leftrightarrow}$$
$$\frac{1}{1 - e^{-j\Omega}} X(\Omega) + \pi X(0) \sum_{i=-\infty}^{\infty} \delta(\Omega - 2\pi\, i) \tag{3.2.11}$$

3.2.8 Frequency Differentiation

By differentiating both sides of the DTFT formula (3.1.1) w.r.t. Ω, we obtain

$$\frac{dX(\Omega)}{d\Omega} \overset{(3.1.1)}{=} -\sum_{n=-\infty}^{\infty} j\, n\, x[n]\, e^{-j\Omega n}$$

which yields the frequency-differentiation property of the DTFT as

$$n\, x[n] \overset{\mathcal{F}}{\leftrightarrow} j \frac{dX(\Omega)}{d\Omega} \tag{3.2.12}$$

This means that multiplication by n in the time domain results in differentiation w.r.t. Ω and multiplication by j in the frequency domain.

3.2.9 Time and Frequency Scaling

In the continuous-time case, we have Eq. (2.5.21) as

$$x(at) \overset{\mathcal{F}}{\leftrightarrow} \frac{1}{|a|} X\left(\frac{\omega}{a}\right)$$

However, it is not so simple to define $x[an]$ because of the following reasons:

- If a is not an integer, say, $a = 0.5$, then $x[0.5\, n]|_{n=1} = x[0.5]$ is indeterminate.
- If a is an integer, say, $a = 2$, then it does not merely speed up $x[n]$, but takes the even-indexed samples of $x[n]$.

Thus we define a time-scaled version of $x[n]$ as

$$x_{(K)}[n] = \begin{cases} x[n/K] & \text{for } n = Kr \text{ (a multiple of } K\text{) with some integer } r \\ 0 & \text{elsewhere} \end{cases} \tag{3.2.13}$$

which is obtained by placing $(K-1)$ zeros between successive samples of $x[n]$. Then we apply the DTFT formula (3.1.1) to get

$$X_{(K)}(\Omega) \overset{(3.1.1)}{=} \sum_{n=-\infty}^{\infty} x_{(K)}[n]e^{-j\Omega n} \underset{\text{with } n=Kr}{\overset{(3.2.13)}{=}} \sum_{r=-\infty}^{\infty} x[r]e^{-j\Omega K r} = X(K\Omega) \tag{3.2.14}$$

3.2.10 Parseval's Relation (Rayleigh Theorem)

If $x[n]$ has finite energy and the DTFT $X(\Omega) = \mathcal{F}\{x[n]\}$, then we have

$$\sum_{n=-\infty}^{\infty} |x[n]|^2 = \frac{1}{2\pi}\int_{2\pi} |X(\Omega)|^2 d\,\Omega \tag{3.2.15}$$

where $|X(\Omega)|^2$ is called the *energy-density spectrum* of the signal $x[n]$.

3.3 Polar Representation and Graphical Plot of DTFT

Similarly to the continuous-time case, we can write the polar representation of the DTFT as

$$X(\Omega) = |X(\Omega)|\angle X(\Omega)$$

If $x[n]$ is real, then its DTFT $X(\Omega) = \mathcal{F}\{x[n]\}$ has the following properties:

- $X(\Omega)$ is periodic with period 2π in Ω.
- The magnitude $|X(\Omega)|$ is an even function of Ω and is symmetric about $\Omega = m\pi$ (m: an integer).
- The phase $\angle X(\Omega)$ is an odd function of Ω and is anti-symmetric about $\Omega = m\pi$ (m: an integer).

Note that all the information about the DTFT of a real-valued sequence is contained in the frequency range $[0, \pi]$ since the portion for other ranges can be determined from that for $[0, \pi]$ using symmetry and periodicity. Consequently, we usually plot the spectrum for $0 \le \Omega \le \pi$ only.

Remark 3.3 Phase Jumps in the DTFT Phase Spectrum

From the phase spectra shown in Fig. 3.1(c1)–(c2) and 3.7(c1)–(c2), it can be observed that there are two occasions for which the phase spectrum has discontinuities or jumps:

- A jump of $\pm 2\pi$ occurs to maintain the phase value within the principal range of $[-\pi, +\pi]$.
- A jump of $\pm\pi$ occurs when the sign of $X(\Omega)$ changes.

The sign of phase jump is chosen in such a way that the resulting phase spectrum is odd or antisymmetric and lies in the principal range $[-\pi, +\pi]$ after the jump.

Remark 3.4 The DTFT Magnitude/Phase Spectra of Symmetric Sequences

(1) Especially for anti-symmetric sequences, their magnitude spectrum is zero at $\Omega = 0$ (see Fig. 3.7(b1)–(b2)). This implies that the DC gain of digital filters having an anti-symmetric impulse response is zero so that they cannot be used as a lowpass filter.
(2) As for the sequences that are even/odd about some point, their DTFTs have linear phase $-M\Omega$ (proportional to the frequency) except for the phase jumps so that the DTFT phase spectra are piecewise linear. Also, symmetric/anti-symmetric sequences, that are just shifted versions of even/odd ones, preserve linear phase because shifting does not impair the linear phase characteristic (see Eq. (3.2.6)). This is illustrated by Examples 3.1, 3.11, and 3.12 and will be restated in Remark 4.8, Sect. 4.6.

Example 3.11 DTFT of an Odd Sequence
For an odd sequence

$$\begin{array}{lccccccccccc} n= & & -4 & -3 & -2 & -1 & 0 & 1 & 2 & 3 & 4 & 5 \\ x_1[n] = & \cdots & 0 & 1 & 2 & 1 & 0 & -1 & -2 & -1 & 0 & 0 & \cdots \end{array} \tag{E3.11.1}$$

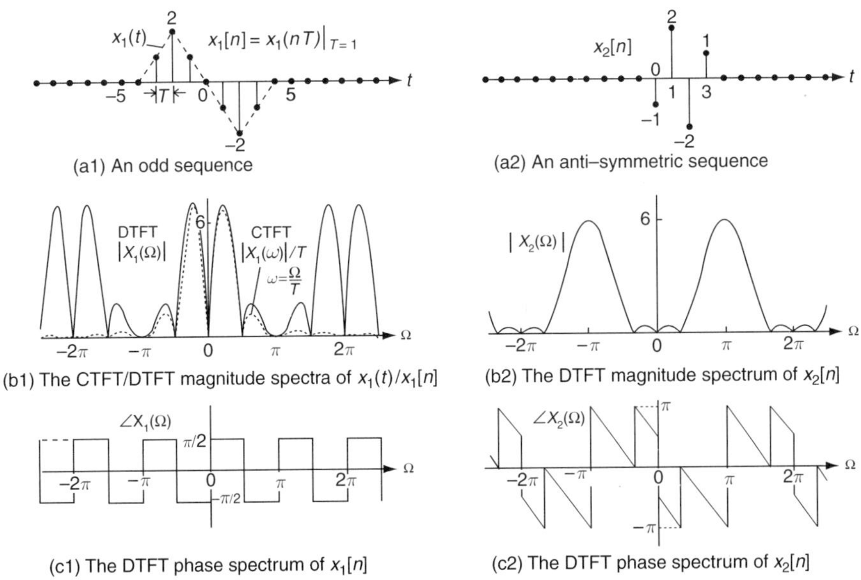

Fig. 3.7 The DTFT spectra of an odd sequence and an anti-symmetric sequence

we have its DTFT as

$$X_1(\Omega) = \sum\nolimits_{n=-\infty}^{\infty} x_1[n]\, e^{-j\Omega n}$$
$$= e^{j3\Omega} - e^{-j3\Omega} + 2(e^{j2\Omega} - e^{-j2\Omega}) + e^{j\Omega} - e^{-j\Omega} \overset{(D.22)}{=} j2\sin(3\Omega)$$
$$+ j4\sin(2\Omega) + j2\sin(\Omega) \overset{(D.12)}{=} 4\sin(2\Omega)(1+\cos\Omega)\, e^{j\pi/2} \quad \text{(E3.11.2)}$$

Noting that the continuous-time version of this sequence is apparently the sum of two opposite-sign shifted triangular pulses, we can compute its CTFT spectrum as

$$x_1(t) = 2(\lambda_D(t+2) - \lambda_D(t-2)) \underset{(2.5.6)\text{ with } t_1=2}{\overset{(E2.3.4)\text{ with } A=2, D=2}{\leftrightarrow}}$$

$$X_1(\omega) = 4\,\mathrm{sinc}^2\left(\frac{\omega}{\pi}\right)(e^{j2\omega} - e^{-j2\omega}) = j8\sin(2\omega)\,\mathrm{sinc}^2\left(\frac{\omega}{\pi}\right) \quad \text{(E3.11.3)}$$

This sequence and its DTFT magnitude/phase spectra are depicted in Fig. 3.7(a1) and (b1)/(c1), respectively. The CTFT magnitude spectrum (E3.11.3) (divided by sampling interval or period T) is plotted (in a dotted line) together with the DTFT magnitude spectrum in Fig. 3.7(b1) for comparison.

Example 3.12 DTFT of an Anti-Symmetric Sequence

For the sequence which is anti-symmetric about $n = 1.5$

$$\begin{array}{lccccccccc} n = & & -2 & -1 & 0 & 1 & 2 & 3 & 4 & 5 \\ x_2[n] = & \cdots & 0 & 0 & -1 & 2 & -2 & 1 & 0 & 0 \cdots \end{array} \quad \text{(E3.12.1)}$$

we have its DTFT as

$$X_2(\Omega) = \sum\nolimits_{n=-\infty}^{\infty} x_2[n]\, e^{-j\Omega n} = -1 + 2e^{-j\Omega} - 2e^{-j2\Omega} + e^{-j3\Omega}$$
$$= -(1 - e^{-j3\Omega}) + 2(e^{-j\Omega} - e^{-j2\Omega})$$
$$= -e^{-j3\Omega/2}(e^{j3\Omega/2} - e^{-j3\Omega/2}) + 2e^{-j3\Omega/2}(e^{j\Omega/2} - e^{-j\Omega/2})$$
$$\overset{(D.22)}{=} 2j\, e^{-j3\Omega/2}\left(2\sin\left(\frac{\Omega}{2}\right) - \sin\left(3\frac{\Omega}{2}\right)\right)$$
$$= 2\left(2\sin\left(\frac{\Omega}{2}\right) - \sin\left(3\frac{\Omega}{2}\right)\right) \angle\left(-3\frac{\Omega}{2} + \frac{\pi}{2}\right) \quad \text{(E3.12.2)}$$

This sequence and its magnitude/phase spectra are depicted in Fig. 3.7(a2) and (b2)/(c2), respectively.

3.4 Discrete Fourier Transform (DFT)

Before going into the DFT, let us consider the computational difference between the CTFS or CTFT for a periodic or an aperiodic continuous-time function $x(t)$ and the DTFT for the discrete-time sequence $x[n]$ consisting of the samples of $x(t)$. Since the CTFS (2.1.5b) and CTFT (2.2.1a) involving a continuous-time integral can be computed analytically for only $x(t)$ described by some basic functions and require a numerical integration method for general finite-duration time functions, they are not practical when we need to find the spectra of real-world signals. The DTFT (3.1.1) is computationally advantageous over the CTFS or CTFT since it does not involve any continuous-time integration. However, it can be computed, say, by using the MATLAB routine "`DTFT()`", only for finite-duration sequences unless the sequence is expressed by some basic sequences such as an exponential or sinc one. Therefore in practice, signals are truncated in various ways to obtain a set of finite number of samples [S-2]:

- A transient signal is assumed to be zero after it decays to a negligible amplitude.
- A periodic signal is sampled over an integral number of periods.
- A random signal is multiplied by a "window" of finite duration (short-time Fourier transform).

In any case, suppose we have a causal, finite-duration sequence $x[n]$ containing M samples. Then the DTFT formula (3.1.1) becomes

$$X(\Omega) = X[e^{j\Omega}] = \mathcal{F}\{x[n]\} = \sum\nolimits_{n=0}^{N-1} x[n]e^{-j\Omega n} \tag{3.4.1}$$

where N is set to be greater than or equal to M (the number of samples). Although it is easier to compute than the CTFS, CTFT, or the original DTFT, it may still annoy the computer because the frequency Ω takes continuous values for all that it is conventionally called the "digital" frequency. That is why we define and use another Fourier technique, called the N-point DFT (*discrete Fourier transform*), as follows:

$$X(k) = \mathrm{DFT}_N\{x[n]\} = \sum\nolimits_{n=0}^{N-1} x[n]e^{-j2\pi kn/N} = \sum\nolimits_{n=0}^{N-1} x[n]W_N^{kn} \tag{3.4.2}$$

$$\text{with } W_N = e^{-j2\pi/N} \text{ for } k = 0 : N-1$$

This is an N-point discrete-frequency sequence obtained by sampling one period of the finite version of DTFT, Eq. (3.4.1), conventionally at

$$\Omega = k\Omega_0 = k\frac{2\pi}{N} \text{ for } 0 \le k \le N-1$$

where N is called the DFT size and $\Omega_0 = 2\pi/N$ the *digital fundamental* or *resolution frequency* just like $\omega_0 = 2\pi/P$ in the CTFS. Note that these N frequency points are equally-spaced over the digital frequency range $[0, 2\pi)$ (excluding the upperbound 2π). Also we can write the IDFT (*inverse discrete Fourier transform*) formula as

$$x[n] = \text{IDFT}_N\{X(k)\} = \frac{1}{N}\sum_{k=0}^{N-1} X(k)\, e^{j2\pi k n/N}$$

$$= \frac{1}{N}\sum_{k=0}^{N-1} X(k) W_N^{-kn} \text{ for } n = 0 : N-1 \qquad (3.4.3)$$

Here, by substituting Eq. (3.4.2) into Eq. (3.4.3), we are going to demonstrate the validity of the IDFT formula, which amounts to verifying that the set of DFT samples $\{X(k), k = 0 : N-1\}$ conveys all the information contained in the set of time-domain samples $\{x[n], n = 0 : N-1\}$, that is, $x[n]$ can be recovered perfectly from $X(k)$:

$$x[n] \underset{(3.4.3)}{\overset{?}{=}} \frac{1}{N}\sum_{k=0}^{N-1} X(k)\, W_N^{-kn} \overset{(3.4.2)}{=} \frac{1}{N}\sum_{k=0}^{N-1}\sum_{m=0}^{N-1} x[m]\, W_N^{km}\, W_N^{-kn}$$

$$= \sum_{m=0}^{N-1} x[m]\left(\frac{1}{N}\sum_{k=0}^{N-1} W_N^{k(m-n)}\right)$$

$$= \sum_{m=0}^{N-1} x[m]\delta[(m-n) \bmod N] \overset{\text{O.K.}}{=} x[n]$$

(a mod b or a modulo b : the remainder after division of a by b)

where we have used the fact that

$$\frac{1}{N}\sum_{k=0}^{N-1} W_N^{k(m-n)} = \frac{1}{N}\sum_{k=0}^{N-1} e^{-j2\pi k(m-n)/N}$$

$$= \begin{cases} 1 & \text{if } m-n = i\,N (i : \text{an integer}) \\ 0 & \text{otherwise} \end{cases} \qquad (3.4.4)$$

There are two observations about the DFT pair (3.4.2) and (3.4.3):

- The DFT sequence $X(k)$ is periodic in k with period N:

$$X(k+mN) \overset{(3.4.2)}{=} \sum_{n=0}^{N-1} x[n]\, e^{-j2\pi(k+mN)n/N}$$

$$= \sum_{n=0}^{N-1} x[n]\, e^{-j2\pi k n/N} e^{-j2\pi m n} \overset{(3.4.2)}{=} X(k) \qquad (3.4.5)$$

- The IDFT sequence $x[n]$ is periodic in n with period N:

$$x\,[n+mN] \overset{(3.4.3)}{=} \frac{1}{N}\sum\nolimits_{k=0}^{N-1} X(k)\, e^{\,j2\pi k(n+mN)/N}$$

$$= \frac{1}{N}\sum\nolimits_{k=0}^{N-1} X(k)\, e^{\,j2\pi k\,n/N} e^{\,j\,2\pi k\,m} \overset{(3.4.3)}{=} x[n] \qquad (3.4.6)$$

The first observation (3.4.5) is no wonder since the DFT $X(k)$ inherits the periodicity of the DTFT $X(\Omega)$ as $X(k)$ originates from the samples of $X(\Omega)$ over its period. In contrast, the second observation is very surprising since there is no reason to believe that any sequence $x[n]$ is periodic and that with period equal to the DFT size N, which can be arbitrarily chosen as long as it covers the duration of $x[n]$. Because of this illusory periodicity, the shift involved with the time shifting property and the convolution related with the real convolution property of the DFT turn out to be not linear, but circular as will be discussed subsequently in Sects. 3.4.1 and 3.4.2. This headache is a kind of cost that we should pay in return for the computational convenience of DFT, which is discrete in both time and frequency so that it needs only multiplication and summation instead of integration. If only the sequence $x[n]$ is really periodic with period N, the DFT is a perfect Fourier technique for analyzing the frequency characteristic and thus it might well be referred to as the *discrete-time Fourier series* (DTFS). In this context, when we want to emphasize the periodicity of DFT, we call it DFS (*discrete Fourier series*) pair and write the DFS pair as

$$\tilde{X}(k) = \text{DFS}_N\{\tilde{x}\,[n]\} = \sum\nolimits_{n=0}^{N-1} \tilde{x}[n]\, e^{-j\,2\pi k\,n/N} = \sum\nolimits_{n=0}^{N-1} \tilde{x}[n]\, W_N^{kn} \qquad (3.4.7)$$

$$\tilde{x}\,[n] = \text{IDFS}_N\{\tilde{X}(k)\} = \frac{1}{N}\sum\nolimits_{k=0}^{N-1} \tilde{X}(k)\, e^{\,j\,2\pi k\,n/N} = \frac{1}{N}\sum\nolimits_{k=0}^{N-1} \tilde{X}(k)\, W_N^{-k\,n} \qquad (3.4.8)$$

3.4.1 Properties of the DFT

The DFT has properties similar to those of the CTFS such as the real convolution property (2.5.13) since the DFT regards a time-domain sample sequence $x[n]$ as one period of its periodic extension (with period equal to the DFT size N) that can be described by

$$x\,[n] = \tilde{x}_N[n]\, r_N[n] \qquad (3.4.9)$$

where $\tilde{x}_N[n]$ is the *periodic extension* of $x[n]$ with period N and $r_N[n]$ the rectangular pulse sequence with duration of $n = 0 : N-1$:

$$\tilde{x}\,[n] = \sum\nolimits_{m=-\infty}^{\infty} x[n+m\,N] = x\,[n \mod N] \qquad (3.4.10a)$$

$$r_N[n] = u_s[n] - u_s[n-N] = \begin{cases} 1 & \text{for } 0 \le n \le N-1 \\ 0 & \text{elsewhere} \end{cases} \qquad (3.4.10b)$$

On the other hand, the DFT sequence is born from the samples of the DTFT and thus it inherits most of the DTFT properties except that the real translation and convolution in the time domain are not linear, but circular as with the CTFS. Therefore we discuss only the two properties and summarize all the DFT properties in Appendix B.

3.4.1.1 Real/Complex Translation – Circular Shift in Time/Frequency

Figure 3.8(a), (b1), (b2), and (d) show a sequence $x[n]$ of finite duration $N = 4$, its periodic extension $\tilde{x}_N[n]$, its shifted version $\tilde{x}_N[n-M]$ $(M = 2)$, and its rectangular-windowed version $\tilde{x}_N[n - M]r_N[n]$, respectively. You can see more clearly from Fig. 3.8(a), (c1), (c2), and (d) that $\tilde{x}_N[n - M]r_N[n]$ is the circular-shifted (rotated) version of $x[n]$ where Fig. 3.8(c1) and (c2) show the circular representation of the finite-duration sequence $x[n]$ and its shifted or rotated version displayed around the circle with a circumference of N points. With this visual concept of *circular shift*, we can write the (circular) time-shifting property of DFS and DFT as

$$\tilde{x}_N[n - M]\,(\text{circular shift}) \overset{\text{DFS}}{\leftrightarrow} W_N^{Mk}\tilde{X}(k) \tag{3.4.11a}$$

$$\tilde{x}_N[n - M]r_N[n]\,(\text{one period of circular shift}) \overset{\text{DFT}}{\leftrightarrow} W_N^{Mk}X(k) \tag{3.4.11b}$$

We can also apply the duality between the time and frequency domains to write the (circular) frequency-shifting property of DFS/DFT:

$$W_N^{-Ln}\tilde{x}_N[n] \overset{\text{DFS}}{\leftrightarrow} \tilde{X}(k - L)\,(\text{circular shift}) \tag{3.4.12a}$$

$$W_N^{-Ln}x[n] \overset{\text{DFT}}{\leftrightarrow} \tilde{X}(k - L)r_N[k]\,(\text{one period of circular shift}) \tag{3.4.12b}$$

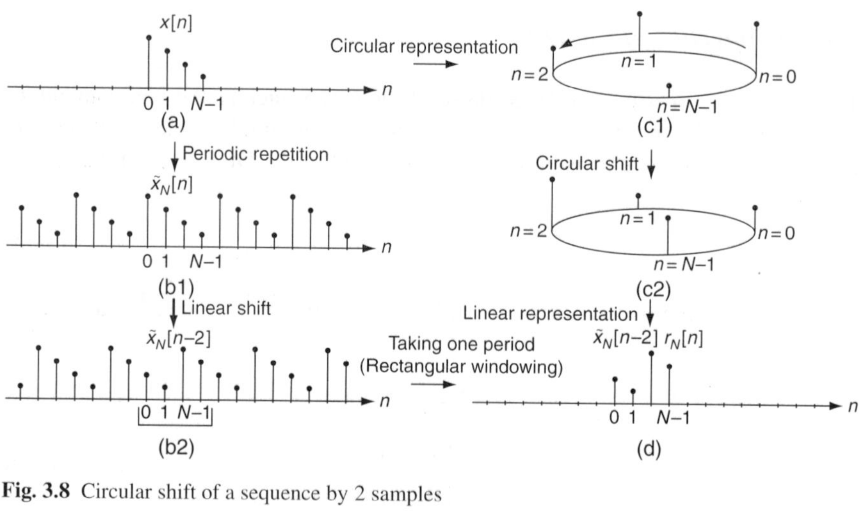

Fig. 3.8 Circular shift of a sequence by 2 samples

3.4.1.2 Real/Complex Convolution – Circular Convolution in Time/Frequency

Let us consider two sequences $x[n]$ and $y[n]$, both of duration $M \le N$ and each with DFT $X(k)$ and $Y(k)$, respectively. Also let us denote the periodic extensions of the two sequences and their DFSs by $\tilde{x}[n]$, $\tilde{y}[n]$, $\tilde{X}(k)$, and $\tilde{Y}(k)$, respectively. Then we can write the product of the two DFSs $\tilde{X}(k)$ and $\tilde{Y}(k)$ as

$$\tilde{X}(k) = \text{DFS}_N\{\tilde{x}[n]\} = \sum_{m=0}^{N-1} \tilde{x}[m]\, W_N^{km}$$

$$\tilde{Y}(k) = \text{DFS}_N\{\tilde{y}[n]\} = \sum_{r=0}^{N-1} \tilde{y}[r]\, W_N^{kr}$$

$$;\ \tilde{X}(k)\tilde{Y}(k) = \sum_{m=0}^{N-1} \sum_{r=0}^{N-1} \tilde{x}[m]\, \tilde{y}[r]\, W_N^{k(m+r)} \tag{3.4.13}$$

and compute its IDFS (inverse discrete Fourier series) as

$$\begin{aligned}
\text{IDFS}_N\left\{\tilde{X}(k)\tilde{Y}(k)\right\} &\overset{(3.4.8)}{=} \frac{1}{N}\sum_{k-0}^{N-1} \tilde{X}(k)\tilde{Y}(k)\, W_N^{-kn} \\
&\overset{(3.4.13)}{=} \sum_{m=0}^{N-1} \tilde{x}[m] \sum_{r=0}^{N-1} \tilde{y}[r]\left(\frac{1}{N}\sum_{k=0}^{N-1} W_N^{k(r-n+m)}\right) \\
&\overset{(3.4.4)}{=} \sum_{m=0}^{N-1} \tilde{x}[m] \sum_{r=0}^{N-1} \tilde{y}[r]\delta[r-(n-m)] \\
&\overset{(1.1.32)}{=} \sum_{m=0}^{N-1} \tilde{x}[m]\, \tilde{y}[n-m] = \tilde{x}[n] \underset{N}{*} \tilde{y}[n]
\end{aligned} \tag{3.4.14}$$

where $\underset{N}{*}$ denotes the circular or periodic convolution sum with period N. This implies the real convolution property of DFS/DFT as

$$\tilde{x}[n] \underset{N}{*} \tilde{y}[n]\,(\text{circular convolution}) \overset{\text{DFS}}{\leftrightarrow} \tilde{X}(k)\tilde{Y}(k) \tag{3.4.15a}$$

$$(\tilde{x}[n] \underset{N}{*} \tilde{y}[n])\, r_N[n]\,(\text{one period of circular convolution}) \overset{\text{DFT}}{\leftrightarrow} X(k)\,Y(k) \tag{3.4.15b}$$

In duality with this, we can interchange the time and frequency indices to write the complex convolution property as

$$\tilde{x}[n]\,\tilde{y}[n] \overset{\text{DFS}}{\leftrightarrow} \frac{1}{N}\sum_{i=0}^{N-1} \tilde{X}(i)\tilde{Y}(k-i) = \frac{1}{N}\tilde{X}(k) \underset{N}{*} \tilde{Y}(k)\,(\text{circular convolution}) \tag{3.4.16a}$$

$$x[n]\,y[n] \overset{\text{DFT}}{\leftrightarrow} \frac{1}{N}\left(\tilde{X}(k) \underset{N}{*} \tilde{Y}(k)\right) r_N[k]\,(\text{one period of circular convolution}) \tag{3.4.16b}$$

Note that the shift of a periodic sequence along the time/frequency axis actually causes a rotation, making the sequence values appear to wrap around from the beginning of the sequence to the end.

3.4.2 Linear Convolution with DFT

As discussed in the previous section, circular translation/convolution is natural with the DFT/DFS. However, in most applications of digital signal processing, linear convolution is necessary. As will be seen later, very efficient algorithms called FFT are available for computing the DFT of a finite-duration sequence. For this reason, it can be computationally advantageous to use the FFT algorithm for computing a convolution in such a way that the linear convolution corresponds to a single period of a circular convolution.

Specifically, suppose the signals $x[n]$ and $y[n]$ are of finite duration N_x and N_y, and defined on $n = 0 : N_x - 1$ and on $n = 0 : N_y - 1$, respectively. Then, the procedure for computing the linear convolution $z[n] = x[n] * y[n]$ is as below (see Fig. 3.9):

0) Extend the given sequences $x[n]$ and $y[n]$ to the duration of $N \geq N_x + N_y - 1$ by zero-padding where $N_x + N_y - 1$ is the expected length of the linear convolution of the two sequences (see Remark 1.5(1)).
   (cf.) Zero-padding a sequence means appending zeros to the end of the sequence.)
1) Compute $X(k) = \text{DFT}_N\{x[n]\}$ and $Y(k) = \text{DFT}_N\{y[n]\}$. (3.4.17a)
2) Multiply the two DFT sequences $Z(k) = X(k)Y(k)$ for $k = 0 : N - 1$. (3.4.17b)
3) Compute $z[n] = \text{IDFT}_N\{Z(k)\}$ where $z[n] = x[n] * y[n]$. (3.4.17c)

Example 3.13 Linear Convolution Using the DFT

We want to find the linear convolution $z[n] = x[n] * y[n]$ of the two sequences depicted in Fig. 3.10(a), where

$$x[n] = \begin{cases} 1 & \text{for } n = 0 \\ 0.5 & \text{for } n = 1 \\ 0 & \text{elsewhere} \end{cases} \quad \text{and} \quad y[n] = \begin{cases} 0.5 & \text{for } n = 0 \\ 1 & \text{for } n = 1 \\ 0 & \text{elsewhere} \end{cases} \tag{E3.13.1}$$

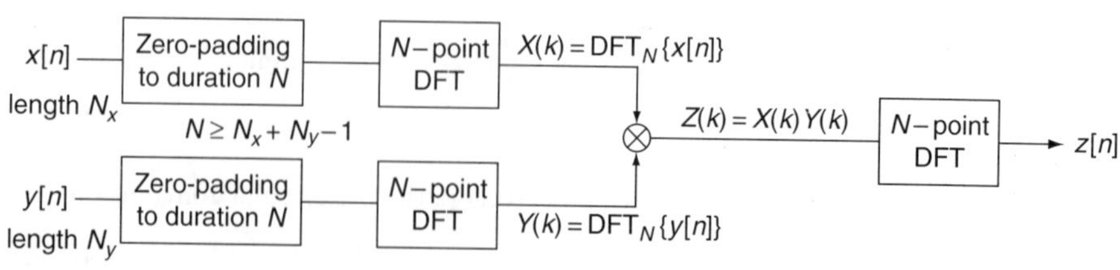

Fig. 3.9 Linear convolution using the DFT

(a) Time-domain approach (Fig. 3.10(a))

$$\begin{aligned} z[n] &= 0 \text{ for } n < 0 \text{ or } n \geq 3 (= N_x + N_y - 1 = 2 + 2 - 1) \\ z[0] &= x[0]\,y[0] + x[1]\,y[-1] = 0.5 \\ z[1] &= x[0]\,y[1] + x[1]\,y[0] = 1.25 \\ z[2] &= x[0]\,y[2] + x[1]\,y[1] = 0.5 \\ z[3] &= x[0]\,y[3] + x[1]\,y[2] = 0 \end{aligned} \quad \text{(E3.13.2)}$$

(b) Frequency-domain approach with 2-point DFT (Fig. 3.10(b))
We can use the DFT formula (3.4.2) with $N = 2$ to get the DFTs of the two sequences as

$$X_2(k) = \sum_{n=0}^{2-1} x[n] e^{-j2\pi kn/2} = 1 + 0.5(-1)^k = 1.5, \quad 0.5 \quad \text{for} \quad k = 0,\ 1$$

$$Y_2(k) = \sum_{n=0}^{2-1} y[n] e^{-j2\pi kn/2} = 0.5 + (-1)^k = 1.5, \quad -0.5 \quad \text{for} \quad k = 0,\ 1$$

so that

$$Z_2(k) = X_2(k)\,Y_2(k) = 2.25,\ -0.25 \quad \text{for} \quad k = 0,\ 1 \quad \text{(E3.13.3)}$$

Then we use the IDFT formula (3.4.3) to get

$$\begin{aligned} z_2[n] &= \frac{1}{2} \sum_{k=0}^{2-1} Z_2(k) e^{j2\pi kn/2} \\ &= \frac{1}{2} \left(Z_2(0) + (-1)^n Z_2(1) \right) = 1, \quad 1.25 \quad \text{for} \quad n = 0, 1 \end{aligned} \quad \text{(E3.13.4)}$$

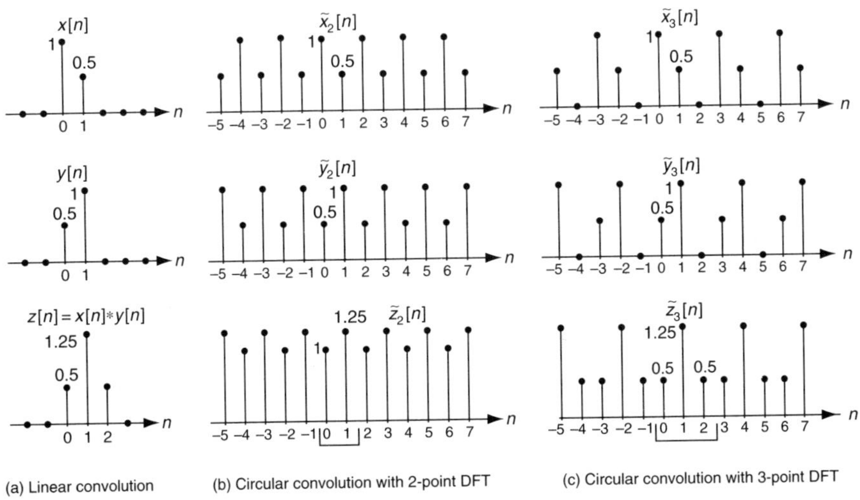

Fig. 3.10 Linear and circular convolutions with DFT

This corresponds to one period of the periodic extension of $z[n]$ (obtained in part (a)) with period $N = 2$ and is not equal to $z[n]$. This illustrates that *time-aliasing* effect might occur if the DFT size N is not sufficiently large.

(c) Frequency-domain approach with 3-point DFT (Fig. 3.10(c))
We can use the DFT formula (3.4.2) with $N = 3$ to get the DFTs of the two sequences as

$$X_3(k) = \sum_{n=0}^{3-1} x[n]e^{-j2\pi kn/3} = 1 + 0.5e^{-j2\pi k/3}$$
$$= 1.5, \quad 1 + 0.5e^{-j2\pi/3}, \ 1 + 0.5e^{-j4\pi/3} \quad \text{for } k = 0,\ 1,\ 2$$
$$Y_3(k) = \sum_{n=0}^{3-1} y[n]e^{-j2\pi kn/3} = 0.5 + e^{-j2\pi k/3}$$
$$= 1.5, \quad 0.5 + e^{-j2\pi/3}, \ 0.5 + e^{-j4\pi/3} \quad \text{for } k = 0,\ 1,\ 2$$

so that

$$Z_3(k) = X_3(k)Y_3(k) = 0.5 + 1.25e^{-j2\pi k/3} + 0.5e^{-j4\pi k/3} \tag{E3.13.5}$$

We can match this with the IDFT formula (3.4.3) to get

$$z_3[n] = 0.5,\ 1.25,\ 0.5 \quad \text{for } n = 0,\ 1,\ 2 \tag{E3.13.6}$$

This agrees with the linear convolution $z[n]$ obtained in part (a) for $n = 0 \sim 2$.

(d) MATLAB Program
We can run the following program "sig03e13.m" to apply the above procedure to get the linear convolution using the DFT. Note the following:

- The built-in MATLAB functions "`fft(x,N)`" and "`ifft(X,N)`" are used to compute the N-point DFT $X(k) = \text{DFT}_N\{x[n]\}$ and IDFT $x[n] = \text{IDFT}_N\{X(k)\}$, respectively.
- Depending on whether the length of the sequence given as the first input argument of "`fft(x,N)`" or "`ifft(X,N)`" is less or greater than the second input argument N, the sequence will be zero-padded or truncated so that its resulting length will be N.
- Note that, to exploit the computational efficiency of FFT (fast Fourier transform), the DFT size N should be chosen as a power of two greater than or equal to the length of the sequence.

```
%sig03e13.m
% Use 2-point/3-point DFT for computing a (linear) convolution
clear, clf
x= [1 0.5]; y= [0.5 1];
z= conv(x,y) % Linear convolution
N= 3; % DFT size
```

```
XN= fft(x,N); YN=fft(y,N);
ZN= XN.*YN;
zN= ifft(ZN) % Circular convolution expected to agree with the linear one
% Sometimes, you had better take the real part of IDFT result
% by using real(ifft(ZN)) to make sure of its being real-valued.
```

(cf.) The above procedure, which is depicted in Fig. 3.9 and illustrated in Example 3.13, paves the way to use the DFT for computing the linear convolution of two finite-duration sequences.

Remark 3.5 How to Choose the DFT Size N in Connection with Zero Padding

(1) In computing the DFT of a given sequence $x[n]$ of length M, we are free to choose the DFT size N, i.e., the number of sampled frequency points over $[0, 2\pi)$ (one period of digital frequency) as long as it is greater than or equal to M. Choosing $N > M$ corresponds to increasing the length of $x[n]$ by appending it with additional zero-valued samples – padding with zeros. This procedure called *zero padding* may be used to fill out the sequence length so that an $N = 2^L$-point FFT (fast Fourier transform) algorithm could be applied (see Sect. 3.6) or a linear convolution of two sequences could be performed without causing a time-aliasing problem (see Example 3.13). It can also be used to provide a better-looking display of the frequency contents of a finite-duration sequence by decreasing the digital frequency spacing (resolution frequency) $2\pi/N$ so that the discrete DFT spectrum $X(k)$ can look close to the continuous DTFT spectrum $X(\Omega)$ (see Example 3.14).
(2) As can be seen from the DTFT formula (3.1.1), zero padding does not alter the continuous spectrum $X(\Omega)$, but just decreases the interval width between successive discrete frequencies in the DFT spectrum. However, when the signal is not of finite duration, zero padding can lead to erroneous results. Conclusively, zero padding is justified only when a signal is of finite duration and has already been sampled over the entire range where it is nonzero.

3.4.3 DFT for Noncausal or Infinite-Duration Sequence

Let us consider the DFT formula (3.4.2):

$$X(k) = \text{DFT}_N\{x[n]\} = \sum_{n=0}^{N-1} x[n]e^{-j2\pi kn/N} \text{ for } k = 0 : N-1$$

This is defined for a causal, finite-duration sequence $x[n]$ so that we can cast it into a computer program without taking much heed of the negative indices. (C language prohibits using the negative indices for array and MATLAB does not accept even zero index.) Then, how can we get the DFT spectrum of noncausal or infinite-duration sequences? The answer is as follows:

- For a noncausal sequence, append the noncausal part to the end of the sequence after any necessary zero-padding is done so that the resulting sequence can be causal.
- For an infinite-duration sequence, set the time-duration $[0, N-1]$ so that the most significant part of the sequence can be covered. To increase the DFT size is also helpful in making the DFT close to the DTFT. Compare Fig. 3.11(b2)/(c2) with (b3)/(c3). Also compare Fig. 3.12.1(b1) with Fig. 3.12.2(b1).

Example 3.14 DFT of a Non-causal Pulse Sequence

Consider a sequence which is depicted in Fig. 3.11(a1). Shifting the noncausal part $x[-1] = 1/3$ into the end of the sequence yields a causal sequence

$$x_N[n] = \begin{cases} 1/3 & \text{for } n = 0,\ 1,\ N-1 \\ 0 & \text{elsewhere} \end{cases} \tag{E3.14.1}$$

where N is the DFT size. This corresponds to one period of the periodic extension $\tilde{x}_N[n]$ (see Fig. 3.11(a2) with $N = 4$ and (a3) with $N = 8$).

Then we can compute the N-point DFT of $x_N[n]$ as

$$\begin{aligned} X_N(k) &\overset{(3.4.2)}{=} \sum_{n=0}^{N-1} x_N[n]e^{-j2\pi kn/N} = \frac{1}{3}\overset{n=0}{e^{-j2\pi k0/N}} + \frac{1}{3}\overset{n=1}{e^{-j2\pi k1/N}} + \frac{1}{3}\overset{n=N-1}{e^{-j2\pi k(N-1)/N}} \\ &= \frac{1}{3}\left(1 + e^{-j2\pi k/N} + e^{-j2\pi k(N-1)/N}\right) = \frac{1}{3}\left(1 + e^{-j2\pi k/N} + e^{j2\pi k/N}\right) \\ &= \frac{1}{3}(1 + \cos(2\pi k/N)) \quad \text{for } k = 0 : N-1 \end{aligned} \tag{E3.14.2}$$

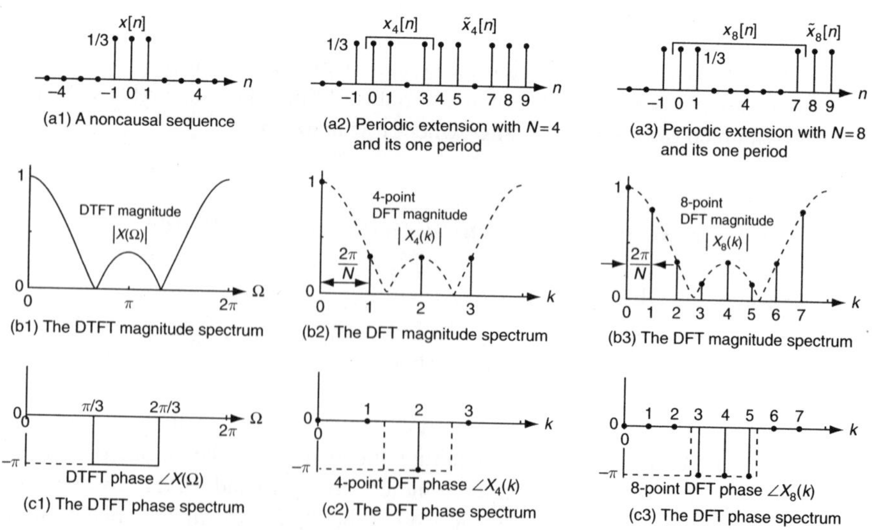

Fig. 3.11 DTFT and DFT for a noncausal sequence

Figure 3.11(b2)/(c2) and (b3)/(c3) show the magnitude/phase of this DFT with $N = 4$ and those with $N = 8$, respectively. Note the following:

- The overlapped DTFT spectra in dotted lines, obtained in Example 3.1 and plotted in Figs. 3.1 and 3.11(b1)/(c1), illustrate the fact that the DFT $X_N(k)$ are just the samples of the DTFT $X(\Omega)$ at $\Omega = k\Omega_0 = 2\pi k/N$ for $0 \le k \le N-1$ as long as the whole duration of the sequence is covered by $[0 : N-1]$.
- Figure 3.11(a2) and (a3) differ in the length of zero padding performed before appending the noncausal part to the end of the sequence. Comparing the corresponding DFT spectra in Figure 3.11(b2)/(c2) and (b3)/(c3), we see that longer zero-padding increases the DFT size and thus decreases the digital resolution frequency $\Omega_0 = 2\pi/N$ so that the DFT looks closer to the DTFT.

Example 3.15 DFT of an Infinite-Duration Sequence

Consider a real exponential sequence of infinite duration described by

$$x[n] = a^n u_s[n] \;(\,|a| < 1\,) \tag{E3.15.1}$$

This sequence is shown partially for $n = 0 : 7$ in Fig. 3.12.1(a1) and for $n = 0 : 15$ in Fig. 3.12.2(a1).

(a) The DTFT of this infinite-duration sequence was obtained in Example 3.2 as

$$X(\Omega) \overset{\text{(E3.2.1)}}{=} \frac{1}{1 - a\,e^{-j\Omega}} \tag{E3.15.2}$$

Thus we get the samples of the DTFT at $\Omega = k\Omega_0 = 2\pi k/N$ as

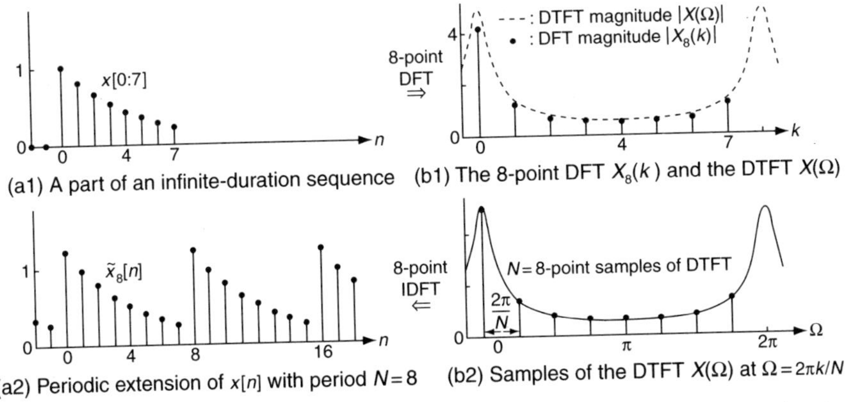

Fig. 3.12.1 Relationship between the 8-point DFT and the DTFT for an infinite-duration sequence

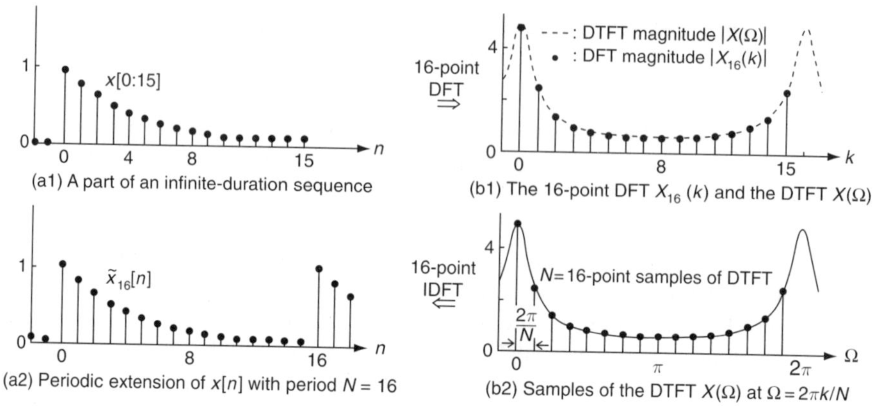

Fig. 3.12.2 Relationship between the 16-point DFT and the DTFT for an infinite-duration sequence

$$X(k\Omega_0) \overset{(E3.15.2)}{=} \frac{1}{1-ae^{-jk\Omega_0}} = \frac{1}{1-ae^{-j2\pi k/N}}$$

$$= \frac{1}{1-aW_N^k} \text{ with } W_N = e^{-j2\pi/N} \tag{E3.15.3}$$

(b) We can use the DFT formula (3.4.2) to find the N-point DFT of $x[n]$ for $n = 0 : N-1$:

$$X(k) \overset{(3.4.2)}{=} \sum_{n=0}^{N-1} x[n]W_N^{nk}$$

$$\overset{(E3.15.1)}{=} \sum_{n=0}^{N-1} a^n W_N^{nk} = \frac{1-a^N W_N^{kN}}{1-aW_N^k} = \frac{1-a^N}{1-aW_N^k} \text{ for } k = 0 : N-1 \tag{E3.15.4}$$

Comparing this with (E3.15.3) reveals the following:

- The DFT and the DTFT samples are not exactly the same for an infinite-duration sequence, while they conform with each other for a finite-duration sequence whose duration can somehow be covered in $[0 : N-1]$
- Larger DFT size N will make the DFT closer to the sampled DTFT. This can also be seen visually by comparing Figs. 3.12.1(b1) and 3.12.2(b1).

(c) One might wonder what the time-domain relationship between the DFT $X(k)$ and the sampled DTFT $X(k\Omega_0)$ is for the infinite-duration sequence. This curiosity seduces us to find the IDFT of $X(k\Omega_0)$ as

$$x_N[n] = \text{IDFT}\{X(k\Omega_0)\} \overset{(3.4.3)}{=} \frac{1}{N}\sum_{k=0}^{N-1} X(k\Omega_0)W_N^{-kn}$$

$$\overset{(E3.15.3)}{=} \frac{1}{N}\sum_{k=0}^{N-1} W_N^{-kn}\frac{1}{1-aW_N^k} \overset{(D.23)}{=} \frac{1}{N}\sum_{k=0}^{N-1} W_N^{-kn}\left\{\sum_{r=0}^{\infty} a^r W_N^{kr}\right\}$$

$$= \sum_{r=0}^{\infty} a^r \left(\frac{1}{N}\sum_{k=0}^{N-1} W_N^{k(r-n)}\right) \overset{(3.4.4)}{=} \sum_{i=0}^{\infty} a^{n+iN}$$

$$\overset{(D.23)}{=} \frac{a^n}{1-a^N} \text{ for } n=0:N-1 \underset{|a|<1}{\overset{N\to\infty}{\to}} x[n] = a^n u_s[n] \tag{E3.15.5}$$

This corresponds to one period of the periodic extension $\tilde{x}_N[n]$ of $x[n]$ with period N and it becomes closer to $x[n]$ as the DFT size N increases. Note the following:

- Just as the sampling of continuous signal $x(t)$ in the time domain results in the periodic extension of $X(\omega) = \mathcal{F}\{x(t)\}$ in the frequency domain (Eq. (E2.13.3)), so the sampling of continuous spectrum $X(\omega)$ or $X(\Omega)$ in the frequency domain results in the periodic extension $\tilde{x}_P(t)$ (of $x(t) = \mathcal{F}^{-1}\{X(\omega)\}$) or $\tilde{x}_N[n]$ (of $x[n] = \mathcal{F}^{-1}\{X(\Omega)\}$) in the time domain.
- Besides, just as shorter time-domain sampling interval (T) in the sampling of $x(t)$ to make $x[n] = x(nT)$ increases the frequency band on which the CTFT $X(\omega) = \mathcal{F}\{x(t)\}$ is close to the DTFT $X(\Omega) = \mathcal{F}\{x[n]\}$, so narrower frequency-domain sampling interval (ω_0/Ω_0) in the sampling of $X(\omega)/X(\Omega)$ to make $X(k\omega_0)/X(k\Omega_0)$ expands the time range on which $\tilde{x}_P(t)$ (with $P = 2\pi/\omega_0$) or $\tilde{x}_N[n]$ (with $N = 2\pi/\Omega_0$) is close to $x(t)$ or $x[n]$.
- However short the time-domain sampling interval in the sampling of $x(t)$ to make $x[n]$ may be, $X(\Omega) = \mathcal{F}\{x[n]\}$ for $-\pi \le \Omega \le \pi$ cannot be exactly the same as $X(\omega) = \mathcal{F}\{x(t)\}$ for $-\pi/T \le \omega \le \pi/T$ due to the frequency-aliasing unless $X(\omega)$ is strictly bandlimited. Likewise, however narrow the frequency-domain sampling interval in the sampling of $X(\omega)/X(\Omega)$ to make $X(k\omega_0)/X(k\Omega_0)$ may be, the corresponding periodic extension $\tilde{x}_P(t)/\tilde{x}_N[n]$ cannot be exactly the same as $x(t)/x[n]$ for one period of length P/N due to the time-aliasing unless $x(t)/x[n]$ is strictly time-limited.

Remark 3.6 The DFT for Noncausal/Infinite-Duration Sequences

The DFT pair (3.4.2) and (3.4.3) can be used to analyze the frequency characteristic of causal, finite-duration sequences. Then, how do we deal with noncausal or infinite-duration sequences?

(1) For a noncausal sequence, append the noncausal part to the end of the sequence after any necessary zero-padding is done so that the resulting sequence can be causal.
(2) For an infinite-duration sequence, set the time-duration $[0, N-1]$ so that the most significant part of the sequence can be covered. If the duration is shifted,

apply the time-shifting property to the DFT for obtaining the right phase spectrum. You can also increase the DFT size to make the DFT close to the DTFT, which accommodates infinite-duration sequences.

3.5 Relationship Among CTFS, CTFT, DTFT, and DFT

3.5.1 Relationship Between CTFS and DFT/DFS

To investigate the relationship between CTFS and DFT/DFS, suppose we have a continuous-time periodic signal $\tilde{x}_P(t)$ with period P and its discrete-time version $\tilde{x}_N[n] = \tilde{x}_P(nT)$, which is obtained by sampling $\tilde{x}_P(t)$ at the rate of N times per period P. Since $\tilde{x}_P(t)$ is a continuous-time periodic signal with period P, we can use Eq. (2.1.5a) to write its CTFS representation as

$$\tilde{x}_P(t) \overset{(2.1.5a)}{=} \frac{1}{P}\sum_{k=-\infty}^{\infty} X_k\, e^{jk\omega_0 t} \underset{\omega_0=2\pi/P=2\pi/NT}{\overset{P=NT}{=}} \frac{1}{NT}\sum_{k=-\infty}^{\infty} X_k\, e^{j2\pi kt/NT} \tag{3.5.1}$$

Substituting $t = nT$ into this equation and using the fact that $e^{j2\pi kn/N}$ is unique only for n mod N yields

$$\begin{aligned}\tilde{x}_P(nT) &\overset{(3.5.1)}{=} \frac{1}{NT}\sum_{k=-\infty}^{\infty} X_k e^{j2\pi knT/NT} = \frac{1}{NT}\sum_{k=-\infty}^{\infty} X_k e^{j2\pi kn/N} \\ &\quad (\text{since } e^{j2\pi kn/N} \text{ is unique only for } n \bmod N) \\ &= \frac{1}{NT}\sum_{k=0}^{N-1}\sum_{m=-\infty}^{\infty} X_{k+mN} e^{j\,2\pi(k+mN)n/N} \\ &= \frac{1}{N}\sum_{k=0}^{N-1}\left(\frac{1}{T}\sum_{m=-\infty}^{\infty} X_{k+mN}\right)e^{j\,2\pi kn/N}\end{aligned} \tag{3.5.2}$$

We can match this equation with the IDFS/IDFT formula (3.4.8)/(3.4.3) for $\tilde{x}_N[n] = \tilde{x}_P(nT)$

$$\tilde{x}_N[n] = \frac{1}{N}\sum_{k=0}^{N-1} \tilde{X}_N(k)\, e^{j\,2\pi k\,n/N}$$

to write the relation between the CTFS coefficients X_k of a periodic signal $\tilde{x}_P(t)$ (with period $P = NT$) and the N-point DFT/DFS coefficients $\tilde{X}_N(k)$ of $\tilde{x}_N[n] = \tilde{x}_P(nT)$ as

$$\tilde{X}_N(k) = \frac{1}{T}\sum_{m=-\infty}^{\infty} X_{k+mN} \tag{3.5.3}$$

This implies that the DFT/DFS of $\tilde{x}_N[n] = \tilde{x}_P(nT)$ is qualitatively the periodic extension of the CTFS of $\tilde{x}_P(t)$ (with period N in frequency index k), i.e., the sum of infinitely many shifted version of CTFS. This explains how the DFT/DFS strays from the CTFS because of *frequency-aliasing* unless the CTFS spectrum X_k is strictly limited within the low-frequency band of $(-(N/2)\omega_0, (N/2)\omega_0) = (-\pi/T, \pi/T)$ where the DFT size N equals the number of samples per period P and the fundamental frequency is $\omega_0 = 2\pi/P = 2\pi/NT$.

3.5.2 Relationship Between CTFT and DTFT

To investigate the relationship between CTFT and DTFT, suppose we have a continuous-time signal $x(t)$ and its discrete-time version $x[n] = x(nT)$. As a bridge between $x(t)$ and $x[n]$, let us consider the sampled version of $x(t)$ with sampling interval or period T as

$$x_*(t) = x(t)\delta_T(t)\ (\delta_T(t) = \sum\nolimits_{n=-\infty}^{\infty} \delta(t - nT) : \text{the impulse train}) \tag{3.5.4}$$

Noting that $x_*(t)$ is still a continuous-time signal, we can use Eq. (2.2.1a) to write its CTFT as

$$\begin{aligned} X_*(\omega) &\overset{(2.2.1a)}{=} \int_{-\infty}^{\infty} x_*(t)e^{-j\omega t}\,dt \overset{(3.5.4)}{=} \int_{-\infty}^{\infty} x(t)\sum\nolimits_{n=-\infty}^{\infty} \delta(t-nT)e^{-j\omega t}dt \\ &= \sum\nolimits_{n=-\infty}^{\infty} \int_{-\infty}^{\infty} x(t)e^{-j\omega t}\delta(t-nT)\,dt \overset{(1.1.25)}{=} \sum\nolimits_{n=-\infty}^{\infty} x(nT)e^{-j\omega nT} \\ &= \sum\nolimits_{n=-\infty}^{\infty} x[n]\,e^{-j\Omega n}\Big|_{\Omega=\omega T} \overset{(3.1.1)}{=} X_d(\Omega)|_{\Omega=\omega T} \end{aligned} \tag{3.5.5}$$

This implies that $X_d(\Omega) = \text{DTFT}\{x[n]\}$ and $X_*(\omega) = \text{CTFT}\{x_*(t)\}$ are essentially the same and that $X_d(\Omega)$ can be obtained from $X_*(\omega)$ via a variable substitution $\omega = \Omega/T$. On the other hand, we recall from Eq. (E2.13.3) that $X_*(\omega) = \text{CTFT}\{x_*(t)\}$ is expressed in terms of $X(\omega) = \text{CTFT}\{x(t)\}$ as

$$X_*(\omega) \overset{(E2.13.3)}{=} \frac{1}{T}\sum\nolimits_{m=-\infty}^{\infty} X(\omega + m\omega_s) \text{ with } \omega_s = \frac{2\pi}{T} \tag{3.5.6}$$

Combining these two equations (3.5.5) and (3.5.6), we can write the relation between the CTFT and the DTFT as

$$X_d(\Omega) \overset{(3.5.5)}{=} X_*(\omega)|_{\omega=\Omega/T} \overset{(3.5.6)}{=} \frac{1}{T}\sum\nolimits_{m=-\infty}^{\infty} X\left(\frac{\Omega}{T} + m\frac{2\pi}{T}\right) \tag{3.5.7}$$

where ω and Ω are called the analog and digital frequency, respectively.

This implies that the DTFT of $x[n] = x(nT)$ is qualitatively the periodic extension of the CTFT of $x(t)$ (with period $2\pi/T$ in analog frequency ω or 2π in digital frequency Ω), i.e., the sum of infinitely many shifted version of CTFT. This explains how the DTFT strays from the CTFT because of *frequency-aliasing* unless the CTFT spectrum $X(\omega)$ is strictly limited within the low-frequency band of $(-\pi/T, \pi/T)$ where T is the sampling interval of $x[n] = x(nT)$. Fig. 3.7(b1) illustrates the deviation of the DTFT spectrum from the CTFT spectrum caused by frequency-aliasing.

3.5.3 Relationship Among CTFS, CTFT, DTFT, and DFT/DFS

As stated in Remark 2.7(2) and illustrated in Fig. 2.8, the CTFS X_k's of a periodic function $\tilde{x}_P(t)$ are the samples of the CTFT $X(\omega)$ of the one-period function $x_P(t)$ at $k\omega_0 = 2\pi k/P$:

$$X_k \overset{(2.2.4)}{=} X(\omega)|_{\omega=k\omega_0=2\pi k/P} \tag{3.5.8}$$

Likewise, as discussed in Sect. 3.4 and illustrated in Fig. 3.11, the DFT/DFS $X(k)$'s of a periodic sequence $\tilde{x}_P[n]$ are the samples of the DTFT $X_d(\Omega)$ of the one-period sequence $x_P[n]$ at $k\Omega_0 = 2\pi k/N$:

$$X(k) \overset{(3.4.1)\&(3.4.2)}{=} X_d(\Omega)|_{\Omega=k\Omega_0=2\pi k/N} \tag{3.5.9}$$

Figure 3.13 shows the overall relationship among the CTFS, CTFT, DTFT, and DFT/DFS based on Eqs. (3.5.3), (3.5.7), (3.5.8), and (3.5.9). Figure 3.14 shows the CTFT, DTFT, DFT/DFS, and CTFS spectra for a continuous-time/discrete-time rectangular pulse or wave, presenting us with a more specific view of their relationship. Some observations are summarized in the following remark:

Remark 3.7 Relationship among the CTFS, CTFT, DTFT, and DTFS (DFT/DFS)

Figures 3.13 and 3.14 shows the overall relationship among the CTFS, CTFT, DTFT, and DTFS (DFT/DFS) from a bird's-eye point of view. The following observations and comparisons are made.

(1) Among the four Fourier spectra, the CTFS and CTFT are more desired than the DTFS and DTFT since all physical signals are continuous-time signals.

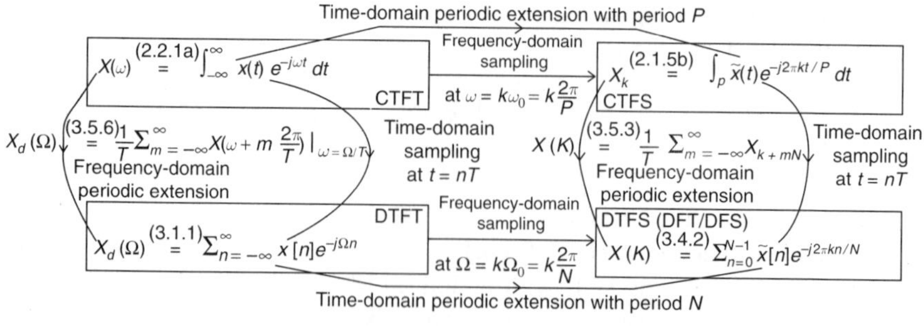

Fig. 3.13 Overall relationship among the CTFT, CTFS, DTFT, and DTFS (DFT/DFS)

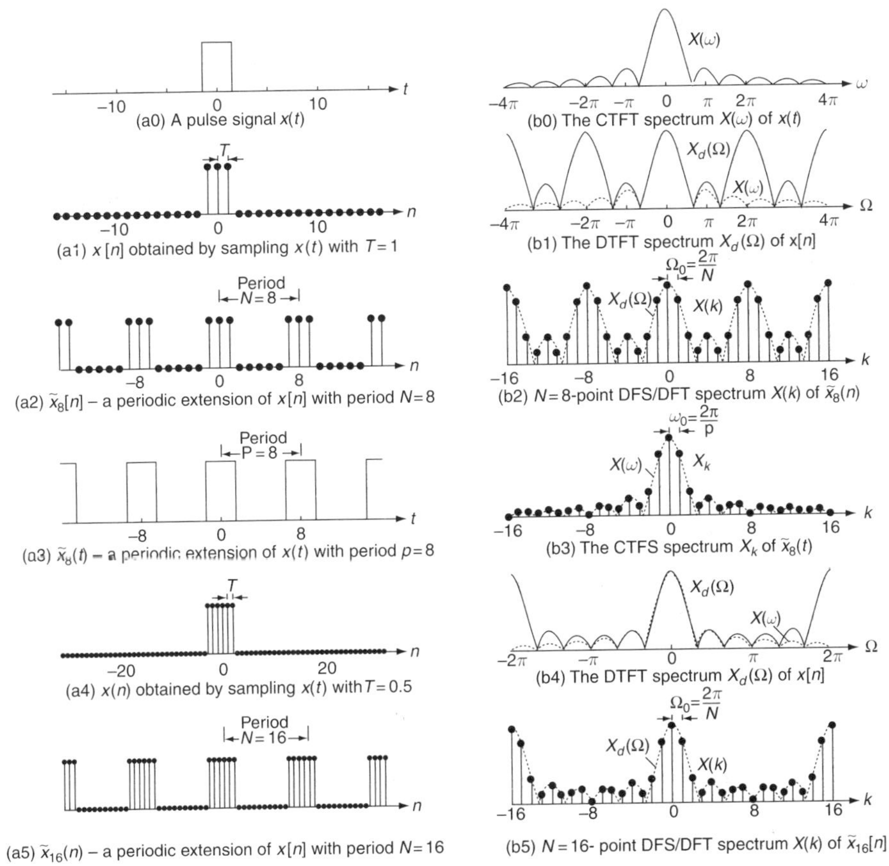

Fig. 3.14 Examples of CTFT, DTFT, DFS/DFT, and CTFS spectra

Between the CTFS and CTFT, we prefer to have the CTFT because it has all the information contained in the CTFS on the assumption that the CTFS consists of the samples of CTFT (Eq. (3.5.8) and Fig. 3.14(b3)). Besides, the CTFS is not so practical because it is hard to find the period or even periodicity of periodic signals due to a noise. Therefore, we think of the CTFT as a standard when we need to compare the spectra in terms of how faithfully they describe the frequency contents of a given signal.

(2) The problem with the CTFS and CTFT is that they are difficult to compute due to the integration. Compared with them, the DTFT $X_d(\Omega)$ is easier to deal with since it has only multiplications and additions. However, the sampling of $x(t)$ (with sampling interval T) to make $x[n] = x(nT)$ produces the periodic extension of the CTFT spectrum $X(\omega)$ with period $2\pi/T$ in ω, causing *frequency-aliasing* in the case of non-zero frequency components outside the *principal analog frequency band* $[-\pi/T, \pi/T]$. This is the cost we pay in return for the computational convenience of the DTFT. This frequency-aliasing

can be reduced by decreasing the sampling interval T so that more frequency components can be contained in $[-\pi/T, \pi/T]$. (Compare the DTFT spectra in Fig. 3.14(b1) (for $T = 1$) and (b4) (for $T = 0.5$) with the CTFT plotted in dotted lines.) Refer to the sampling theorem to be discussed in Sect. 5.3, which presents a criterion for selecting the sampling interval.

(cf.) To compare the DTFT $X_d(\Omega)$ with the CTFT $X(\omega)$, we should divide $X(\omega)$ by the sampling interval T (refer to Eq. (3.5.7)).

(3) The DTFT $X_d(\Omega)$ of $x[n]$ is computationally advantageous over the CTFS or CTFT, but is still not so handy since it is continuous in the frequency Ω and thus requires an integration for IDTFT (inverse DTFT). That is why we sample the DTFT in the frequency domain at $k\Omega_0 = 2\pi k/N$ for $k = 0 : N-1$ to make an N-point DFT $X(k)$ for more computational efficiency. However, it also costs us the (illusory) periodic extension of $x[n]$ with period N (the DFT size) irrespective of whether $x[n]$ is originally periodic or not and no matter what the real period is even if $x[n]$ is really periodic. This causes *time-aliasing* if the original signal is not sufficiently covered within the whole time interval $[0, N-1]$ (Example 3.15) and *spectral leakage* problem when the DFT size does not conform to the real period of the signal (Example 3.16).

(4) The analog resolution frequency $\omega_0 = \Omega_0/T = 2\pi/NT = 2\pi/P$ can be improved by increasing the whole time interval $P = NT$. Increasing the DFT size N (, say, by zero-padding) helps the DFT spectrum to become close to the DTFT spectrum. Decreasing the sampling interval T increases the period $2\pi/T$ of periodic extension of the CTFT spectrum (Eq. (3.5.7)) or equivalently, expands the principal analog frequency band $[-\pi/T, \pi/T]$ so that the chance and degree of frequency aliasing can be reduced.

(5) Generally, we can choose the DFT size N and sampling interval T and thus, eventually $P = NT$ (the product of N and T) during which a signal is to be measured, sampled, and collected as a set of N data points. Therefore, it is hard to imagine that NT happens to be identical with the real period of the signal. For this reason, it will be reasonable to call $P = NT$ the *whole time interval* rather than the period that was originated from the definition of the CTFS.

3.6 Fast Fourier Transform (FFT)

In this section we discuss the FFT algorithm that exploits the periodicity and symmetry of the discrete-time complex exponentials $e^{j2\pi nk/N}$ to reduce significantly the number of multiplications for the DFT computation. The FFT algorithm discussed here achieves its efficiency when N is a power of 2, i.e., $N = 2^{NLOG2}$ for some integer *NLOG*2. This makes no practical problem since the length of $x[n]$ can be increased to the next power of 2 by zero-padding.

To get some understanding of the steps in the FFT algorithm, let us consider a sequence $x[n]$ for $0 \le n \le N-1$ with $N = 2^{NLOG2}$. There are two approaches, each of which is based on the decimation process in the time and frequency domain, respectively.

3.6.1 Decimation-in-Time (DIT) FFT

In this approach, we break the N-point DFT into two $N/2$-point DFTs, one for even-indexed subsequence $x[2r]$ and the other for odd-indexed subsequence $x[2r+1]$, then break each $N/2$-point DFT into two $N/4$-point DFTs and continue this process until 2-point DFTs appear. Specifically, we can write the N-point DFT of $x[n]$ as

$$
\begin{aligned}
X(k) &\overset{(3.4.2)}{=} \sum_{n=0}^{N-1} x[n]W_N^{kn} = \sum_{n=2r(\text{even})} x[n]W_N^{kn} + \sum_{n=2r+1(\text{odd})} x[n]W_N^{kn} \\
&= \sum_{r=0}^{N/2-1} x[2r]W_N^{2rk} + \sum_{r=0}^{N/2-1} x[2r+1]W_N^{(2r+1)k} \\
&\qquad (W_N^{2rk} = e^{-j2\pi(2r)k/N} = e^{-j2\pi rk/(N/2)} = W_{N/2}^{rk}) \\
&= \sum_{r=0}^{N/2-1} x_e[r]W_{N/2}^{rk} + W_N^k \sum_{r=0}^{N/2-1} x_o[r]W_{N/2}^{rk} \\
&\overset{(3.4.2)}{=} X_e(k) + W_N^k X_o(k) \quad \text{for } 0 \le k \le N-1
\end{aligned} \tag{3.6.1}
$$

so that

$$
X(k) \overset{(3.6.1)}{=} X_e(k) + W_N^k X_o(k) \quad \text{for} \quad 0 \le k \le N/2-1 \tag{3.6.2a}
$$

$$
\begin{aligned}
X(k) &\overset{(3.6.1)}{=} X_e(k) + W_N^k X_o(k) \quad \text{for} \quad N/2 \le k \le N-1; \\
X(k+N/2) &\overset{(3.6.1)}{=} X_e(k+N/2) + W_N^{k+N/2} X_o(k+N/2) \\
&= X_e(k) - W_N^k X_o(k) \text{ for } 0 \le k \le N/2-1
\end{aligned} \tag{3.6.2b}
$$

where $X_e(k)$ and $X_o(k)$ are $N/2$ -point DFTs that are periodic with period $N/2$ in k. If $N/2$ is even, then we can again break $X_e(k)$ and $X_o(k)$ into two $N/4$ -point DFTs in the same way:

$$
\begin{aligned}
X_e(k) &\underset{\text{with } N\to N/2}{\overset{(3.6.1)}{=}} X_{ee}(k) + W_{N/2}^k X_{eo}(k) \\
&= X_{ee}(k) + W_N^{2k} X_{eo}(k) \quad \text{for } 0 \le k \le N/2-1
\end{aligned} \tag{3.6.3a}
$$

$$
\begin{aligned}
X_o(k) &\underset{\text{with } N\to N/2}{\overset{(3.6.1)}{=}} X_{oe}(k) + W_{N/2}^k X_{oo}(k) \\
&= X_{oe}(k) + W_N^{2k} X_{oo}(k) \quad \text{for } 0 \le k \le N/2-1
\end{aligned} \tag{3.6.3b}
$$

If $N = 2^{NLOG2}$, we repeat this procedure $NLOG2-1$ times to obtain $N/2$ 2-point DFTs, say, for $N = 2^3$, as

$$X_{ee}(k) = \sum_{n=0}^{N/4-1} x_{ee}[n]W_{N/4}^{kn} = x[0] + (-1)^k x[4] \tag{3.6.4a}$$

with $x_{ee}[n] = x_e[2n] = x[2^2 n]$

$$X_{eo}(k) = \sum_{n=0}^{N/4-1} x_{eo}[n]W_{N/4}^{kn} = x[2] + (-1)^k x[6] \tag{3.6.4b}$$

with $x_{eo}[n] = x_e[2n+1] = x[2^2 n + 2]$

$$X_{oe}(k) = \sum_{n=0}^{N/4-1} x_{oe}[n]W_{N/4}^{kn} = x[1] + (-1)^k x[5] \tag{3.6.4c}$$

with $x_{oe}[n] = x_o[2n] = x[2^2 n + 1]$

$$X_{oo}(k) = \sum_{n=0}^{N/4-1} x_{oo}[n]W_{N/4}^{kn} = x[3] + (-1)^k x[7] \tag{3.6.4d}$$

with $x_{oo}[n] = x_o[2n+1] = x[2^2 n + 2 + 1]$

Along with this procedure, we can draw the signal flow graph for an 8-point DFT computation as shown in Fig. 3.15(a). By counting the number of branches with a gain W_N^r (representing multiplications) and empty circles (denoting additions), we see that each stage needs N complex multiplications and N complex additions. Since there are $\log_2 N$ stages, we have a total of $N \log_2 N$ complex multiplications and additions. This is a substantial reduction of computation compared with the direct DFT computation, which requires N^2 complex multiplication and $N(N-1)$ complex additions since we must get N values of $X(k)$ for $k = 0 : N-1$, each $X(k)$ requiring N complex multiplications and $N-1$ complex additions.

Remark 3.8 Data Rearrangement in "Bit Reversed" Order

The signal flow graph in Fig. 3.15 shows that the input data $x[n]$ appear in the bit reversed order:

Position		Binary equivalent		Bit reversed		Sequence index
3	→	011	→	110	→	6
4	→	100	→	001	→	1

Remark 3.9 Simplified Butterfly Computation

(1) The basic computational block in the flow graph (Fig. 3.15(b)), called a *butter fly*, for stage $m+1$ represents the following operations such as Eq. (3.6.2):

$$X_{m+1}(p) = X_m(p) + W_N^r X_m(q), \text{ with } q = p + 2^m \tag{3.6.5a}$$

$$X_{m+1}(q) = X_m(p) + W_N^{r+N/2} X_m(q) = X_m(p) - W_N^r X_m(q) \tag{3.6.5b}$$

$$(\because W_N^{r+N/2} = -W_N^r)$$

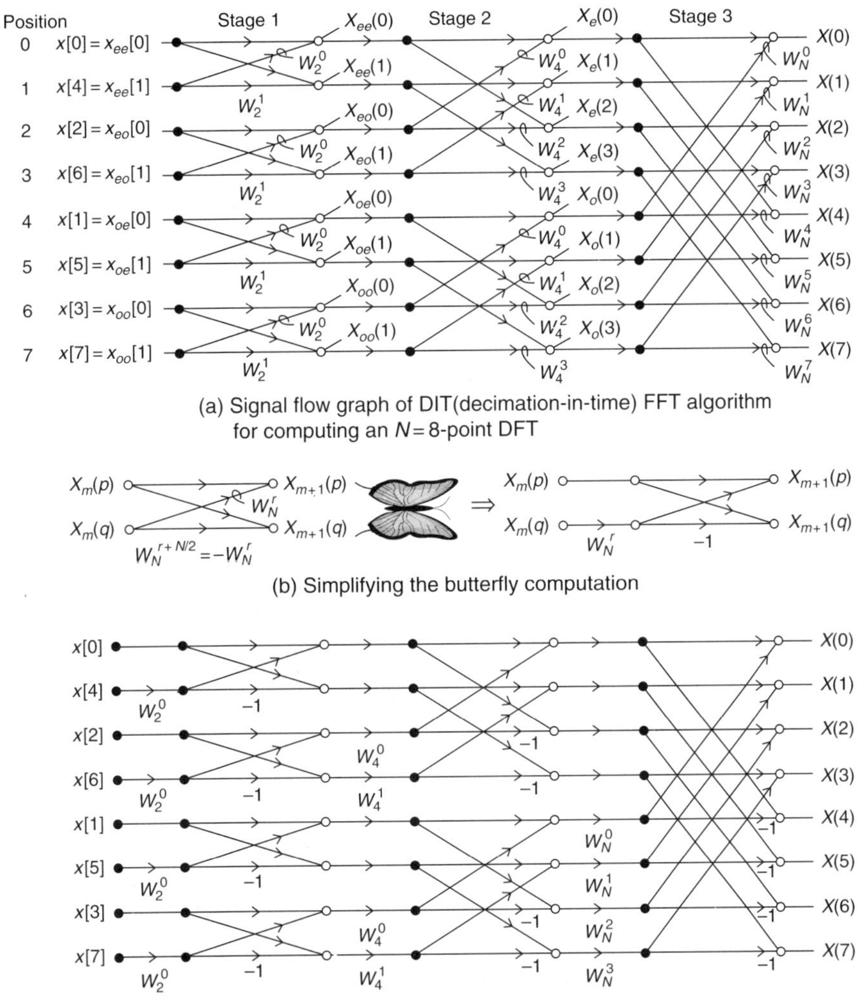

(a) Signal flow graph of DIT(decimation-in-time) FFT algorithm for computing an $N = 8$-point DFT

(b) Simplifying the butterfly computation

(c) Signal flow graph of DIT(decimation-in-time) FFT algorithm (with simplified butterfly pattern) for computing an $N = 8$-point DFT

Fig. 3.15 Signal flow graph of DIT(decimation-in-time) FFT algorithm

where p and q are the position indices in stage $m + 1$. Notice that $X_{m+1}(p)$ and $X_{m+1}(q)$, the outputs of the butterfly at stage $m + 1$ are calculated in terms of $X_m(p)$ and $X_m(q)$, the corresponding values from the previous stage and no other inputs. Therefore, if a scratch memory for storing some intermediate results is available, $X_{m+1}(p)$ and $X_{m+1}(q)$ can be calculated and be placed back into the storage registers for $X_m(p)$ and $X_m(q)$. Thus only N registers for storing one complex array are needed to implement the complete computation. This kind of memory-efficient procedure is referred to as an *in-place* computation.

(2) Noting that the horizontal multiplier is just the negative of the diagonal multiplier, the basic butterfly configuration can be further simplified. Specifically, with $T = W_N^r X_m(q)$, we have the simplified butterfly computation

$$X_{m+1}(p) = X_m(p) + T$$
$$X_{m+1}(q) = X_m(p) - T$$

which is depicted in Fig.3.15(b). This reduces the number of complex multiplications for the DFT computation to half of that with the original butterfly computation (3.6.5a) (Fig. 3.15(c)):

$$N \log_2 N \quad \rightarrow \quad \frac{N}{2} \log_2 N$$

3.6.2 Decimation-in-Frequency (DIF) FFT

An alternative algorithm for computing the DFT results from breaking up or decimating the transform sequence $X(k)$. It begins by breaking up $X(k)$ as follows:

$$X(k) \overset{(3.4.2)}{=} \sum_{n=0}^{N-1} x[n]W_N^{kn} = \sum_{n=0}^{N/2-1} x[n]W_N^{kn} + \sum_{n=N/2}^{N-1} x[n]W_N^{kn}$$
$$= \sum_{n=0}^{N/2-1} x[n]W_N^{kn} + W_N^{kN/2} \sum_{n=0}^{N/2-1} x[n+N/2]W_N^{kn}$$
$$= \sum_{n=0}^{N/2-1} (x[n] + (-1)^k x[n+N/2])W_N^{kn} \tag{3.6.6}$$

We separate this into two groups, i.e., one group of the even-indexed elements (with $k = 2r$) and the other group of the odd-indexed elements ($k = 2r + 1$) of $X(k)$;

$$X(2r) = \sum_{n=0}^{N/2-1} (x[n] + x[n+N/2])W_N^{2rn} \tag{3.6.7a}$$
$$= \sum_{n=0}^{N/2-1} (x[n] + x[n+N/2])W_{N/2}^{rn} \quad \text{for } 0 \le r \le N/2-1$$
$$X(2r+1) = \sum_{n=0}^{N/2-1} (x[n] - x[n+N/2])W_N^{(2r+1)n} \tag{3.6.7b}$$
$$= \sum_{n=0}^{N/2-1} (x[n] - x[n+N/2])W_N^n W_{N/2}^{rn} \quad \text{for } 0 \le r \le N/2-1$$

These are $N/2$ -point DFTs of the sequences $(x[n]+x[n+N/2])$ and $(x[n]-x[n+N/2])W_N^n$, respectively. If $N = 2^{NLOG2}$, we can proceed in the same way until it ends up with $N/2$ 2-point DFTs. The DIF FFT algorithm with simplified butterfly computation and with the output data in the bit reversed order is illustrated for a 8-point DFT in Fig. 3.16.

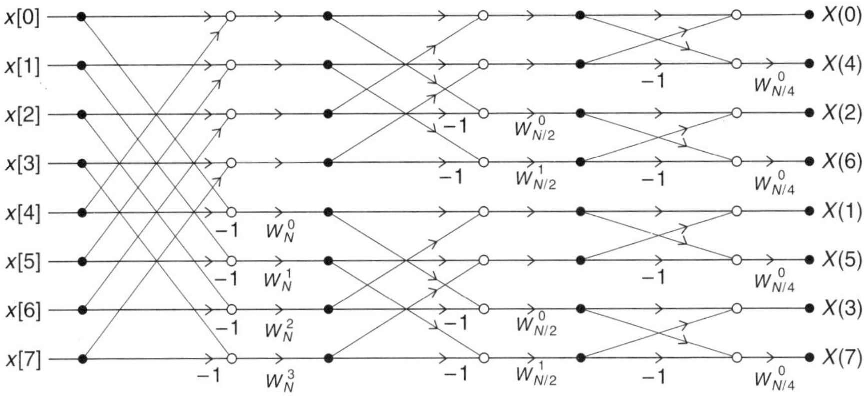

Fig. 3.16 Signal flow graph of DIF(decimation-in-frequency) FFT algorithm (with simplified butterfly pattern) for computing an $N = 8$-point DFT

3.6.3 Computation of IDFT Using FFT Algorithm

The inverse DFT is given by Eq. (3.4.3) as

$$x[n] = \text{IDFT}_N\{X(k)\} = \frac{1}{N}\sum\nolimits_{k=0}^{N-1} X(k)e^{j2\pi kn/N}$$
$$= \frac{1}{N}\sum\nolimits_{k=0}^{N-1} X(k)W_N^{-kn} \text{ for } n = 0 : N-1 \qquad (3.6.8)$$

Comparing this with the DFT formula (3.4.2), we see that the computational procedure remains the same except that the *twiddle* factors are negative powers of W_N and the output is scaled by $1/N$. Therefore, an inverse fast Fourier transform (IFFT) program can be obtained from an FFT algorithm by replacing the input data $x[n]$'s with $X(k)$'s, changing the exponents of W_N to negative values, and scaling the last output data by $1/N$.

An alternative way to get the IDFT by using the FFT algorithm is derived by taking the complex conjugate of the IDFT formula (3.4.3) or (3.6.8) as follows:

$$x^*[n] = \frac{1}{N}\sum\nolimits_{k=0}^{N-1} X^*(k)W_N^{kn} = \frac{1}{N}\text{FFT}\{X^*(k)\};$$
$$x[n] = \frac{1}{N}\left(\text{FFT}\{X^*(k)\}\right)^* \qquad (3.6.9)$$

It is implied that we can get $x[n] = \text{IDFT}_N\{X(k)\}$ by taking the complex conjugate of $X(k)$, applying the FFT algorithm for $X^*(k)$, taking the complex conjugate of the output, and scaling by $1/N$.

The MATLAB built-in functions "`fft(x,N)`"/"`ifft(X,N)`" implement the N-point FFT/IFFT algorithm for the data $x[n]/X(k)$ given as their first input argument if the second input argument N is given as a power of 2 or if N is not given and the length of the given data is a power of 2. If the length of the data sequence differs from the second input argument N, it will be zero-padded or truncated so that the resulting length will be N.

3.7 Interpretation of DFT Results

In general, the DFT takes a complex-valued sequence $\{x[n],\ n = 0 : N-1\}$ and produces another complex-valued sequence $\{X(k),\ k = 0 : N-1\}$. Then, what is the physical meaning or practical usage of it? To understand it, let us take some examples where we think of a discrete sequence $x[n]$ as the discrete-time version of an analog signal $x(t)$ such that $x[n] = x(nT)$.

Example 3.16 DFT Spectrum of a Single-Tone Sinusoidal Wave [L-1]

Suppose we have a continuous-time signal

$$x(t) = \cos(\omega_1 t) = \cos\left(\frac{15}{4}\pi t\right) \quad \text{with} \quad \omega_1 = \frac{15}{4}\pi \tag{E3.16.1}$$

Let us use the MATLAB function "`fft()`" to find the DFT spectrum of $\{x[n] = x(nT),\ n = 0 : N-1\}$ with different sampling interval T and DFT size or number of sample points N. Then we will discuss how we can read the frequency information about $x(t)$ from the DFT spectrum, which has different shape depending on the sampling interval T and DFT size N.

(a) Let the sampling interval and number of samples (DFT size) be $T = 0.1$ and $N = 16$, respectively so that the discrete-time sequence will be

$$x_a[n] = \cos\left(\frac{15}{4}\pi t\right)\bigg|_{t=nT=0.1n} = \cos\left(\frac{3}{8}\pi n\right), \quad n = 0, 1, \cdots, 15 \tag{E3.16.2}$$

The magnitude spectrum of this sequence, i.e., the magnitude of $X_a(k) = \text{DFT}_{16}\{x_a[n]\}$ is depicted in Fig. 3.17(a). It has a spike at the frequency index $k = 3$ corresponding to the digital and analog frequencies as

$$\begin{aligned} \Omega_3 &= 3\Omega_0 = 3\frac{2\pi}{N} = \frac{6\pi}{16} = \frac{3\pi}{8}[\text{rad/sample}] \\ \to \omega &\overset{(1.1.15)}{=} \frac{\Omega_3}{T} = \frac{3\pi}{8 \times 0.1} = \frac{15\pi}{4}[\text{rad/s}] \end{aligned} \tag{E3.16.3}$$

This analog frequency agrees exactly with the real frequency of the original signal (E3.16.1). Note that $|X_a(k)|$ has another spike at the frequency index $k = 13$, which virtually corresponds to $k = 13 - N = -3$, but it is just like the

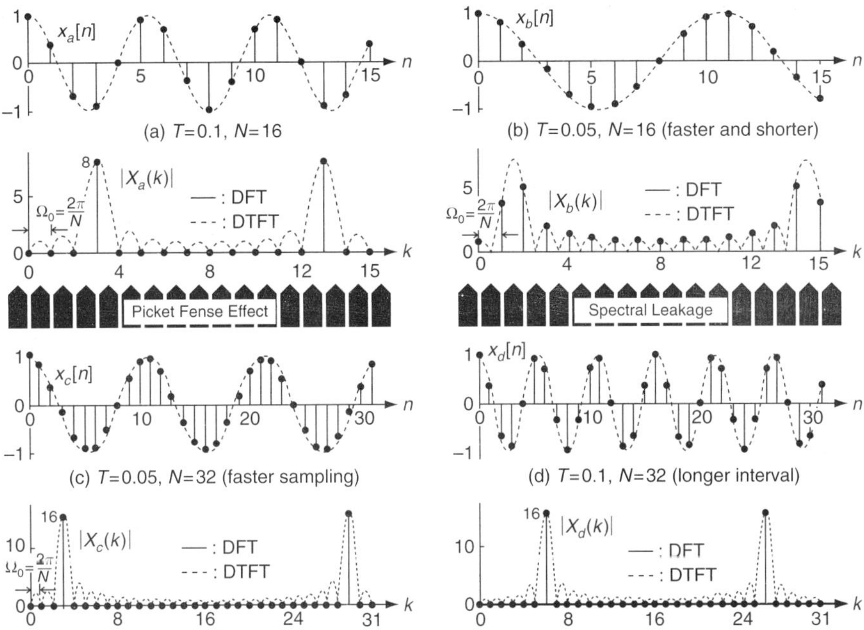

Fig. 3.17 DFT spectra of discrete-time sinusoidal sequences having a single tone

mirror image of $k = 3$. It can be assured by writing the discrete-time sequence (E3.16.2) as

$$x[n] = \frac{1}{2}\left(e^{j3\pi n/8} + e^{-j3\pi n/8}\right)$$

$$\overset{\text{IDFT (3.4.3)}}{=} \frac{1}{N}\left(\frac{N}{2}e^{j2\pi 3n/N} + \frac{N}{2}e^{-j2\pi 3n/N}\right)\Bigg|_{N=16} \tag{E3.16.4}$$

$$= \frac{1}{16}\left(8\,e^{j2\pi 3n/16} + 8\,e^{j2\pi(13)n/16}\right) = \frac{1}{N}\sum\nolimits_{k=0}^{N-1} X(k)\,e^{j2\pi kn/N}$$

This explains why the magnitude of the DFT spectrum has two spikes of 8 at $k = 3$ and $N - 3 = 13$.

(b) Let the sampling interval and number of samples (DFT size) be $T = 0.05$ (*up-sampling*) and $N = 16$, respectively so that the discrete-time sequence will be

$$x_b[n] = \cos\left(\frac{15}{4}\pi t\right)\Bigg|_{t=nT=0.05n} = \cos\left(\frac{3}{16}\pi n\right), \quad n = 0, 1, \cdots, 15 \tag{E3.16.5}$$

The magnitude spectrum of this sequence, i.e., the magnitude of $X_b(k) = \text{DFT}_{16}\{x_b[n]\}$ is depicted in Fig. 3.17(b), which looks quite different from

Fig. 3.17(a). That is, so many nonzero values show up here and there in the magnitude spectrum, though the highest two values occur at $k = 1$ and 2. It is because the *picket fence* effect (due to the frequency domain sampling of DFT) hides the spectral peak, allowing the many side lobes (resulting from the *spectral leakage* due to the time domain windowing) to be seen. This is a contrast to the other cases where the picket fence effect coincidentally hides the side lobes except the main peak as if by magic. Still we might fake a "far-fetched" story about the frequency corresponding to $k = 1.5$ (between 1 and 2) as

$$\Omega_{1.5} = 1.5\Omega_0 = 1.5\frac{2\pi}{N} = \frac{3\pi}{16}[\text{rad/sample}]$$
$$\rightarrow \omega \overset{(1.1.15)}{=} \frac{\Omega_{1.5}}{T} = \frac{3\pi}{16 \times 0.05} = \frac{15\pi}{4}[\text{rad/s}] \tag{E3.16.6}$$

This agrees with the result of (a), showing the right frequency information.

(c) Let the sampling interval and DFT size be $T = 0.05$ (*up-sampling*) and $N = 32$ (more sampling), respectively so that the discrete-time sequence will be

$$x_c[n] = \cos\left(\frac{15}{4}\pi t\right)\bigg|_{t=nT=0.05n} = \cos\left(\frac{3}{16}\pi n\right), \quad n = 0, 1, \cdots, 31 \tag{E3.16.7}$$

The magnitude spectrum of this sequence, i.e., the magnitude of $X_c(k) = \text{DFT}_{32}\{x_c[n]\}$ is depicted in Fig. 3.17(c). Like Fig. 3.17(a), it has two spikes at $k = 3$ and $N - 3 = 29$ with only a difference in their amplitude (16), which is two times as large as those in (a) because of the increased number of samples ($N = 32$). Thus we can read the digital and analog frequencies as

$$\Omega_3 = 3\Omega_0 = 3\frac{2\pi}{N} = \frac{6\pi}{32} = \frac{3\pi}{16}[\text{rad/sample}]$$
$$\rightarrow \omega \overset{(1.1.15)}{=} \frac{\Omega_3}{T} = \frac{3\pi}{16 \times 0.05} = \frac{15\pi}{4}[\text{rad/s}] \tag{E3.16.8}$$

This analog frequency also agrees exactly with the real frequency of the original signal (E3.16.1).

(d) Let the sampling interval and DFT size be $T = 0.1$ and $N = 32$ (longer interval/more sampling), respectively so that the discrete-time sequence will be

$$x_d[n] = \cos\left(\frac{15}{4}\pi t\right)\bigg|_{t=nT=0.1n} = \cos\left(\frac{3}{8}\pi n\right), \quad n = 0, 1, \cdots, 31 \tag{E3.16.9}$$

The magnitude spectrum of this sequence, i.e., the magnitude of $X_d(k) = \text{DFT}_{32}\{x_d[n]\}$ is depicted in Fig. 3.17(d). It has two spikes at $k = 6$ and $N - 6 = 26$, which tells us about the existence of the digital and analog frequencies as

$$\Omega_6 = 6\Omega_0 = 6\frac{2\pi}{N} = \frac{12\pi}{32} = \frac{3\pi}{8}\,[\text{rad/sample}]$$
$$\rightarrow \omega \overset{(1.1.15)}{=} \frac{\Omega_6}{T} = \frac{3\pi}{8 \times 0.1} = \frac{15\pi}{4}\,[\text{rad/s}] \qquad \text{(E3.16.10)}$$

This analog frequency also agrees exactly with the real frequency of the original signal (E3.16.1).

Example 3.17 DFT Spectrum of a Multi-Tone Sinusoidal Wave

Suppose we have a continuous-time signal

$$x(t) = \sin(\omega_1 t) + 0.5\cos(\omega_2 t) = \sin(1.5\pi t) + 0.5\cos(3\pi t) \qquad \text{(E3.17.1)}$$
$$\text{with } \omega_1 = 1.5\pi \text{ and } \omega_2 = 3\pi$$

Figure 3.18 shows the four different discrete-time versions of this signal and their DFT spectra, which will be discussed below.

(a) With sampling interval $T = 0.1$ and DFT size $N = 32$, the discrete-time sequence will be

$$x_a[n] = x(nT)|_{T=0.1} = \sin(0.15\pi n) + 0.5\cos(0.3\pi n),\ n = 0,\ 1, \cdots,\ 31 \qquad \text{(E3.17.2)}$$

The magnitude spectrum depicted in Fig. 3.18(a) is large at $k = 2$ & 3 and 4 & 5 and they can be alleged to represent two tones, one between $k\omega_0 = k\Omega_0/T = 2\pi k/NT = 2\pi k/3.2 \overset{k=2}{=} 1.25\,\pi$ and $k\omega_0 \overset{k=3}{=} 1.875\,\pi$ and the other between $k\omega_0 \overset{k=4}{=} 2.5\,\pi$ and $k\omega_0 \overset{k=5}{=} 3.125\,\pi$. Of these two tones, the former corresponds to $\omega_1 = 1.5\pi$ (with larger amplitude) and the latter to $\omega_2 = 3\pi$ (with smaller amplitude).

(b) With shorter sampling interval $T = 0.05$ (*up-sampling*) and larger DFT size $N = 64$, the discrete-time sequence will be

$$x_b[n] = x(nT)|_{T=0.05} = \sin(0.075\pi n) + 0.5\cos(0.15\pi n), \qquad \text{(E3.17.3)}$$
$$n = 0,\ 1, \cdots,\ 63$$

For all the up-sampling by a factor of 2, the magnitude spectrum depicted in Fig. 3.18(b) does not present us with any more information than (a). Why? Because all the frequency components ($\omega_1 = 1.5\pi$ and $\omega_2 = 3\pi$) are already covered within the principal analog frequency range $[-\pi/T,\ \pi/T]$ with $T = 0.1$, there is nothing to gain by expanding it via shorter sampling interval.

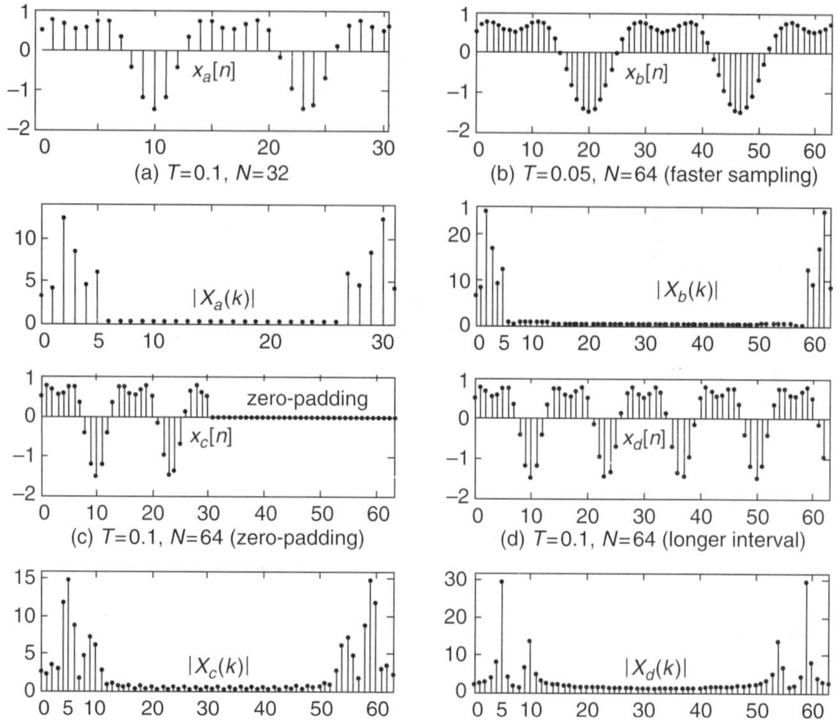

Fig. 3.18 DFT spectra of a two-tone sinusoidal wave

(c) With the same sampling interval $T = 0.1$, but with larger DFT size $N = 64$ (by zero-padding), the discrete-time sequence will be

$$x_c[n] = \begin{cases} x_a[n] & \text{for } n = 0,\ 1,\ \cdots,\ 31 \\ 0 & \text{for } n = 32,\ \cdots,\ 63 \end{cases} \tag{E3.17.4}$$

The magnitude spectrum depicted in Fig. 3.18(c) is large at $k = 4$ & 5 and 9 & 10 and they can be alleged to represent two tones, one between $k\omega_0 = k\Omega_0/T = 2\pi k/NT = 2\pi k/6.4 \overset{k=4}{=} 1.25\,\pi$ and $k\omega_0 \overset{k=5}{=} 1.5625\,\pi$ and the other between $k\omega_0 \overset{k=9}{=} 2.8125\,\pi$ and $k\omega_0 \overset{k=10}{=} 3.125\,\pi$. Comparing this with the result obtained in (a), we see that larger DFT size by zero-padding can yield better resolution for the spectrum.

(d) With the same sampling interval $T = 0.1$, but with larger DFT size $N = 64$ (by more sampling), the discrete-time sequence will be

$$x_d[n] = x(nT)|_{T=0.1} = \sin(0.15\pi n) + 0.5\cos(0.3\pi n), \quad n = 0,\ 1, \cdots,\ 63 \tag{E3.17.5}$$

The magnitude spectrum depicted in Fig. 3.18(d) is strikingly large at $k = 5$ and 10, which can be alleged to represent two tones of $k\omega_0 = 2\pi k/NT = 2\pi k/6.4 \overset{k=5}{=} 1.5625\ \pi$ and $k\omega_0 \overset{k=10}{=} 3.125\ \pi$. Comparing this with the result obtained in (a), we see that larger DFT size by more sampling can improve the spectrum resolution. Comparing this with (c), we see that more sampling can yield better resolution than zero-padding as it collects more information about the signal.

Example 3.18 DFT Spectrum of a Triangular Pulse

Consider the following sequences that are obtained by sampling the continuous-time triangular wave $\tilde{x}_8(t) = 2(\tilde{\lambda}_2(t) - \tilde{\lambda}_2(t-4))$ of period 8 and with peak-to-peak range between -2 and $+2$. Let us consider the DFT magnitude spectra of the sequences that are depicted in Fig. 3.18.

(a) With sampling interval $T = 1$ and DFT size $N = 8$, the discrete-time sequence will be

$$x_a[n] = \tilde{x}_8\ (nT)|_{T=1}\,, \quad n = 0,\ 1, \cdots, 7 \tag{E3.18.1}$$

The 8-point DFT magnitude spectrum of $x_a[n]$ together with the CTFT magnitude spectrum of the single-period triangular pulse $x_8(t)$ (Eq. (E3.11.3)) is depicted in Fig. 3.19(a).

(b) With shorter sampling interval $T = 0.5$ (up-sampling) and larger DFT size $N = 16$, the discrete-time sequence will be

$$x_b[n] = \tilde{x}_8\ (nT)|_{T=0.5}\,, \quad n = 0,\ 1, \cdots, 15 \tag{E3.18.2}$$

The 16-point DFT magnitude spectrum of $x_b[n]$ together with the CTFT magnitude spectrum of the single-period triangular pulse $x_8(t)$ is depicted in Fig. 3.19(b). It shows that the DFT is similar to the CTFT for the expanded principal analog frequency range $[-\pi/T, \pi/T]$.

(cf.) Note that to be compared with the DFT, the CTFT spectrum has been scaled not only vertically in consideration for the sampling interval T (Eq. (3.5.7) or Remark 3.7(2)) and the number of sample points, but also horizontally in consideration for the relationship between the digital frequency Ω and analog frequency $\omega = \Omega/T$.

(c) With the same sampling interval $T = 1$, but with larger DFT size $N = 16$ (by zero-padding), the discrete-time sequence will be

$$x_c[n] = \begin{cases} x_a[n] & \text{for } n = 0,\ 1, \cdots, 7 \\ 0 & \text{for } n = 8, \cdots, 15 \end{cases} \tag{E3.18.3}$$

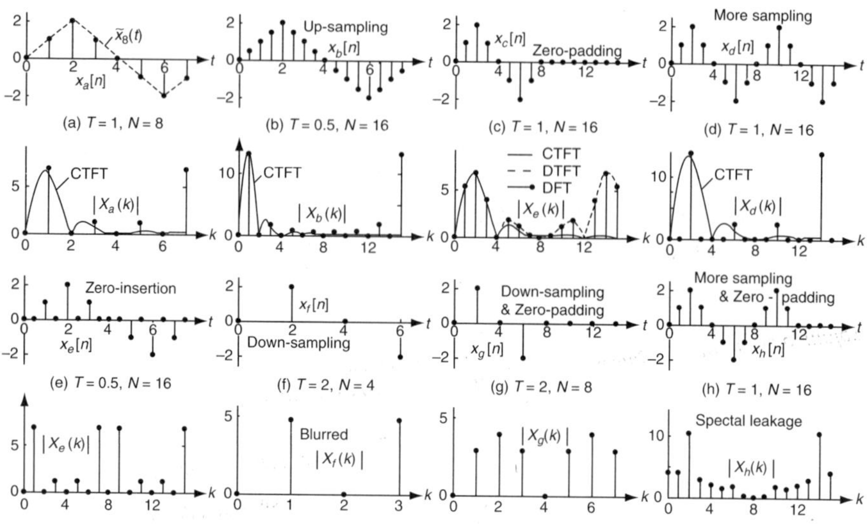

Fig. 3.19 DFT spectra of various discrete-time versions of a triangular wave

Figure 3.19(c) shows that the DFT becomes closer to the DTFT spectrum, while it has the same principal analog frequency range for which it is similar to the CTFT.

(d) With the same sampling interval $T = 1$, but with larger DFT size $N = 16$ (by more sampling), the discrete-time sequence will be

$$x_d[n] = \begin{cases} x_a[n] & \text{for } n = 0,\ 1,\ \cdots,\ 7 \\ x_a[n-8] & \text{for } n = 8,\ \cdots,\ 15 \end{cases} \tag{E3.18.4}$$

Figure 3.19(d) shows that we gained nothing with more sampling and larger DFT size in contrast with the case of Example 3.17(d). However, we lost nothing with more sampling.

(e) With larger DFT size $N = 16$ by *zero-insertion*, i.e., inserting one zero between the samples, the discrete-time sequence will be

$$x_e[n] = \begin{cases} x_a[n/2] & \text{for } n = 0,\ 2,\cdots,\ 14 \text{ (even)} \\ 0 & \text{for } n = 1,\ 3,\cdots,\ 15 \text{ (odd)} \end{cases} \tag{E3.18.5}$$

Figure 3.19(e) shows that zero-insertion results in the periodic extension of the DFT spectrum. This has an interpretation that zero-insertion increases the variation rate of the signal, producing the high frequency components.

(f) With longer sampling interval $T = 2$ (*down-sampling* or *subsampling*) and smaller DFT size $N = 4$, the discrete-time sequence will be

$$x_f[n] = x_a[2n], \quad n = 0,\ 1,\cdots,\ 3 \tag{E3.18.6}$$

Figure 3.19(f) shows that down-sampling compresses the principal analog frequency range and besides, smaller DFT size harms the spectrum resolution.

(g) With longer sampling interval $T = 2$ (*down-sampling* or *subsampling*), but with the same DFT size $N = 8$ (by zero-padding), the discrete-time sequence will be

$$x_g[n] = \begin{cases} x_f[n] & \text{for } n = 0, 1, \cdots, 3 \\ 0 & \text{for } n = 4, 5, \cdots, 7 \end{cases} \tag{E3.18.7}$$

Figure 3.19(g) shows that zero-padding may help the spectrum to look better, but it does not recover the spectrum damaged by down-sampling.

(h) With the same sampling interval $T = 1$, but with larger DFT size $N = 16$ (by more sampling partly and zero-padding partly), the discrete-time sequence will be

$$x_h[n] = \begin{cases} x_d[n] & \text{for } n = 0, 1, \cdots, 11 \\ 0 & \text{for } n = 12, \cdots, 15 \end{cases} \tag{E3.18.8}$$

Figure 3.19(h) shows that zero-padding may help the spectrum to look better compared with (d).

Remark 3.10 DFS/DFT (Discrete Fourier Series/Transform) and Spectral Leakage

(1) Generally, the DFT $X(k)$ is complex-valued and denotes the magnitude & phase of the signal component having the digital frequency $\Omega_k = k\Omega_0 = 2\pi k/N$ [rad/sample], which corresponds to the analog frequency $\omega_k = k\omega_0 = k\Omega_0/T = 2\pi k/NT$ [rad/s]. We call $\Omega_0 = 2\pi/N$ and $\omega_0 = \Omega_0/T$ (N: DFT size) the digital/analog fundamental or resolution frequency since it is the minimum digital/analog frequency difference that can be distinguished by the N-point DFT. Note that the frequency indices $k = 0$ and $N/2$ represent the DC component ($\Omega = 0$) and the virtually highest digital frequency component ($\Omega_{N/2} = N/2 \times 2\pi/N = \pi$), respectively.

(2) As illustrated in Figs. 3.17(b) and 3.18(a)-(d), if a periodic signal does not go through a multiple of periods within the sampled signal range $[0, NT)$, its DFT spectrum is dirty. This is because the *spectral leakage* is not hidden by the picket fence effect. It seems that we might avoid this problem by setting the sampled signal range so that it covers exactly a multiple of the period of the signal. However, it is only our desire because we hardly know in advance the frequency contents of the signal and besides, most signals have many frequency components.

(3) The spectral leakage problem is always in the midst of DFT because it is inherently due to the time-domain windowing as can be observed in the DTFT spectrum of Fig. 3.5(b3). Then, how could the DFT spectra of Figs. 3.17(a), (c), and (d) be free from the spectral leakage? The fact is that we could not

see the existing spectral leakage (as shown by the DTFT spectrum plotted in dotted lines) because the *picket fence* effect [D-1] of DFT (attributed to the frequency-domain sampling of DTFT) coincidentally happens to hide the many side lobe ripples except the main peaks.

(4) From another point of view, we might put the responsibility for spectral leakage on the assumption of DFT that every signal is periodic with period equal to the DFT size, which is hard to be true. As a measure to alleviate the spectral leakage problem, there is a smooth windowing technique [W-1] that can reduce the "faked" abrupt change of the signal (at both ends of the signal range) caused by the rectangular windowing. Interested readers can see Problem 3.14.

Through the examples given above, we have already observed how the DFT spectrum is affected by the sampling interval, DFT size, zero-padding, and so on. The observations are summarized as follows:

Remark 3.11 The Effects of Sampling Period (Interval) T and DFT Size N on DFT

(1) Shorter sampling interval expands the principal analog frequency range $[-\pi/T, \pi/T]$ so that it can help higher frequency components to be reflected on the DFT spectrum if every frequency component is not covered within the present principal analog frequency range.
(2) Larger DFT size (by zero-padding or more sampling) may help the (discrete) DFT spectrum to become closer to the (continuous) DTFT spectrum and that with better frequency resolution.

3.8 Effects of Signal Operations on DFT Spectrum

In this section, we summarize the formulas for finding the DFTs of different versions of a signal from the DFT of the original one. They will be of help in understanding the effect of self-repetition, zero-padding, zero-insertion, and so on.

Let the DFT of a finite-duration sequence $x[n]$ be

$$X(k) = \sum_{n=0}^{N-1} x[n]e^{-j2\pi kn/N}$$

$$= \sum_{n=0}^{N-1} x[n]W_N^{kn} \quad \text{with} \quad W_N = e^{-j2\pi/N} \text{ for } k = 0 : N-1 \qquad (3.8.1)$$

(a) Flipping: $x_a[n] = x[N-1-n]$

$$X_a(k) = \sum_{n=0}^{N-1} x[N-1-n]W_N^{kn} \overset{n\to N-1-m}{=} \sum_{m=N-1}^{0} x[m]W_N^{k(N-1-m)}$$

$$\overset{m\to n}{=} \sum_{n=0}^{N-1} x[n]W_N^{-kn}W_N^{-k} = X(N-k)W_N^{-k} \qquad (3.8.2)$$

(b) Frequency-shifting: $x_b[n] = (-1)^n x[n]$

$$X_b(k) = \sum_{n=0}^{N-1} (-1)^n x\,[n]\; W_N^{k\,n}$$

$$= \sum_{n=0}^{N-1} x[n]\; W_N^{k\,n}\; W_N^{(N/2)\,n} = X(k + N/2) \qquad (3.8.3)$$

(c) Zero-Padding: $x_c[n] = \begin{cases} x[n] & \text{for } n = 0, 1, \cdots, N-1 \\ 0 & \text{for } n = N, \cdots, 2N-1 \end{cases}$

$$X_c(k) = \sum_{n=0}^{2N-1} x_c[n] W_{2N}^{kn} = \sum_{n=0}^{N-1} x[n] W_N^{(k/2)n} = X(k/2) \text{ (interpolation)} \qquad (3.8.4)$$

(d) Self-Repetition: $x_d[n] = \begin{cases} x[n] & \text{for } n = 0, 1, \cdots, N-1 \\ x[n-N] & \text{for } n = N, \cdots, 2N-1 \end{cases}$

$$X_d(k) = \sum_{n=0}^{2N-1} x_d[n] W_{2N}^{kn} = \sum_{n=0}^{N-1} x[n] W_N^{(k/2)n} + \sum_{n=N}^{2N-1} x[n-N] W_N^{(k/2)n}$$

$$\overset{n \to m+N}{=} \sum_{n=0}^{N-1} x[n] W_N^{(k/2)n} + \sum_{m=0}^{N-1} x[m] W_N^{(k/2)(m+N)}$$

$$= \sum_{n=0}^{N-1} x[n] W_N^{(k/2)n} + \sum_{n=0}^{N-1} x[n] W_N^{(k/2)n} W_N^{(N/2)k} \qquad (3.8.5)$$

$$= X(k/2) + (-1)^k X(k/2) = \begin{cases} 2X(k/2) & \text{for } k \text{ even} \\ 0 & \text{for } k \text{ odd} \end{cases} \quad \text{(zero-insertion)}$$

(e) Zero-Insertion: $x_e[n] = \begin{cases} x[n/2] & \text{for } n \text{ even} \\ 0 & \text{for } n \text{ odd} \end{cases}$

$$X_e(k) = \sum_{n=0}^{2N-1} x_e[\,n]\; W_{2N}^{k\,n} = \sum_{n=\text{even}}^{2N-1} x\,[\,n/2]\; W_{2N}^{kn}$$

$$\overset{n=2m}{=} \sum_{m=0}^{N-1} x[m]\; W_{2N}^{2km} \overset{m \to n}{=} \sum_{n=0}^{N-1} x[n]\; W_N^{k\,n} = \tilde{X}(k)$$

$$= \begin{cases} X(k) & \text{for } k = 0,\ 1,\ \cdots,\ N-1 \\ X(k-N) & \text{for } k = N,\ \cdots,\ 2N-1 \end{cases} \quad \text{(self-repetition)} \qquad (3.8.6)$$

(f) Down-sampling, Subsampling, or Decimation: $x_f[n] = x[2n],\ n = 0, 1, \cdots, \frac{N}{2} - 1$

$$X_f(k) = \sum_{n=0}^{N/2-1} x[2n]\, W_{N/2}^{kn} \overset{2n=m}{=} \sum_{m=0}^{N-1} x[m] \frac{1+(-1)^m}{2}\, W_{N/2}^{km/2}$$

$$\overset{m\to n}{=} \frac{1}{2} \sum_{n=0}^{N-1} x[n]\, W_N^{kn} + \frac{1}{2} \sum_{n=0}^{N-1} x[n]\, W_N^{(k+N/2)n} \tag{3.8.7}$$

$$= \frac{1}{2}(X(k) + X(k+N/2)) \quad \text{for} \quad k = 0, 1, \cdots, \frac{N}{2} - 1 \text{ (blurring)}$$

3.9 Short-Time Fourier Transform – Spectrogram

The short-time or short-term Fourier transform (STFT) breaks a long signal into small segments, optionally overlapped and/or windowed, and finds the DFT of each (windowed) segment separately to record the local spectra in a matrix with frequency/time indices:

$$\text{STFT}\{x[n]\} = X[k,n] = \sum_{m=0}^{N-1} x[m+n]w[m]W_N^{km} \tag{3.9.1a}$$

$$\overset{m\to m-n}{=} \sum_{m=n}^{n+N-1} x[m]w[m-n]W_N^{k(m-n)} \text{ for } k = 0, 1, \cdots, N-1 \tag{3.9.1b}$$

```
%sig03f20.m : to plot Fig.3.20
clear, clf
T=0.1; Fs=1/T; % Sampling period and Sampling frequency
w1=25*pi/16; w2=30*pi/16; w3=40*pi/16; w4=60*pi/16;
n=[0:31]; x=[cos(w1*T*n) sin(w2*T*n) cos(w3*T*n) sin(w4*T*n)];
Nx=length(x); nn=0:Nx-1; % Length and duration (period) of the signal
N=32; kk=0:N/2; ff=kk*Fs/N; % DFT size and frequency range
wnd= hamming(N).'; % Hamming window of length N
Noverlap=N/4; % the number of overlap
M=N-Noverlap; % the time spacing between DFT computations
for i=1:fix((Nx-Noverlap)/M)
   xiw= x((i-1)*M+[1:N]).*wnd; % ith windowed segment
   Xi= fft(xiw); % DFT X(k,i) of ith windowed segment
   X(:,i)= Xi(kk+1).'; % insert X(0:N/2,i) into the ith column
   tt(i)=(N/2+(i-1)*M)*T;
end
% Use the MATLAB signal processing toolbox function specgram()
[X_sp,ff1,tt1] = spectrogram(x,wnd,Noverlap,N,Fs,'yaxis');
% Any discrepancy between the above result and spectrogram()?
discrepancy= norm(X-X_sp)/norm(X_sp)
figure(2), clf, colormap(gray(256));
subplot(221), imagesc(tt,ff,log10(abs(X))), axis xy
subplot(222), imagesc(tt1,ff1,log10(abs(X_sp))), axis xy
% specgram(x,N,Fs,wnd,noverlap) in MATLAB of version 6.x
```

where the window sequence $w[m]$ is used to select a finite-length (local) segment from the sliding sequence $x[m+n]$ and possibly to reduce the spectral leakage. Note that the frequency and time denoted by frequency index k and time index n of $X[k,n]$ are

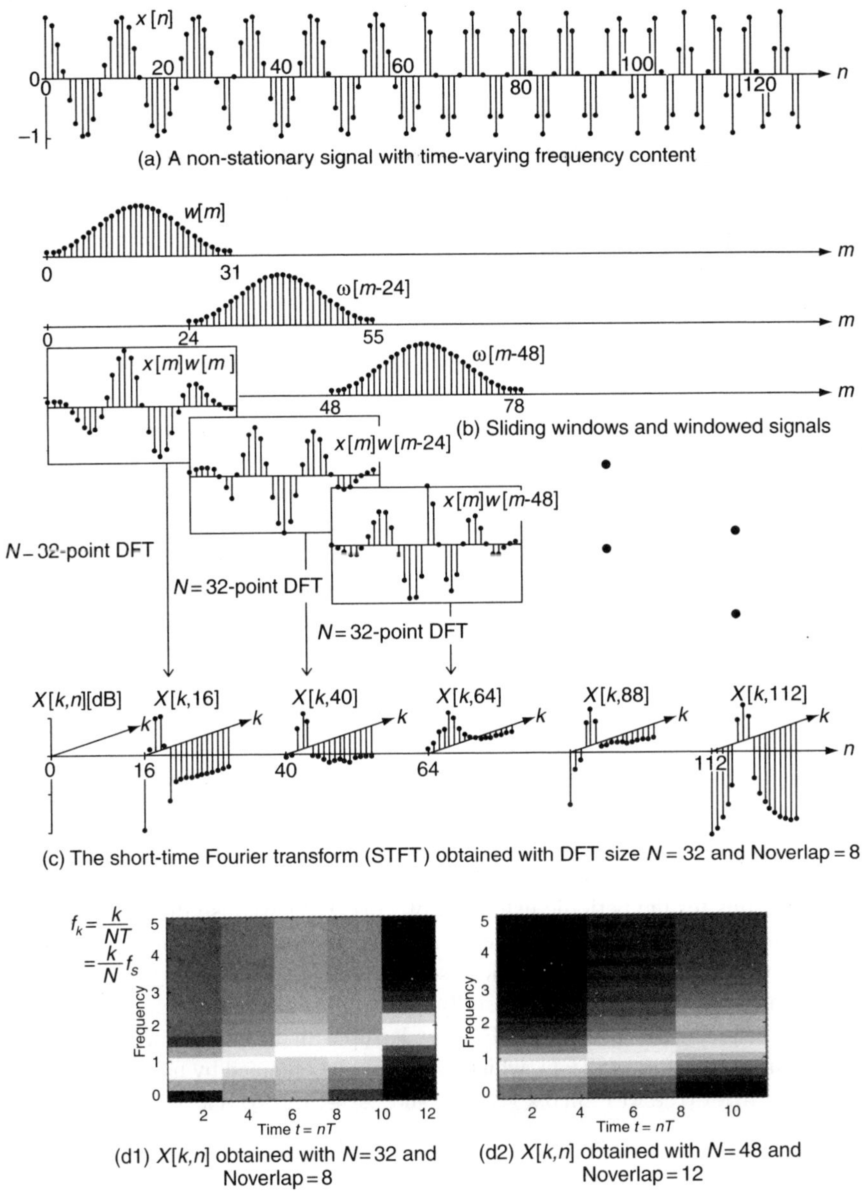

Fig. 3.20 Short-time Fourier transforms with different time/frequency resolution

$$f_k = \frac{k}{NT} = \frac{k}{N} f_s[\text{Hz}] \quad \text{and} \quad t_n = \left(\frac{N}{2} + n\right) T[\text{sec}] \tag{3.9.2}$$

where t_n is set to the center of the sliding window (Fig. 3.20(b)). If we take the STFT not for every n but only for $n = iM$ (a multiple of M), each segment gets overlapped with its neighboring one(s) by $Noverlap = N - M$ samples.

Note that M determines the width of interval time between the DFT computations. The MATLAB function "`spectrogram()`" can be used to find the STFT (see Appendix E.8).

We run the above MATLAB program "sig03f20.m" to get the STFT $X[k, n]$ of $x[n]$ (Fig. 3.20(a)) as depicted in Fig. 3.20(d1) (with DFT size $N = 32$ and $Noverlap = 8$) and (d2) (with $N = 48$ and $Noverlap = 12$), which show that the frequency becomes higher as time goes by. Such a STFT can be used to determine and display the spectrum of a non-stationary signal (such as speech or music) whose frequency content changes with time.

Now, comparing Fig. 3.20(d1) and (d2), let us consider the effect of segment or window duration NT on the time and frequency resolutions of the STFT. We might say that the time resolution is NT [sec] (the segment duration) and the frequency resolution is $1/NT$ [Hz]. This implies that if we increase or decrease the segment duration, the time/frequency resolution gets poorer/better or better/poorer, showing the trade-off between the time and frequency resolutions. It is an illustration of uncertainty principle (Remark 2.9(1)) that both time resolution and frequency resolution cannot be improved. Although the boundary of the uncertainty principle (best simultaneous resolution of both) is reached with a Gaussian window function, the wavelet transform or multi-resolution analysis was devised, which can give a good time/frequency resolution for high/low-frequency components.

3.10 Summary

In this chapter we have defined the discrete-time Fourier transform (DTFT) and the discrete Fourier series(DFS)/transform(DFT) for discrete-time signals and examined their properties together with their relationship. The DFT of a discrete-time sequence $x[n]$ has been shown to correspond to one period of the discrete-time Fourier series for the periodic extension of $x[n]$. We have also discussed the FFT algorithms that deserve the overwhelming popularity owing to their computational efficiency. This chapter also introduced some examples, which give you some interpretation of the DFT and will hopefully help you to gain an insight into physical meaning or practical usage of the DFT. It ended up with the introduction of the short-time Fourier transform (STFT), which can easily be obtained by using a MATLAB function "`specgram()`" in 6.x versions and "`spectrogram()`" in 7.x versions.

Problems

3.1 Properties of DTFT (Discrete-Time Fourier Transform)

(a) Prove the frequency-shifting property (3.2.7) by using the DTFT formula (3.1.1) or IDTFT formula (3.1.3).

$$x[n]e^{j\Omega_1 n} \overset{\mathcal{F}}{\leftrightarrow} X(\Omega - \Omega_1) \tag{P3.1.1}$$

(b) Prove the convolution property (3.2.8).

$$y[n] = x[n]*g[n] \overset{\mathcal{F}}{\leftrightarrow} Y(\Omega) = X(\Omega)G(\Omega) \tag{P3.1.2}$$

(c) Prove the modulation property (3.2.9).

$$y[n] = x[n]\, m[n] \overset{\mathcal{F}}{\leftrightarrow} Y(\Omega) = \frac{1}{2\pi} X(\Omega)*M(\Omega) \tag{P3.1.3}$$

(d) Prove the summation property (3.2.11) by using the convolution property (3.2.8) together with Eq. (E3.8.6).

$$\sum\nolimits_{m=-\infty}^{n} x[m] = x[n] * u_s[n] \overset{\mathcal{F}}{\leftrightarrow} \frac{1}{1-e^{-j\,\Omega}} X(\Omega) + \pi X(0) \sum\nolimits_{i=-\infty}^{\infty} \delta(\Omega - 2\pi i) \tag{P3.1.4}$$

3.2 Discrete-Time Hilbert Transformer

(a) Prove that the impulse response of a system, called the Hilbert transformer, with frequency response

$$H(\Omega) = \begin{cases} -j & \text{for } 0 < \Omega < \pi \\ +j & \text{for } -\pi < \Omega < 0 \end{cases} \tag{P3.2.1}$$

is

$$h[n] = \begin{cases} 2/n\pi & \text{for } n: \text{ odd} \\ 0 & \text{for } n: \text{ even} \end{cases} \tag{P3.2.2}$$

```
%sig03p_02.m
EPS=1e-10; Nx=200; nx=[-Nx:Nx]; W1=0.1*pi; xn= cos(W1*nx);
Nh=50; nh=[-Nh:Nh]; hn= (mod(nh,2)==1).*(2/pi./(nh+EPS));
yn= conv(hn,xn); ny=Nh+Nx+1+nx; yn1= yn(ny);
W=[-50:-1 1:50]*(pi/50);
X= DTFT(xn,W); Xmag= abs(X); Xphase= angle(X);
Xphase= (abs(imag(X))>=EPS).*Xphase;
H= DTFT(hn,W); Hmag= abs(H); Hphase= angle(H);
Y= DTFT(yn,W); Ymag= abs(Y); Yphase= angle(Y);
subplot(331), plot(nx,xn), axis([nx([1 end]) -1.5 1.5])
subplot(332), plot(W,Xmag), subplot(333), plot(W,Xphase)
subplot(334), plot(nh,hn), axis([nh([1 end]) -1 1])
subplot(335), plot(W,Hmag), subplot(336), plot(W,Hphase)
subplot(337), plot(nx,yn1, nx,sin(W1*nx),'r')
subplot(338), plot(W,Ymag), subplot(339), plot(W,Yphase)
```

(b) By taking the IDTFT of $Y(\Omega) = H(\Omega)X(\Omega)$, prove that the output of the Hilbert transformer to a sinusoidal input $x[n] = \cos(\Omega_0 n)$ is

$$y[n] = \sin(\Omega_0 n) \tag{P3.2.3}$$

(c) We can run the above program "sig03p_02.m" to rest assured of the fact that the output of the Hilbert transformer to $x[n] = \cos(0.1\pi n)$ is $y[n] = \sin(0.1\pi n)$ where $y[n] = h[n]*x[n]$ has been obtained from the time-domain input-output relationship. Identify the statements yielding the impulse response $h[n]$, output $y[n]$, and frequency response $H(\Omega)$.

3.3 Discrete-Time Differentiator

(a) Prove that the impulse response of a discrete-time differentiator with frequency response

$$H(\Omega) = j\Omega \tag{P3.3.1}$$

is

$$h[n] = \frac{(-1)^n}{n} \tag{P3.3.2}$$

(b) By taking the IDTFT of $Y(\Omega) = H(\Omega)X(\Omega)$, prove that the output of the differentiator to a sinusoidal input $x[n] = \sin(\Omega_0 n)$ is

$$y[n] = \Omega_0 \cos(\Omega_0 n) \tag{P3.3.3}$$

(c) By reference to the program "sig03p_02.m", compose a program and run it to rest assured of the fact that the output of the differentiator to $x[n] = \sin(0.1\pi n)$ is $y[n] = 0.1\pi \cos(0.1\pi n)$ where $y[n] = h[n] * x[n]$ has been obtained from the time-domain input-output relationship.

3.4 BPF Realization via Modulation-LPF-Demodulation – Frequency Shifting Property of DTFT
Consider the realization of BPF of Fig. P3.4(a), which consists of a modulator (multiplier), an LPF, and a demodulator (multiplier). Assuming that the spectrum of $x[n]$ is as depicted in Fig. P3.4(b), sketch the spectra of the signals $s[n]$, $v[n]$, and $y[n]$.

3.5 Quadrature Multiplexing – Modulation (Complex Convolution) Property of DTFT
Consider the quadrature multiplexing system depicted in Fig. P3.5 where the two signals are assumed to be bandlimited, i.e.,

$$X_1(\Omega) = X_2(\Omega) = 0 \text{ for } \Omega > \Omega_M \tag{P3.5.1}$$

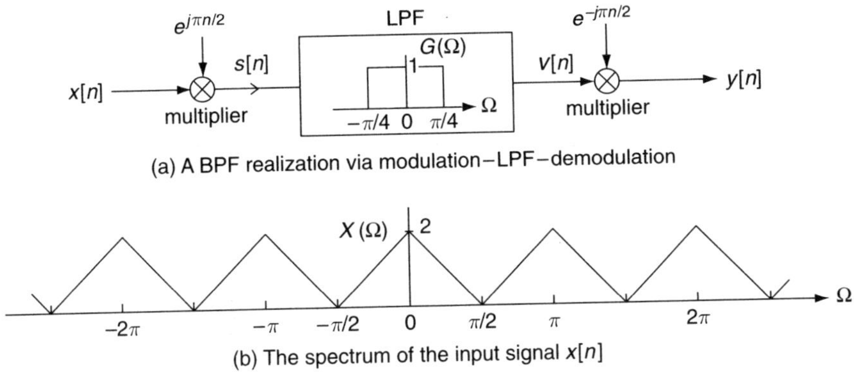

Fig. P3.4

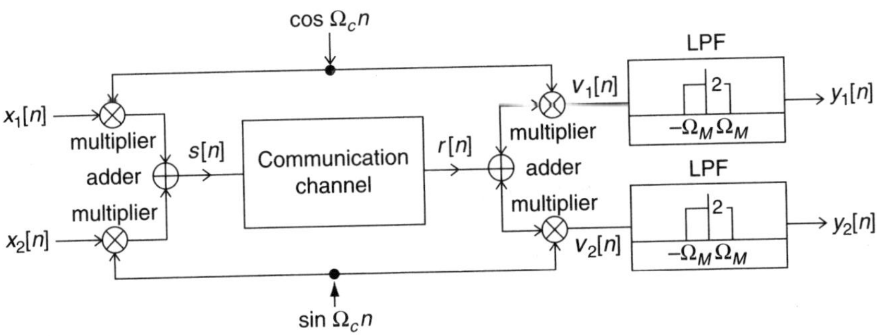

Fig. P3.5 Quadrature multiplexing

Assuming that the channel is ideal so that $r[n] = s[n]$, express the spectra $S(\Omega)$, $V_1(\Omega)$, $V_2(\Omega)$, $Y_1(\Omega)$, and $Y_2(\Omega)$ of $s[n]$, $v_1[n]$, $v_2[n]$, $y_1[n]$, and $y_2[n]$ in terms of $X_1(\Omega)$ and $X_2(\Omega)$.

3.6 Windowing (Complex Convolution) Property of DTFT
Referring to Example 3.10, answer the following questions:

(a) In the program "sig03e10.m", increase the (rectangular) window size parameter M and run the program to see the window spectrum and the frequency response of the designed FIR filter. Based on the results, describe the shape of the filter frequency response when M is set to infinity.

(b) With the window size parameter M set to 3 as it was, change the shape of the window into, say, one of the following: Bartlett (triangular), Hamming, Hanning, Blackman, and Kaiser. Note that the window sequences can be generated (as column vectors) by the MATLAB commands, `bartlett()`, `hamming()`, `hanning()`, `blackman()`, and `kaiser()`, respectively. Has the ripple of the frequency response been reduced?

3.7 DTFT and CTFT of a Finite-Duration Signal

```
%sig03p_07.m
% Fig. 3.7(To tell the relationship between CTFT and DTFT)
EPS=1e-10; T=1; t =[-10:T:10];
N=400; f=[-500:500]/N + EPS; % Normalized frequency range
W=2*pi*f; w=W/T; % Digital/Analog frequency range
xn= [zeros(1,7) 1 2 1 0 -1 -2 -1 zeros(1,7)];
Xw= inline('j*8*sin(2*w).*sinc(w/pi).^2','w');
X_CTFT= Xw(w); X_CTFT_mag= abs(X_CTFT);
Sum_of_X_CTFTs= (Xw(w+2*pi/T)+Xw(w)+Xw(w-2*pi/T))/T;
X_DTFT=DTFT(xn,W); X_DTFT_mag= abs(X_DTFT); X_DTFT_ph= angle(X_DTFT);
Discrepancy_between_CTFT_DTFT= norm(X_DTFT-Sum_of_X_CTFTs)/norm(X_DTFT)
subplot(321), stem(t,xn,'.'), hold on, plot(t,xn,':')
subplot(323), plot(W,X_DTFT_mag,'b', W,X_CTFT_mag/T,'k:')
hold on, plot(W,abs(Sum_of_X_CTFTs),'r:')
```

The objective of the above program "sig03p_07.m" is to plot the CTFT spectrum $X_1(\omega)/T$ of the unsampled dual triangular pulse signal $x_1(t)$ given by Eq. (E3.11.3) and the DTFT spectrum $X_1(\Omega)$ of the triangular pulse sequence $x_1[n] = x_1(nT)$ with $T = 1$. Also, it plots the sum of three CTFTs shifted by $-2\pi/T$, 0, and $+2\pi/T$ along the analog frequency $\omega = \Omega/T$ and scaled by $1/T$ to check the validity of the relationship between the CTFT and DTFT described by Eq. (3.5.7). Modify the program so that it does the same job with sampling interval $T = 0.5$. What can we get by shortening the sampling interval? Refer to Remark 3.7(2).

3.8 DFT/DFS, DTFT, and CTFS of a Noncausal Signal
Consider the dual triangular pulse sequence which would be obtained by sampling $x_1(t)$ in Fig. 3.7(a1) with sampling interval $T = 0.5$.

(a) Use the MATLAB function "`fft()`" to get the $N = 16$-point DFT $X_N(k)$ of the sequence and plot its magnitude together with the DTFT magnitude against $\Omega_k = [0 : N-1] \times (2\pi/N)$ and $\Omega = [-500 : 500] \times (2\pi/400)$, respectively. Note that the DTFT might have been obtained in Problem 3.7.

(b) Noting that the CTFS coefficients X_k are the samples of the CTFT $X(\omega)$ (Eq. (E3.11.3)) at $\omega = k\omega_0 = 2\pi k/P = 2\pi k/NT$ as described by Eq. (2.2.4) or stated in Remark 2.7(2), find the relative errors between the DFT $X_N(k)$ and two partial sums of the shifted CTFSs and compare them to argue for the relationship (3.5.3) between the DFT and CTFS.

$$E_1 = \frac{||X_N(k) - X_k/T||}{||X_N(k)||} \quad \text{and}$$

$$E_2 = \frac{||X_N(k) - (X_{k+N} + X_k + X_{k-N})/T||}{||X_N(k)||} \tag{P3.8.1}$$

<Hint> You might need to combine some part of the program "sig03p_07.m" with the following statements:

```
xn_causal= [0 -0.5 -1 -1.5 -2 -1.5 -1 -0.5 0 0.5 1 1.5 2 1.5 1 0.5];
X_DFT= fft(xn_causal); X_DFT_mag= abs(X_DFT);
N=length(xn_causal); k=[0:N-1];
f0=1/N; fk=k*f0; W0=2*pi*f0; Wk=k*W0; w0=W0/T;
stem(Wk,X_DFT_mag,'m.')
% (b): Partial sum of CTFSs by Eq.(3.5.3)
Sum_of_Xks= (Xw((k+N)*w0)+Xw(k*w0)+Xw((k-N)*w0))/T;
Discrepancy_between_CTFS_and_DFT= norm(X_DFT-Xw(k*w0)/T)/norm(X_DFT)
Discrepancy_between_CTFS_and_DFT3= norm(X_DFT-Sum_of_Xks)/norm(X_DFT)
```

3.9 CTFS, CTFT, and DFT/DFS of a Noncausal Sequence

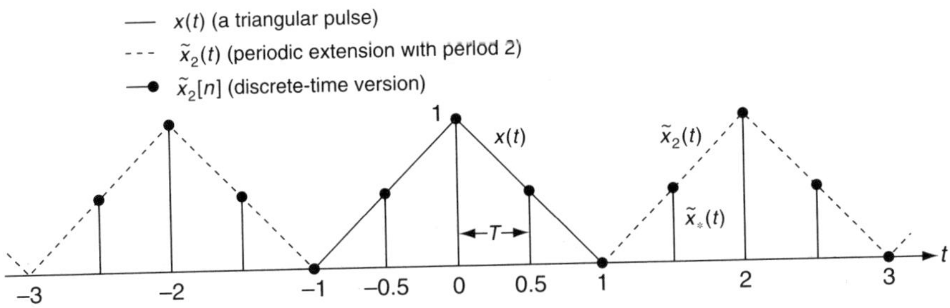

Fig. P3.9

Consider the triangular pulse signal $x(t)$ of Fig. P3.9, its periodic extension $\tilde{x}_2(t)$ with period $P = 2$, and its sampled periodic version $\tilde{x}_*(t) = \tilde{x}_2(t)\delta_T(t)$ with sampling interval $T = 0.5$.

(a) Find the CTFT $X(\omega)$ of $x(t)$. You can use Eq. (E2.3.4).
(b) Find the CTFS X_k of $\tilde{x}_2(t)$. You can use the result of (a) and Eq. (2.2.4) stating that the CTFS coefficients are the samples of the CTFT at $\omega = k\omega_0 = 2\pi k/P$.
(c) Show that the $N = 4$-point DFT of the discrete-time version $\tilde{x}[n]$ is

$$X(k) = [2\ 1\ 0\ 1] \tag{P3.9.1}$$

(d) Show the validity of the relationship between the CTFS and DFT described by Eq. (3.5.3) for this case, i.e.,

$$X(k) \overset{(P3.9.1)}{=} [2\ 1\ 0\ 1] \overset{?}{=} \frac{1}{T}\sum_{m=-\infty}^{\infty} X_{k+mN} \tag{P3.9.2}$$

<Hint> You might need the following formula

$$\sum_{m=0}^{\infty} \frac{1}{(2m+1)^2} = \frac{\pi^2}{8} \tag{P3.9.3}$$

which can be derived by substituting $A = 1$, $D/P = 1/2$ into Eq. (E2.1.6) as

$$A\tilde{\lambda}_{D/P}(t) \overset{(E2.1.6)}{=} \frac{1}{P}\sum_{k=-\infty}^{\infty} AD\text{sinc}^2\left(k\frac{D}{P}\right) \overset{A=1,\ P=2D}{\rightarrow}$$

$$1 = \frac{1}{2}\sum_{k=-\infty}^{\infty} \text{sinc}^2\left(\frac{k}{2}\right)$$

$$= \frac{1}{2}\left(\overset{\text{for } k=0}{1} + \overset{\text{for } k=\text{even}(\neq 0)}{0} + \overset{\text{for } k=odd}{\sum_{k=2m+1}^{\infty} \frac{1}{(k\pi/2)^2}}\right);$$

$$\frac{1}{2} = \frac{1}{2}\sum_{m=-\infty}^{\infty} \frac{1}{(2m+1)^2\pi^2/2^2} = 2\sum_{m=-\infty}^{\infty} \frac{1}{(2m+1)^2\pi^2}$$

$$= 4\sum_{m=0}^{\infty} \frac{1}{(2m+1)^2\pi^2}$$

3.10 DTFT and DFT of Discrete-Time Periodic Sequences

When we need to find the DTFT of a periodic sequence $\tilde{x}_N[n]$ with period N, we first get its DFS representation

$$\tilde{x}_N[n] \underset{\text{IDFT}}{\overset{(3.4.3)}{=}} \frac{1}{N}\sum_{k=0}^{N-1} \tilde{X}_N(k)e^{j2\pi kn/N} \tag{P3.10.1}$$

$$\text{with } \tilde{X}_N(k) \underset{\text{DFT}}{\overset{(3.4.2)}{=}} \sum_{n=0}^{N-1} \tilde{x}_N[n]e^{-j2\pi kn/N}$$

and then use Eq. (3.1.5) to obtain the DTFT of $\tilde{x}_N[n]$ as

$$\tilde{x}_N[n] = \frac{1}{N}\sum_{k=0}^{N-1} \tilde{X}_N(k)e^{j2\pi kn/N} \overset{\mathcal{F}}{\leftrightarrow}$$

$$X(\Omega) \underset{(3.1.5)}{\overset{(P3.10.1)}{=}} \frac{2\pi}{N}\sum_{k=-\infty}^{\infty} \tilde{X}_N(k)\delta\left(\Omega - \frac{2\pi k}{N}\right) \tag{P3.10.2}$$

(a) Verify the validity of this DTFT pair by showing that the IDTFT of (P3.10.2) is $\tilde{x}_N[n]$.

$$F^{-1}\{X(\Omega)\} \overset{(3.1.3)}{=} \frac{1}{2\pi}\int_{2\pi} X(\Omega)\, e^{j\Omega n}\, d\Omega$$

$$\underset{(1.1.25)}{\overset{(P3.10.2)}{=}} \frac{1}{N}\sum_{k=0}^{N-1} \tilde{X}_N(k)\, e^{j\,2\pi kn/N} \overset{(P3.10.1)}{=} \tilde{x}_N[n] \tag{P3.10.3}$$

(b) Use the DTFT relation (P3.10.2) to show that the DTFT of the discrete-time impulse train

$$\delta_N[n] = \sum_{m=-\infty}^{\infty} \delta[n-mN] \text{ (also called the } \textit{comb sequence}) \quad \text{(P3.10.4)}$$

is

$$D_N(\Omega) = \frac{2\pi}{N}\sum_{k=-\infty}^{\infty} \delta\left(\Omega - \frac{2\pi k}{N}\right) = \frac{2\pi}{N}\sum_{k=-\infty}^{\infty} \delta\left(\Omega + \frac{2\pi k}{N}\right) \quad \text{(P3.10.5)}$$

3.11 Effect of Discrete-Time Sampling on DTFT

As a continuous-time sampling with sampling interval T can be described by

$$x_*(t) \overset{\text{(E2.13.1)}}{=} x(t)\delta_T(t) \quad \left(\delta_T(t) = \sum_{m=-\infty}^{\infty} \delta(t-mT) : \text{ the impulse train}\right)$$

so can a discrete-time sampling with sampling interval N be described by

$$x_*[n] = x[n]\delta_N[n] \quad \left(\delta_N[n] = \sum_{m=-\infty}^{\infty} \delta[n-mN] : \text{the impulse train}\right) \quad \text{(P3.11.1)}$$

(a) Use the DTFT modulation property (3.2.9) or B.4(9) and Eq. (P3.10.5) to show that the DTFT of this sampled sequence $x_*[n]$ can be expressed in terms of that of the original sequence $x[n]$ as

$$X_*(\Omega) = \frac{1}{N}\sum_{m=-\infty}^{\infty} X\left(\Omega + \frac{2\pi m}{N}\right) = \frac{1}{N}\sum_{m=-\infty}^{\infty} X(\Omega + m\Omega_s) \text{ with } \Omega_s = \frac{2\pi}{N} \quad \text{(P3.11.2)}$$

(b) Show that the DTFT of a decimated sequence $x_d[n]$ obtained by removing the zeros from the sampled sequence $x_*[n] = x[n]\delta_N[n]$ is

$$X_d(\Omega) = X_*\left(\frac{\Omega}{N}\right) \quad \text{(P3.11.3)}$$

(cf.) This conforms with Eq. (3.2.14), which is the scaling property of DTFT.

(c) Substitute Eq. (P3.11.2) into Eq. (P3.11.3) to show that

$$X_d(\Omega) = \frac{1}{N}\sum_{m=0}^{N-1} X\left(\frac{\Omega}{N} + \frac{2\pi m}{N}\right) \quad \text{(P3.11.4)}$$

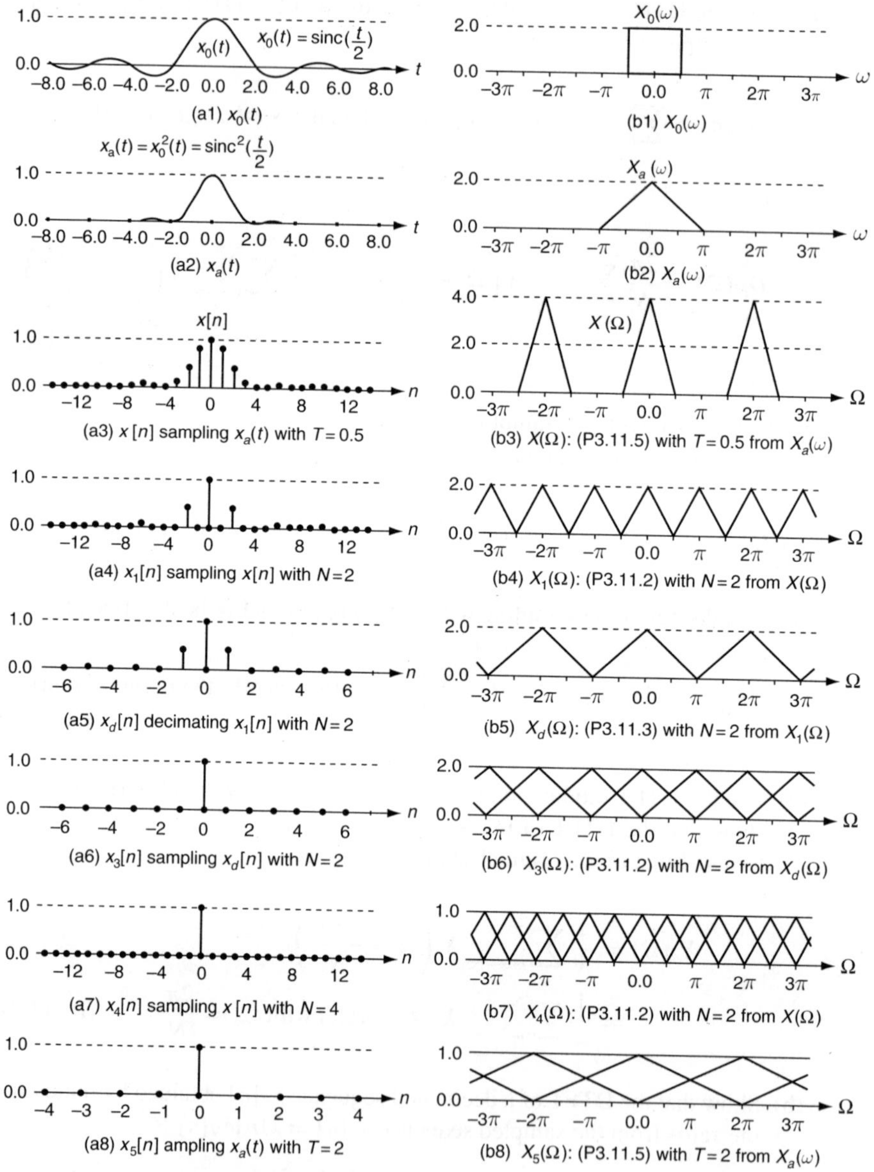

Fig. P3.11 Effect of sampling on the spectrum

(d) Let $x[n]$ be obtained from the samples of $x_a(t)$ with sampling interval T. Then, from Eq. (3.5.7), we can write

$$X(\Omega) = X_*(\omega)|_{\omega=\Omega/T} \overset{(3.5.7)}{=} \frac{1}{T} \sum\nolimits_{k=-\infty}^{\infty} X_a\left(\frac{\Omega}{T} + k\frac{2\pi}{T}\right) \quad \text{(P3.11.5)}$$

Substitute this into Eq. (P3.11.4) to show that

$$X_d(\Omega) = \frac{1}{NT}\sum_{k=-\infty}^{\infty} X_a\left(\frac{\Omega}{NT} + \frac{2\pi k}{NT}\right) \tag{P3.11.6}$$

(e) Verify the CTFT relation between the signal and its spectrum depicted in Figs. P3.11(a2) and (b2):

$$x_a(t) = x_0^2(t) = \mathrm{sinc}^2\left(\frac{t}{2}\right) \overset{\mathcal{F}}{\leftrightarrow} \frac{1}{2\pi}2r_\pi(\omega) * 2r_\pi(\omega) \text{ (triangular pulse)} \tag{P3.11.7}$$

Also verify the DTFT relation between the signal and its spectrum depicted in Fig. P3.11(a3) and (b3):

$$x[n] = x_0^2[n] = \mathrm{sinc}^2\left(\frac{n}{4}\right) \overset{\mathcal{F}}{\leftrightarrow} \frac{1}{2\pi}4r_{\pi/2}(\Omega) * 4r_{\pi/2}(\Omega) \text{ (triangular pulse)} \tag{P3.11.8}$$

(f) Show that the DTFT spectrum $X(\Omega)$ of $x[n]$ (Fig. P3.11(a3)) can be expressed in terms of the CTFT spectrum $X_a(\omega)$ of $x_a(t)$ (Fig. P3.11(a2)) as

$$X(\Omega) = 2\sum_{k=-\infty}^{\infty} X_a(2(\Omega - 2\pi k)) \tag{P3.11.9}$$

Show that the DTFT spectrum $X_1(\Omega)$ of $x_1[n]$ (Fig. P3.11(a4)) can be expressed in terms of the DTFT spectrum $X(\Omega)$ of $x[n]$ (Fig. P3.11(a3)) as

$$X_1(\Omega) = \frac{1}{2}(X(\Omega) + X(\Omega + \pi)) \tag{P3.11.10}$$

Show that the DTFT spectrum $X_d(\Omega)$ of $x_d[n]$ (Fig. P3.11(a5)) can be expressed in terms of the DTFT spectrum $X(\Omega)$ of $x[n]$ (Fig. P3.11(a3)) as

$$X_d(\Omega) = \frac{1}{2}\left(X\left(\frac{\Omega}{2}\right) + X\left(\frac{\Omega + 2\pi}{2}\right)\right) \tag{P3.11.11}$$

Show that the DTFT spectrum $X_3(\Omega)$ of $x_3[n]$ (Fig. P3.11(a6)) can be expressed in terms of the DTFT spectrum $X_d(\Omega)$ of $x[n]$ (Fig. P3.11(a5)) as

$$X_3(\Omega) = \frac{1}{2}(X_d(\Omega) + X_d(\Omega + \pi)) \tag{P3.11.12}$$

Show that the DTFT spectrum $X_4(\Omega)$ of $x_4[n]$ (Fig. P3.11(a7)) can be expressed in terms of the DTFT spectrum $X_d(\Omega)$ of $x_d[n]$ (Fig. P3.11(a5)) as

$$X_4(\Omega) = \frac{1}{4}\left(X_d(\Omega) + X_d\left(\Omega + \frac{\pi}{2}\right) + X_d(\Omega + \pi) + X_d\left(\Omega + \frac{3}{2}\pi\right)\right) \quad \text{(P3.11.13)}$$

Show that the DTFT spectrum $X_5(\Omega)$ of $x_5[n]$ (Fig. P3.11(a8)) can be expressed in terms of the CTFT spectrum $X_a(\omega)$ of $x_a(t)$ (Fig. P3.11(a2)) as

$$X_5(\Omega) = \frac{1}{2}\sum_{m=-\infty}^{\infty} X_a\left(\frac{\Omega + 2m\pi}{2}\right) \quad \text{(P3.11.14)}$$

(cf.) The sampled sequences $x_3[n]$, $x_4[n]$, and $x_5[n]$ are all identical to the impulse sequence and their flat spectra illustrate the extreme case of a spectrum blurred by downsampling.

3.12 Linear Convolution and Correlation Using DFT
Consider the following two sequences:

$$x[n] = [1\ \ 2\ \ 3\ \ 4\ \ 5\ \ 6\ \ 7\ \ 8] \ \text{and}\ \ y[n] = [1\ \ -2\ \ 3] \quad \text{(P3.12.1)}$$

(a) Referring to Remark 1.5, determine the length of the linear convolution or correlation of the two sequences. Also use the MATLAB command "`nextpow2()`" to find the smallest power of 2 that is greater than or equal to the expected length of the linear convolution or correlation. What is the minimum size N of DFT that can make the FFT operation most efficient for computing the linear convolution or correlation?

(b) Referring to Sect. 3.4.2, use the N-point FFT to compute the linear convolution of $x[n]$ and $y[n]$. Check if the result conforms with that obtained by using "`conv(x,y)`".

(c) It seems that we can use the N-point FFT to compute the correlation of $x[n]$ and $y[n]$ based on the following facts:

$$\circ\ \tilde{\phi}_{xy}[n] \overset{(1.4.12\text{b})}{=} \tilde{x}[n] * \tilde{y}[-n] \underset{\text{with period } N}{\overset{\text{periodicity of } \tilde{y}[n]}{=}} \tilde{x}[n] * \tilde{y}[N-n] \quad \text{(P3.12.2)}$$

$$\circ\ x[-n] \underset{(3.2.3)}{\overset{\mathcal{F}}{\leftrightarrow}} X(-\Omega) \text{ implies that } \mathrm{DFT}_N\{\tilde{y}[N-n]\} = \tilde{Y}(N-k) \quad \text{(P3.12.3)}$$

Noting that $\tilde{y}[N-n]$ can be programmed as "`[y(1) fliplr(y(2:end))]`", compose a program to compute the correlation and run it to get $\phi_{xy}[n]$. Check if the result conforms with that obtained by using "`xcorr(x,y)`".

3.13 DFT Spectra of Sampled Signals

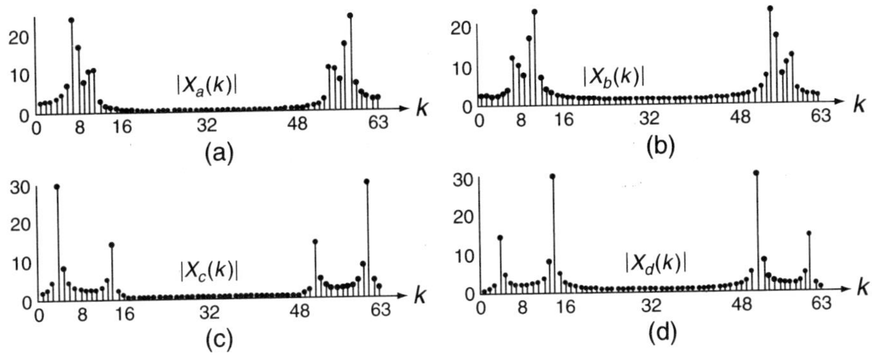

Fig. P3.13 DFT spectra of several discrete-time signals

With the sampling period $T = 0.1$ [s], we have sampled the following signals

(1) $0.5\cos 3\pi t + \sin 2\pi t$
(2) $\cos 3\pi t + 0.5\sin 2\pi t$
(3) $0.5\cos 4\pi t + \sin \pi t$
(4) $\cos 4\pi t + 0.5\sin \pi t$
(5) $0.5\cos 23\pi t + 0.5\sin 2\pi t - 0.5\sin 18\pi t$
(6) $0.5\sin 21\pi t + 0.5\cos 24\pi t + 0.5\cos 16\pi t$

and obtained their DFT spectra as depicted in Fig. P3.13.

(a) Find the corresponding spectrum for each signal from the DFT spectra depicted in Fig. P3.13.
(b) If these signals have been through an anti-aliasing prefilter, which one is expected to have the DFT magnitude spectrum like $|X_a(k)|$?

3.14 Spectral Leakage and Windowing Technique
Let us think about the cause of spectral leakage illustrated in Fig. 3.17(b) and some measure against it. As discussed in Remark 3.10(2), the spectral leakage stems from that any sequence is supposed to have periodicity with period N (equal to the DFT size), regardless of its peridocity, just because it is an object sequence of an N-point DFT. More specifically, the sequence $x_b[n]$ is supposed to be periodic with period $N = 16$ as depicted in Fig. P3.14(a), which shows such abrupt changes as discontinuities at both ends. A measure to reduce this abrupt change is to multiply a window sequence whose magnitude decreases gradually at both ends rather than cuts off abruptly as the rectangular window does. Figure P3.14(b), (c), (d1), and (d2) show a triangular (Bartlett) window $w_T[n]$ of length $N = 16$, the windowed sequence $x_b[n]w_T[n]$, and the DFT spectra of $x_b[n]$ and $x_b[n]w_T[n]$. Comparing the DFT spectra of unwindowed and windowed sequences, we can see that the windowing has

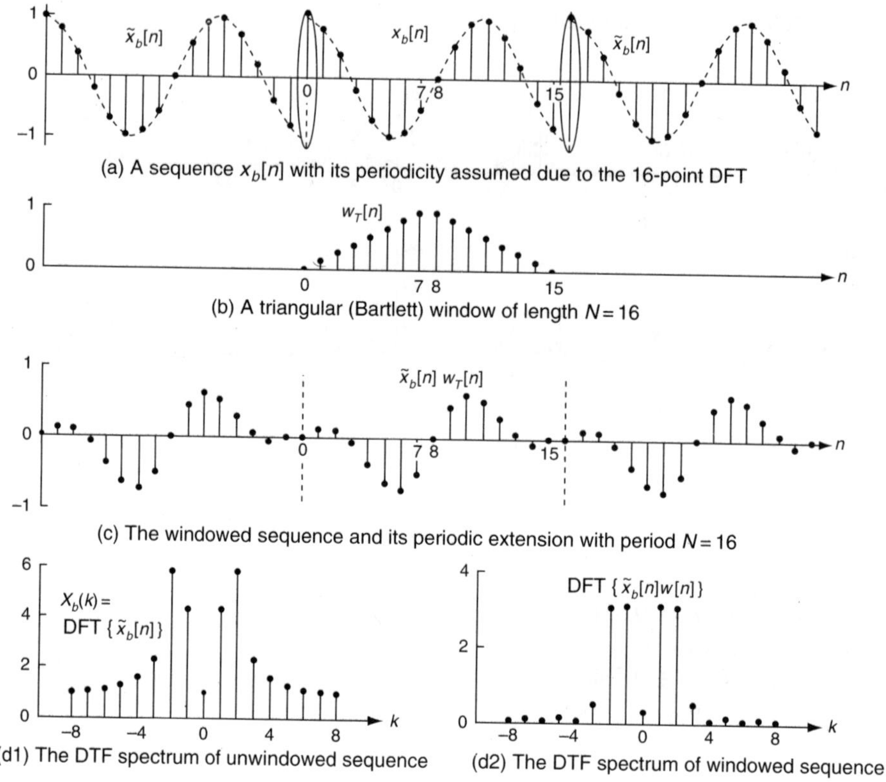

Fig. P3.14 The effect of windowing on the DFT spectrum

the effect of reducing the spectral leakage. You can try with another window such as the Hamming window or Hanning (raised-cosine) window.

3.15 Spectrum Blurring due to Down/Sub-sampling (Slower Sampling)
As can be seen in Fig. 3.19(f) or Eq. (3.8.7), slower sampling (sub-sampling) may make the spectrum blurred.

(a) Noting that $x_f[n]$ in Fig. 3.19(f) has been obtained from downsampling $x_a[n]$ by a factor of 2, sketch the DFT spectrum of the sequence which will be obtained from downsampling $x_f[n]$ by a factor of 2.
(b) Referring to Eq. (3.8.7), determine the condition on which the DFT spectrum is not blurred by downsampling it by a factor of 2.

3.16 Parallel Computation of DFTs for Two Real-valued Sequences (Problem 4.10 of [K-2])
We can compute the DFTs of two real-valued sequences $x_1[n]$ and $x_2[n]$ from the DFT $Y(k)$ of a complex-valued sequence

$$y[n] = x_1[n] + j\, x_2[n] \tag{P3.16.1}$$

where the DFT of each sequence can be written as

$$X_1(k) = X_{1R}(k) + j\,X_{1I}(k) = \sum_{n=0}^{N-1} x_1[n]\cos\left(\frac{2\pi kn}{N}\right) - j\sum_{n=0}^{N-1} x_1[n]\sin\left(\frac{2\pi kn}{N}\right) \quad \text{(P3.16.2a)}$$

$$X_2(k) = X_{2R}(k) + j\,X_{2I}(k) = \sum_{n=0}^{N-1} x_2[n]\cos\left(\frac{2\pi kn}{N}\right) - j\sum_{n=0}^{N-1} x_2[n]\sin\left(\frac{2\pi kn}{N}\right) \quad \text{(P3.16.2b)}$$

(a) Show that, since $x_i[n]$, $i = 1$ & 2, is real, $X_{iR}(k) = \mathrm{Re}\{X_i(k)\}$ is even and $X_{iI}(k) = \mathrm{Im}\{X_i(k)\}$ is odd via the conjugate symmetry (3.2.5) or Table B.5(4) so that

$$X_{iR}(k) = \tilde{X}_{iR}(-k)r_N[k] = \tilde{X}_{iR}(N-k) = X_{iR}(N-k) \quad \text{for} \quad k = 1,\cdots,N-1 \quad \text{(P3.16.3a)}$$

$$X_{iI}(k) = -\tilde{X}_{iI}(-k)r_N[k] = -\tilde{X}_{iI}(N-k) = -X_{iI}(N-k) \quad \text{for} \quad k = 1,\cdots,N-1 \quad \text{(P3.16.3b)}$$

where $r_N[k] = u_s[k] - u_s[N-k]$ is a rectangular pulse sequence.

(b) Show that the real and imaginary parts of $Y(k) = \mathrm{DFT}_N\{y[n]\}$ can be expressed in terms of $X_{1R}(k)$, $X_{1I}(k)$, $X_{2R}(k)$, and $X_{2I}(k)$ as

$$Y(k) = \mathrm{DFT}_N\{y[n] = x_1[n] + j\,x_2[n]\} = X_1(k) + jX_2(k) = X_{1R}(k) + j\,X_{1I}(k) + j\,(X_{2R}(k) + j\,X_{2I}(k));$$

$$Y_R(k) + j\,Y_I(k) = X_{1R}(k) - X_{2I}(k) + j\,(X_{1I}(k) + X_{2R}(k)) \quad \text{(P3.16.4)}$$

(c) Show that we can use the above relations (P3.16.3) and (P3.16.4) to express the DFTs $X_1(k)$ and $X_2(k)$ in terms of the real and imaginary parts of $Y(k) = \mathrm{DFT}_N\{y[n]\}$ as

$$X_1(k) = X_{1R}(k) + j\,X_{1I}(k) = \frac{1}{2}(Y_R(k) + \tilde{Y}_R(N-k)) + j\frac{1}{2}(Y_I(k) - \tilde{Y}_I(N-k)) = \frac{1}{2}(Y(k) + \tilde{Y}^*(N-k)) \quad \text{(P3.16.5a)}$$

$$X_2(k) = X_{2R}(k) + j\,X_{2I}(k) = \frac{1}{2}(Y_I(k) + \tilde{Y}_I(N-k)) - j\frac{1}{2}(Y_R(k) - \tilde{Y}_R(N-k)) = -j\frac{1}{2}(Y(k) - \tilde{Y}^*(N-k)) \quad \text{(P3.16.5b)}$$

(d) Compose a MATLAB function "[X1,X2]=fft_2real(x1,x2,N)", which uses Eq. (P3.16.5) to compute the DFTs of two given real-valued sequences x1 and x2. Run it to get the DFTs of the two sequences $x_a[n]$ and $x_b[n]$ of Fig. 3.17(a) and (b) simultaneously and plot them to check for its legitimacy.

3.17 Half-Size DFT Computation of Real-valued Sequences in OFDM Communication

Suppose we have an OFDM (Orthogonal Frequency Division Multiplexing) communication system in which the transmitter converts a length $N-1$ complex data sequence $\{X(k), k = 1 : N-1\}$ into a length $2N$ conjugate symmetric sequence

$$Y(k) = \begin{cases} 0 & \text{for } k = 0 \\ X(k) & \text{for } k = 1, \cdots, N-1 \\ 0 & \text{for } k = N \\ X^*(2N-k) & \text{for } k = N+1, \cdots, 2N-1 \end{cases} \quad \text{(P3.17.1)}$$

and performs a $2N$-point IFFT operation on $\{Y(k), k = 0 : 2N-1\}$ to make a length $2N$ real-valued sequence $\{y[n], n = 0 : 2N-1\}$ to be transmitted. With a noiseless channel and an exact synchronization, the receiver performs a $2N$-point FFT operation on the received sequence $y[n]$ to get $Y(k)$, and then restore $X(k)$. Let us consider how to exploit the conjugate symmetry to cut the DFT size by half.

(a) Referring to Problem 3.16, we let the receiver construct a complex-valued sequence from the received sequence $y[n]$ as

$$\begin{aligned} w[n] &\overset{\text{(P3.16.1)}}{=} w_1[n] + jw_2[n] \\ &= y[2n] + j\,y[2n+1] \quad \text{for } k = 0 : N-1 \end{aligned} \quad \text{(P3.17.2)}$$

and take the N-point FFT operation on $w[n]$ to compute $W(k) = \text{DFT}_N \{w[n]\}$. Then we can use Eq. (P3.16.5) to express the DFTs of $w_1[n]$ and $w_2[n]$ in terms of $W(k)$ as

$$W_1(k) = \text{DFT}_N\{w_1[n]\} \overset{\text{(P3.16.5a)}}{=} \frac{1}{2}(W(k) + \tilde{W}^*(N-k)) \quad \text{(P3.17.3a)}$$

$$W_2(k) = \text{DFT}_N\{w_2[n]\} \overset{\text{(P3.16.5b)}}{=} -j\frac{1}{2}(W(k) - \tilde{W}^*(N-k)) \quad \text{(P3.17.3b)}$$

Note that the $2N$-point DFT of the even-indexed sequence $y[2n] = w_1[n]$, i.e., $y[n] = w_1[n/2]$ (with zero-insertion) is the same as the repetition of $W_1(k)$ (Eq. (3.8.6)) and the $2N$-point DFT of the odd-indexed sequence $y[2n+1] = w_2[n]$ is the same as the repetition of $W_2(k)e^{-j2\pi k/2N}$ (Eq. (3.8.6)) and the time-shifting property B.5(7)) since deploying $w_2[n]$ as $y[2n+1]$ amounts to a zero-insertion followed by time-shifting by

$n = -1$. Therefore the DFT of the original received sequence $y[n] = y[2n] + y[2n+1]$ can be written as

$$Y(k) = W_1(k) + W_2(k)\, e^{-j\pi k/N} \overset{\text{(P3.17.3a,b)}}{=} \frac{1}{2}(W(k) + \tilde{W}^*(N-k)) \\ - j\frac{1}{2}e^{-j\pi k/N}(W(k) - \tilde{W}^*(N-k)) \qquad \text{(P3.17.4)}$$

This equation is used by the receiver for post-processing the FFT result.

(b) On the other hand, the transmitter is required to perform a pre-processing operation on $Y(k)$ to construct $W(k)$ whose IDFT is the complex-valued sequence as

$$w[n] \overset{\text{(P3.17.2)}}{=} w_1[n] + jw_2[n] \\ = y[2n] + j\, y[2n+1] \quad \text{for } k = 0 : N-1 \qquad \text{(P3.17.5)}$$

and take the N-point IFFT operation on $W(k)$ to compute $w[n] = \text{IDFT}_N\{W(k)\}$. Then the transmitter is supposed to construct a length $2N$ real-valued sequence $y[n] = y[2n] + y[2n+1]$ (to be transmitted) such that $y[2n] = w_1[n]$ and $y[2n+1] = w_2[n]$. To determine the pre-processing operation (at the transmitter) which is reverse to the post-processing one (at the receiver), let us write Eq. (P3.17.4) together with its Hermitian in matrix form as

$$\begin{bmatrix} Y(k) \\ \tilde{Y}^*(N-k) \end{bmatrix} \\ \underset{\text{with its Hermitian}}{\overset{\text{(P3.17.4)}}{=}} \frac{1}{2}\begin{bmatrix} 1 - j\,e^{-j\pi k/N} & 1 + j\,e^{-j\pi k/N} \\ 1 - j\,e^{j\pi(N-k)/N} & 1 + j\,e^{j\pi(N-k)/N} \end{bmatrix} \begin{bmatrix} W(k) \\ \tilde{W}^*(N-k) \end{bmatrix} \\ = \frac{1}{2}\begin{bmatrix} 1 - j\,e^{-j\pi k/N} & 1 + j\,e^{-j\pi k/N} \\ 1 + j\,e^{-j\pi k/N} & 1 - j\,e^{-j\pi k/N} \end{bmatrix} \begin{bmatrix} W(k) \\ \tilde{W}^*(N-k) \end{bmatrix} \qquad \text{(P3.17.6)}$$

This implies that the pre-processing operation on $Y(k)$ to construct $W(k)$ can be performed by

$$\begin{bmatrix} W(k) \\ \tilde{W}^*(N-k) \end{bmatrix} = 2\begin{bmatrix} 1 - j\,e^{-j\pi k/N} & 1 + j\,e^{-j\pi k/N} \\ 1 + j\,e^{-j\pi k/N} & 1 - j\,e^{-j\pi k/N} \end{bmatrix}^{-1} \begin{bmatrix} Y(k) \\ \tilde{Y}^*(N-k) \end{bmatrix} \\ = \frac{2}{-j\,4e^{-j\pi k/N}} \begin{bmatrix} 1 - j\,e^{-j\pi k/N} & -(1 + j\,e^{-j\pi k/N}) \\ -(1 + j\,e^{-j\pi k/N}) & 1 - j\,e^{-j\pi k/N} \end{bmatrix} \begin{bmatrix} Y(k) \\ \tilde{Y}^*(N-k) \end{bmatrix}$$

```
%sig03p_17.m
clear, clf
N=1024; N2=N*2; kk=[-N:N];
n=[0:N-1]; k=[0:N-1]; WN_k=exp(j*pi*k/N); WNk=exp(-j*pi*k/N);
Y= [0 2*(randint(1,N-1)+j*randint(1,N-1))-1]; % assumed data
% Transmitter
Y2= [Y 0 conj(fliplr(Y(2:N)))]; y2= ifft(Y2); % using 2N-point IDFT
Yc= conj([Y(1) fliplr(Y(2:N))]);
W= (Y+Yc + j*WN_k.*(Y-Yc))/2; w= ifft(W); % using N-point IDFT
y2_from_w= [real(w); imag(w)]; y2_from_w= y2_from_w(:).';
error_between_y2_and_y2_from_w= norm(y2-y2_from_w)/norm(y2)
% Receiver
y2_r= y2; Y2_r= fft(y2_r); % received sequence and its 2N-point DFT
y_r= reshape(y2_r,2,N); w_r= y_r(1,:)+j*y_r(2,:);
W_r= fft(w_r); Wc_r= conj([W_r(1) fliplr(W_r(2:N))]);
Y_from_W= (W_r+Wc_r - j*WNk.*(W_r-Wc_r))/2;
error_between_Y2_r_and_Y_from_W=norm(Y2_r(1:N)-Y_from_W)/norm(Y2_r)
```

$$= \frac{1}{2}\begin{bmatrix} 1 + j\,e^{j\pi k/N} & 1 - j\,e^{j\pi k/N} \\ 1 - j\,e^{j\pi k/N} & 1 + j\,e^{j\pi k/N} \end{bmatrix} \begin{bmatrix} Y(k) \\ \tilde{Y}^*(N-k) \end{bmatrix} \tag{P3.17.7}$$

$$W(k) = \frac{1}{2}(Y(k) + \tilde{Y}^*(N-k)) + j\frac{1}{2}e^{j\pi k/N}(Y(k) - \tilde{Y}^*(N-k)) \tag{P3.17.8}$$

The objective of the above program "sig03p_17.m" is to check the validity of this scheme by seeing if the result of using the N -point DFT agrees with that of using the $2N$ -point DFT. Identify the statements corresponding to Eqs. (P3.17.4) and (P3.17.8).

3.18 On-line Recursive Computation of DFT

We can compute the N -point DFT $X(k, m)$ of a long sequence $x[n]$ multiplied by a sliding rectangular window $r_N[n-m]$ of length N in the following recursive way where we call each windowed sequence $x[n]r_N[n-m]$ the mth data segment.

Let us define the 0th, m^{th}, and $(m+1)^{\text{th}}$ data segments as

$$\{x_0[0],\ x_0[1], \cdots, x_0[N-1]\} = \{0,\ 0, \cdots, 0\} \tag{P3.18.1a}$$

$$\{x_m[0],\ x_m[1], \cdots, x_m[N-1]\}$$
$$= \{x[m],\ x[m+1], \cdots, x[m+N-1]\} \tag{P3.18.1b}$$

$$\{x_{m+1}[0],\ x_{m+1}[1], \cdots, x_{m+1}[N-1]\}$$
$$= \{x[m+1],\ x[m+2], \cdots, x[m+N]\} \tag{P3.18.1c}$$

Thus we can express the N -point DFT $X(k, m+1)$ of $(m+1)^{\text{th}}$ data segment in terms of $X(k, m)$ (i.e., the DFT of the previous segment) as

$$\begin{aligned}
X(k, m+1) &= \sum_{n=0}^{N-1} x_{m+1}[n]\, W_N^{kn} = \sum_{n=0}^{N-1} x_m\,[n+1]\, W_N^{kn} \\
&= \sum_{n=0}^{N-1} x_m\,[n+1]\, W_N^{k(n+1)} W_N^{-k} \overset{n+1\to n}{=} \sum_{n=1}^{N} x_m[n]\, W_N^{kn}\, W_N^{-k} \\
&= \left(\sum_{n=0}^{N-1} x_m\,[n]\, W_N^{kn} + x[N] - x\,[0]\right) W_N^{-k} \\
&= (X(k, m) + x\,[N] - x\,[0])\, W_N^{-k} \qquad \text{(P3.18.2)}
\end{aligned}$$

Complete the following program "sig03p_18.m" to check the validity. What is the meaning of "discrepancy" in the program?

```
%sig03p_18.m
% Recursive computation of DFT
clear, clf
N=64; n=0:N-1; k=0:N-1;
W1=2*pi*8/N; x=sin(W1*n);
X=fft(x); %FFT
xa=[zeros(1,N) x]; % augment x with N zeros
Xm=zeros(1,N); WN_k=exp(j*2*pi*k/N);
for m=1:N
  Xm=(Xm+xa(N+m)-xa(m)).*????; %RDFT
end
discrepancy= norm(Xm-X)/norm(X)
stem(k,abs(Xm)), hold on, pause, stem(k,abs(X),'r')
title('RDFT and FFT')
```

3.19 Two-Dimensional (2-D) DFT
Two-dimensional DFT $X(k, l)$ of a 2-D sequence $x[m, n]$ and its IDFT $x[m, n]$ are defined as

$$X(k, l) = \sum_{m=0}^{M-1} \sum_{n=0}^{N-1} x\,[m, n]\, W_M^{km}\, W_N^{ln} \qquad \text{(P3.19.1a)}$$

for $0 \le k \le M-1$ and $0 \le l \le N-1$

$$x\,[m, n] = \frac{1}{MN} \sum_{k=0}^{M-1} \sum_{l=0}^{N-1} X(k, l)\, W_M^{-km}\, W_N^{-ln} \qquad \text{(P3.19.1b)}$$

for $0 \le m \le M-1$ and $0 \le n \le N-1$

The 2-D DFT can be computed by using a MATLAB built-in function `fft2()`. On the other hand, the 2-D DFT (P3.19.1a) can be written as two cascaded 1-D DFTs:

$$X(k,l)=\sum_{m=0}^{M-1} P(m,l)\, W_M^{km} \tag{P3.19.2a}$$

$$\text{for } 0\le k\le M-1 \quad \text{and} \quad 0\le l\le N-1$$

with
$$P(m,l)=\sum_{n=0}^{N-1} x\,[m,n]\, W_N^{ln} \tag{P3.19.2b}$$

$$\text{for } 0\le m\le M-1 \quad \text{and} \quad 0\le l\le N-1$$

This implies that the 2-D DFT can be computed by taking the 1-D DFT for each row of x to get $P(m,l)$ and then taking the 1-D DFT for each column of $P(m,l)$. Noting that the MATLAB function `fft()` computes the DFTs of each column of the input matrix, compose a program to check the validity of Eq. (P3.19.2a–b) for a 2-D rectangular pulse $x[m,n]$ that is produced and 2-D or 3-D plotted (in black and white) by the following MATLAB statements:

```
>>N=64; x=zeros(N,N); x(N/2-3:N/2+3,N/2-5:N/2+5)=255;
>>mm=[-N/2:N/2-1]; nn=[-N/2:N/2-1]; image(nn,mm,x) % 2-D plot
>>mesh(nn,mm,x), colormap(gray(256)); % Alternatively, 3-D plot
```

In fact, it does not matter whichever of the column and row is taken the DFT of first.

3.20 Short-Time Fourier Transform (STFT) Using Spectrogram
Let us use the MATLAB function spectrogram() to find the STFT for some signals.

```
%play_music_wave.m
clear, clf
Fs=10000; Ts=1/Fs; % 10kHz Sampling Frequency and Sampling period
Tw=2; % Duration of a whole note
melody_rhythm= [40 42 44 45 47 49 51 52; 1/4 1/4 1/4 1/4 1/8 1/8 1/8 1/8];
[x,tt]= music_wave(melody_rhythm,Ts,Tw); sound(x,Fs)
N=256; wnd=N; Noverlap= N/2;
subplot(221), spectrogram(x,wnd,Noverlap,N,Fs,'yaxis'); % specgram(x)
colormap(gray(256)) % colormap('default')
```

```
function [wave,tt]=music_wave(melody_rhythm,Ts,Tw)
% Ts: Sampling period, Tw: duration for a whole note
% Copyleft: Won Y. Yang, wyyang53@hanmail.net, CAU for academic use only
if nargin<3, Tw=2; end
if nargin<2, Ts=0.0001; end
[M,N]= size(melody_rhythm);
wave= []; tt=[]; pi2= 2*pi; phase= 0;
for i=1:N
   t= [Ts:Ts:melody_rhythm(2,i)*Tw];
   if i==1, tt=[tt t]; else tt=[tt tt(end)+t]; end
   w= pi2*440*2^((melody_rhythm(1,i)-49)/12); angle= w*t + phase;
   wave= [wave sin(angle)]; phase= angle(end);
end
```

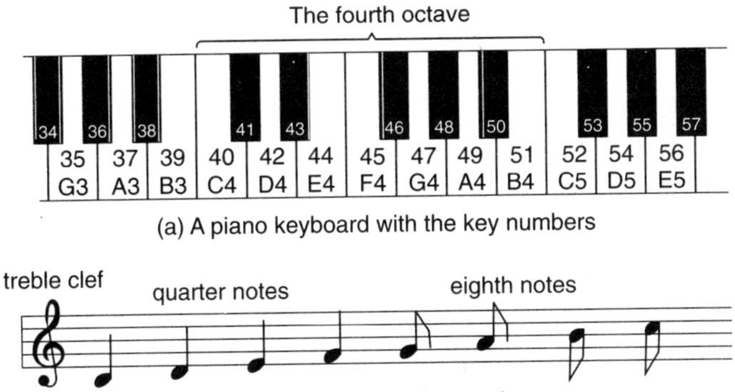

(a) A piano keyboard with the key numbers

(b) A melody written on the music score

Fig. P3.20.1

(a) Referring to the above program "play_music_wave.m" and the layout of a piano keyboard depicted in Fig. P3.20.1(a), use the MATLAB routine "music_wave" to produce a sinusoidal wave for a series of musical notes shown in Fig. P3.20.1(b) where the frequency for the central note A4 is 440 Hz and that for any other note is

$$f_n = 440 \times 2^{(n-49)/12}[\text{Hz}] \text{ where } n \text{ is the key number.} \tag{P3.20.1}$$

(b) Use the MATLAB function "specgram" (in MATLAB of version 6.x) or "spectrogram" (in MATLAB of version 7.x) to find the STFT for the sinusoidal wave produced in (a). You can change the DFT size from its default value N = 256 to 512 and then to 1024 and localize the frequency range if you think it to be of help in distinguishing the distinct frequency components. Which of the frequency and time resolutions do you gain or lose by increasing the DFT size?

(c) Beyond the current limit of your ability and the scope of this book, you can use the RDFT (recursive DFT) or any other technique such as the wavelet transform to improve the time and frequency resolution.

(d) Produce a beat-note signal by summing two sinusoids of slightly different frequencies 495 Hz and 505 Hz as

$$\begin{aligned} x(t) &= \cos(2\pi \times 495t) + \cos(2\pi \times 505t) \\ &= 2\cos(2\pi \times 500t)\cos(2\pi \times 5t) \end{aligned} \tag{P3.20.2}$$

Determine the FFT size (as a power of 2) such that the two frequencies are distinguishable in the spectrogram. How about the DFT size N which makes the frequency resolution

$$\frac{2}{NT} \cong \frac{505-495}{2}\text{Hz} \tag{P3.20.3}$$

where the sampling period T is chosen as 0.0001 s?

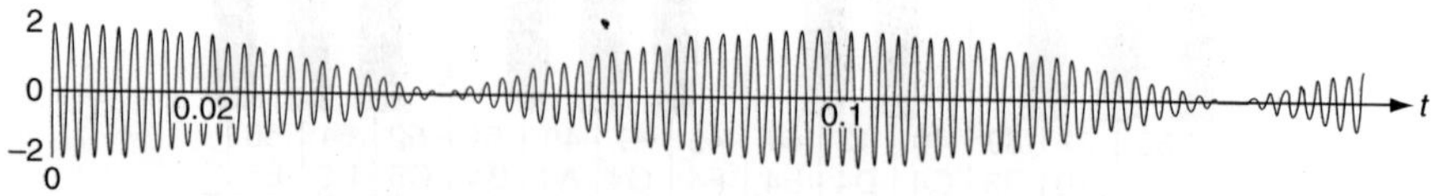

Fig. P3.20.2 A beat-note signal

(cf.) For example, the spectrogram with $N = 512$ shows a single frequency of around 500 Hz varying in amplitude with time, while that with $N = 4096$ reveals the two different frequencies with constant amplitude.

3.21 Design of DTMF (Dual-Tone Multi-Frequency) Signal Decoder
DTMF system transmits keys pressed on a keypad (Fig. P3.21(a)) through an audio channel such as a telephone line. Every time a key is pressed, it transmits two frequencies, each corresponding to the row and column in which the key is in.

```
%sig03p_21.m
% DTMF encoding and decoding
clear, clf
Fs=10000; Ts=1/Fs; % 10kHz Sampling Frequency and Sampling period
keypad.keys= ['1' '2' '3' 'A';
              '4' '5' '6' 'B';
              '7' '8' '9' 'C';
              '*' '0' '#' 'D'];
keypad.row_freqs= [697 770 852 941];
keypad.col_freqs= [1209 1336 1477 1633];
w=2*pi*[keypad.row_freqs keypad.col_freqs];
[x,tt]= dtmf_generator('159D',[0.1 0.2 0.1 0.2],Ts,keypad);
soundsc(x,Fs)
[keys,B,A]= dtmf_decoder(x,Ts,keypad);
keys
sim('dtmf',tt(end)); % run the Simulink model file 'dtmf' for tt(end)[sec]
```

```
function [wave,tt]=dtmf_generator(keys,durations,Ts,keypad)
% keys, durations : keys pressed and their durations in vectors
% Ts              : Sampling period
% Copyleft: Won Y. Yang, wyyang53@hanmail.net, CAU for academic use only
Nkey= length(keys); Nduration= length(durations);
if Nduration<Nkey
   durations= [durations durations(1)*ones(1,Nkey-Nduration)];
end
wave= []; tt=[]; pi2= 2*pi;
Nzero= ceil(0.1/Ts); zero_between_keys=zeros(1,Nzero);
tzero= [1:Nzero]*Ts;
for i=1:Nkey
   t= [Ts:Ts:durations(i)];
   if i==1, tt=[tt t];
     else tt=[tt tt(end)+t];
   end
   [m,n]= find(keys(i)==keypad.keys);
   if isempty(m), error('Wrong key in dtmf_generator'); end
   w2= pi2*[keypad.row_freqs(m); keypad.col_freqs(n)];
   wave= [wave sum(sin(w2*t)) zero_between_keys];
   tt=[tt tt(end)+tzero];
end
```

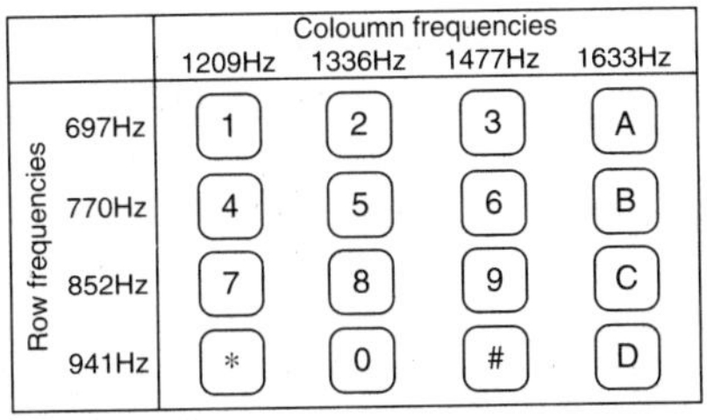

(a) A DTMF keypad with the row and column frequencies

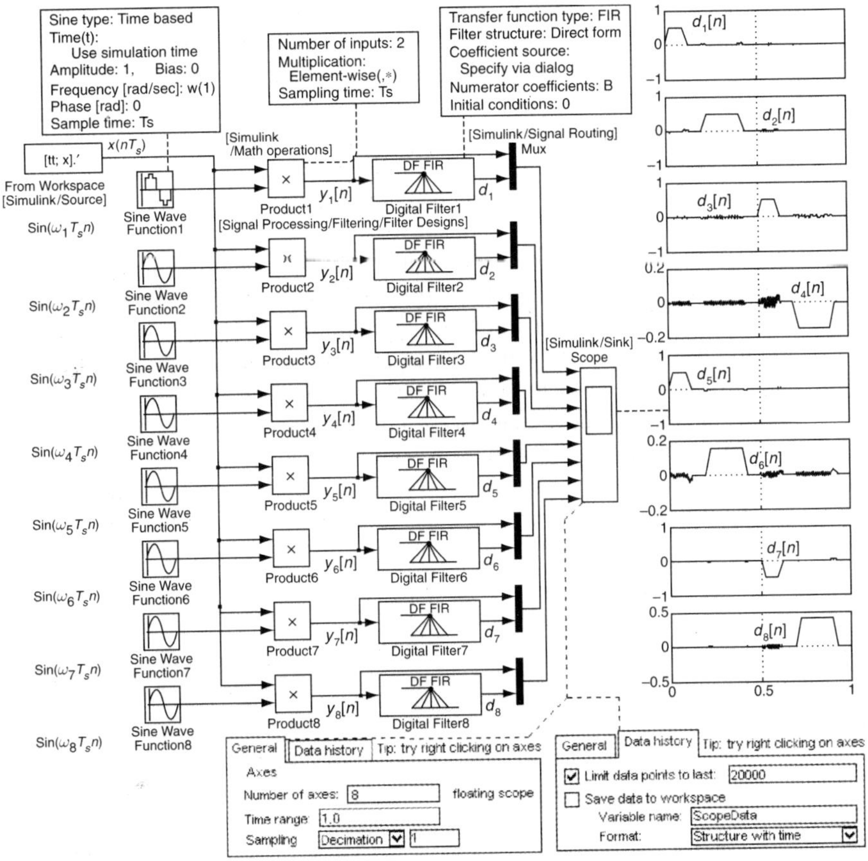

(b) The Simulink block diagram for a DTMF decoder

Fig. P3.21

(a) The program "dtmf_generator()" listed below generates a DTMF signal (wave) together with the time vector (tt) for given vector (keys) of keys pressed and vector (durations) of key pressing duration. A zero period of 0.1 s is inserted between the key signals so that the decoder can distinguish between a single long-pressed key and multiple repeated keys by accepting a new key only after detecting a zero period. Referring to the main program “sig03p_21.m”, make a DTMF signal for the leading three digits of your phone number and listen to its sound.

(b) Figure 3.21(b) shows a Simulink block diagram for a filter bank which can be used to detect the DTMF signal. In each bank, $\sin(\omega_i T_s n) = \sin(2\pi f_i T_s n)$ is multiplied with the input signal where $f_i = 697, 770, 852,$ 941, 1209, 1336, 1477, and 1633 Hz and the modulated signal is passed through a moving or running average filter (MAF). The DTMF signal x together with the time vector tt, the sampling period Ts, the analog frequency vector w ($\omega_i = 2\pi f_i$'s), and the running average filter coefficient vector B are supposed to be supplied via the MATLAB workspace. Note that the Simulink model file “dtmf.mdl” can be run directly or indirectly from a MATLAB program by using the MATLAB command “`sim()`” as illustrated in the program “sig03p_21.m”. Based on the simulation results observed through the scope, determine the common threshold value which can be used for detecting any one of the eight frequencies from the output ($d_i[n]$) of each MAF.

(c) Complete the following program “dtmf_decoder()” and run it on the DTMF signal generated in (a) for demonstration. Note that freq_durations(i) for each frequency is accumulated every sample time the corresponding MAF output is larger than the threshold value (Thd) and zero_duration is increased by one every time no frequency is detected. Whenever the accumulated zero duration exceeds some threshold value (Zero_duration) and the frequency durations for at least one of the row frequencies and at least one of the column frequencies exceed some threshold (Freq_duration), the key corresponding to the row/column frequencies is determined.

```
function [keys,B,A]=dtmf_decoder(x,Ts,keypad)
% <Inputs>    x    : Signal vector of DTMF tones
%             Ts   : Sampling period
% <Outputs>   Keys: Detected key vector
%             B,A : Numerator/denominator of the filter transfer function
% Copyleft:  Won Y. Yang, wyyang53@hanmail.net, CAU for academic use only
w=2*pi*[keypad.row_freqs keypad.col_freqs]; wTs=w*Ts;
NB=ceil(0.03/Ts); B=ones(1,NB)/NB; A=1; % FIR filter for running average
DCgain= sum(B)/sum(A); B= B/DCgain;
NB=length(B); NA=length(A);
M=length(wTs); % the input dimension of the filter
%wi=zeros(max(NA,NB)-1,M); % the initial state
Nx= length(x); tt=[1:Nx]*Ts; % the length of signal and the time vector
y=[]; d=[]; keys= [];
freq_durations=zeros(1,M); zero_duration=0; Thd=0.1;
Zero_duration= 0.08/Ts; Freq_duration= 0.09/Ts;
for n=1:Nx
   y= [y; x(n)*sin(wTs*n)];
   if n==1, [dn,wi]= filter(B,A,[zeros(1,M); y(end,:)]); % column-wise
    else [dn,wi]= filter(B,A,y([end-1 end],:),wi); % multi-filtering
   end
   d = [d; dn(end,:)]; tmp = (abs(dn(end,:))>Thd);
   freq_durations = freq_durations + tmp;
   zero_duration = zero_duration + (sum(tmp)==0);
   cond1= (zero_duration>Zero_duration);
   cond2= sum(freq_durations(1:4)>Freq_duration)>0;
   cond3= sum(freq_durations(5:8)>Freq_duration)>0;
   if cond1&cond2&cond3
    ???????????????????????????????????????
    ????????????????????????????????????????
    ?????????????????????????????????????????
    ??????????????????????????
    ???????????????????
   end
end
```

Chapter 4
The z-Transform

Contents

The z-transform is the discrete-time counterpart of the Laplace transform. It can be viewed as a generalization of the DTFT (discrete-time Fourier transform) just as the Laplace transform can be viewed as a generalization of the CTFT (continuous-time Fourier transform). It plays a vital role in the analysis and design of discrete-time linear time-invariant (LTI) systems that are described by difference equations. In this chapter we introduce the z-transform and discuss some of its properties and applications to LTI system analysis and representations.

W.Y. Yang et al., *Signals and Systems with MATLAB®*,
DOI 10.1007/978-3-540-92954-3_4,

4.1 Definition of the z-Transform

For a discrete-time sequence $x[n]$, the bilateral (two-sided) and unilateral (one-sided) z-transforms are defined as

$$\mathbb{X}[z] = \mathrm{Z}\{x[n]\} = \sum_{n=-\infty}^{\infty} x[n]z^{-n} \quad (4.1.1a)$$

$$X[z] = \mathcal{Z}\{x[n]\} = \sum_{n=0}^{\infty} x[n]z^{-n} \quad (4.1.1b)$$

where z is a complex variable. For convenience, the z-transform relationship will sometimes be denoted as

$$x[n] \overset{\mathrm{Z}}{\leftrightarrow} \mathbb{X}[z] \qquad\qquad x[n] \overset{\mathcal{Z}}{\leftrightarrow} X[z]$$

Note that the unilateral z-transform of $x[n]$ can be thought of as the bilateral z-transform of $x[n]u_s[n]$ ($u_s[n]$: the unit step sequence) and therefore the two definitions will be identical for causal sequences. The unilateral z-transform is particularly useful in analyzing causal systems described by linear constant-coefficient difference equations with initial conditions.

The z-transform does not necessarily exist for all values of z in the complex plane. For the z-transform to exist, the series in Eq. (4.1.1) must converge. For a given sequence $x[n]$, the domain of the z-plane within which the series converges is called the *region of convergence* (ROC) of the z-transform $\mathbb{X}[z]$ or $X[z]$. Thus, the specification of the z-transform requires both the algebraic expression in z and its region of convergence.

Note that a sequence $x[n]$ is called

- a *right-sided* sequence if $x[n] = 0\ \forall\, n < n_0$ for some finite integer n_0,
- a *causal* sequence if $x[n] = 0\ \forall\, n < n_0$ for some nonnegative integer $n_0 \geq 0$,
- a *left-sided* sequence if $x[n] = 0\ \forall\, n \geq n_0$ for some finite integer n_0, and
- a *anti-causal* sequence if $x[n] = 0\ \forall\, n \geq n_0$ for some non-positive integer $n_0 \leq 0$,

respectively.

Example 4.1 The z-Transform of Exponential Sequences

(a) For a right-sided and causal sequence $x_1[n] = a^n u_s[n]$, we can use Eq. (4.1.1a) to get its bilateral z-transform as

$$\mathbb{X}_1[z] = \mathrm{Z}\{x_1[n]\} = \sum_{n=-\infty}^{\infty} a^n u_s[n]z^{-n} = \sum_{n=0}^{\infty} a^n z^{-n} \overset{(D.23)}{=} \frac{1}{1 - az^{-1}} = \frac{z}{z - a} \quad (E4.1.1)$$

This geometric sequence converges for $|az^{-1}| < 1$, which implies that the ROC of $\mathbb{X}_1[z]$ is $\mathcal{R}_1 = \{z : |z| > |a|\}$ (see Fig. 4.1(a)).

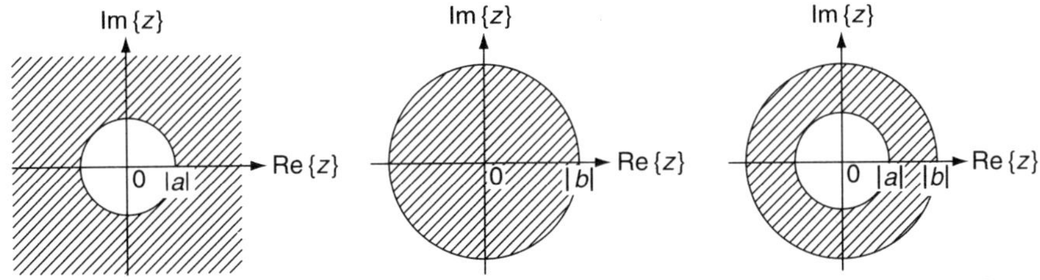

$\mathcal{R}_1 = \{z : |z| > |a|\}$
(a) ROC for a right-sided sequence

$\mathcal{R}_2 = \{z : |z| < |b|\}$
(b) ROC for a left-sided sequence

$\mathcal{R} = \mathcal{R}_1 \cap \mathcal{R}_2 = \{z : |a| < |z| < |b|\}$
(c) ROC for a both-sided sequence

Fig. 4.1 Three forms of ROC (region of convergence)

(b) For a left-sided and anti-causal sequence $x_2[n] = -b^n u_s[-n-1]$, we can use Eq. (4.1.1a) to get its bilateral z-transform as

$$\begin{aligned}\mathbb{X}_2[z] = \mathrm{Z}\{x_2[n]\} &= \sum_{n=-\infty}^{\infty} -b^n u_s[-n-1] z^{-n} = -\sum_{n=-\infty}^{-1} b^n z^{-n} \\ &= -\sum_{n=1}^{\infty} (b^{-1}z)^n \overset{(D.23)}{=} \frac{-b^{-1}z}{1-b^{-1}z} = \frac{z}{z-b} \qquad \text{(E4.1.2)}\end{aligned}$$

This geometric sequence converges for $|b^{-1}z| < 1$, which implies that the ROC of $\mathbb{X}_2[z]$ is $\mathcal{R}_2 = \{z : |z| < |b|\}$ (see Fig. 4.1(b)).

(c) For the both-sided sequence

$$y[n] = x_1[n] + x_2[n] = a^n u_s[n] - b^n u_s[-n-1] \qquad \text{(E4.1.3)}$$

we can combine the above results of (a) and (b) to obtain its bilateral z-transform as

$$\mathbb{Y}[z] = \mathbb{X}_1[z] + \mathbb{X}_2[z] = \frac{z}{z-a} + \frac{z}{z-b} \qquad \text{(E4.1.4)}$$

For this series to converge, both of the two series must converge and therefore, its ROC is the intersection of the two ROCs $\mathcal{R}_1$ and $\mathcal{R}_2 : \mathcal{R} = \mathcal{R}_1 \cap \mathcal{R}_2 = \{z : |a| < |z| < |b|\}$. This is an annular region $|a| < |z| < |b|$ (Fig. 4.1(c)) if $|b| > |a|$ and it will be an empty set if $|b| \le |a|$.

(cf) This example illustrates that different sequences may have the same z-transform, but with different ROCs. This implies that a z-transform expression may correspond to different sequences depending on its ROC.

Remark 4.1 Region of Convergence (ROC)

(1) The ROC for $X[z]$ is an annular ring centered at the origin in the z-plane of the form

$$r^- < |z| < r^+ \tag{4.1.2}$$

where r^+ can be as large as ∞ (Fig. 4.1(a)) for causal sequences and r^- can be as small as zero (Fig. 4.1(b)) for anti-causal sequences.

(2) As illustrated in Example 4.1, the three different forms of ROC shown in Fig. 4.1 can be associated with the corresponding three different classes of discrete-time sequences; $0 < r^- < |z|$ for right-sided sequences, $|z| < r^+ < \infty$ for left-sided sequences, and $r^- < |z| < r^+$ (annular ring) for two-sided sequences.

Example 4.2 A Causal Sequence Having Multiple-Pole z-transform

For a sum of two causal sequences

$$x[n] = x_1[n] + x_2[n] = 2u_s[n] - 2(1/2)^n u_s[n] \tag{E4.2.1}$$

we can use Eq. (E4.1.1) to get the z-transform for each sequence and combine the z-transforms as

$$\mathbb{X}[z] = Z\{x[n]\} = \mathbb{X}_1[z] + \mathbb{X}_2[z] \underset{\text{with } a=1 \text{ and } a=1/2}{\overset{\text{(E4.1.1)}}{=}} 2\frac{z}{z-a}\bigg|_{a=1} -2\frac{z}{z-a}\bigg|_{a=1/2}$$
$$= \frac{2z}{z-1} - \frac{2z}{z-1/2} = \frac{z}{(z-1)(z-1/2)} \tag{E4.2.2}$$

Since both $\mathbb{X}_1[z] = Z\{x_1[n]\}$ and $\mathbb{X}_2[z] = Z\{x_2[n]\}$ must exist for this z-transform to exist, the ROC is the intersection of the ROCs for the two z-transforms:

$$\mathcal{R} = \mathcal{R}_1 \cap \mathcal{R}_2 = \{z : |z| > 1\} \cap \{z : |z| > 1/2\} = \{z : |z| > 1\} \tag{E4.2.3}$$

(cf) Figure 4.2 shows the pole-zero pattern of the z-transform expression (E4.2.2) where a pole and a zero are marked with x and o, respectively. Note that the ROC of the z-transform $\mathbb{X}[z]$ of a right-sided sequence is the exterior of the circle which is centered at the origin and passes through the pole farthest from the origin in the z-plane.

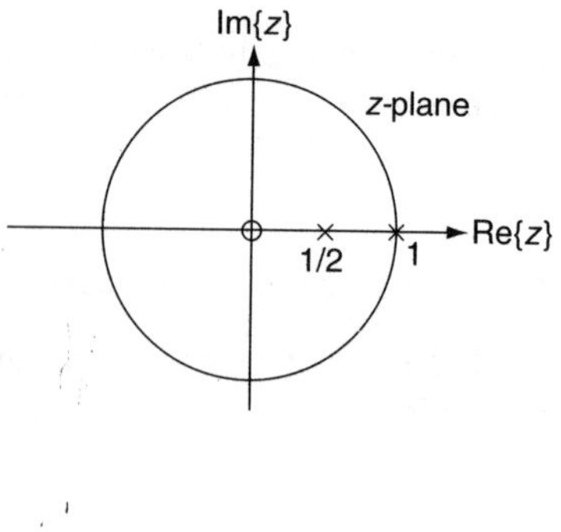

Fig. 4.2 Pole-zero pattern of the z-transform (E4.2.2)

Before working further with more examples, let us see the relationship between the z-transform and DTFT (discrete-time Fourier transform), which is summarized in the following remark:

Remark 4.2 The z-Transform and the DTFT (Discrete-Time Fourier Transform)

Comparing the bilateral z-transform (4.1.1a) with the DTFT (3.1.1), we can see their relationship that the DTFT can be obtained by substituting $z = e^{j\Omega}$ into the (bilateral) z-transform:

$$\begin{aligned}\mathcal{F}\{x[n]\} &= X(j\Omega) \overset{(3.1.1)}{=} \sum\nolimits_{n=-\infty}^{\infty} x[n]e^{-j\Omega n} = X[e^{j\Omega}] \\ &= X[z]|_{z=e^{j\Omega}} = Z\{x[n]\}|_{z=e^{j\Omega}} \end{aligned} \tag{4.1.3}$$

This implies that the evaluation of the z-transform along the unit circle in the z-plane yields the DTFT. Therefore, in order for the DTFT to exist, the ROC of the z-transform must include the unit circle (see Fig. 4.3).

Example 4.3 The z-transform of a Complex Exponential Sequence

Let us consider a causal sequence

$$x[n] = e^{j\Omega_1 n}u_s[n] = \cos(\Omega_1 n)u_s[n] + j\sin(\Omega_1 n)u_s[n] \tag{E4.3.1}$$

We can find the z-transform of this exponential sequence by substituting $a = e^{j\Omega_1}$ into Eq. (E4.1.1) as

$$\mathbb{X}[z] = \mathrm{Z}\{e^{j\Omega_1 n}u_s[n]\} \underset{a=e^{j\Omega_1}}{\overset{(E4.1.1)}{=}} \frac{z}{z - e^{j\Omega_1}} = \frac{z(z - e^{-j\Omega_1})}{(z - e^{j\Omega_1})(z - e^{-j\Omega_1})},\ |z| > |e^{\pm j\Omega_1}| = 1;$$

$$e^{j\Omega_1 n}u_s[n] \overset{(D.20)}{=} \cos(\Omega_1 n)u_s[n] + j\sin(\Omega_1 n)u_s[n]$$

$$\overset{Z}{\leftrightarrow} \frac{z(z - \cos\Omega_1)}{z^2 - 2z\cos\Omega_1 + 1} + j\frac{z\sin\Omega_1}{z^2 - 2z\cos\Omega_1 + 1}$$

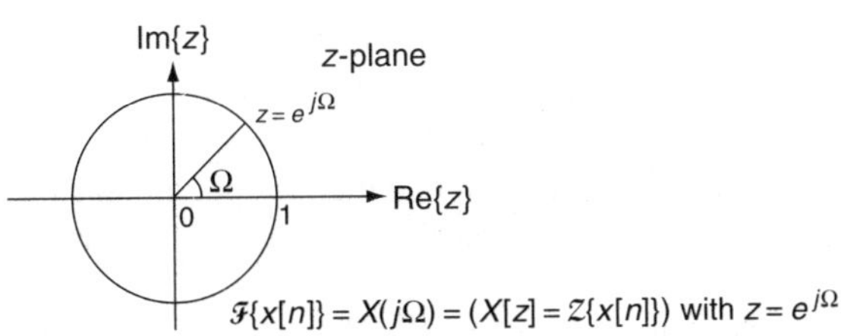

Fig. 4.3 The relationship between the DTFT and z-transform

This implies the following two z-transform pairs:

$$\cos(\Omega_1 n)u_s[n] \overset{Z}{\leftrightarrow} \frac{z(z-\cos\Omega_1)}{z^2-2z\cos\Omega_1+1} = \frac{z(z-\cos\Omega_1)}{(z-\cos\Omega_1)^2+\sin^2\Omega_1} \quad \text{with } \mathcal{R}=\{z:|z|>1\} \tag{E4.3.2}$$

$$\sin(\Omega_1 n)u_s[n] \overset{Z}{\leftrightarrow} \frac{z\sin\Omega_1}{z^2-2z\cos\Omega_1+1} = \frac{z\sin\Omega_1}{(z-\cos\Omega_1)^2+\sin^2\Omega_1} \quad \text{with } \mathcal{R}=\{z:|z|>1\} \tag{E4.3.3}$$

Example 4.4 The z-transform of an Exponentially Decreasing Sinusoidal Sequence
For a causal sequence

$$x[n] = r^n\cos(\Omega_1 n)u_s[n] \overset{(D.21)}{=} \frac{1}{2}r^n(e^{j\Omega_1 n}+e^{-j\Omega_1 n})u_s[n] \tag{E4.4.1}$$

we can use Eq. (E4.1.1) with $a = e^{\pm j\Omega_1}$ to find the z-transform of this sequence as

$$\mathbb{X}[z] \underset{a=re^{j\Omega_1} \text{ and } a=re^{-j\Omega_1}}{\overset{(E4.1.1)}{=}} \frac{1}{2}\left(\frac{z}{z-re^{j\Omega_1}}+\frac{z}{z-re^{-j\Omega_1}}\right) = \frac{z(z-r\cos\Omega_1)}{z^2-2zr\cos\Omega_1+r^2} \tag{E4.4.2}$$

where the ROC is $\mathcal{R} = \{z : |z| > |r\,e^{\pm j\Omega_1}| = |r|\}$.

(cf) Note that if $|r| \geq 1$, the ROC does not contain the unit circle and consequently, the DTFT does not exist in the strict sense.

We could elaborate on more examples of the z-transform computation. However, we instead list a number of useful z-transform pairs in Table B.9 in Appendix B so that the readers can use them for finding the z-transform or inverse z-transform.

In many cases, we deal with the z-transform that is expressed in the form of a rational function, i.e., a ratio of polynomials in z. For such a rational z-transform $\mathbb{X}[z] = \mathbb{Q}[z]/\mathbb{P}[z]$, we define its singularities, i.e., poles and zeros as follows:

Remark 4.3 Poles and Zeros of a Rational z-Transform Expression $\mathbb{X}[z] = \mathbb{Q}[z]/\mathbb{P}[z]$

The roots of $\mathbb{Q}[z] = 0$, or the values of z at which $\mathbb{X}[z] = 0$ are called the *zeros* of $\mathbb{X}[z]$. The *poles* are the roots of $\mathbb{P}[z] = 0$, or the values of z for which $\mathbb{X}[z] = \infty$. If we count the singularities at $z = \infty$, the number of poles will be equal to that of zeros. Notice that, for polynomials with real-valued coefficients, the complex-valued singularities must occur in complex-conjugate pairs.

The ROC of a z-transform $\mathbb{X}[z]$ can be described in terms of its poles as follows:

Remark 4.4 Pole Locations and Region of Convergence (ROC)

(1) It is obvious that there can be no poles of a z-transform $\mathbb{X}[z]$ inside the ROC since the z-transform does not converge at its poles. In fact, the ROC is bounded by poles or infinity as illustrated by Examples 4.1 and 4.2.
(2) For a right-sided sequence $x_a[n]$ and left-sided sequence $x_b[n]$, the ROCs for their z-transforms $\mathbb{X}_a[z]$ and $\mathbb{X}_b[z]$ are outside/inside the outermost/innermost pole, respectively:

$$\mathcal{R}_1 = \{z : |z| > \text{the maximum of } |a_i|'\text{s}\}(z = a_i : \text{the poles of } \mathbb{X}_a[z]) \quad (4.1.4a)$$

(as illustrated by Example 4.2) or

$$\mathcal{R}_2 = \{z : |z| < \text{the minimum of } |b_i|'\text{s}\}(z = b_i : \text{the poles of } \mathbb{X}_b[z]) \quad (4.1.4b)$$

For a two-sided sequence $x[n] = x[n] + x_b[n]$, the ROC is an annular region given by

$$\mathcal{R} = \mathcal{R}_1 \cap \mathcal{R}_2 = \{z : \max_i |a_i| < |z| < \min_j |b_j|\} \quad (4.1.4c)$$

4.2 Properties of the z-Transform

In this section we discuss the properties of the z-transform that are useful in obtaining z-transform pairs and in applying the transform to the analysis of discrete-time linear time-invariant systems. These properties are very similar to those of the Laplace transform and Fourier transform.

Most of the properties are commonly applicable to the bilateral and unilateral z-transforms. There are, however, a few exceptions such as the time shifting property, time reversal property, and initial/final value theorems. In particular, it is the time shifting property that makes the unilateral z-transform very useful for solving difference equations with initial conditions.

4.2.1 Linearity

Let the z-transforms of two time sequences $x[n]$ and $y[n]$ be $\mathrm{Z}\{x[n]\} = \mathbb{X}[z]$ with ROC $\mathcal{R}_x$ and $\mathrm{Z}\{y[n]\} = \mathbb{Y}[z]$ with ROC $\mathcal{R}_y$. Then we can express the z-transform of their linear combination in terms of $\mathbb{X}[z]$ and $\mathbb{Y}[z]$ as

$$ax[n] + \beta y[n] \overset{\mathrm{Z}}{\leftrightarrow} \alpha\mathbb{X}[z] + \beta\mathbb{Y}[z] \text{ with ROC } \mathcal{R} \supset (\mathcal{R}_1 \cap \mathcal{R}_2) \quad (4.2.1)$$

where the ROC is generally the intersection of the two ROCs $\mathcal{R}_x$ and $\mathcal{R}_y$, but it can be larger when any pole of $\mathbb{X}[z]$ and $\mathbb{Y}[z]$ is cancelled by the zeros resulting from the addition.

4.2.2 Time Shifting – Real Translation

Let the bilateral z-transform of $x[n]$ be $\mathrm{Z}\{x[n]\} = \mathbb{X}[z]$ with ROC $\mathcal{R}_x$. Then the bilateral z-transform of a time-shifted version $x[n-n_1]$ is

$$x[n-n_1] \overset{z}{\leftrightarrow} z^{-n_1}\mathbb{X}[z],\ \text{ROC } \mathcal{R}_x \text{ (possibly with } z=0 \text{ or } \infty \text{ added or removed)} \tag{4.2.2}$$

The addition/removal of $z=\infty$ to/from the ROC occurs when a causal/non-causal sequence is shifted to become a non-causal/causal sequence. Similarly, the addition/removal of the origin $z=0$ to/from the ROC is owing to the possible transition between anticausal and non-anticausal sequences resulting from the shift operation.

On the other hand, the time-shifting property of the unilateral z-transform is as follows:

<Case 1> $x[n-n_1],\ n_1>0$ (delayed sequence)

$$\begin{aligned}
\mathcal{Z}\{x[n-n_1]\} &= \sum_{n=0}^{\infty} x[n-n_1]z^{-n} \\
&= \sum_{n=0}^{n_1-1} x[n-n_1]z^{-n} + \sum_{n=n_1}^{\infty} x[n-n_1]z^{-n} \\
&\overset{n-n_1=m,n=m+n_1}{=} \sum_{m=-n_1}^{-1} x[m]z^{-(m+n_1)} + \sum_{m=0}^{\infty} x[m]z^{-(m+n_1)} \\
&= z^{-n_1}\left(\sum_{m=0}^{\infty} x[m]z^{-m} + \sum_{m=-n_1}^{-1} x[m]z^{-m}\right);
\end{aligned}$$

$$x[n-n_1] \overset{z}{\leftrightarrow} z^{-n_1}\left(X[z] + \sum_{m=-n_1}^{-1} x[m]z^{-m}\right) \tag{4.2.3a}$$

where the second term in the RHS will disappear when $x[n]$ is causal.

<Case 2> $x[n+n_1],\ n_1>0$ (advanced sequence)

$$\begin{aligned}
\mathcal{Z}\{x[n+n_1]\} &= \sum_{n=0}^{\infty} x[n+n_1]z^{-n} \overset{n+n_1=m,n=m-n_1}{=} \sum_{n=n_1}^{\infty} x[m]z^{-(m-n_1)} \\
&= z^{n_1}\left(\sum_{m=0}^{\infty} x[m]z^{-m} - \sum_{m=0}^{n_1-1} x[m]z^{-m}\right);
\end{aligned}$$

$$x[n+n_1] \overset{z}{\leftrightarrow} z^{n_1}\left(X[z] -- \sum_{m=0}^{n_1-1} x[m]z^{-m}\right) \tag{4.2.3b}$$

Example 4.5 Applying Linearity and Time Shifting Properties of the z-Transform

For a rectangular pulse sequence of duration N given by

$$x[n] = u_s[n] - u_s[n-N] = \{\overset{n=\cdots-1}{\cdots 0}\ \overset{0}{1}\ \overset{1}{1}\cdots\ \overset{\cdots N-1}{1}\ \overset{N\cdots}{0}\cdots\} \tag{E4.5.1}$$

we can use Eq. (E4.1.1) (with $a = 1$) together with the linearity (4.2.1) and time-shifting property (4.2.2) to get its z-transform as

$$\mathbb{X}[z] \underset{(4.2.1)\&(4.2.2)}{\overset{\text{(E4.1.1) with } a=1}{=}} \frac{z}{z-1} - z^{-N}\frac{z}{z-1} = \frac{z^N - 1}{z^{N-1}(z-1)} \tag{E4.5.2}$$

Note that $\mathbb{X}[z]$ has multiple pole of order $N-1$ at $z = 0$ and $N-1$ zeros at $z = e^{j2m\pi/N}$ (for $m = 1, 2, \cdots, N-1$) on the unit circle where the zero at $z = 1$ (resulting from the addition) cancels the pole (of $\mathrm{Z}\{u_s[n]\} = z/(z-1)$) at that point. Due to this pole-zero cancellation, the ROC of $\mathbb{X}[z]$ becomes $|z| > 0$ (the entire z-plane except the origin $z = 0$), which is larger than that of $\mathrm{Z}\{u_s[n]\} = z/(z-1)$, i.e., $|z| > 1$.

4.2.3 Frequency Shifting – Complex Translation

Let the bilateral z-transform of $x[n]$ be $\mathrm{Z}\{x[n]\} = \mathbb{X}[z]$ with ROC $\mathcal{R}_x = \{z : r_x^- < |z| < r_x^+\}$. Then we have the z-transform of a frequency-shifted or modulated version $z_1^n x[n]$ as

$$e^{j\Omega_1 n}x[n] \overset{\mathrm{Z}}{\leftrightarrow} \mathbb{X}[e^{-j\Omega_1}z] \text{ with ROC}: \ \mathcal{R}_x = \{z : r_x^- < |z| < r_x^+\} \tag{4.2.4a}$$

$$z_1^n x[n] \overset{\mathrm{Z}}{\leftrightarrow} \mathbb{X}[z/z_1] \text{ with ROC}: |z_1|\mathcal{R}_x = \{z : |z_1|r_x^- < |z| < |z_1|r_x^+\} \tag{4.2.4b}$$

4.2.4 Time Reversal

Let the bilateral z-transform of $x[n]$ be $\mathrm{Z}\{x[n]\} = \mathbb{X}[z]$ with ROC $\mathcal{R}_x = \{z : r_x^- < |z| < r_x^+\}$. Then we have the bilateral z-transform of a time-reversed version $x[-n]$ as

$$x[-n] \overset{\mathrm{Z}}{\leftrightarrow} \mathbb{X}[z^{-1}] \text{ with ROC}: 1/\mathcal{R}_x = \{z : 1/r_x^+ < |z| < 1/r_x^-\} \tag{4.2.5}$$

where if $z \in \mathcal{R}_x$, then $1/z \in 1/\mathcal{R}_x$.

4.2.5 Real Convolution

Let the z-transforms of two time sequences $g[n]$ and $x[n]$ be $\mathrm{Z}\{g[n]\} = \mathbb{G}[z]$ with ROC $\mathcal{R}_g$ and $\mathrm{Z}\{x[n]\} = \mathbb{X}[z]$ with ROC $\mathcal{R}_x$. Then we can express the z-transform of their convolution in terms of $\mathbb{G}[z]$ and $\mathbb{X}[z]$ as

$$y[n] = g[n] * x[n] \overset{\mathrm{Z}}{\leftrightarrow} \mathbb{Y}[z] = \mathbb{G}[z]\,\mathbb{X}[z] \text{ with ROC } \mathcal{R} \supset (\mathcal{R}_g \cap \mathcal{R}_x) \tag{4.2.6}$$

(proof)

$$\mathsf{Z}\{g[n]*x[n]\} \overset{(4.1.1a)}{=} \sum_{n=-\infty}^{\infty}(g[n]*x[n])z^{-n}$$

$$\overset{(1.2.9)}{=} \sum_{n=-\infty}^{\infty}\left(\sum_{m=-\infty}^{\infty} g[m]x[n-m]\right)z^{-n}$$

$$= \sum_{m=-\infty}^{\infty}\left(g[m]\sum_{n=-\infty}^{\infty} x[n-m]z^{-(n-m)}z^{-m}\right)$$

$$\overset{n-m\to n,\ n\to n+m}{=} \sum_{m=-\infty}^{\infty}\left(g[m]z^{-m}\sum_{n=-\infty}^{\infty} x[n]z^{-n}\right) \overset{(4.1.1a)}{=} \mathbb{G}[z]\mathbb{X}[z]$$

This convolution property holds for causal sequences, which implies that the unilateral z-transform also has the same property. It is very useful for describing the input-output relationship of a discrete-time LTI system with the input $x[n]$, output $y[n]$, and impulse response $g[n]$ where the transform of the impulse response, $\mathsf{Z}\{g[n]\} = \mathbb{G}[z]$, is called the *system* or *transfer function* (see Sects. 1.2.3 and 1.2.4).

4.2.6 Complex Convolution

Let the z-transforms of $x[n]$ and $y[n]$ be $\mathsf{Z}\{x[n]\} = \mathbb{X}[z]$ with ROC $\mathcal{R}_x = \{z : r_x^- < |z| < r_x^+\}$ and $\mathsf{Z}\{y[n]\} = \mathbb{Y}[z]$ with ROC $\mathcal{R}_y = \{z : r_y^- < |z| < r_y^+\}$, respectively. Then the z-transform of the product of the two sequences can be expressed in terms of $\mathbb{X}[z]$ and $\mathbb{Y}[z]$ as

$$x[n]y[n] \overset{\mathsf{Z}}{\leftrightarrow} \frac{1}{2\pi j}\oint_{C_1} \mathbb{X}[z/v]\mathbb{Y}[v]v^{-1}dv \text{ with ROC } \mathcal{R} = \{z : r_x^- r_y^- < |z| < r_x^+ r_y^+\} \tag{4.2.7a}$$

$$x[n]y[n] \overset{\mathsf{Z}}{\leftrightarrow} \frac{1}{2\pi j}\oint_{C_2} \mathbb{X}[v]\mathbb{Y}[z/v]v^{-1}dv \text{ with ROC } \mathcal{R} = \{z : r_x^- r_y^- < |z| < r_x^+ r_y^+\} \tag{4.2.7b}$$

where $\oint_{C_i}$ means the complex integral along a closed contour C_i within the intersection of the ROCs of $\mathbb{X}[z/v]$ and $\mathbb{Y}[v]$ or $\mathbb{X}[v]$ and $\mathbb{Y}[z/v]$ (see [O-2], Sect. 2.3.9 for its proof).

4.2.7 Complex Differentiation

Let the z-transform of $x[n]$ be $\mathsf{Z}\{x[n]\} = \mathbb{X}[z]$ with ROC $\mathcal{R}_x$. Then the z-transform of $nx[n]$ can be expressed in terms of $\mathbb{X}[z]$ as

$$nx[n] \overset{\mathsf{Z}}{\leftrightarrow} -z\frac{d}{dz}\mathbb{X}[z] \text{ with ROC } \mathcal{R}_x \tag{4.2.8}$$

This can be proved by differentiating the definition of the z-transform (4.1.1) w.r.t. z.

4.2.8 Partial Differentiation

If a z-transform pair is given in the form of $Z\{x(nT,a)\} = \mathbb{X}(z,a)$ with a parameter a, we can differentiate it w.r.t. a to obtain another z-transform pair as

$$\frac{\partial}{\partial a}x(nT,a) \overset{Z}{\leftrightarrow} \frac{\partial}{\partial a}X(z,a) \tag{4.2.9}$$

Example 4.6 Complex Differentiation and Partial Differentiation

For $y[n] = n^2 a^n u_s[n]$, we can apply the complex differentiation property (4.2.8) twice for Eq. (E4.1.1) to write

$$na^n u_s[n] \overset{Z}{\leftrightarrow} -z\frac{d}{dz}\left(\frac{z}{z-a}\right) = -z\frac{(z-a)-z}{(z-a)^2} = \frac{az}{(z-a)^2} \tag{E4.6.1}$$

$$n^2 a^n u_s[n] \overset{Z}{\leftrightarrow} -z\frac{d}{dz}\left(\frac{az}{(z-a)^2}\right) = -z\frac{a(z-a)^2 - 2az(z-a)}{(z-a)^4} = \frac{az(z+a)}{(z-a)^3} \tag{E4.6.2}$$

Alternatively, we can apply the partial differentiation property for Eq. (E4.1.1) to get the same results:

$$na^{n-1} u_s[n] \overset{Z}{\leftrightarrow} \frac{\partial}{\partial a}\left(\frac{z}{z-a}\right) = \frac{z}{(z-a)^2} \tag{E4.6.3}$$

$$n^2 a^{n-1} u_s[n] \overset{Z}{\leftrightarrow} \frac{\partial}{\partial a}\left(\frac{az}{(z-a)^2}\right) = \frac{z(z-a)^2 + 2az(z-a)}{(z-a)^4} = \frac{z(z+a)}{(z-a)^3} \tag{E4.6.4}$$

Multiplying both sides of these equations by a yields Eqs. (E4.6.1) and (E4.6.2), which are listed in Table B.9(10) and (12).

4.2.9 Initial Value Theorem

For a causal sequence $x[n]$ such that $x[n] = 0 \; \forall \; n < 0$, we can get its initial value $x[0]$ from its z-transform as

$$x[0] = \lim_{z\to\infty} X[z] \tag{4.2.10}$$

This can easily be shown by substituting $z = \infty$ into the z-transform definition (4.1.1b).

4.2.10 *Final Value Theorem*

For a causal sequence $x[n]$ such that $x[n] = 0 \ \forall \ n < 0$, we can get its final value $x[\infty]$ from its z-transform as

$$x[\infty] = \lim_{z \to 1}(z-1)X[z] = \lim_{z \to 1}(1-z^{-1})X[z] \tag{4.2.11}$$

This requires that $x[n]$ should converge or equivalently, all the poles of its z-transform $X[z]$ should lie inside the unit circle possibly with the exception of a simple pole at $z = 1$.
(proof)
We can use the z-transform definition (4.1.1b) to write

$$\begin{aligned}\mathcal{Z}\{x[n+1]-x[n]\} &= \lim_{k \to \infty}\left(\sum\nolimits_{n=0}^{k} x[n+1]z^{-n} - \sum\nolimits_{n=0}^{k} x[n]z^{-n}\right) \\ &= \lim_{k \to \infty}\left(\sum\nolimits_{n=1}^{k} x[n]z^{-(n-1)} + x[k+1]z^{-k}\right. \\ &\qquad \left. - x[0] - \sum\nolimits_{n=1}^{k} x[n]z^{-n}\right)\end{aligned}$$

On the other hand, from the time shifting property (4.2.3b), we can write

$$\begin{aligned}\mathcal{Z}\{x[n+1]-x[n]\} &\overset{(4.2.1)}{=} \mathcal{Z}\{x[n+1]\} - \mathcal{Z}\{x[n]\} \\ &\overset{(4.2.3)}{=} z(X[z]-x[0]) - X[z] = (z-1)X[z] - zx[0]\end{aligned}$$

Noting that these two equations are commonly the z-transform of $(x[n+1]-x[n])$, we can equate their RHSs and substitute $z = 1$ to get the desired result.

4.3 The Inverse z-Transform

In this section we consider how to find $x[n]$ for a given z-transform $X[z]$ with its ROC. From the complex variable theory, the inverse z-transform formula can be derived (see [O-2], Sect. 2.2) as

$$x[n] = \frac{1}{2\pi j}\oint_C \mathbb{X}[z]z^{-n}dz \tag{4.3.1}$$

where $\oint_C$ means the complex integral along a closed contour C within the ROC of $\mathbb{X}[z]$ encircling the origin of the z-plane in the counterclockwise direction. It is, however, difficult to directly evaluate this integral and therefore we make resort to alternative procedures for obtaining the inverse z-transform.

4.3.1 Inverse z-Transform by Partial Fraction Expansion

Noting that the complex variable z appears in the numerator of almost every basic z-transform listed in Table B.9, we apply the same procedure as with the inverse Laplace transform (Sect. A.4) to get the partial fraction expansion on $X[z]/z$ and then multiply both sides by z so that we can directly use the z-transform table to get the inverse z-transform.

More specifically, let $X[z]/z$ be rational as

$$\frac{X[z]}{z} = \frac{Q_1[z]}{P[z]} = \frac{b_M z^M + \ldots + b_1 z + b_0}{a_N z^N + \ldots + a_1 z + a_0} \tag{4.3.2}$$

where M and N are the degrees of the numerator and denominator polynomials, respectively. If $M \geq N$, we divide $Q_1[z]$ by $P[z]$ starting with the highest powers of z to produce the remainder polynomial of degree less than N:

$$\frac{X[z]}{z} = \frac{Q[z]}{P[z]} + c_{M-N} z^{M-N} + \ldots + c_1 z + c_0 \tag{4.3.3}$$

If $M < N$, we have only the first term on the RHS where $c_i = 0$ for all i. Now, for the purpose of illustration, we assume that all the poles of $Q[z]/P[z]$ are simple except one multiple pole of order L at $z = p$ so that we can write $Q[z]/P[z]$ in the following form:

$$\frac{Q[z]}{P[z]} = \left(\sum_{i=1}^{N-L} \frac{r_i}{z - p_i} \right) + \frac{r_{N-L+1}}{z - p} + \ldots + \frac{r_N}{(z-p)^L} + K \tag{4.3.4}$$

where

$$r_i = (z - p_i) \left. \frac{Q[z]}{P[z]} \right|_{z = p_i}, \quad i = 1, 2, \ldots, N - L \tag{4.3.5a}$$

$$r_{N-l} = \frac{1}{l!} \frac{d^l}{dz^l} \left\{ (z-p)^L \frac{Q[z]}{P[z]} \right\} \bigg|_{z = p}, \quad l = 0, 1, \ldots, L - 1 \tag{4.3.5b}$$

Now, substituting Eq. (4.3.4) into Eq. (4.3.3), multiplying the both sides by z, and using the z-transform Table B.9, we can obtain the inverse z-transform of $X[z]$ as

$$x[n] = \left\{ \sum_{i=1}^{N-L} r_i p_i^n + r_{N-L+1} p^n + r_{N-L+2} n p^{n-1} + \ldots \right.$$
$$\left. + r_N \frac{n!}{(L-1)!(n-L+1)!} p^{n-L+1} \right\} u_s[n] + \sum_{i=0}^{M-N} c_i \delta[n+i+1] \tag{4.3.6}$$

Example 4.7 The Inverse z-Transform by Partial Fraction Expansion

(a) Let us find the inverse z-transform of

$$X[z] = \frac{z}{z^3 - 2z^2 + (5/4)z - 1/4} = \frac{z}{(z-1)(z-1/2)^2} \quad \text{with ROC } \mathcal{R} = \{z : |z| > 1\} \tag{E4.7.1}$$

We first divide this by z and then expand it into partial fractions:

$$\frac{X[z]}{z} = \frac{r_1}{z-1} + \frac{r_2}{z-1/2} + \frac{r_3}{(z-1/2)^2} \overset{(4.7.3)}{=} \frac{4}{z-1} - \frac{4}{z-1/2} - \frac{2}{(z-1/2)^2} \tag{E4.7.2}$$

where the coefficient of each term can be found from Eq. (4.3.5) as

$$r_1 \overset{(4.3.5a)}{=} (z-1)\frac{X[z]}{z}\bigg|_{z=1} = 4 \tag{E4.7.3a}$$

$$r_2 \underset{l=1}{\overset{(4.3.5b)}{=}} \frac{d}{dz}\left((z-1/2)^2\frac{X[z]}{z}\right)\bigg|_{z=1/2} = \frac{d}{dz}\left(\frac{1}{z-1}\right)\bigg|_{z=1/2} = -\frac{1}{(z-1)^2}\bigg|_{z=1/2} = -4 \tag{E4.7.3b}$$

$$r_3 \underset{l=0}{\overset{(4.3.5b)}{=}} (z-1/2)^2\frac{X[z]}{z}\bigg|_{z=1/2} = \frac{1}{z-1}\bigg|_{z=1/2} = -2 \tag{E4.7.3c}$$

Now, moving the z (which we have saved) from the LHS back into the RHS yields

$$X[z] = 4\frac{z}{z-1} - 4\frac{z}{z-1/2} - 4\frac{(1/2)z}{(z-1/2)^2} \tag{E4.7.4}$$

Then we can use Table B.9(3), (5), and (10) to write the inverse z-transform as

$$x[n] = \left(4 - 4\left(\frac{1}{2}\right)^n - 4n\left(\frac{1}{2}\right)^n\right)u_s[n] \tag{E4.7.5}$$

where the right-sided sequences are chosen over the left-sided ones since the given ROC is not the inside, but the outside of a circle.

We can use the MATLAB command 'residue( )' or 'residuez( )' to get the partial fraction expansion and 'iztrans( )' to obtain the whole inverse z-transform.

It should, however, be noted that 'iztrans()' might not work properly for high-degree rational functions.

```
>>Nz=[0 1]; Dz=poly([1 1/2 1/2]), [r,p,k]=residue(Nz,Dz); [r p],k %(E4.7.2)
   r =  4.0000     p =  1.0000    % (E4.7.3a)  4/(z-1)
       -4.0000          0.5000    % (E4.7.3b) -4/(z-0.5)
       -2.0000          0.5000    % (E4.7.3c) -2/(z-0.5)^2
   k = []
>>syms z, x=iztrans(z/(z^3-2*z^2+1.25*z-0.25)) % (E4.7.1)
   x = 4-4*(1/2)^n-4*(1/2)^n*n    % (E4.7.5)
```

(b) Let us find the inverse z-transform of

$$X[z] = \frac{3z}{z^2 - (1/4)z - 1/8} = \frac{3z}{(z-1/2)(z+1/4)} \tag{E4.7.6}$$

with one of the following three ROCs:

$$\mathcal{R}_1 = \left\{z : |z| > \frac{1}{2}\right\}, \mathcal{R}_2 = \left\{z : |z| < \frac{1}{4}\right\}, \text{ and}$$
$$\mathcal{R}_3 = \left\{z : \frac{1}{4} < |z| < \frac{1}{2}\right\} \tag{E4.7.7}$$

We first divide this by z and then expand it into partial fractions:

$$\frac{X[z]}{z} = \frac{4}{z-1/2} - \frac{4}{z+1/4}; \quad X[z] = 4\frac{z}{z-1/2} - 4\frac{z}{z+1/4} \tag{E4.7.8}$$

Now, depending on the ROC, we use Table B.9(5) or (6) to write the inverse z-transform as follows:

$$\text{(i) } \mathcal{R}_1 = \left\{z : |z| > \frac{1}{2}\right\}: \quad x[n] = 4\left(\frac{1}{2}\right)^n u_s[n] - 4\left(-\frac{1}{4}\right)^n u_s[n] \tag{E4.7.9a}$$

$$\text{(ii) } \mathcal{R}_2 = \left\{z : |z| < \frac{1}{4}\right\}:$$

$$x[n] = -4\left(\frac{1}{2}\right)^n u_s[-n-1] + 4\left(-\frac{1}{4}\right)^n u_s[-n-1] \tag{E4.7.9b}$$

$$\text{(iii) } \mathcal{R}_3 = \left\{z : \frac{1}{4} < |z| < \frac{1}{2}\right\}:$$

$$x[n] = -4\left(\frac{1}{2}\right)^n u_s[-n-1] - 4\left(-\frac{1}{4}\right)^n u_s[n] \tag{E4.7.9c}$$

```
>>syms z, x=iztrans(3*z/(z^2-0.25*z-1/8)) % (E4.7.6)
   x = 4*(1/2)^n-4*(-1/4)^n  % (E4.7.9a) just a right-sided sequence
```

Example 4.8 The Inverse z-Transform by Partial Fraction Expansion

Let us find the inverse z-transform of

$$X[z] = \frac{2z^2}{z^2 - z + 1/2} = \frac{2z^2}{(z-0.5-j0.5)(z-0.5+j0.5)} \quad \text{(E4.8.1)}$$
$$\text{with ROC } \mathcal{R} = \{z : |z| > 1\}$$

We first divide this by z and then expand it into partial fractions:

$$\frac{X[z]}{z} = \frac{r_R + jr_I}{z-0.5-j0.5} + \frac{r_R - jr_I}{z-0.5+j0.5}$$
$$= \frac{1-j}{z-0.5-j0.5} + \frac{1+j}{z-0.5+j0.5} \quad \text{(E4.8.2)}$$

Now, moving the z (which we have saved) from the LHS back into the RHS yields

$$X[z] = \frac{(1-j)z}{z-0.5-j0.5} + \frac{(1+j)z}{z-0.5+j0.5} \quad \text{(E4.8.3)}$$

We can use Table B.9(5) to write the inverse z-transform as

$$x[n] = (1-j)(0.5+j0.5)^n u_s[n] + (1+j)(0.5-j0.5)^n u_s[n]$$
$$= \sqrt{2}\left(e^{-j\pi/4}\sqrt{2}^{-n} e^{jn\pi/4} + e^{j\pi/4}\sqrt{2}^{-n} e^{-jn\pi/4}\right) u_s[n]$$
$$= \sqrt{2}^{-n+3} \cos((n-1)\frac{\pi}{4}) u_s[n] \quad \text{(E4.8.4)}$$

As a nice alternative for $X[z]$ having complex conjugate poles like (E4.8.1), we can decompose it into the following form, which can be matched exactly with some element of the z-transform table:

$$X[z] = \frac{2z(z-1/2)}{(z-1/2)^2 + (1/2)^2} + \frac{2(1/2)z}{(z-1/2)^2 + (1/2)^2}$$
$$= 2\frac{z(z - r\cos\Omega_1)}{(z - r\cos\Omega_1)^2 + r^2\sin^2\Omega_1} + 2\frac{zr\sin\Omega_1}{(z - r\cos\Omega_1)^2 + r^2\sin^2\Omega_1} \quad \text{(E4.8.5)}$$

where $r = 1/\sqrt{2}$, $\cos\Omega_1 = 1/\sqrt{2}$, $\sin\Omega_1 = 1/\sqrt{2}$, and $\Omega_1 = \pi/4$. Then we can use B.9(18) and (17) to obtain the same result as (E4.8.3):

$$x[n] = 2\left(r^n\cos(\Omega_1 n) + r^n\sin(\Omega_1 n)\right) u_s[n]$$
$$= 2\left(\sqrt{2}^{-n}\cos(\frac{\pi}{4}n) + \sqrt{2}^{-n}\sin(\frac{\pi}{4}n)\right) u_s[n]$$
$$= \sqrt{2}^{-n+3}\cos((n-1)\frac{\pi}{4})u_s[n] = \sqrt{2}^{-n+3}\sin((n+1)\frac{\pi}{4})u_s[n] \quad \text{(E4.8.6)}$$

(cf) It seems that the MATLAB command 'iztrans()' does not work properly for this problem:

```
>>syms z, x=iztrans(2*z^2/(z^2-z+1/2))  % (E4.8.1)
>>n=1:10; xn=2.^(-(n-3)/2).*cos((n-1)*pi/4); stem(n,xn), hold on %(E4.8.6)
>>[r,p,k]=residuez([2 0 0],[1 -1 1/2]) % Partial fraction expansion
>>xn1=real(r.'*[p(1).^n; p(2).^n]); stem(n,xn1,'r') % (E4.8.3) Alternative
```

4.3.2 Inverse z-Transform by Long Division

Noting that the inverse z-transform can rarely be found in an elegant form like Eq. (E4.8.6), we may think of it as an alternative to expand $\mathbb{X}[z] = Q_1[z]/P[z]$ into a polynomial in powers of z^{-1} and equate each coefficient of z^{-n} to $x[n]$. More specifically, starting with the highest/lowest powers of z depending on the shape of the ROC (for a right/left-sided sequence), we divide $Q_1[z]$ by $P[z]$ to expand $\mathbb{X}[z]$ into the power series form of the z-transform definition (4.1.1). For example, let us consider $\mathbb{X}[z]$ given by Eq. (E4.7.6) in Example 4.7.

$$\mathbb{X}[z] = \frac{3z}{z^2 - (1/4)z - 1/8} = \frac{3z}{(z - 1/2)(z + 1/4)} \tag{4.3.7}$$

(Case 1) If the ROC is given as $\{z : |z| > 1/2\}$, we perform the long division as

$$\begin{array}{r|l}
 & 3z^{-1} + (3/4)z^{-2} + (9/16)z^{-3} + \cdots \\
\hline
z^2 - (1/4)z - 1/8 & 3z \\
 & 3z - 3/4 \quad - (3/8)z^{-1} \\
\hline
 & 3/4 \quad + (3/8)z^{-1} \\
 & 3/4 \quad - (3/16)z^{-1} - (3/32)z^{-2} \\
\hline
 & (9/16)z^{-1} - (3/32)z^{-2} \\
 & \cdots\cdots\cdots\cdots
\end{array}$$

Then each coefficient of the quotient polynomial in z^{-n} is equated with $x[n]$, yielding

$$x[n] = [\overset{n=0}{0} \quad \overset{1}{3} \quad \overset{2}{3/4} \quad \overset{3}{9/16} \cdots \quad] : \text{the same as Eq. (E4.7.9a)} \tag{4.3.8a}$$

(Case 2) If the ROC is given as $\{z : |z| < 1/4\}$, we perform the long division as

$$
\begin{array}{r l}
 & -24z \;+\; 48z^2 \;-\; 288z^3 \;+\cdots \\
\cline{2-2}
-1/8-(1/4)z+z^2 \,\big) & 3z \\
 & \underline{3z \;+\; 6z^2 \;-\; 24z^3} \\
 & \qquad -6z^2 \;+\; 24z^3 \\
 & \qquad \underline{-6z^2 \;-\; 12z^3 \;+\; 48z^4} \\
 & \qquad\qquad 36z^3 \;-\; 48z^4 \\
 & \cdots\cdots\cdots\cdots\cdots\cdots
\end{array}
$$

Then each coefficient of the quotient polynomial in z^n is equated with $x[-n]$, yielding

$$x[n] = [\overset{-3}{-288} \;\; \overset{-2}{48} \;\; \overset{-1}{-24} \;\; \overset{n=0}{0} \;\cdots\;\;] : \text{the same as Eq. (E4.7.9b)} \tag{4.3.8b}$$

(Case 3) If the ROC is given as $\{z : r^- = 1/4 < |z| < r^+ = 1/2\}$, $\mathbb{X}[z]$ should be separated into two parts, one having the poles on or inside the circle of radius r^- and the other having the poles on or outside the circle of radius r^+. Then, after performing the long division as in case 1/2 for the former/latter, we add the two quotients and equate each coefficient of the resulting polynomial in $z^{\pm n}$ with $x[\mp n]$.

4.4 Analysis of LTI Systems Using the z-Transform

So far we have seen that the z-transform is a general way of describing and analyzing discrete-time sequences. Now we will see that the z-transform also plays a very important role in the description and analysis of discrete-time linear time(shift)-invariant (LTI) systems. This stems from the fact that an LTI system can be characterized by the impulse response. Since the impulse response itself is a discrete-time signal, its z-transform, referred to as the system or transfer function, provides another way to characterize discrete-time LTI systems both in the time domain and in the frequency domain.

Let us consider a discrete-time causal LTI system with the impulse response $g[n]$ and input $x[n]$. Then the output $y[n]$ is the convolution of $g[n]$ and $x[n]$ given by Eq. (1.2.9) as

$$y[n] = g[n] * x[n] \tag{4.4.1}$$

so that, from the convolution property (4.2.6),

$$Y[z] = G[z]X[z] \tag{4.4.2}$$

where $X[z]$, $Y[z]$, and $G[z]$ are the z-transforms of the input $x[n]$, output $y[n]$, and impulse response $g[n]$, respectively. Note that $G[z]$ is referred to as the *system* or *transfer function*.

Remark 4.5 System Function, Pole Location, ROC, Causality, and Stability

(1) Eqs. (4.4.2) and (3.2.8) have an interpretation of describing the input-output relationship of a discrete-time LTI system in the z-domain and in the frequency domain, respectively. Comparing these two equations, we can state that the system function $G[z]$, evaluated on the unit circle $z = e^{j\Omega}$, yields the frequency response $G[e^{j\Omega}] = G(\Omega)$ of the system (Remark 4.2) if $G(\Omega) = \mathcal{F}\{g[n]\}$ exists, or equivalently, the ROC of $G[z]$ includes the unit circle. This is analogous to the continuous-time case where the frequency response $G(\omega)$ can be obtained by evaluating the system function $G(s)$ on the imaginary axis $s = j\omega$.
(2) Characteristics of a system such as stability and causality can be associated with the ROC and pole location of the system function $G[z]$. For example, if a system is causal, its impulse response $g[n]$ is a right-sided sequence and therefore, the ROC of $G[z] = Z\{g[n]\}$ must be the outside of the outermost pole (see Remark 4.4). If a system is stable, the ROC of $G[z]$ includes the unit circle so that the frequency response $G[e^{j\Omega}]$ can be defined (see Remark 4.2). If a system is both causal and stable, then the ROC of $G[z]$ must include the unit circle and be outside the outermost pole. It is implied that for a causal system to be stable, all the poles of its system function $G[z]$ must be inside the unit circle (Fig. 4.4(a) vs. (b)).

In particular, for systems characterized by linear constant-coefficient difference equations, the z-transform provides a very convenient procedure for obtaining the system function, frequency response, or time response. Consider a causal linear time-invariant (LTI) system (in Fig. 4.5) whose input-output relationship is described by the following difference equation

$$\sum_{i=0}^{NA-1} a_i y[n-i] = \sum_{j=0}^{NB-1} b_j x[n-j] \tag{4.4.3}$$

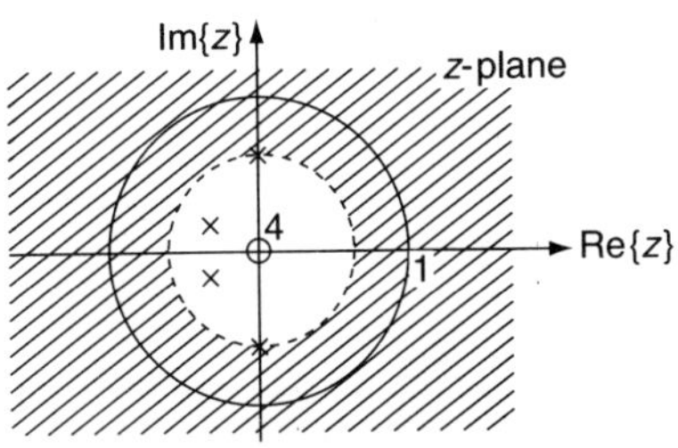

(a) All poles inside the unit circle

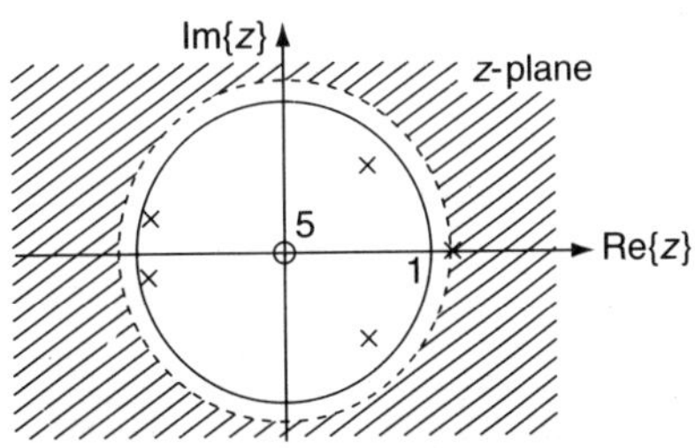

(b) Not all poles inside the unit circle

Fig. 4.4 Pole locations, ROC, unit circle, causality, and stability

Input $x[n]$	Impulse response $g[n]$ $\sum_{i=0}^{NA-1} a_i y[n-i] = \sum_{j=0}^{NB-1} b_j x[n-j]$	Output $y[n] = g[n] * x[n]$
z-Transform of input $X[z]$	$A[z]Y[z] = B[z]X[z]$ System or transfer function $G[z] = \frac{Y[z]}{X[z]} = \frac{B[z]}{A[z]}$	z-Transorm of output $Y[z] = G[z]X[z]$
$X[\Omega]$ Input spectrum	Frequency response $G(\Omega) = \frac{Y(\Omega)}{X(\Omega)} = \frac{B(\Omega)}{A(\Omega)}$	$Y(\Omega) = G(\Omega)X(\Omega)$ Output spectrum

Fig. 4.5 The input–output relationship, system function, and frequency response of a discrete–time LTI system

where the initial conditions are given as $y[n_0],\ y[n_0-1],\ \cdots,\ y[n_0-NA+2]$. This can be solved iteratively for the time response $y[n]$ to an input $x[n]$ starting from $n = n_0 + 1$:

$$y[n] = \frac{1}{a_0}\left(-a_1 y[n-1] - \cdots - a_{NA-1} y[n-NA+1] + \sum_{j=0}^{NB-1} b_j x[n-j]\right) \tag{4.4.4}$$

With zero initial conditions and the unit impulse input $x[n] = \delta[n]$, this yields the impulse response $y[n] = g[n]$. To find the system function as the z-domain input-output relationship, we assume zero initial conditions and use the linearity and time-shifting properties to take the z-transform of Eq. (4.4.3) as

$$\sum_{i=0}^{NA-1} a_i z^{-i} Y[z] = \sum_{j=0}^{NB-1} b_j z^{-j} X[z]; \quad A[z]Y[z] = B[z]X[z];$$

$$G[z] = \frac{Y[z]}{X[z]} = \frac{B[z]}{A[z]} \text{ with } A[z] = \sum_{i=0}^{NA-1} a_i z^{-i} \text{ and } B[z] = \sum_{j=0}^{NB-1} b_j z^{-j} \tag{4.4.5}$$

This is referred to as the system function. Substituting $z = e^{j\Omega}$ into the system function or taking the DTFT of the impulse response $g[n]$, we can obtain the frequency response of the system. Figure 4.6 shows the overall relationship among the time-domain relationship (in the form of difference equation), the system (or transfer) function, the impulse response, and the frequency response.

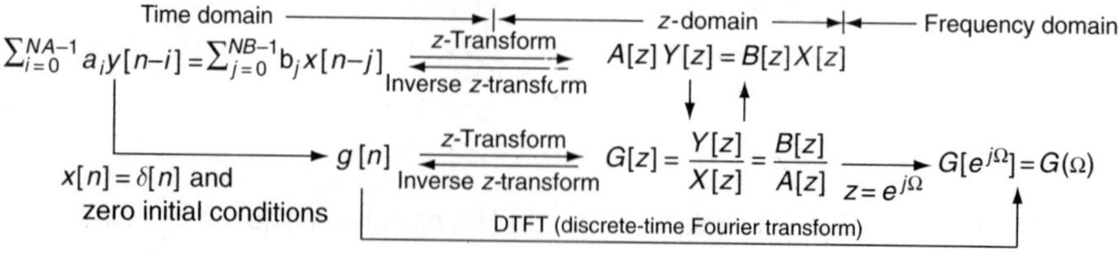

Fig. 4.6 The relationship between the impulse response, system function, and frequency response

Especially when we are interested only in right-sided sequences and causal linear systems, it is sensible to use the unilateral z-transform instead of the bilateral z-transform. It is the time shifting property that makes the unilateral z-transform particularly useful in analyzing causal systems described by difference equations with initial conditions.

Example 4.9 Difference Equation, System Function, and Impulse Response

Consider a discrete-time causal LTI (linear time-invariant) system whose input-output relationship is described by the following difference equation:

$$y[n] - \frac{1}{4}y[n-1] - \frac{1}{8}y[n-2] = x[n-1] \tag{E4.9.1}$$

(a) Find the system function $G[z]$.

Applying the linearity and time-shifting properties of the z-transform or using Eq. (4.4.5), we can obtain the system function as

$$G[z] - \frac{Y[z]}{X[z]} = \frac{z^{-1}}{1-(1/4)z^{-1}-(1/8)z^{-2}} = \frac{z}{(z-1/2)(z+1/4)} \tag{E4.9.2}$$

(b) Find the impulse response $g[n]$.

Noting that the system is causal and accordingly, the ROC of $G[z]$ is $z > 1/2$ (the outside of the circle passing through the outermost pole), we obtain the inverse z-transform of $G[z]$ as

$$g[n] = \mathcal{Z}^{-1}\left\{\frac{z}{(z-1/2)(z+1/4)}\right\} \overset{\text{partial fraction expansion}}{=} \mathcal{Z}^{-1}\left\{\frac{4}{3}\left(\frac{z}{z-1/2} - \frac{z}{z+1/4}\right)\right\} \overset{\text{B.9(5)}}{=} \frac{4}{3}\left(\left(\frac{1}{2}\right)^n - \left(-\frac{1}{4}\right)^n\right)u_s[n] \tag{E4.9.3}$$

Alternatively, the impulse response can be obtained directly from the difference equation, which can be solved iteratively with the unit impulse input $x[n] = \delta[n]$ and zero initial conditions:

$$y[n] = \frac{1}{4}y[n-1] + \frac{1}{8}y[n-2] + x[n-1] \tag{E4.9.4}$$

where $y[-1] = y[-2] = 0$ and $x[n-1] = \delta[n-1] = 1$ only for $n = 1$.

$$n = 0: \quad y[0] = (1/4)y[-1] + (1/8)y[-2] + x[-1] = 0 - 0 + 0 = 0$$
$$n = 1: \quad y[1] = (1/4)y[0] + (1/8)y[-1] + x[0] = 0 - 0 + 1 = 1$$
$$n = 2: \quad y[2] = (1/4)y[1] + (1/8)y[0] + x[1] = 1/4 - 0 + 0 = 1/4$$
$$n = 3: \quad y[3] = (1/4)y[2] + (1/8)y[1] + x[2] = 1/16 + 1/8 + 0 = 3/16$$

....

```
%sig04e09.m
syms z, Gz=z/(z^2-(1/4)*z-1/8); %system function
g = iztrans(Gz)  % symbolic inverse z transform
N=16; nn=[0:N-1];
for n=0:N-1, gn(n+1) = eval(g); end
% Solving the difference equation with the unit impulse input
B= [1 0]; A= [1 -1/4 -1/8]; %numerator/denominator
NB=length(B);  NA=length(A);
xn = [0 1 zeros(1,N-1+NB)]; % x[n-1] impulse input delayed by one sample
y = zeros(1,NA-1); % Initial condition
for m=NA:NA+N-1  % To solve the difference equation iteratively
   y(m)= -A(2:NA)*y(m-[1:NA-1]).' +B*xn(m-NA+[1:NB]).';
end
y = y(NA:NA+N-1);
% Using filter()
yp=[0 0]; xp=0; w0=filtic(B,A,yp,xp) %Initial condition from past history
yn = filter(B,A,xn,w0) % With zero initial condition by default
subplot(211)
stem(nn,gn), hold on, pause, stem(nn,y(1:N),'r'), stem(nn,yn(1:N),'k.')
% To plot the frequency response
N=64; dW=2*pi/N; W=[0:N]*dW; % frequency range
GW = DTFT(gn,W); %DTFT of the impulse response
GW1 = freqz(B,A,W); %substitute z=exp(j*W) into the system ftn B(z)/A(z)
subplot(212), plot(W,abs(GW),'b', W,abs(GW1),'r')
```

The objective of the above program "sig04e09.m" is as follows:

- Find the impulse response $g[n]$ in two ways, that is, by taking the inverse z-transform of the system function $G[z]$ and by solving the difference equation for the output $y[n]$ to the impulse input $x[n] = \delta[n]$. The MATLAB built-in function '`filter()`' together with '`filtic()`' (Sect. E.12) can also be used to obtain the output to any input and any initial condition. Also check if the two results conform to each other.
- Find the frequency response $G(\Omega)$ in two ways, that is, by taking the DTFT of the impulse response $g[n]$ and by substituting $z = e^{j\Omega}$ into the system function $G[z] = B[z]/A[z]$, where the latter job is done by using the MATLAB built-in function '`freqz()`'.

(cf) Comparing Eq. (E4.9.2) with Eq. (E4.9.3), we can tell that the poles of the system function, say, $p_1 = 1/2$ and $p_2 = -1/4$ yield the modes of the system, each of which determines how the corresponding output term evolves with time. See the stability theorem A.1.

Remark 4.6 Computational Method for Inverse z-Transform

Example 4.9 suggests another way of obtaining the inverse z-transform. That is, we can regard a rational z-transform expression $G[z]$ as a system function and set it equal to $Y[z]/X[z]$. Then, cross multiplying yields the z-domain input-output relationship

$$\frac{Y[z]}{X[z]} = G[z] = \frac{B[z]}{A[z]}; \quad A[z]Y[z] = B[z]X[z] \tag{4.4.6}$$

with $A[z] = \sum_{i=0}^{NA-1} a_i z^{-i}$ and $B[z] = \sum_{j=0}^{NB-1} b_j z^{-j}$. We can write the corresponding difference equation

$$\sum_{i=0}^{NA-1} a_i y[n-i] = \sum_{j=0}^{NB-1} b_j x[n-j] \tag{4.4.7}$$

and solve it iteratively with $x[n] = \delta[n]$ and zero initial conditions for $y[n]$ in the forward/backward direction to get a right/left-sided sequence $g[n] = \mathcal{Z}^{-1}\{G[z]\}$. Just as with the long division method, this gives us no analytical solution in a closed form. Note the following fact:

$$\begin{aligned} Y[z]|_{X[z]=1} &= G[z]X[z]|_{X[z]=1} = G[z] \\ y[n]|_{x[n]=\delta[n]} &= \mathcal{Z}^{-1}\{G[z]\} = g[n] \end{aligned} \tag{4.4.8}$$

So far, we have never felt the necessity of the unilateral z-transform over the bilateral one. Now, we are about to look at an initial value problem for which the unilateral transform is indispensable.

Example 4.10 Different Difference Equations Describing the Same System

(a) Find the output $y_1[n]$ of the causal system whose input-output relationship is described by

$$y_1[n] - a\, y_1[n-1] = x_1[n-1], \; n \ge 0 \tag{E4.10.1}$$

where $y_1[-1] = y_0$ and

$$x_1[n] = b^n u_s[n] \text{ with } b \ne a \tag{E4.10.2}$$

(Solution)
To solve this difference equation for $y_1[n]$, we apply the time shifting property (4.2.3a) for Eq. (E4.10.1) to write its z-transform as

$$\begin{aligned} Y_1[z] - a(z^{-1}Y_1[z] + y_1[-1]) &= z^{-1}X_1[z] + x_1[-1]; \\ (1 - az^{-1})Y_1[z] &= a\, y_0 + z^{-1}X_1[z] \end{aligned} \tag{E4.10.3}$$

since $x_1[-1] \overset{\text{(E4.10.2)}}{=} b^n u_s[n]|_{n=-1} = 0$. We can solve this algebraic equation for $Y_1[z]$ as

$$Y_1[z] = \frac{1}{1 - a\, z^{-1}}(a\, y_0 + z^{-1}X_1[z]) = \frac{z}{z-a}\left(a\, y_0 + z^{-1}\frac{z}{z-b}\right) \tag{E4.10.4}$$

To take the inverse z-transform of this expression, we divide its both sides by z and perform the partial fraction expansion as

$$\frac{Y_1[z]}{z} = \frac{a\,y_0}{z-a} + \frac{1}{(z-a)(z-b)} = \frac{a\,y_0}{z-a} + \frac{1}{a-b}\left(\frac{1}{z-a} - \frac{1}{z-b}\right)$$

$$Y_1[z] = \frac{a\,y_0 z}{z-a} + \frac{1}{a-b}\left(\frac{z}{z-a} - \frac{z}{z-b}\right) \quad \text{(E4.10.5)}$$

Now we take the inverse z-transform to get the output as

$$y_1[n] = a y_0 a^n u_s[n] + \frac{1}{a-b}(a^n - b^n)u_s[n]$$
$$= y_0 a^{n+1} u_s[n+1] + \frac{1}{a-b}(a^n - b^n)u_s[n] \quad \text{(E4.10.6)}$$

Here, we replaced $u_s[n]$ by $u_s[n+1]$ to express the existence of the given initial condition $y_1[-1] = y_0$.

(b) Find the output $y_2[n]$ of the causal system whose input-output relationship is described by

$$y_2[n+1] - a y_2[n] = x_2[n],\, n \geq 0 \quad \text{(E4.10.7)}$$

where $y_2[0] = y_0$ and

$$x_2[n] = x_1[n-1] = b^{n-1}u_s[n-1] \text{ with } b \neq a \quad \text{(E4.10.8)}$$

(Solution)

To solve this difference equation for $y_2[n]$, we apply the time shifting property (4.2.3b,a) for Eq. (E4.10.7) to write its z-transform as

$$z(Y_2[z] - y_2[0]) - aY_2[z] = X_2[z];$$
$$(z-a)Y_2[z] = z y_0 + X_2[z] = z y_0 + z^{-1}X_1[z] \quad \text{(E4.10.9)}$$

since $x_1[-1] \overset{\text{(E4.10.2)}}{=} b^n u_s[n]|_{n=-1} = 0$. We can solve this algebraic equation for $Y_2[z]$ as

$$Y_2[z] = \frac{1}{z-a}(z y_0 + z^{-1}X_1[z]) = \frac{y_0 z}{z-a} + \frac{1}{(z-a)(z-b)} \quad \text{(E4.10.10)}$$

To take the inverse z-transform of this expression, we divide its both sides by z and perform the partial fraction expansion as

$$\frac{Y_2[z]}{z} = \frac{y_0}{z-a} + \frac{1}{z(z-a)(z-b)}$$
$$= \frac{y_0}{z-a} + \frac{1/ab}{z} + \frac{1/a(a-b)}{z-a} + \frac{1/b(b-a)}{z-b};$$
$$Y_2[z] = \frac{y_0 z}{z-a} + \frac{1}{ab} + \frac{1}{a-b}\left(\frac{1}{a}\frac{z}{z-a} - \frac{1}{b}\frac{z}{z-b}\right) \quad \text{(E4.10.11)}$$

Now we take the inverse z-transform to get the output as

$$y_2[n] = y_0 a^n u_s[n] + \frac{1}{ab}\delta[n] + \frac{1}{a-b}(a^{n-1} - b^{n-1})u_s[n]$$
$$= y_0 a^n u_s[n] + \frac{1}{a-b}(a^{n-1} - b^{n-1})u_s[n-1] \quad \text{(E4.10.12)}$$

Here, $u_s[n]$ is replaced by $u_s[n-1]$ to show that the 2nd and 3rd terms cancel each other at $n = 0$.

(cf) Comparing Eqs. (E4.10.6) and (E4.10.12), we see that $y_1[n-1] = y_2[n]$. This can be verified by showing that

$$\mathcal{Z}\{y_1[n-1]\} = Y_2[z] = \mathcal{Z}\{y_2[n]\} \quad \text{(E4.10.13)}$$

(Proof)

$$\mathcal{Z}\{y_1[n-1]\} \overset{(4.2.3a)}{=} z^{-1}(Y_1[z] + y_1[-1]z) = z^{-1}(Y_1[z] + y_0 z)$$
$$\overset{(4.10.4)}{=} z^{-1}\left(\frac{1}{z-a}(z\,a\,y_0 + X_1[z]) + y_0 z\right) = \frac{1}{z-a}(z\,y_0 + z^{-1}X_1[z]) \overset{(4.10.10)}{=} Y_2[z] \quad \text{(E4.10.14)}$$

The justification is as follows: The two LTI systems of (a) and (b) are inherently the same. Compared with (a), the initial conditions and input of (b) are delayed by $n_1 = 1$ and consequently, the output of the system (b) is also delayed by $n_1 = 1$.

4.5 Geometric Evaluation of the z-Transform

In this section we discuss a geometrical method to evaluate a rational function in z at any point in the z-plane, particularly on the unit circle $z = e^{j\Omega}$ for obtaining the frequency response $G(\Omega)$ from the pole-zero plot of the system function $G[z]$.

Let us consider a system function $G[z]$ given in a rational form as

$$G[z] = K\frac{\prod_{j=1}^{M}(z - z_j)}{\prod_{i=1}^{N}(z - p_i)} = K\frac{(z-z_1)(z-z_2)\cdots(z-z_M)}{(z-p_1)(z-p_2)\cdots(z-p_N)} \quad (4.5.1)$$

where z_j's and p_i's are finite zeros and poles of $G[z]$, respectively. The value of $G[z]$ at some point $z = z_0$ in the z-plane is a complex number that can be expressed in the polar form as

$$G[z_0] = |G[z_0]|\angle G[z_0] \tag{4.5.2}$$

where

$$\text{Magnitude : } |G[z_0]| = |K| \frac{\prod_{j=1}^{M} |z_0 - z_j|}{\prod_{i=1}^{N} |z_0 - p_i|} \tag{4.5.3a}$$

$$\text{Phase : } \angle G[z_0] = \sum_{j=1}^{M} \angle(z_0 - z_j) - \sum_{i=1}^{N} \angle(z_0 - p_i)(\pm\pi)$$

$$\text{with } \pm\pi \text{ only for } K < 0 \tag{4.5.3b}$$

$(z_0 - z_j)$'s and $(z_0 - p_i)$'s in the above equations are complex numbers, each of which can be represented by a vector in the z-plane from z_j or p_i to z_0. They can be easily constructed from the pole/zero plot where $|z_0 - z_j|$ and $|z_0 - p_i|$ are the distances, while $\angle z_0 - z_j$ and $\angle z_0 - p_i$ are the phase angles.

As mentioned in Remark 4.5(1), the frequency response $G(\Omega)$ can be obtained by evaluating the system function $G[z]$ on the unit circle if it exists. Keeping this in mind, we can use the *pole-zero plot* to get the approximate shape of the *magnitude response* $|G(\Omega)|$ without computing the frequency response from Eq. (4.5.2). Here is the overall feature of frequency response related with the pole-zero pattern:

Remark 4.7 Frequency (Magnitude/Phase) Response and *Pole-Zero Pattern*

(1) For a pole/zero near the unit circle, the magnitude response curve tends to have a sharp peak/valley and the phase changes rapidly by about 180° at the corresponding frequency. As the pole/zero moves to the unit circle, the peak/valley becomes sharper. On the other hand, as the pole/zero moves to the origin, the peak/valley becomes smoother. Such a tendency can be observed from Fig. 3.2 where the frequency responses of $G[z] = z/(z-a)$ are depicted for $a = 0.8$ and 0.5. Also, the phase jump of π [rad] occurring at the frequency corresponding to a zero on the unit circle is illustrated by Figs. 3.1, 3.7, 3.11, and 4.8.

(2) Generally speaking, if a singularity (pole/zero) is located close to the unit circle, it will dominate the frequency response in the frequency range adjacent to that location. This idea of dominant singularity is helpful not only for getting an approximate frequency response, but also for the pole/zero placement design to achieve a desired frequency response. For example, the magnitude response (Fig. 4.7(b)) of the system described by (E4.11.1) has the maximum around $\Omega = \pm\pi/4$ due to the pole at $z = 0.5\sqrt{2}e^{\pm j\pi/4}$ and becomes zero at $\Omega = \pi$ because of the zero at $z = -1$. In contrast, the zero at $z = 0$ (which is far from the unit circle) has no influence on the magnitude response.

Example 4.11 Pole-Zero Pattern and Frequency Response

For the system function

$$G[z] = \frac{z(z+1)}{z^2 - z + 0.5} = \frac{z(z+1)}{(z-0.5-j0.5)(z-0.5+j0.5)} \tag{E4.11.1}$$

we have the pole-zero plot and the frequency response magnitude and phase curves depicted in Fig. 4.7(a), (b), and (c), respectively. As shown in Fig. 4.7(b), the magnitude of $G(\Omega) = G[e^{j\Omega}]$ becomes zero at $\Omega = \pi$ corresponding to the zero $z = -1$, i.e.,

$$|G[e^{j\Omega}]|\big|_{\Omega=\pi} = |G[z]|\big|_{z=-1} = 0 \tag{E4.11.2}$$

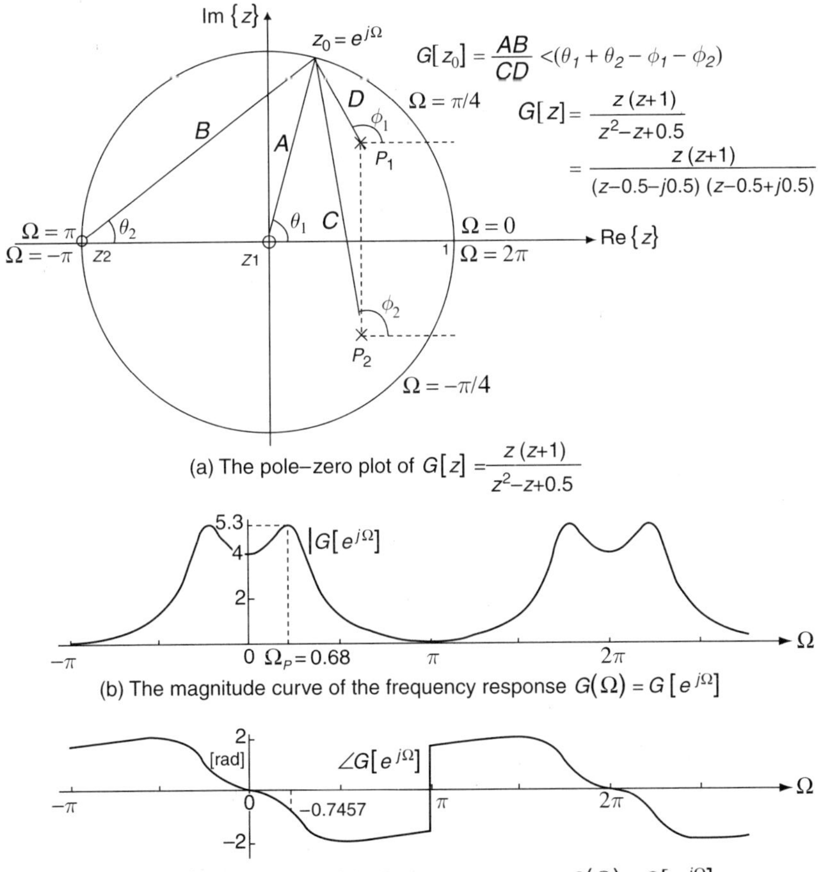

(a) The pole–zero plot of $G[z] = \dfrac{z(z+1)}{z^2-z+0.5}$

(b) The magnitude curve of the frequency response $G(\Omega) = G[e^{j\Omega}]$

(c) The phase curve of the frequency response $G(\Omega) = G[e^{j\Omega}]$

Fig. 4.7 The pole–zero plot and frequency response for Example 4.11

and it reaches the peak around $\Omega = \pm\pi/4$ adjacent to the phases of the poles

$$z = 0.5 \pm j0.5 = 0.5\sqrt{2}e^{\pm j\pi/4} \tag{E4.11.3}$$

(a) Let us find the response of the system $G[z]$ to a sinusoidal input $x_1[n] = \sin(\Omega_p n)$ with $\Omega_p = 0.68$.

$$Y_1[z] = X_1[z]G[z] = \frac{z\sin\Omega_p}{(z-\cos\Omega_p)^2+\sin^2\Omega_p}\frac{z(z+1)}{(z-0.5)^2+0.5^2} \tag{E4.11.4}$$

$$= \frac{-3.60\times z(z-\cos\Omega_p)+3.89\times z\sin\Omega_p}{(z-\cos\Omega_p)^2+\sin^2\Omega_p} + \left.\frac{3.60\times z(z-R\cos\Omega_s)-1.65\times zR\sin\Omega_s}{(z-R\cos\Omega_s)^2+(R\sin\Omega_s)^2}\right|_{R=\frac{1}{\sqrt{2}},\Omega_s=\frac{\pi}{4}}$$

$$y_1[n] = \mathcal{Z}^{-1}\{Y_1[z]\} = -3.6\cos(\Omega_p n) + 3.89\sin(\Omega_p n) + \sqrt{2}^{-n}(3.6\cos(n\pi/4) - 1.65\sin(n\pi/4))$$

$$= (5.3\sin(\Omega_p n - 0.7457) + \sqrt{2}^{-n}(3.6\cos(n\pi/4) - 1.65\sin(n\pi/4)))u_s[n] \tag{E4.11.5}$$

Note that the sinusoidal steady-state response $y_{1,ss}[n] = 5.3\sin(\Omega_p n - 0.7457)$ has the same frequency with the input $x_1[n]$, but its amplitude is $|G(\Omega_p)| = 5.3$ times that of the input and its phase is $\angle G(\Omega_p) = -0.7457 \simeq -1.1\Omega_p$ plus that of the input, as can be seen from Fig. 4.8(a).

(b) Let us find the response of the system $G[z]$ to a unit step (DC) input $x_2[n] = u_s[n]$.

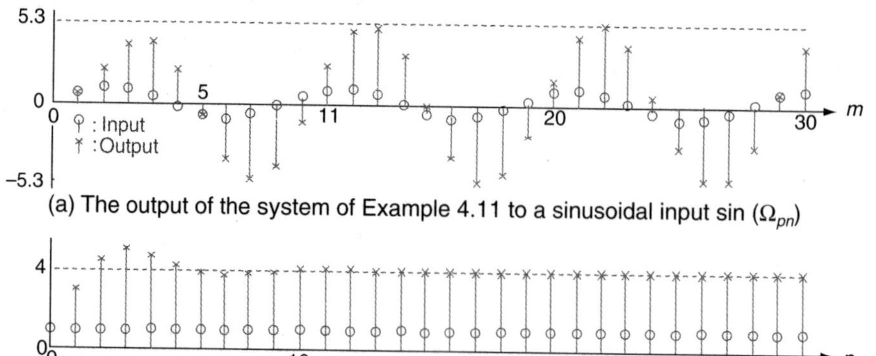

(a) The output of the system of Example 4.11 to a sinusoidal input sin (Ω_{pn})

(b) The output of the system of Example 4.11 to a unit step (DC) input

Fig. 4.8 The output of the system of Example 4.11 to sinusoidal/DC inputs

$$Y_2[z] = X_2[z]G[z] = \frac{z}{z-1}\frac{z(z+1)}{(z-0.5)^2+0.5^2}$$

$$= \frac{4\times z}{z-1} + \frac{-3\times z(z-0.5)}{(z-0.5)^2+0.5^2} + \frac{1\times z0.5}{(z-0.5)^2+0.5^2} \quad \text{(E4.11.6)}$$

where the partial fraction expansion is found as follows:

$$Y_2[z] = \frac{K_1\times z}{z-1} + \frac{K_2\times z(z-0.5)}{(z-0.5)^2+0.5^2} + \frac{K_3\times z0.5}{(z-0.5)^2+0.5^2}$$

$$K_1 = (z-1)\frac{Y_1[z]}{z} = \left.\frac{z(z+1)}{(z-0.5)^2+0.5^2}\right|_{z=1} = 4$$

$$K_1 + K_2 = 1, -K_1 - 0.5K_2 + 0.5K_3 = 1, 0.5K_1 + 0.5K_2 - 0.5K_3 = 0$$

$$K_2 = 1 - K_1 = 1 - 4 = -3,\ K_3 = K_1 + K_2 = 4 - 3 = 1$$

Thus the response to $x_2[n] = u_s[n]$ can be obtained from the inverse z-transform as

$$y_2[n] = \mathcal{Z}^{-1}\{Y_2[z]\} = \left(4 + \sqrt{2}^{-n}(-3\cos(n\pi/4) + \sin(n\pi/4))\right)u_s[n] \quad \text{(E4.11.7)}$$

Note that the DC steady-state response $y_{2,ss}[n] = 4$ is the DC gain $|G(0)| = 4$ times that of the input, as can be seen from Fig. 4.8(b).

In fact, the inverse z-transform of $Y_1[z]$ as well as Figs. 4.7 and 4.8 is obtained by running the following MATLAB program "sig04e11.m", where we managed to get the coefficients of the partial fraction expansion using residue() since iztrans() does not work properly for this case.

```
%sig04e11.m
clear, clf
B=[1 1 0]; A=[1 -1 0.5]; %numerator/denominator of system function (E4.11.1)
figure(1), zplane(roots(B),roots(A)) % pole-zero plot in Fig. 4.7(a)
% To get the frequency response
N=360; dW=2*pi/N; k0=-200; W=[k0:500]*dW;
GW= freqz(B,A,W); % frequency response
GW_mag= abs(GW); GW_ph= angle(GW); % magnitude/phase of frequency response
[GW_mag_peak,i]= max(GW_mag(-k0+[1:N/2])) % peak frequency response magnitude
ip=-k0+i-1; Wp=W(ip); GW_ph_peak=GW_ph(ip); % peak frequency
GW_dc= GW(-k0+1); % DC gain
figure(2), subplot(411), plot(W,GW_mag) % Fig. 4.7(b)
subplot(412), plot(W,GW_ph)             % Fig. 4.7(c)
% To get the time response from filtering of a sine-wave input signal
nn=[0:30]; % time index vector
xn= sin(Wp*nn); % A sinusoidal input of peak frequency
yn= filter(B,A,xn); % With zero initial condition by default
% plot the time response
subplot(413), stem(nn,xn,'Markersize',5), hold on
```

```
stem(nn,yn,'x','Markersize',5)
% Try with the inverse z-transform
syms z
y1=iztrans(sin(Wp)*z^2*(z+1)/((z-cos(Wp))^2+sin(Wp)^2)/((z-0.5)^2+0.5^2))
[r,p,k]=residue(sin(Wp)*[1 1 0],conv([1 -2*cos(Wp) 1],[1 -1 0.5]))
R = sqrt(real(p(3)*p(4))); Ws= angle(p(3));
% numerator of the 1st&2nd terms reduced to a common denominator
n1= r(1)*[1 -p(2)]+r(2)*[1 -p(1)];
K1= n1(1); K2= (n1(2)+K1*cos(Wp))/sin(Wp);
% numerator of the 3rd&4th terms reduced to a common denominator
n2= r(3)*[1 -p(4)]+r(4)*[1 -p(3)];
K3= n2(1); K4= (n2(2)+K3*R*cos(Ws))/(R*sin(Ws));
y1n= K1*cos(Wp*nn) + K2*sin(Wp*nn) + R.^nn.*(K3*cos(Ws*nn)+K4*sin(Ws*nn));
stem(nn,y1n,'rx','Markersize',5)
%filtering of a DC input signal
nn=[0:30]; xn= ones(size(nn)); % A DC input signal (of zero frequency)
yn_DC= filter(B,A,xn); % With zero initial condition by default
subplot(414)
stem(nn,xn,'Markersize',5) % plot the time response together with the input
hold on, stem(nn,yn_DC,'m','Markersize',5)
% Try with the inverse z-transform
y2=iztrans(z^2*(z+1)/(z-1)/((z-0.5)^2+0.5^2))
% A nice alternative for the case of all simple poles
[r,p,k]=residue([1 1 0],conv([1 -1],[1 -1 0.5]))
y2n= 0;
for i=1:length(r)
   y2n= y2n + r(i)*p(i).^nn;
end
stem(nn,real(y2n),'m^','Markersize',5)
```

4.6 The z-Transform of Symmetric Sequences

In this section we explore some features of the phase characteristic and pole-zero pattern for systems having (anti-)symmetric impulse responses of finite duration.

4.6.1 Symmetric Sequences

Let us consider a symmetric sequence $g[n]$ of duration $N+1$ such that

$$g[n] = g[N-n] \text{ for } n = 0 : N \tag{4.6.1}$$

<Case 1> If N is even, i.e., $N = 2M$ for some integer M, then the z-transform of $g[n]$ is

$$G[z] = \sum_{n=0}^{M-1} g[n]z^{-n} + g[M]z^{-M} + \sum_{n=M+1}^{N} g[n]z^{-n}$$

$$\overset{(4.6.1)}{=} \sum_{n=0}^{M-1} g[n](z^{-n} + z^{-(N-n)}) + g[M]z^{-M}$$

$$= z^{-M}\left(g[M] + \sum_{n=0}^{M-1} g[n](z^{-n+M} + z^{n-M})\right) \text{ with } M = \frac{N}{2} \tag{4.6.2a}$$

which, with $z = e^{j\Omega}$, yields the frequency response as

$$G(\Omega) = \left(g[\frac{N}{2}] + \sum_{n=0}^{N/2-1} 2g[n]\cos\left(\left(\frac{N}{2} - n\right)\Omega\right)\right) \angle - \frac{N}{2}\Omega \tag{4.6.2b}$$

<Case 2> If N is odd, i.e., $N = 2M - 1$ for some integer M, then the z-transform of $g[n]$ is

$$G[z] = \sum_{n=0}^{M-1} g[n]z^{-n} + \sum_{n=M}^{N} g[n]z^{-n}$$

$$\overset{(4.6.1)}{=} \sum_{n=0}^{M-1} g[n]z^{-n} + \sum_{n=0}^{M-1} g[N-n]z^{-(N-n)}$$

$$= \sum_{n=0}^{M-1} g[n](z^{-n} + z^{n-N})$$

$$\overset{M=(N+1)/2}{=} z^{-N/2} \sum_{n=0}^{M-1} g[n](z^{-n+N/2} + z^{n-N/2}) \tag{4.6.3a}$$

which, with $z = e^{j\Omega}$, yields the frequency response as

$$G(\Omega) = \sum_{n=0}^{(N-1)/2} 2g[n]\cos\left(\left(\frac{N}{2} - n\right)\Omega\right) \angle - \frac{N}{2}\Omega \tag{4.6.3b}$$

Note that $G[z]|_{-1} = G[e^{j\Omega}]|_{\Omega=\pi} = 0$.

4.6.2 Anti-Symmetric Sequences

Let us consider an anti-symmetric sequence $g[n]$ of duration $N + 1$ such that

$$g[n] = -g[N-n] \text{ for } n = 0 : N \tag{4.6.4}$$

<Case 1> If N is even, i.e., $N = 2M$ for some integer M, then we have $g[M] = -g[M]$, which implies $g[M] = 0$. The z-transform of $g[n]$ is

$$G[z] = z^{-M} \sum_{n=0}^{M-1} g[n](z^{-n+M} - z^{n-M}) \text{ with } M = \frac{N}{2} \tag{4.6.5a}$$

which, with $z = e^{j\Omega}$, yields the frequency response as

$$G(\Omega) = \sum_{n=0}^{N/2-1} 2g[n]\sin\left(\left(\frac{N}{2} - n\right)\Omega\right) \angle -\frac{N}{2}\Omega + \frac{\pi}{2} \tag{4.6.5b}$$

Note that $G[e^{j\Omega}]\big|_{\Omega=0} = 0$ and $G[e^{j\Omega}]\big|_{\Omega=\pi} = 0$.

<Case 2> If N is odd, i.e., $N = 2M - 1$ for some integer M, then the z-transform of $g[n]$ is

$$G[z] = z^{-N/2} \sum_{n=0}^{M-1} g[n](z^{-n+N/2} - z^{n-N/2}) \tag{4.6.6a}$$

which, with $z = e^{j\Omega}$, yields the frequency response as

$$G(\Omega) = \sum_{n=0}^{(N-1)/2} 2g[n]\sin\left(\left(\frac{N}{2} - n\right)\Omega\right) \angle -\frac{N}{2}\Omega + \frac{\pi}{2} \tag{4.6.6b}$$

Note that $G[e^{j\Omega}]\big|_{\Omega=0} = 0$.

Remark 4.8 Pole-Zero Pattern and Linear Phase of (Anti-)Symmetric Sequences

(1) From Eqs. (4.6.2a)/(4.6.5a) and (4.6.3a)/(4.6.6a), we can see that if $G[z_0] = 0$, then $G[z_0^{-1}] = 0$, which implies that real zeros occur in reciprocal pairs (z_0 and z_0^{-1}) and complex zeros occur in reciprocal, complex-conjugate quadruplets ($z_0 = r_0\angle \pm \Omega_0$ and $z_0^{-1} = r_0^{-1}\angle \pm \Omega_0$). Note that zeros on the unit circle form their own reciprocal pairs and real zeros on the unit circle, i.e., $z = 1$ or $z = -1$, form their own reciprocal, complex conjugate pairs. Note also that all the poles are located at $z = 0$ or ∞. See Fig. 4.9(a1) and (a2).

(2) From Eqs. (4.6.2b)/(4.6.5b) and (4.6.3b)/(4.6.6b), we can see that they have *linear phase*, i.e., their phases are (piecewise) linear in Ω except for phase jumps of $\pm\pi$ or $\pm 2\pi$ (see Remark 3.3 and Fig. 4.9(c1) and (c2)).

(3) If a system has the impulse response represented by a symmetric or anti-symmetric sequence of finite duration, such a system has *linear phase shifting* property so that it will reproduce the input signals falling in the passband with a delay equal to the slope of the phase curve. That is why such a system is called the linear phase FIR filter.

Example 4.12 Pole-zero Pattern of Symmetric or Anti-symmetric Sequences

(a) Consider a system whose impulse response is

$$g_1[n] = [\cdots \overset{n=-1}{0} \ \overset{0}{1} \ \overset{1}{-2.5} \ \overset{2}{5.25} \ \overset{3}{-2.5} \ \overset{4}{1.0} \ \overset{5}{0} \cdots]. \tag{E4.12.1}$$

This system has the system function $G_1[z] = \mathcal{Z}\{g_1[n]\}$ as

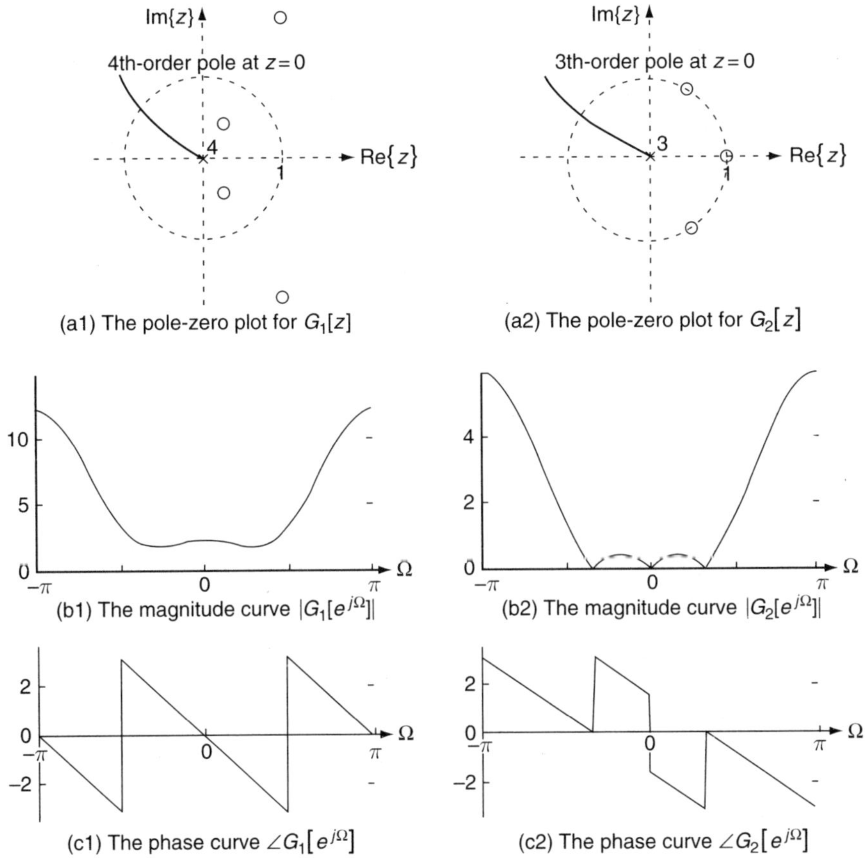

Fig. 4.9 The pole–zero plots and frequency responses for Example 4.12

$$
\begin{aligned}
G_1[z] &= 1 - 2.5z^{-1} + 5.25z^{-2} - 2.5z^{-3} + z^{-4} \\
&= z^{-4}(z - 0.5e^{j\pi/3})(z - 0.5e^{-j\pi/3})(z - 2e^{j\pi/3})(z - 2e^{-j\pi/3})
\end{aligned} \tag{E4.12.2}
$$

whose pole-zero pattern and frequency response magnitude/phase curves are plotted in Fig. 4.9(a1) and (b1)/(c1), respectively.

(b) Consider a system whose impulse response is

$$
g_2[n] = [\cdots \overset{n=-1}{0} \ \overset{0}{-1} \ \overset{1}{2} \overset{2}{-2} \ \overset{3}{1} \ \overset{4}{0} \cdots]. \tag{E4.12.3}
$$

This system has the system function $G_2[z] = \mathcal{Z}\{g_2[n]\}$ as

$$
\begin{aligned}
G_2[z] &= -1 + 2z^{-1} - 2z^{-2} + z^{-3} = -z^{-3}(z-1)(z^2 - z + 1) \\
&= -z^{-3}(z-1)(z - e^{j\pi/3})(z - e^{-j\pi/3})
\end{aligned} \tag{E4.12.4}
$$

whose pole-zero pattern and frequency response magnitude/phase curves are plotted in Fig. 4.9(a2) and (b2)/(c2), respectively.

The following program "sig04e12.m" is run to yield Fig. 4.9, where zplane(B,A) is used to create the pole-zero plot of a system with system function $G[z] = B[z]/A[z]$.

```
%sig04e12.m
clear, clf
N=360; W=pi/N*[-N:N]; % frequency range
for i=1:2
   if i==1, B=[1 -2.5 5.25 -2.5 1]; A=1;
    else  B=[-1 2 -2 1]; A=1; %numerator/denominator of system function
   end
   figure(1), subplot(220+i), zplane(B,A)
   GW= freqz(B,A,W); % frequency response
   GW_mag= abs(GW); % magnitude of frequency response
   GW_phase= angle(GW); % phase of frequency response
   figure(2)
   subplot(220+i), plot(W,GW_mag)
   set(gca,'fontsize',9, 'Xlim',[-pi pi], 'xtick',[-pi 0 pi], ...
        'xticklabel',{'-pi' '0' 'pi'})
   subplot(222+i), plot(W,GW_phase)
   set(gca,'fontsize',9, 'Xlim',[-pi pi], 'xtick',[-pi 0 pi], ...
        'xticklabel',{'-pi' '0' 'pi'})
end
```

4.7 Summary

In this chapter we have defined the bilateral/unilateral z-transform and derived their basic properties. We have also presented the several methods to find the inverse z-transform of a given z-transform expression. We have explored the relationship among the system function, frequency response, and difference equation and discussed how to utilize the z-transform for analyzing discrete-time LTI systems.

Problems

4.1 z-Transform

(a) Find the bilateral z-transform $X[z]$ (with its ROC) of

$$x[n] = a^{|n|} = a^n u_s[n] + a^{-n} u_s[-n-1] \text{ with } |a| < 1 \tag{P4.1.1}$$

(b) Could you get the bilateral z-transform of

$$x[n] = 1 = u_s[n] + u_s[-n-1] \tag{P4.1.2}$$

If your answer is yes, find it. Otherwise, state why it is not possible.

(c) Find the bilateral z-transform $X[z]$ (with its ROC) of

$$x[n] = u_s[-n] \tag{P4.1.3}$$

in the following three ways:

(i) Noting that $u_s[-n]$ is the one-sample delayed version of $u_s[-n-1]$, use the time shifting property.
(ii) Noting that $u_s[-n]$ is the time-reversal version of $u_s[n]$, use the time reversal property.
(iii) Noting that $u_s[-n]$ can be expressed as $u_s[-n-1]+\delta[n]$, use B.9(6) & (1).

4.2 Inverse z-Transform

(a) Find the inverse z-transform $x[n]$ of

$$X[z] = \frac{z^2}{(z+1)^2} = z \times \frac{z}{(z+1)^2} = \frac{z}{z+1} \times \frac{z}{z+1} \tag{P4.2.1}$$
with ROC $\mathcal{R} = \{z : |z| > 1\}$

in the following two different ways:

(i) Use the z-transform table B.9(10) and the time shifting property 4.2.3(b) or B.7(2).
(ii) Use the z-transform table B.9(5) and the convolution property 4.2.6 or B.7(4).

(b) Use the z-transform table B.9(17) and the time shifting property 4.2.3(a) or B.7(2) to find the inverse z-transform of

$$X[z] = \frac{a}{z^2+a^2} = z^{-1} \times \frac{az}{z^2+a^2} \text{ with ROC } \mathcal{R} = \{z : |z| > a\} \tag{P4.2.2}$$

(c) Find the inverse z-transform $x[n]$ of

$$X[z] = \frac{2z-1}{(z-1)(z^2-z+1)} \text{ with ROC } \mathcal{R} = \{z : |z| > 1\} \tag{P4.2.3}$$

(d) Find the inverse z-transform $x[n]$ of

$$X[z] = \frac{1}{(z-1)^2 z^2} = \frac{1}{(z-1)^2} - \frac{2}{z-1} + \frac{1}{z^2} + \frac{2}{z} \tag{P4.2.4}$$

with ROC $\mathcal{R} = \{z : |z| > 1\}$

4.3 Inverse z-Transform, Causality, Convergence, and Stability

(a) Find the inverse z-transforms $g_{a1}[n]$, $g_{a2}[n]$, and $g_{a3}[n]$ of

$$G_a[z] = \frac{3(1-z^{-1})}{1-(5/2)z^{-1}+z^{-2}} = \frac{3z(z-1)}{(z-2)(z-1/2)} \tag{P4.3.1}$$

with one of the following ROCs:

(i) $\mathcal{R}_1 = \{z : |z| > 2\}$, (ii) $\mathcal{R}_2 = \{z : |z| < 1/2\}$, and (iii) $\mathcal{R}_3 = \{z : 1/2 < |z| < 2\}$. Then determine the causality and convergence (to zero) for each of $g_i[n]$'s.

(b) Find the inverse z-transforms $g_{b1}[n]$, $g_{b2}[n]$, and $g_{b3}[n]$ of

$$G_b[z] = \frac{2z(z+1)}{z^2 + (3/4)z + 1/8} = \frac{2z(z+1)}{(z+1/4)(z+1/2)} \tag{P4.3.2}$$

with one of the following ROCs: (i) $\mathcal{R}_1 = \{z : |z| > 1/2\}$, (ii) $\mathcal{R}_2 = \{z : |z| < 1/4\}$, and (iii) $\mathcal{R}_3 = \{z : 1/4 < |z| < 1/2\}$. Then determine the causality and convergence (to zero) for each of $g_i[n]$'s.

(c) Suppose $G_a[z]$ and $G_b[z]$ are the system functions of discrete-time systems. Then the causality and stability of the systems depends on the causality and geometric convergence (to zero) of the impulse response $g_i[n]$'s. Referring to the stability criterion that the ROC should contain the unit circle (Remark 4.5(2)), determine the stability of the above six cases. For each case, plot the impulse response for $n = -20 : 20$ and complete the following table:

(d) What do you think about the stability of a system having a pole on the unit circle, which is the boundary between the stability and instability regions? Let us consider a system with the system function

$$G_c[z] = \frac{1}{z-1} \text{ with ROC } \mathcal{R} = \{z : |z| > 1\} \tag{P4.3.3}$$

Suppose a bounded input such as the unit step sequence $x[n] = u_s[n]$ is applied to this system and find the output. Is it bounded? Is the system (BIBO) stable? Can you find any (bounded) input such as a sine-wave which makes the output of the system blow up?

(cf) If a discrete-time LTI system has all its poles inside or on the unit circle and the poles on the unit circle are simple (distinct), the system is said to be *marginally stable*. Likewise, if a continuous-time LTI system has all its poles inside the LHP (left-half s -plane) or on the $j\omega$-axis and the poles

Table P4.3 Pole location, ROC, causality, and stability of a system

	All poles inside the unit circle	The unit circle inside ROC	Causality	Stablility
(a)–(i)				
(a)–(ii)				
(a)–(iii)				
(b)–(i)				
(b)–(ii)				
(b)–(iii)				

on the $j\omega$-axis are simple (distinct), the system is said to be *marginally stable*.

4.4 To Solve Difference Equation Using the z-Transform
Consider the following difference quation

$$y[n] - \frac{1}{3}y[n-1] = x[n] \text{ with } x[n] = 2^{-n}u_s[n] \qquad \text{(P4.4.1)}$$

(a) Solve it with the initial condition $y[-1] = -3$.
(b) Solve it with the initial condition $y[-1] = 6$.

(cf) This problem implies that the output of an LTI system has generally the system mode (in the natural response) and the input mode (in the forced response), but the system mode may disappear depending on the initial conditions.

4.5 Forward/Backward Solution of Difference Equation by Iteration
Consider the following difference quation:

$$y[n] - \frac{1}{4}y[n-1] - \frac{1}{8}y[n-2] = 3x[n-1] \text{ with } x[n] = \delta[n] \qquad \text{(P4.5.1)}$$

(a) Solve it forward with no initial condition ($y[-2] = y[-1] = 0$) to find $y[n]$ for $n = 0 : 3$. To do this, you should rewrite the difference equation as

$$y[n] = \frac{1}{4}y[n-1] + \frac{1}{8}y[n-2] + 3x[n-1] \text{ with } x[n] = \delta[n] \quad \text{(P4.5.2)}$$

(b) Solve it backward with no initial condition ($y[1] = y[2] = 0$) to find $y[n]$ for $n = -3 : 0$. To do this, you should rewrite the difference equation as

$$y[n-2] = -2y[n-1] + 8y[n] - 24x[n-1] \text{ with } x[n] = \delta[n] \quad \text{(P4.5.3)}$$

(cf) You can use the following program "sig04p_05.m". Do the results agree with the solutions (4.3.8a,b) obtained using the long division in Sec. 4.3.2 or (E4.7.9a,b)?

```
%sig04p_05.m
clear
% To solve a difference equation forward by iteration
ya(1)=0; ya(2)=0; % zero initial conditions
xa= zeros(1,10); xa(3)=1; % unit impulse input
for n=3:11
   ya(n)= (1/4)*ya(n-1) + (1/8)*ya(n-2) +3*xa(n-1);
end
nn=0:8; ya1= 4*(1/2).^nn - 4*(-1/4).^nn;
[ya(3:end); ya1]
% To solve a difference equation backward by iteration
```

```
yb(10)=0; yb(11)=0; % zero initial conditions
xb= zeros(1,10); xb(9)=1; % unit impulse input
for n=11:-1:3
   yb(n-2)= -2*yb(n-1) + 8*yb(n) -24*xb(n-1);
end
yb
nn=-8:0; yb1= -4*(1/2).^nn + 4*(-1/4).^nn;
[yb(1:end-2); yb1]
```

4.6 Difference Equation, System Function, Impulse Response, and Frequency Response

Consider a causal system whose input-output relationship is described by the following difference quation:

$$y[n] - y[n-1] + 0.5y[n-2] = x[n] + x[n-1] \text{ with } y[-2] = y[-1] = 0 \tag{P4.6.1}$$

(a) Find the system function $G[z] = Y[z]/X[z]$ and determine the stability based on its pole locations.
(b) Find the frequency response $G(\Omega) = G[e^{j\Omega}]$ at $\Omega = 0,\ \pi/2$, and π or $e^{j\Omega} = 1,\ j$, and -1 where $G(0) = G[e^{j0} = 1]$ is referred to as the *DC* (direct current) *gain*.
(c) Find the impulse response $g[n]$, i.e., the output of the system to the unit impulse input $x[n] = \delta[n]$ in two ways; once using the z-transform and once using the iterative method for $n = 0 : 4$. Does the DTFT of $\{g[n] : n = 0 : 100\}$ agree with $G(\Omega)$ for $\Omega = 0 : 0.01 : 2\pi$? You can use the following program "sig04p_06.m".
(d) Find the step response, i.e., the output of the system to the unit step input $x[n] = u_s[n]$ in two ways; once using the z-transform and once using the iterative method for $n = 0 : 4$. Is the steady-state value the same as the DC gain obtained in (b) and the result of applying the final value theorem?
(e) Find the z-transform $Y[z]$ of the output to a sinusoidal input $x[n] = \sin(n\pi/2)u_s[n]$. Find the steady-state response of the output by taking the inverse z-transform of $Y[z]$ or using the frequency response obtained in (b). Support your analytical solution by plotting it together with the iterative (numerical) solution for $n = 0 : 100$. Could you apply the final value theorem? If not, state why it is not possible.

```
%sig04p_06.m
clear, clf
disp('(a)')
syms z
% Gz= z*(z+1)/(z^2-z+0.5);
Gz= z*(z+1)/(z-0.5+0.5i)/(z-0.5-0.5i);
B=[1 1]; A=[1 -1 0.5];
```

```
disp('(b)')
DC_gain=subs(Gz,'z',exp(j*0))
[DC_gain subs(Gz,'z',exp(j*pi/2)) subs(Gz,'z',exp(j*pi))]
disp('(c)') % Impulse response and frequency response
g= iztrans(Gz)
% To solve a difference equation by iteration
gn(1)=0; gn(2)=0; % zero initial conditions
xc= zeros(1,103); xc(3)=1; % unit impulse input
r=1/sqrt(2); W1=pi/4;
for m=3:103
   gn(m)= gn(m-1) - 0.5*gn(m-2) + xc(m) + xc(m-1);
   n=m-3; gn_iztrans(m-2)= eval(g);
   gn_iztrans_hand(m-2)= r^n*(cos(W1*n)+3*sin(W1*n));
end
gn=gn(3:end); gn(1:4)
discrepancy_gn1= norm(gn-gn_iztrans)
discrepancy_gn2= norm(gn-gn_iztrans_hand)
W=0:0.01:2*pi;
GW_DTFT= DTFT(gn,W,0);
GW= freqz(B,A,W);
discrepancy_GW= norm(GW-GW_DTFT)
```

4.7 Sinusoidal Steady-State Response from z-Transform or Frequency Response
Consider a causal system which has the following system function:

$$H[z] = \frac{1-e^{-1}}{z-e^{-1}} \tag{P4.7.1}$$

Find the output of the system to a sinusoidal input $x[n] = \sin(0.2\pi n)u_s[n]$ in the following two ways:

(a) Using the z-transform, show that the sinusoidal response is

$$\begin{aligned} y[n] &= (0.688e^{-n} + 0.5163\sin(0.2\pi n) - 0.688\cos(0.2\pi n))u_s[n] \\ &= (0.688e^{-n} + 0.8602\sin(0.2\pi n - 53.1^{\circ}))u_s[n] \\ &= (0.688e^{-n} + 0.8602\sin(0.2\pi(n - 1.4754)))u_s[n] \end{aligned} \tag{P4.7.2}$$

(b) Based on the fact that the frequency response at $\Omega = 0.2\pi$ is

$$\begin{aligned} H[e^{j\Omega}]\big|_{\Omega=0.2\pi} &= \frac{1-e^{-1}}{e^{j0.2\pi} - e^{-1}} \\ &= \frac{1-e^{-1}}{\sqrt{(\cos(0.2\pi)-e^{-1})^2 + \sin^2(0.2\pi)}} \angle -\tan^{-1}\frac{\sin(0.2\pi)}{\cos(0.2\pi)-e^{-1}} \\ &= 0.8602\angle -0.927 \end{aligned} \tag{P4.7.3}$$

find the sinusoidal steady-state response excluding the transient response which decays as time goes by.

4.8 System Function and Frequency Response from Input and Output
Consider a causal system which has the following input and output:

$$x[n] = r^n \cos(\Omega_1 n) u_s[n] \tag{P4.8.1}$$

$$y[n] = r^n \sin(\Omega_1 n) u_s[n] \tag{P4.8.2}$$

(a) Find the system function $G[z] = Y[z]/X[z]$ and impulse response of this system.
(b) Find the frequency response $G(\Omega) = G[e^{j\Omega}]$ and its peak frequency.

4.9 Goertzel Algorithm [G-2]
Consider two second-order filters whose impulse responses are

$$g_1[n] = \cos(2\pi kn/N) u_s[n] \tag{P4.9.1a}$$

$$g_2[n] = \sin(2\pi kn/N) u_s[n] \tag{P4.9.1b}$$

(a) Show that for the input of real-valued sequence $\{x[n]; 0 \le n \le N-1\}$, the outputs of the two filters at $n = N$ become the real and imaginary parts of N-point DFT $X(k) = \text{DFT}_N\{x[n]\}$:

$$y_{1k}[n]|_{n=N} = \sum_{m=0}^{N-1} x[m] g_1[N-m] \tag{P4.9.2a}$$

$$= \sum_{m=0}^{N-1} x[m] \cos\left(\frac{2\pi km}{N}\right) = X_R(k)$$

$$y_{2k}[n]|_{n=N} = \sum_{m=0}^{N-1} x[m] g_2[N-m]$$

$$= -\sum_{m=0}^{N-1} x[m] \sin\left(\frac{2\pi km}{N}\right) = X_I(k) \tag{P4.9.2b}$$

where

$$X(k) = \text{DFT}_N\{x[n]\} \overset{(3.4.2)}{=} \sum_{m=0}^{N-1} x[m] W_N^{mk} = \sum_{m=0}^{N-1} x[m] e^{-j2\pi km/N} \tag{P4.9.3}$$

$$X_R(k) + jX_I(k) = \sum_{m=0}^{N-1} x[m] \cos\left(\frac{2\pi km}{N}\right) - j\sum_{m=0}^{N-1} x[m] \sin\left(\frac{2\pi km}{N}\right) \tag{P4.9.4}$$

(b) Taking the z-transform of the impulse responses (P4.9.1a,b), find the system functions of the filters and check if they can be implemented as Fig. P4.9.

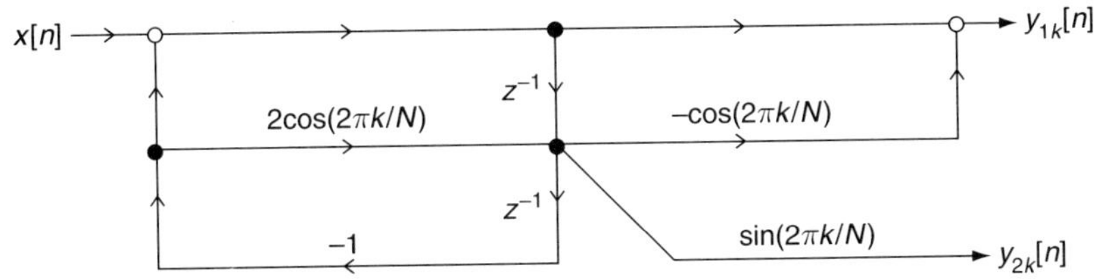

Fig. P4.9 The signal flow graph for Goertzel algorithm

(c) The structure of the filter which generates the DFT sequence of a given input sequence can be obtained from the following equation:

$$y_k[n] = \sum_{m=0}^{N-1} x[m]W_N^{-k(n-m)} = x[n] * W_N^{-kn} \tag{P4.9.5}$$

Comparing this equation with the DFT definition (P4.9.3), we see that the value of the output (P4.9.5) at $n = N$ is the same as the DFT of $x[n]$:

$$\begin{aligned} y_k[n]|_{n=N} &= y_k[N] = \sum_{m=0}^{N-1} x[m]W_N^{-k(N-m)} \\ &= \sum_{m=0}^{N-1} x[m]W_N^{km} = X(k) \end{aligned} \tag{P4.9.6}$$

Noting that (P4.9.5) is the output of a filter with the impulse response

$$g[n] = W_N^{-kn} = e^{j2\pi kn/N} \tag{P4.9.7}$$

show that the system function of the filter is

$$\begin{aligned} G[z] &= \frac{1 - W_N^k z^{-1}}{1 - 2\cos(2\pi k/N)z^{-1} + z^{-2}} \\ &= \frac{1 - \cos(2\pi k/N)z^{-1} + j\sin(2\pi k/N)z^{-1}}{1 - 2\cos(2\pi k/N)z^{-1} + z^{-2}} \end{aligned} \tag{P4.9.8}$$

(cf) This filter is not stable in view of the BIBO stability criterion (1.2.27b). However, we will use them for only a finite number of time points, ensuring that their outputs do not become infinite.

Chapter 5
Sampling and Reconstruction

Contents

In this chapter we are concerned how continuous-time and discrete-time signals are related with each other. We will cover the following:

- Basic functions of D/A (digital-to-analog) converter, A/D (analog-to-digital) converter, and S/H (sample-and-hold) device
- Relationship between the CTFT and DTFT and the effect of sampling on the spectrum of a continuous-time signal
- Sampling theorem
- Reconstruction of the original signal from the sampled signal

These topics will not only give an insight into the relationship between continuous-time signals and discrete-time signals, but also help you to realize what the sampling really means.

Sects. 5.1 and 5.2 are based on the tutorial notes of Richard C. Jaeger ([J-1] and [J-2]).

W.Y. Yang et al., *Signals and Systems with MATLAB®*,
DOI 10.1007/978-3-540-92954-3_5, © Springer-Verlag Berlin Heidelberg 2009

5.1 Digital-to-Analog (DA) Conversion[J-1]

The basic function of the D/A (digital-to-analog) converter (DAC) is to convert a digital (binary) number into its equivalent analog voltage. The output voltage of the DAC can be represented as

$$v_o = V_{FS}(d_1 2^{-1} + d_2 2^{-2} + \cdots + d_N 2^{-N}) + V_{os} \tag{5.1.1}$$

where

v_o: Output voltage
V_{FS}: Full-scale output voltage
V_{os}: Offset voltage (normally to be adjusted to zero)
$d_1 d_2 \cdots d_N$: N-bit input word with the (fictitious) binary point at the left, the most significant bit (MSB) d_1, and the least significant bit (LSB) d_N.

The *resolution* of the converter is the smallest analog change that may be produced by the conversion and is equal to the value of the LSB in volts:

$$\text{Resolution (in volts)} = 2^{-N} V_{FS} = 1\text{LSB value} \tag{5.1.2}$$

It is often stated as a percentage of full scale ($2^{-N} \times 100\%$) or just N-bit resolution.

Figure 5.1 shows an inverted R-$2R$ ladder network, which can be used as a 3-bit DAC where the binary input word controls the switches with a signal of logical value '1'/'0' indicating a transfer to OP Amp/ground side. Taking successive Thevenin equivalent circuits for each stage of the ladder, it can be shown that the input currents are each reduced by a factor of 2 going from the MSB to the LSB so that the analog output voltage to a 3-bit binary input $d_1 d_2 d_3$ becomes

$$v_o = R\,i_o = R\left(d_1 \frac{V_{FS}}{2R} + d_2 \frac{V_{FS}}{4R} + d_3 \frac{V_{FS}}{8R}\right) = \left(d_1 2^{-1} + d_2 2^{-2} + d_3 2^{-3}\right) V_{FS} \tag{5.1.3}$$

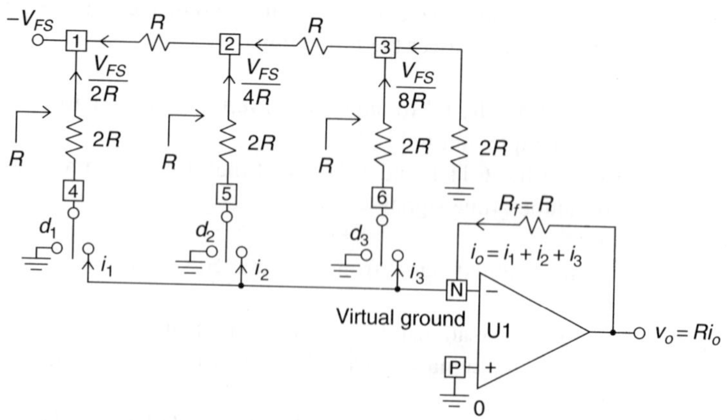

Fig. 5.1 An inverted R-$2R$ ladder network used as a DAC

Note that the currents flowing in the resistors are kept constant since the digital input diverts the current either to ground or to the input (virtual ground) of the operational amplifier functioning as a current-to-voltage converter.

5.2 Analog-to-Digital (AD) Conversion[G-1, J-2, W-2]

An analog-to-digital converter (ADC) takes an unknown analog input signal, most often a voltage V_x, and converts it into an N -bit binary number D representing the ratio of V_x to the converter's full-scale voltage V_{FS}.

Most ADCs use a DAC to vary the reference voltage V_r and use a logic circuit including one or more comparators to deteremine one of the 2^N possible binary numbers $D = d_1 d_2 \cdots d_N$ (d_i's: binary coefficients) which can represent the unknown voltage V_x. The reference voltage V_r can have 2^N different values as

$$V_r = V_{FS} \sum_{i=1}^{N} d_i 2^{-i} \tag{5.2.1}$$

where V_{FS} is the DC reference voltage. The basic difference in converters consists in how to vary V_r to determine the binary coefficients d_i's such that the error $|V_x - V_r|$ is minimized.

5.2.1 Counter (Stair-Step) Ramp ADC

The counter ramp ADC illustrated in Fig. 5.2(a) starts to increment the N -bit counter value from zero by one per clock period on the SOC (start of conversion) pulse till the reference voltage V_r exceeds the unknown input voltage V_x.

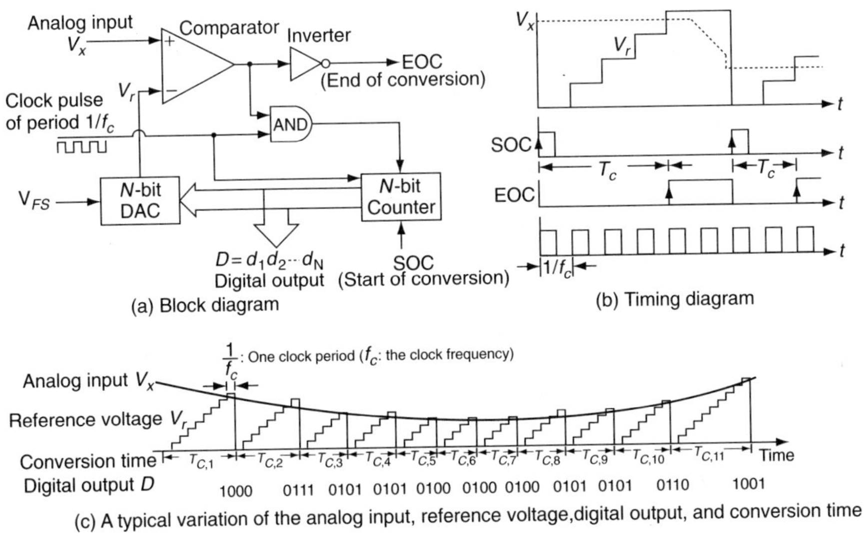

(a) Block diagram (b) Timing diagram

(c) A typical variation of the analog input, reference voltage,digital output, and conversion time

Fig. 5.2 Counter ramp ADC

The sequentially increasing counter output is applied to the N -bit DAC, making its output V_r go up like a staircase as depicted in Fig. 5.2(b). The reference voltage V_r is applied to the $-$ input terminal of the comparator and compared against V_x (applied to the $+$ input terminal) by the comparator. The comparator output keeps to be '1' so that the counter will continue to increase normally till V_r exceeds V_x. When $V_r \geq V_x$, the comparator output will be switched to '0' so that no further clock pulse can increment the counter value and the EOC (end of conversion) signal becomes high to tell other devices that an A/D conversion cycle is completed and the counter value represents the converted (digital) value of the unknown analog voltage V_x. Fig. 5.2(c) shows typical variations of the analog input, reference voltage, digital output, and conversion time.

Some features of this converter should be noted:

<Advantage>

- The simple hardware makes the counter ramp ADC inexpensive to implement.

<Disadvantages>

- The conversion time is proportional to the value of V_x. In the worst case where V_x is equal to or greater than the value of the maximum binary number, i.e., $V_x \geq (1 - 2^{-N})V_{FS}$, it becomes

$$T_C = 2^N / f_c (2^N \text{clock periods}) \tag{5.2.2}$$

where f_c is the clock frequency and V_{FS} is the full-scale DAC output voltage.
- The DAC output is not necessarily the closest to V_x, but the smallest just over V_x among the 2^N possible binary numbers.

5.2.2 *Tracking ADC*

The tracking ADC tries to improve the conversion performance by using an up-down counter with logic to force the DAC output V_r to track changes in the analog input V_x (see Fig. 5.3). Depending on whether $V_r < V_x$ or $V_r > V_x$ (as indicated by the comparator output), the counter value is incremented or decremented by the clock pulse so that the DAC output will alternate between two values differing by one LSB value (2^{-N}) when V_x is constant. When the analog input V_x varies, V_r changes in the proper direction towards V_x so that V_r follows V_x. Consequently, if V_x varies slowly enough, the DAC output V_r is continuously close to V_x and the A/D converted value may be read from the counter at any time. However, if V_x varies too rapidly, the DAC output V_r will not be able to catch up with V_x quickly enough to make the counter value represent V_x closely at any time. The main drawback is the

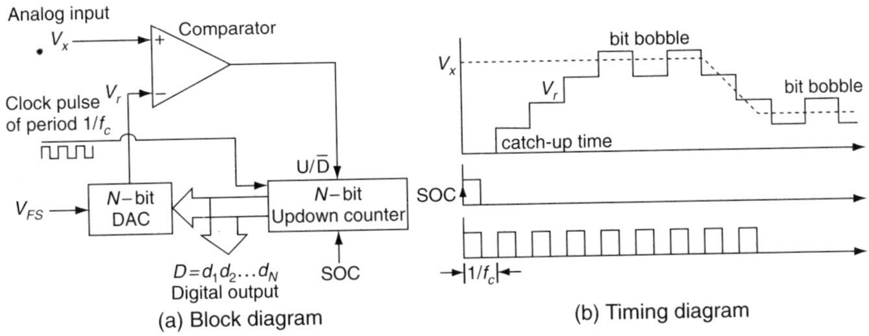

Fig. 5.3 Tracking ADC

phenomenon called 'bit bobble' that the output is never stable since it switches back and forth with every clock pulse even for a constant analog input $V_x = \text{const}$.

5.2.3 Successive Approximation ADC

The successive approximation ADC uses a binary or bisectional search method to determine the best approximation to V_x, requiring only N clock periods to complete an N -bit conversion. Figure 5.4 shows its block diagram and timing diagram. At the start of conversion, the SAR (successive approximation register) is reset with its MSB set to '1', resulting in the DAC output

$$V_r \leftarrow 2^{-1} V_{FS}$$

At the next clock pulse, depending on whether $V_r < V_x$ or $V_r > V_x$ (as indicated by the comparator output), the MSB is left on ('1') or set to '0' and the 2nd MSB is set to '1', resulting in

$$V_r \leftarrow V_r + 2^{-2} V_{FS} \quad \text{or} \quad V_r \leftarrow V_r - 2^{-2} V_{FS}$$

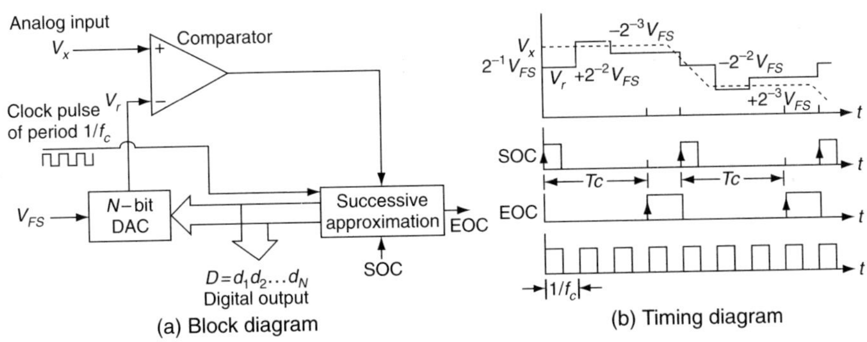

Fig. 5.4 Successive approximation ADC

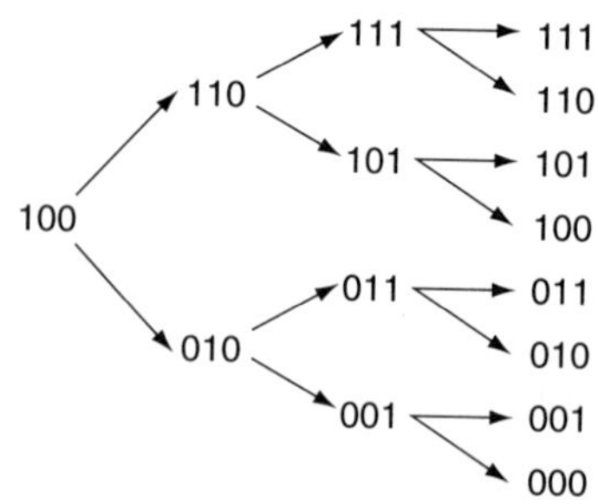

Fig. 5.5 Binary code sequence of a 3-bit successive approximation DAC

Again, depending on $V_r < V_x$ or $V_r > V_x$, the 2nd MSB is left on('1') or set to '0' and the 3rd MSB is set to '1', resulting in

$$V_r \leftarrow V_r + 2^{-3} V_{FS} \quad \text{or} \quad V_r \leftarrow V_r - 2^{-3} V_{FS}$$

When the process has been carried out for every bit, the SAR contains the binary number representing V_x and EOC line indicates that digital output is available. In this way, the 3-bit successive conversion is completed at the end of N clock periods for an N -bit ADC so that we have the A/D conversion time

$$T_C = N/f_c(N \text{ clock periods}) \tag{5.2.3}$$

Figure 5.5 shows the binary code sequence of a 3-bit successive approximation DAC.

This type of converter is very popular due to its fast conversion rate. A problem with the SA ADC is that if the input does not remain constant during the full conversion period, the digital output may not be related to the value of the unknown input voltage V_x. To avoid this problem, sample-and-hold circuits are usually used ahead of the ADC.

5.2.4 Dual-Ramp ADC

Figure 5.6 shows the organization and operation of the dual-ramp ADC. On the SOC pulse, the counter and RC integrator are reset. Then the analog input V_x, connected to the integrator input through switch S_1, is (negatively) integrated during a fixed time interval of $T_1 = 2^N/f_c$. At the end of the integration period, the two switches S_1/S_2 are turned off/on, respectively so that the reference input $-V_{FS}$ are connected to the integrator input through S_2. Then the integrator output v_o increases until it crosses zero to make the comparator output change. The length of the deintegration period will be measured as $T_2 = n_2/f_c$ (n_2 clock periods).

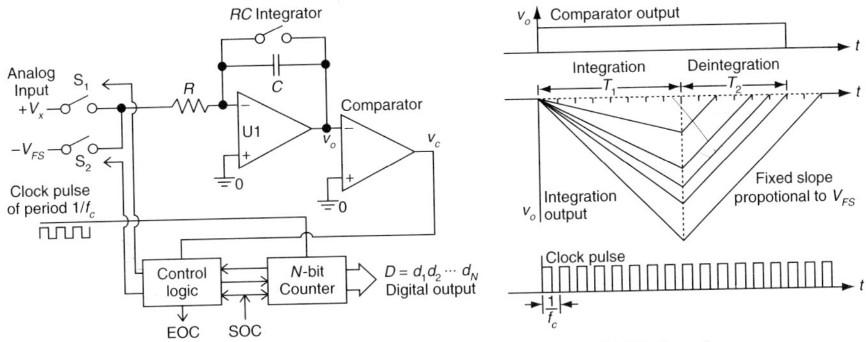

(a) Block diagram (b) Timing diagram

Fig. 5.6 Dual ramp ADC

Noting that the charge accumulated in the capacitor from $t = 0^+$ to T_1 will have been completely discharged at $t = T_1 + T_2$, we can write

$$\frac{1}{RC}\int_0^{T_1} V_x\,dt = \frac{1}{RC}\int_{T_1}^{T_1+T_2} V_{FS}\,dt;\ \frac{\overline{V}_x T_1}{RC} = \frac{V_{FS}T_2}{RC};\ T_2 = T_1\frac{\overline{V}_x}{V_{FS}} \tag{5.2.4}$$

where $T_1 = 2^N/f_c$ and $\overline{V}_x$ is the average of V_x. This implies that the counter value n_2 accumulated during the deintergration period is supposed to represent the average value of the analog input:

$$n_2 = 2^N\frac{\overline{V}_x}{V_{FS}} \tag{5.2.5}$$

The value of RC constant does not matter as long as it remains constant throughout the conversion cycle of duration $T_1 + T_2$.

<Advantage>

- Even if V_x changes during the conversion cycle, the ADC output corresponding to $\overline{V}_x$ is still valid since it represents the average value of V_x during the integration period of duration $T_1 = 2^N/f_c$.
- Any sinusoidal input signals with frequencies $K/T_1 = K2^{-N}f_c$ (K: an integer) will have integrals of zero so that they will not disturb the ADC output. This property is utilized in digital voltmeters which use dual-ramp converters with $T_1 = K/f_o$ where f_o is the power-line frequency (50 or 60Hz), so that harmonic noise at multiples of f_o can be removed ('good rejection of power-line interference').
- Reversed polarity of the analog input V_x can easily be dealt with by reversing the polarity of $-V_{FS}$.

<Disadvantages>

- The conversion time is variable and is as long as

$$T_C = T_1 + T_2 = (2^N + n)/f_c \tag{5.2.6}$$

5.2.5 Parallel (Flash) ADC

Figure 5.7 shows a 3-bit parallel (or flash) ADC in which the analog input V_x is simultaneously compared with $(2^3 - 1)$ different reference values and depending on the comparison results, one of the 2^3 digital values is chosen as the ADC output by the encoding logic circuit.

<Advantage>

- The conversion speed is so fast that the parallel ADC can be thought of as automatically tracking the input signal.
- With the resistors of no-equal values, the parallel ADC can be designed so that it performs a customized, nonlinear A/D conversion. No other ADC design is capable of such a nonlinear AD conversion.

<Disadvantage>

- The cost is expensive and grows rapidly with resolution since $2^N - 1$ comparators and reference voltages are required for an N -bit converter.

(cf.) Visit the web site [W-2] to see the delta-sigma ($\Delta\Sigma$) ADC.

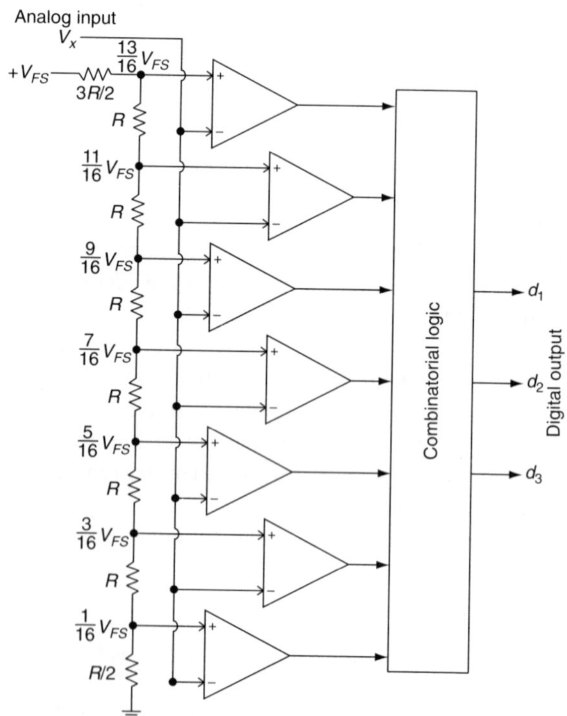

Fig. 5.7 Parallel (or flash) ADC

5.3 Sampling

5.3.1 Sampling Theorem

In Sect. 3.5.2 we derived the relation between the CTFS X_k of a continuous-time periodic signal $\tilde{x}_P(t)$ and the N -point DTFS (DFS/DFT) $\tilde{X}_N(k)$ of the discrete-time version $\tilde{x}_N[n] = \tilde{x}_P(nT)$ (obtained by sampling $\tilde{x}_P(t)$ every T s), which is periodic with period $N = P/T$ in n:

$$\tilde{X}_N(k) \overset{(3.5.3)}{=} \frac{1}{T}\sum_{m=-\infty}^{\infty} X_{k+mN} \tag{5.3.1a}$$

where the frequency components described by the DTFS $\tilde{X}_N(k)$ and CTFS X_k are virtually identical in light of the inter-relationship (1.1.15) between the digital and analog frequencies:

$$k\Omega_0 = k\frac{2\pi}{N}[\text{rad/sample}] \overset{(1.1.15)}{\sim} k\frac{2\pi/N[\text{rad/sample}]}{T\,[\text{sec/sample}]} = k\frac{2\pi}{NT}[\text{rad/sec}] = k\frac{2\pi}{P} = k\omega_0$$

We also showed that the CTFT $X_a(\omega)$ of a continuous-time signal $x_a(t)$ is related with the DTFT $X_d(\Omega)$ of the discrete-time version $x_d[n] = x_a(nT)$ as

$$X_d(\Omega) \overset{(3.5.7)}{=} \frac{1}{T}\sum_{k=-\infty}^{\infty} X_a\left(\omega + k\frac{2\pi}{T}\right)\Big|_{\omega=\Omega/T}$$
$$= \frac{1}{T}\sum_{k=-\infty}^{\infty} X_a\left(\frac{\Omega}{T} + k\frac{2\pi}{T}\right) \tag{5.3.1b}$$

where $\omega = \Omega/T$ [rad/s] and Ω [rad/sample] are virtually identical in light of the inter-relationship (1.1.15) between the digital and analog frequencies.

Equations (5.3.1a) and (5.3.1b) imply that the DTFS/DTFT of a discrete-time sequence $x_d[n] = x_a(nT)$ is qualitatively the periodic extension of the CTFS/CTFT of the continuous-time version $x_a(t)$ (with period $2\pi/T$ in analog frequency ω or 2π in digital frequency Ω), i.e., the sum of infinitely many shifted version of the spectrum of $x_a(t)$. This explains how the DTFS/DTFT spectrum of $x_d[n] = x_a(nT)$ deviates from the CTFS/CTFT spectrum of $x_a(t)$ due to *frequency-aliasing* unless the CTFS/CTFT is strictly limited within the low-frequency band of $(-\pi/T, \pi/T)$ where T is the sampling interval or period of $x_d[n] = x_a(nT)$.

To be more specific, suppose a continuous-time signal $x_a(t)$ has a band-limited CTFT spectrum

$$X_a(\omega) = 0 \quad \text{for} \quad |\omega| > \omega_x \tag{5.3.2}$$

as depicted in Fig. 5.8(a). Then the DTFT spectrum $X_d(\Omega)$ of the discrete-time version $x_d[n] = x_a(nT)$, which is the periodic extension of $X_a(\omega)|_{\omega=\Omega/T}$ with period

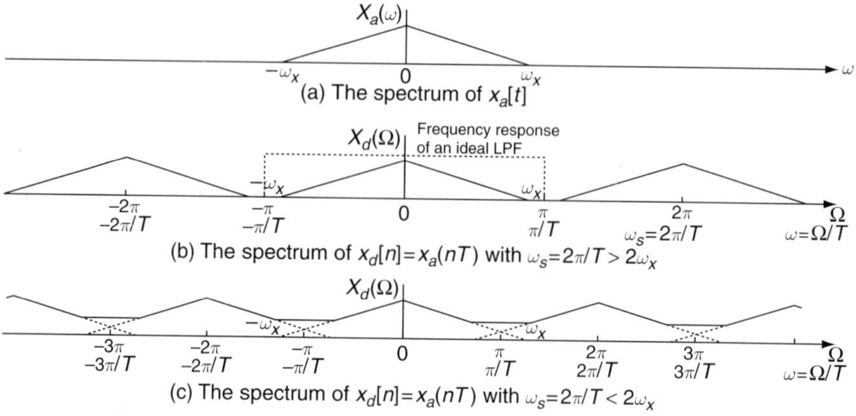

Fig. 5.8 The spectra of discrete–time signals depending on the sampling frequency – the sampling theorem

$\omega_s = 2\pi/T$ in ω or $\Omega_s = 2\pi$ in Ω, is shown in Fig. 5.8(b) and (c) for $\pi/T > \omega_x$ and $\pi/T < \omega_x$, respectively. Notice that $X_d(\Omega)$ in the principal frequency range $(-\pi/T, \pi/T)$ is identical to $X_a(\omega)$ if

$$\omega_s = \frac{2\pi}{T} > 2\omega_x(\text{Nyquist rate}) \text{ or equivalently, } \frac{\pi}{T}(\text{folding frequency}) > \omega_x \tag{5.3.3}$$

where ω_x is the highest frequency in $x_a(t)$ and $\omega_s = 2\pi/T$ and π/T are called the *sampling frequency* and *folding* or *Nyquist frequency*, respectively. In this case (Fig. 5.8(b): oversampling), there is no overlap, called *aliasing*, between adjacent spectral components so that $x_a(t)$ or $X_a(\omega)$ can be extracted exactly from $x_d[n] = x_a(nT)$ or $X_d(\Omega)$ by employing an ideal lowpass filter with the magnitude response depicted (in a dotted line) in Fig. 5.8(b).

On the other hand, in the case of Fig. 5.8(c) (under-sampling) with $\pi/T < \omega_x$ or equivalently, $\omega_s = 2\pi/T < 2\omega_x$, the *frequency aliasing* (or *spectral overlap*) occurs, producing an irretrievable error in the spectrum. Such an aliasing phenomenon makes higher frequencies in $X_a(\omega)$ get reflected into lower frequencies, which presents an interpretation of the *stroboscopic effect*. This effect is occasionally observed in films of wagon wheels or aircraft propellers that appear to be rotating slower than would be consistent with the forward motion and sometimes in the opposite direction. This story can be summarized as follows:

[Sampling Theorem]
In order to retain vital information through sampling with sampling period T, the sampling frequency $\omega_s = 2\pi/T$ must be greater than twice the highest frequency ω_x contained in the continuous-time signal $x_a(t)$ to be sampled, called the Nyquist rate $2\omega_x$. Otherwise, i.e., if $\omega_s \leq 2\omega_x$, the so-called frequency aliasing effect (spectral overlapping phenomenon) results.

When sampling a continuous-time signal $x_a(t)$ every T s, we want to ensure that all the information in the original signal is retained in the samples $x_d[n] = x_a(nT)$ so that we can exactly recover $x_a(t)$ from the discrete-time sequence $x_d[n]$. In this context, the sampling theorem is the most important criterion of the sampling period selection.

To clarify the concept of aliasing, consider two continuous-time signals $x_1(t) = \sin(2\pi t)$ and $x_2(t) = \sin(22\pi t)$ shown in Fig. 5.9(a). Each of these signals is sampled every $T = 0.1$ s to give sequences $x_1[n] = x_1(0.1n) = \sin(0.2\pi n)$ and $x_2[n] = x_2(0.1n) = \sin(2.2\pi n)$, that are identical. Fig. 5.9(b) also shows that two continuous-time signals $x_3(t) = \sin(18\pi t)$ and $x_4(t) = -\sin(2\pi t)$ are sampled every $T = 0.1$ s to yield $x_3[n] = x_3(0.1n) = \sin(1.8\pi n)$ and $x_4[n] = x_4(0.1n) = \sin(-0.2\pi n)$, that are identical. These are examples of frequency aliasing that a higher frequency signal, when sampled, appears like a lower frequency one.

Figure 5.10 shows the output signals $y_1(t)$, $y_2(t)$, and $y_3(t)$ of an A/D- $G[z]$ (digital filter)-D/A structure to the three different input signals where $G[z] = (1 - e^{-1})/(z - e^{-1})$. Note that with the sampling period $T = 0.1$ [s], all the outputs are of the same digital frequency and that $x_2(t) = \sin(22\pi t)$ and $x_3(t) = \sin(18\pi t)$ have been aliased/folded with the phase preserved/reversed, respectively. This result can be obtained by running the following program "sig05f10.m" and predicted from the frequency response of the digital filter $G[z]$, which is shown in Fig. 5.11.

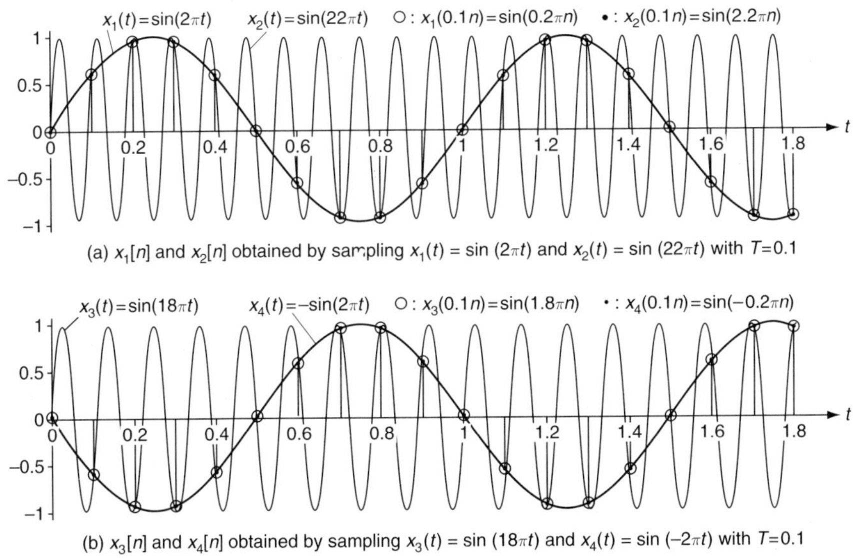

Fig. 5.9 Examples of frequency aliasing/folding or spectral overlap

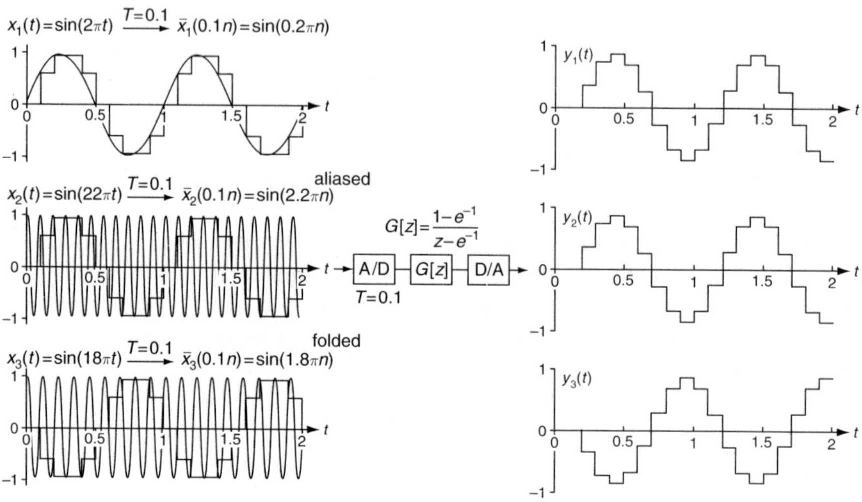

Fig. 5.10 The effect of frequency aliasing on a system of A/D-$G[z]$-D/A structure

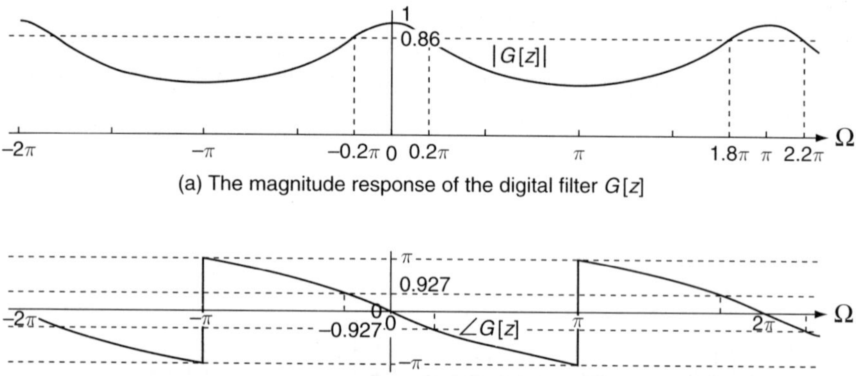

(a) The magnitude response of the digital filter $G[z]$

(b) The phase response of the digital filter $G[z]$

Fig. 5.11 The frequency response of the digital filter $G[z]$

```
%sig05f10.m
clear, clf
B=[0 1-exp(-1)]; A=[1 -exp(-1)]; % system function of the filter
W=[2 22 18]*pi; % the three different frequency of input signal
T=0.1; ts=2e-3; % Sampling period, Segment length
tf=2; t=[0:ts:tf]; n=[0:tf/T]; % Final time, Time ranges
x1t= sin(W(1)*t); x1n=x1t([1:T/ts:end]); y1n= filter(B,A,x1n);
x2t= sin(W(2)*t); x2n=x2t([1:T/ts:end]); y2n= filter(B,A,x2n);
x3t= sin(W(3)*t); x3n=x3t([1:T/ts:end]); y3n= filter(B,A,x3n);
subplot(521), plot(t,x1t), hold on, stairs(n*T,x1n,'r') % Input 1
subplot(522), stairs(n*T,y1n) % Output 1
```

```
subplot(523), plot(t,x2t), hold on, stairs(n*T,x2n,'r') % Input 2
subplot(524), stairs(n*T,y2n) % Output 2
subplot(525), plot(t,x3t), hold on, stairs(n*T,x3n,'r') % Input 3
subplot(526), stairs(n*T,y3n) % Output 3
% The frequency response of the digital filter
W=[-2:0.001:2.3]*pi; GW=freqz(B,A,W);
subplot(514), plot(f,abs(GW))
subplot(515), plot(f,angle(GW))
```

Referring to Remark 1.4 (for the physical meaning of the frequency response) and based on the frequency response of the digital filter $G[z]$ depicted in Fig. 5.11, we can write the steady-state response of the digital filter to the digital inputs $x_1[n] = \sin(0.2\pi n)$ and $x_3[n] = \sin(1.8\pi n)$ as

$$\begin{aligned} y_1[n] &= |\,G[e^{j\Omega}]\,|\big|_{\Omega=0.2\pi} \sin(0.2\pi n + \angle G[e^{j\Omega}]\big|_{\Omega=0.2\pi}) \\ &= 0.86 \sin(0.2\pi n - 0.927) \\ y_3[n] &= |\,G[e^{j\Omega}]\,|\big|_{\Omega=1.8\pi} \sin(1.8\pi n + \angle G[e^{j\Omega}]\big|_{\Omega=1.8\pi}) \\ &= 0.86 \sin(1.8\pi n + 0.927) \\ &= 0.86 \sin(-0.2\pi n + 0.927) = -0.86 \sin(0.2\pi n - 0.927) = -y_1[n] \end{aligned}$$

Here, let us think about how we can interpret the periodicity of the DTFT spectrum of a discrete-time signal and that of the frequency response of a discrete-time system such as $G[z]$. The interpretation of the former is an ambiguity that we cannot tell which of analog frequency components $\{(\Omega_1 \pm 2m\pi)/T,\ m\text{: any integer}\}$ are contained in the original continuous-time signal (before sampling with period T). It should not be interpreted as the existence of infinitely many analog frequency components $\{(\Omega_1 \pm 2m\pi)/T,\ m\text{: any integer}\}$. On the other hand, the interpretation of the latter is an equality that the discrete-time system responds with the same magnitude and phase of frequency response to the discrete-time version of continuous-time signals of different analog frequencies $\{(\Omega_1 \pm 2m\pi)/T,\ m\text{: any integer}\}$.

There are some rules of thumb for choosing the sampling rate, which are listed in the following remark:

Remark 5.1 Rules of Thumb for Sampling Rate Selection

In practice, the sampling frequency is chosen to be much higher than the Nyquist rate. In closed-loop sampled-data systems, low sampling rate has a detrimental effect on stability and therefore, the sampling rate selection is made with stability consideration. The rules of thumb are

- to sample $8 \sim 10$ times during a cycle of damped oscillation in the output if the system is under-damped or during the rise time of the transient response if the system is over-damped, or
- to sample at least 5 times per time constant.

Remark 5.2 Time Aliasing and Frequency Aliasing

In Example 3.15, it can be observed that if the DFT size is not sufficiently large, in other words, if the sampling rate in the frequency domain $[0, \; 2\pi]$ is not sufficiently high, the aliasing problem may occur in the time domain and such a phenomenon is called the *time aliasing*. Now in this section, we can observe that a low sampling rate in the time domain may cause the *frequency aliasing*.

5.3.2 Anti-Aliasing and Anti-Imaging Filters

We know that higher sampling rate satisfies the sampling theorem more easily; in other words, if a continuous-time signal is sampled at a sufficiently high rate, the frequency-aliasing problem can be alleviated. However, the maximum sampling rate of a S/H (sample/hold device) is upper-bounded by the hardware constraints such as the delay time of the S/H and the conversion time of A/D (analog-to-digital) converter. If the total conversion time is 1 μs, the maximum sampling rate is 1 MHz. Therefore we may not be able to make the sampling rate as high as we want. Besides, a higher sampling rate increases the number of calculations needed for implementation.

On the other hand, all physical signals found in the real world are not band-limited and do contain a wide range of frequency components. Besides, an ideal low-pass filter is not realizable. It is therefore impossible to exactly reproduce the original continuous-time signal from its sampled version even though the sampling theorem is satisfied.

Fortunately, most physical transients tend to be smooth so that their spectra are close to zero for high frequencies. Still it may be a problem to deal with the signal corrupted by high-frequency noise. To prevent or reduce aliasing errors caused by undesired high-frequency signals, we must use an analog low-pass filter, called an *anti-aliasing prefilter*. This filter is applied to the continuous-time signal prior to

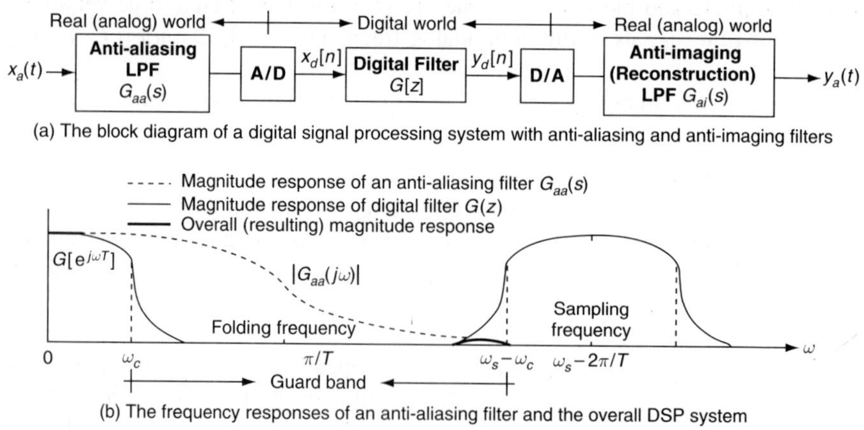

Fig. 5.12 Block diagram and frequency response of a DSP system

sampling and passes the components with frequencies $|\omega| \leq \omega_c$, while attenuating the components with $|\omega| \geq \omega_s - \omega_c$, which would be folded into the range $|\omega| \leq \omega_c$. A digital signal processing system using an anti-aliasing LPF and its typical frequency response are shown in Fig. 5.12. Note that another LPF of similar frequency characteristic, called an *anti-imaging* or *reconstruction filter*, might have to be installed after D/A conversion to remove unwanted spectral components (above the folding or Nyquist frequency $\omega_f = \pi/T$) from the DAC output and construct a smooth analog output signal.

5.4 Reconstruction and Interpolation

In many practical applications, discrete-time sequences are required to be transformed into continuous-time signals. In computer-controlled systems, it is necessary to convert the control actions calculated by the computer as a sequence of numbers into a continuous-time signal that can be applied to the plant or process. In digital audio compact disk system, the audio signals are stored as digital samples on the disk and must ultimately be converted into analog signals to drive the speakers. In these cases, we are faced with the inverse of the sampling operation, which asks us to think about how to reproduce the analog signal $x_a(t)$ from the continuous-time sampled signal $x_*(t) = x_a(t)\delta_T(t)$ or discrete-time sequence $x_d[n] = x_a(nT)$.

5.4.1 Shannon Reconstruction

We begin by considering the un-aliased spectrum $X_*(\omega)$ as shown in Fig. 5.8(b) where $X_*(\omega)$ has the same shape with the original spectrum $X_a(\omega)$ over the principal frequency range $(-\pi/T, \pi/T)$ where T is the sampling interval or period. Suppose we have an analog lowpass filter (LPF) with the ideal 'brick wall' frequency response

$$G_I(\omega) = \begin{cases} T & \text{for } -\pi/T \leq \omega \leq \pi/T \\ 0 & \text{elsewhere} \end{cases} \tag{5.4.1}$$

which passes only frequencies in the range $|\omega| \leq \pi/T$ and masks out all other frequencies. Recall from Eq. (E2.9.2) that

$$g_I(t) = \mathcal{F}^{-1}\{G_I(\omega)\} \overset{\text{(E2.9.2)}}{=} T\frac{B}{\pi}\text{sinc}\left(\frac{Bt}{\pi}\right) \overset{B=\pi/T}{=} \text{sinc}\left(\frac{t}{T}\right) \tag{5.4.2}$$

We can apply this filter to $X_*(\omega)$ to retrieve the original spectrum $X_a(\omega)$ in the frequency domain as

$$X(\omega) = G_I(\omega)X_*(\omega) \tag{5.4.3}$$

or in the time domain as

$$x(t) = \mathcal{F}^{-1}\{X(\omega)\} \overset{(2.5.10)}{=} g_I(t) * x_*(t) \overset{(E2.13.1)}{=} g_I(t) * x_a(t)\delta_T(t)$$

$$\overset{(E2.10.1)}{=} g_I(t) * x_a(t) \sum_{m=-\infty}^{\infty} \delta(t - mT)$$

$$= g_I(t) * \sum_{m=-\infty}^{\infty} x_a(t)\delta(t - mT) = g_I(t) * \sum_{m=-\infty}^{\infty} x_a(mT)\delta(t - mT)$$

$$\overset{(D.37)}{=} \sum_{m=-\infty}^{\infty} x_a(mT) g_I(t - mT);$$

$$\hat{x}(t) \overset{(5.4.2)}{=} \sum_{m=-\infty}^{\infty} x_a(mT)\,\mathrm{sinc}\left(\frac{t - mT}{T}\right) \tag{5.4.4}$$

The above summation (5.4.4), called Whittaker's cardinal interpolation formula or Whittaker-Shannon sampling series [S-1, W-1], suggests a reconstruction formula. It can be modified into a more computationally-efficient form

$$\hat{x}(t) = \frac{\sin(\pi t/T)}{\pi/T} \sum_{m=-\infty}^{\infty} x_a(mT)\frac{(-1)^m}{t - mT} \tag{5.4.5}$$

where we have used the fact that

$$\sin\left(\frac{\pi}{T}(t - mT)\right) = \sin\left(\frac{\pi}{T}t\right)\cos(m\pi) - \cos\left(\frac{\pi}{T}t\right)\sin(m\pi) = (-1)^m \sin\left(\frac{\pi}{T}t\right)$$

Since the sinc function sinc$(t - mT)$ has unity value for $t = mT$ and becomes zero for other Shannon reconstruction sample points, it is obvious that

$$\hat{x}(nT) = x_a(nT) \quad \forall \text{ integer } n \tag{5.4.6}$$

holds for every sample point. The role of $g_I(t)$ is to fill in or interpolate the values of the continuous-time function between the sample points.

```
%sig05f13.m
clear, clf
ts=0.001; T=0.1; tt=-0.5:ts:1.4; t0=0; tf=1; % Time range
fs=[0 0.1 0.2 0.3 0.4 0.5]; ws=2*pi*fs/T; % Frequencies contained in x(t)
Aks= [1 1 1 1 1 1]; phiks= [0.5 -0.5 0.5 -0.5 1 -1];
K=6; xt= Aks(1)*sin(ws(1)*tt + phiks(1));
for k=2:K, xt = xt + Aks(k)*sin(ws(k)*tt + phiks(k)); end
nT= tt([1:T/ts:end]); % Sampling point vector
xn= xt([1:T/ts:end]); % Discrete-time sequence sampled with T=0.1
sincmT= inline('sinc(t/T)','t','T');
plot(tt,xt), hold on, stem(nT,xn)
xht= 0;
for n=1:length(xn)
   xn_sincnT = xn(n)*sincmT(tt-nT(n),T); xht = xht + xn_sincnT; %Eq.(5.4.4)
   plot(tt,xn_sincnT,':')
end
plot(tt,xht,'r'), set(gca,'XLim',[t0 tf],'fontsize',9)
```

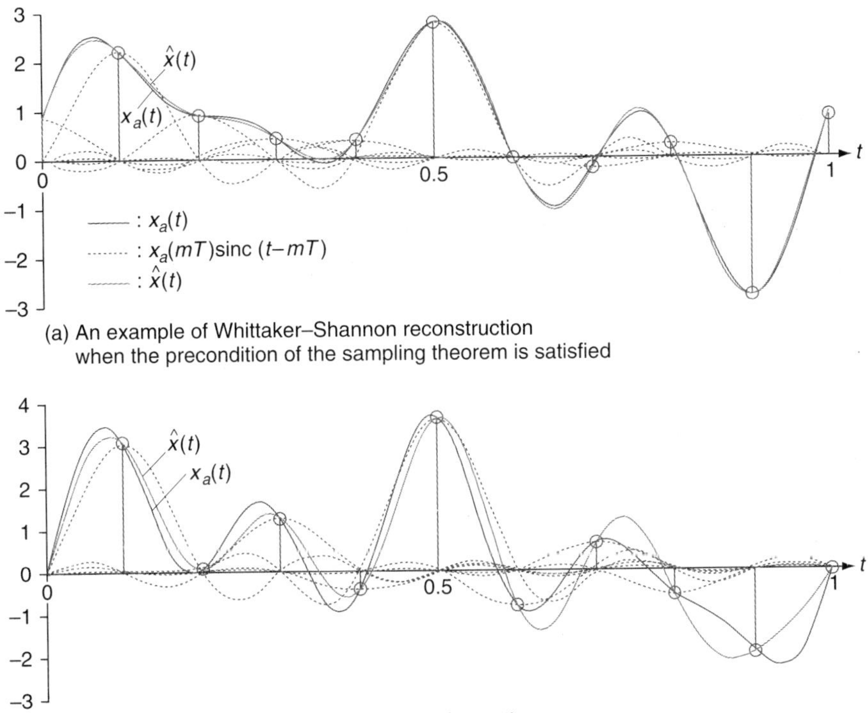

(a) An example of Whittaker–Shannon reconstruction when the precondition of the sampling theorem is satisfied

(b) An example of Whittaker–Shannon reconstruction when the precondition of the sampling theorem is not satisfied

Fig. 5.13 Examples of Whittaker–Shannon reconstruction

Interested readers are recommended to run the above program "sig05f13.m" twice, once with the number of frequencies K = 5 and once with K = 6, to get Fig. 5.13(a) and (b), respectively and then, think about what the difference comes from in connection with the precondition of the sampling theorem.

5.4.2 DFS Reconstruction

Recall again the relationship (5.3.1a) between the CTFS X_k of a continuous-time periodic signal $\tilde{x}_P(t)$ and the N -point DTFS (DFS/DFT) $\tilde{X}_N(k)$ of the discrete-time version $\tilde{x}_N[n] = \tilde{x}_P(nT)$ (obtained by sampling $\tilde{x}_P(t)$ every T s), which is periodic with period $N = P/T$ in n:

$$\tilde{X}_N(k) \overset{(3.5.3) \text{ or } (5.3.1a)}{=} \frac{1}{T} \sum_{m=-\infty}^{\infty} X_{k+mN} \tag{5.4.7}$$

This implies that if $\tilde{x}_P(t)$ does not contain the frequencies above the folding frequency, i.e., half the sampling frequency

$$\omega_f = \frac{\omega_s}{2} = \frac{\pi.}{T} = \frac{N}{2}\frac{2\pi}{NT}$$
$$= \frac{N}{2}\frac{2\pi}{P} = \frac{N}{2}\omega_0 (\text{corresponding to the frequency index } k = \frac{N}{2})$$

so that

$$X_k = 0 \quad \forall \quad |k| \geq \frac{N}{2} \tag{5.4.8}$$

then Eq. (5.4.7) becomes

$$\tilde{X}_N(k) \overset{(5.4.7)\text{ with }(5.4.8)}{=} \frac{1}{T} X_k \text{ for } |k| < \frac{N}{2} \tag{5.4.9}$$

This suggests another reconstruction formula, called the DFS reconstruction, which is similar to the inverse DFS formula (3.4.8):

$$\hat{x}(t) = \frac{1}{N} \sum_{|k|<N/2} \tilde{X}_N(k) e^{j2\pi kt/NT} \tag{5.4.10}$$

```
%sig05f14.m
clear, clf
ts=0.001; T=0.1; tt=-0.5:ts:1.4; t0=0; tf=1;
fs=[0 0.1 0.2 0.3 0.4 0.5]; ws=2*pi*fs/T; % Frequencies contained in x(t)
Aks= [1 1 1 1 1 1]; phiks= [0.5 -0.5 0.5 -0.5 1 -1];
K=5; xt= Aks(1)*sin(ws(1)*tt + phiks(1));
for k=2:K, xt = xt + Aks(k)*sin(ws(k)*tt + phiks(k)); end
nT= tt([1:T/ts:end]); xn= xt([1:T/ts:end]);
xn_causal=[xn(6:end) xn(1:5)]; Xk= fft(xn_causal); N=length(Xk);
plot(tt,xt), hold on, stem(nT,xn)
kk1=[1:N/2].';
xht1_DFS= real(Xk(kk1)*exp(j*2*pi/N/T*(kk1-1)*tt))/N;
kk2=[N/2+2:N].';
xht2_DFS= real(Xk(kk2)*exp(j*2*pi/N/T*(kk2-1-N)*tt))/N;
xht_DFS = xht1_DFS + xht2_DFS;
plot(tt,xht_DFS,'r'), set(gca,'XLim',[t0 tf],'fontsize',9)
```

Interested readers are recommended to run the above program "sig05f14.m" twice, once with the number of frequencies K = 5 and once with K = 6, to get Fig. 5.14(a) and (b), respectively and then, think about what the difference comes from in connection with the precondition of the sampling theorem.

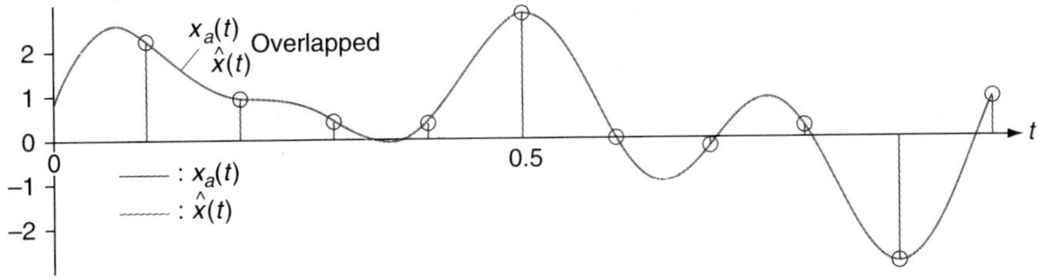

(a) An example of Fourier series reconstruction when the precondition of the sampling theorem is satisfied

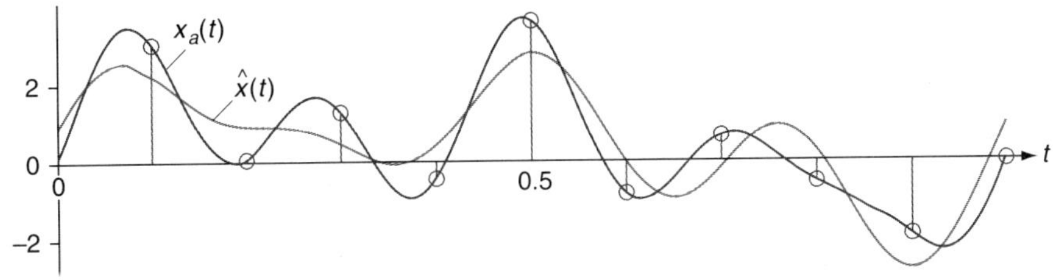

(b) An example of Fourier series reconstruction when the precondition of the sampling theorem is not satisfied

Fig. 5.14 Examples of Fourier series reconstruction

5.4.3 Practical Reconstruction

The problem with the ideal (Whittaker-Shannon) interpolation function $g_I(t) = \text{sinc}(t/T)$ is that it extends in time from $t = -\infty$ to $+\infty$ and accordingly, it incorporates the entire sequence of $x_a(nT)$ including all the future samples (for $n > 0$) as well as all the past samples (for $n < 0$) to find the estimate of $x(t)$. Besides, in control situations we cannot usually wait to observe the entire sequence before an interpolation is performed. In this section we consider a practical interpolator called the zero-order hold (z.o.h.), which is a causal lowpass filter to approximate a signal between two consecutive sampling instants nT and $(n+1)T$ by a constant $x_a(nT)$. This is the beginning of the story about the ideal S/H (sample-and-hold device). As depicted in Fig. 5.15(a), the output of a z.o.h. to an input $x(t)$ can be described as

$$\overline{x}(t) = \sum\nolimits_{n=0}^{\infty} x_a(nT)\,(u_s(t-nT) - u_s(t-nT-T)) \tag{5.4.11}$$

whose Laplace transform is

$$\overline{X}(s) \underset{\text{B.8(3), B.7(2)}}{\overset{(5.4.11)}{=}} \sum\nolimits_{n=0}^{\infty} x_a(nT)\left(\frac{1}{s}e^{-nT s} - \frac{1}{s}\,e^{-nT s - Ts}\right)$$

$$= \sum\nolimits_{n=0}^{\infty} x_a(nT) e^{-nT s}\frac{1}{s}(1 - \, e^{-Ts}); \tag{5.4.12}$$

$$\overline{X}(s) = X_*(s)\,G_{h0}(s) \tag{5.4.13}$$

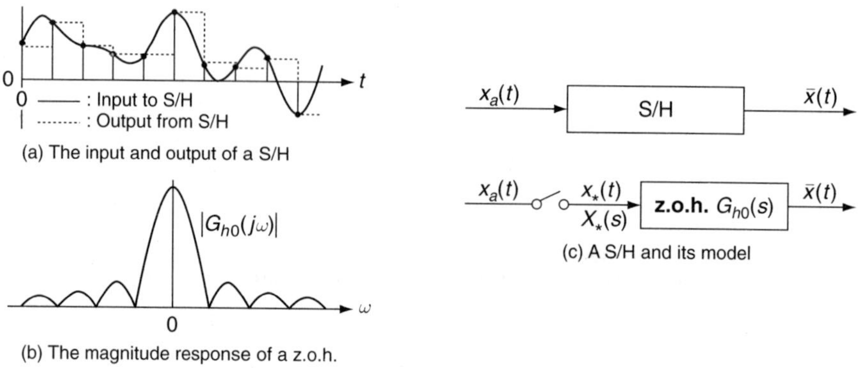

Fig. 5.15 S/H (Sample–and–Hold) modeled by combining an ideal sampler and a z.o.h.

where

$$X_*(s) = \mathcal{L}\{x_*(t)\} \overset{(E2.13.1)}{=} \mathcal{L}\left\{x_a(t)\sum_{n=-\infty}^{\infty}\delta(t-nT)\right\}$$

$$\overset{B.8(2)}{=} \sum_{n=-\infty}^{\infty} x_a(nT)e^{-nT\,s} \overset{(4.1.1)}{=} X_d[z]|_{z=e^{sT}} \tag{5.4.14}$$

$$G_{h0}(s) \overset{(5.4.13)}{=} \frac{\overline{X}(s)}{X_*(s)} \underset{(5.4.14)}{\overset{(5.4.12)}{=}} \frac{1}{s}(1-e^{-Ts}) \tag{5.4.15}$$

Note that in Eq. (5.4.13) describing the Laplace transform of the z.o.h. output, $X_*(s)$ is the Laplace transform of $x_*(t) = x_a(t)\delta_T(t)$ (the output of an ideal sampler to $x_a(t)$), called the *starred transform* of $x_a(t)$. On the other hand, $G_{h0}(s)$ can be regarded as the system or transfer function of the z.o.h. device since it does not depend on $x_a(t)$. Hence the frequency response of the z.o.h. is

$$G_{h0}(j\omega) = G_{h0}(s)|_{s=j\omega} = \frac{1}{j\omega}(1-e^{-j\omega T}) = \frac{e^{j\omega T/2}-e^{-j\omega T/2}}{j\omega}e^{-j\omega T/2}$$
$$= T\frac{\sin(\omega T/2)}{\omega T/2}e^{-j\omega T/2} \tag{5.4.16}$$

which implies that the z.o.h. is a kind of lowpass filter (see Fig. 5.15(b)). Note that $x_*(t)$ is not present in physical systems, but it appears in the mathematical model as a result of factoring and can be considered to be the output of an ideal sampler. Although the ideal sampler and the z.o.h. do not individually model a physical sampler and a physical data hold, their combination does accurately model a physical S/H device (see Fig. 5.15(c)).

5.4.4 Discrete-Time Interpolation

As mentioned in Remark 3.7(4), zero-padding in the time domain can be used to increase the frequency resolution, which is justified for finite-duration signals that have already been sampled over all nonzero portions. In duality with this, zero-padding in the frequency domain can be used to increase the time resolution, which is justified for bandlimited signals having nonzero spectral values only for some finite frequency range. This suggests us a discrete-time interpolation method.

Suppose we are given a sequence $\{x[n],\ n = 0 : N-1\}$ where $X(\Omega) = \mathcal{F}\{x[n]\} \simeq 0$ for the frequency range near $\Omega = \pi$. We first obtain the N-point DFT of $x[n]$, i.e., $X(k) = \text{DFT}_N\{x[n]\}$ for $k = 0 : N-1$ and then pad it with some, say, $(K-1)N$ zeros to make a new KN-point DFT sequence $V(k)$ as follows:

<case 1: N is even>

$$V(k) = \begin{cases} K\tilde{X}(k) & \text{for } \ 0 \le k \le (N/2)-1 \ \text{ and} \\ & \quad KN-(N/2)+1 \le k \le KN-1 \\ (K/2)\tilde{X}(k) & \text{for } \ k = N/2,\ KN-(N/2) \\ 0 & \text{elsewhere (zero-padding)} \end{cases} \tag{5.4.17a}$$

<case 2: N is odd>

$$V(k) = \begin{cases} K\tilde{X}(k) & \text{for } 0 \le k \le (N-1)/2 \text{ and} \\ & \quad KN-(N-1)/2 \le k \le KN-1 \\ 0 & \text{elsewhere (zero-padding)} \end{cases} \tag{5.4.17b}$$

where $\tilde{X}(k) = \text{DFS}_N\{x[n]\}$ is the periodic repetition of $X(k)$ with period N.

Now we compute the KN-point IDFT of $V(k)$ to obtain an interpolation of $x[n]$ as

$$\begin{aligned} v[n] &= \text{IDFT}_{KN}\{V(k)\} \\ &= \frac{1}{KN}\sum\nolimits_{k=0}^{KN-1} V(k)\, e^{j2\pi kn/KN} \text{ for } n = 0,\ 1,\ \cdots,\ KN-1 \end{aligned} \tag{5.4.18}$$

To evaluate this expression at every K th element, we substitute $n = Km$ to get

$$\begin{aligned} v[Km] &= \frac{1}{N}\sum\nolimits_{k=0}^{N-1} \tilde{X}(k)\, e^{j2\pi km/N} \\ &= \text{IDFT}_N\{X(k)\} = x[m] \text{ for } m = 0,\ 1,\ \cdots,\ N-1 \end{aligned} \tag{5.4.19}$$

This implies that the new sequence $v[n]$ has been obtained by placing $(K-1)$ samples between successive samples of the original sequence $x[n]$.

(Q) In the case where $x[n]$ is a real-valued sequence, it seems that we need to ensure the conjugate symmetry of its KN -point DFT $V(k)$ by modifying a part of Eq. (5.4.17a) as

$$V(k) = \begin{cases} (K/2)\tilde{X}(k) & \text{for } k = N/2 \\ (K/2)\tilde{X}^*(k) & \text{for } k = KN - (N/2) \end{cases} \tag{5.4.20}$$

However, we do not have to take the complex conjugate of $\tilde{X}(N/2)$ since it is real for a real-valued sequence $x[n]$.

Remark 5.3 Discrete-Time Interpolation, Zero Insertion, and Lowpass Filtering

The DFT sequence $V(k)$ is the same as the DFT of the output sequence of an ideal lowpass filter (with gain K and bandwidth π/K) to the input signal $x_{(K)}[n]$, which is obtained by inserting $(K-1)$ zeros between successive sample points of $x[n]$. From Eq. (3.2.14), we know that the KN -point DFT of $x_{(K)}[n]$ is

$$X_{KN}(k) = \tilde{X}_N(k) \text{ for } k = 0,\ 1,\ \cdots,\ (KN-1):$$
$$\text{the periodic repetition of } X(k) \text{ with period } N$$

Note that Eq. (5.4.17) is equivalent to taking a scaled version of $X_K(k)$ within the low frequency band, which corresponds to a lowpass filtering (see Fig. 5.16(b2)).

Example 5.1 Discrete-Time Interpolation

Let us find a discrete-time interpolation for the following sequence (see Fig. 5.16(a1)):

$$x[n] = \{\overset{n=0}{2},\ \overset{1}{1},\ \overset{2}{0},\ \overset{3}{1}\} \tag{E5.1.1}$$

We first compute the $N = 4$ -point DFT of $x[n]$ as (see Fig. 5.16(b1))

$$\begin{aligned} X(k) &= \text{DFT}_N\{x[n]\} = \sum_{n=0}^{N-1} x[n]\, e^{-j2\pi nk/N} = \sum_{n=0}^{3} x[n](-j)^{nk} \\ &= x[0] + x[1](-j)^k + x[2](-j)^{2k} + x[3](-j)^{3k} \\ &= 2 + (-j)^k + (-j)^{3k}; \\ X(k) &= \{4,\quad 2,\quad 0,\quad 2\} \end{aligned} \tag{E5.1.2}$$

Then we use Eq. (5.4.17) with $K = 2$ to get $V(k)$ as (see Fig. 5.16(b3))

$$V(k) = \{\overset{k=0}{8},\ \overset{1}{4},\ \overset{2}{0},\ \overset{3}{0},\ \overset{4}{0},\ \overset{5}{0},\ \overset{6}{0},\ \overset{7}{4}\} \tag{E5.1.3}$$

Now, we compute the $KN = 8$ -point IDFT of $V(k)$ to obtain the interpolated sequence $v[n]$ (see Fig. 5.16(a3)):

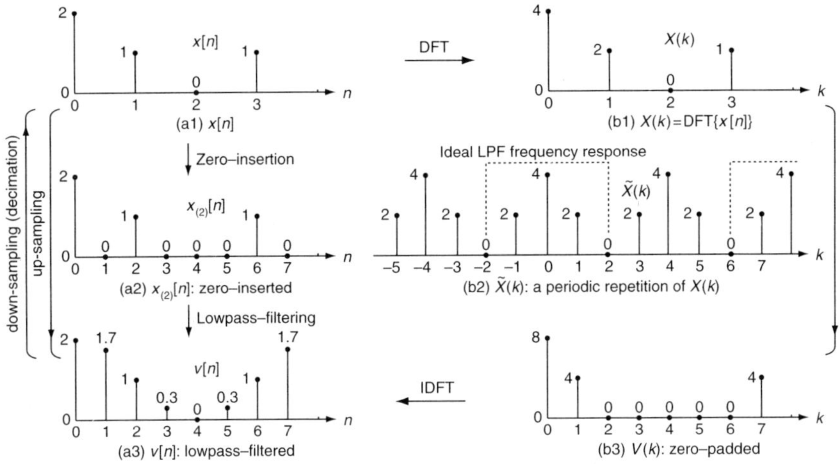

Fig. 5.16 Discrete-time interpolation of a discrete-time sequence

$$
\begin{aligned}
v[n] &= \text{IDFT}_{KN}\{V(k)\} = \frac{1}{8}\sum\nolimits_{k=0}^{7} V(k)\, e^{j2\pi k n/8} \\
&= \frac{1}{8}(8 + 4e^{j\pi kn/4} + 4e^{-j\pi kn/4}) = 1 + \cos\left(\frac{1}{4}k\pi n\right); \qquad \text{(E5.1.4)} \\
v[n] &= \{\overset{n=0}{2},\ \overset{1}{1.707},\ \overset{2}{1},\ \overset{3}{0.293},\ \overset{4}{0},\ \overset{5}{0.293},\ \overset{6}{1},\ \overset{7}{1.707}\}
\end{aligned}
$$

This result can also be obtained by using the MATLAB routine '`interpolation_discrete()`', which implements the procedure described by Eq. (5.4.17) and (5.4.18).

```
>>x=[2 1 0 1]; K=2;
>>v=interpolation_discrete(x,K)
   2.0000 1.7071 1.0000 0.2929 0 0.2929 1.0000 1.7071
```

```
function [xi,Xi]=interpolation_discrete(x,K)
% To find the KN-point discrete-time interpolation xi of an N-point
% discrete-time sequence x[n]
N=length(x); KN=K*N; N1=floor((N+1)/2); KNN1=KN-N1+2;
X= fft(x);
Xi= K*X(1:N1); Xi(KNN1:KN)=K*X(N-N1+2:N); % Eq.(5.4.17)
if mod(N,2)==0, Xi([N/2+1 KN-N/2])=K/2*X(N/2+1)*[1 1]; end % Eq.(5.4.17b)
xi= ifft(Xi); % Eq. (5.4.18)
```

5.5 Sample-and-Hold (S/H) Operation

In a majority of practical digital operations, a S/H function is performed by a single device consisting of a capacitor, an electronic switch, and operational amplifiers, which stores the (sampled) input signal voltage as a charge on the capacitor (Fig. 5.17). OP amps are needed for isolation; capacitor and switch are not connected directly to analog circuitry lest the capacitor should affect the input waveform and should be discharged to the output. The mathematical model for a S/H device or operation has been developed in Sect. 5.4.3.

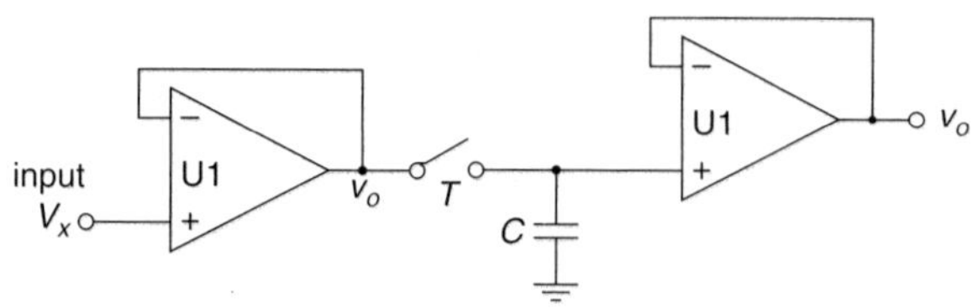

Fig. 5.17 Sample-and-Hold (S/H) device

The DAC described in Sect. 5.1 may be regarded as a device which consists of a decoder and an S/H unit (Fig. 5.18(a)). Note that the sampling operation is not necessary, but it is included since the S/H is usually considered as one unit. The ADC performs the operations of S/H, quantization, and encoding (see Fig. 5.18(b)).

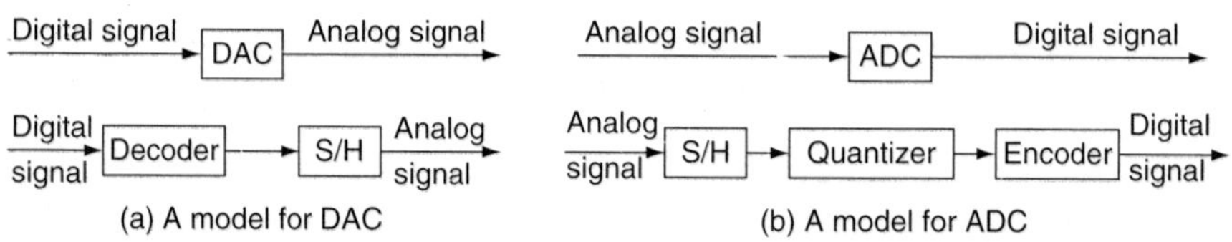

Fig. 5.18 Models for DAC and ADC

5.6 Summary

In this chapter we were concerned how continuous-time and discrete-time signals are and can be related. First, we have included a survey of D/A and A/D conversion methods to let the readers know the relationship between analog signals and the corresponding digital ones and further, to provide them with some practical information that may be useful for the selection of DAC and ADC. Then we discussed the sampling theorem, several reconstruction/interpolation techniques, and the mathematical model of a practical S/H device.

Problems

5.1 CTFS, DTFS, and Sampling Theorem
Consider the following five cosine waves:

$$x_1(t) = \cos(2\pi t) = \frac{1}{2}(e^{j2\pi t} + e^{-j2\pi t})$$
$$x_2(t) = \cos(8\pi t) = \frac{1}{2}(e^{j8\pi t} + e^{-j8\pi t})$$
$$x_3(t) = \cos(10\pi t) = \frac{1}{2}(e^{j10\pi t} + e^{-j10\pi t})$$
$$x_4(t) = \cos(18\pi t) = \frac{1}{2}(e^{j18\pi t} + e^{-j18\pi t})$$
$$x_5(t) = \cos(22\pi t) = \frac{1}{2}(e^{j22\pi t} + e^{-j22\pi t})$$

(a) Referring to the Fourier representation formula (2.1.5a), verify that the Fourier series coefficients of the two-tone signal $x_a(t) = x_1(t) + x_2(t)$ with period $P_a = \max\{2\pi/2\pi,\ 2\pi/8\pi\} = 1$ and fundamental frequency $\omega_0 = 2\pi$ are

$$X_{a,k} = \frac{1}{2}, \frac{1}{2}, \frac{1}{2}, \text{ and } \frac{1}{2} \text{ for } k = -4, -1, +1, \text{ and } +4, \text{ respectively.} \tag{P5.1.1}$$

Also for the sequence $\{x_a(nT),\ n = 0 : 9\}$ obtained by sampling one period of the signal with sampling period $T = 0.1$, find the $N = 10$ -point DFT $X(k)$ and discuss how it is related with $X_{a,k}$ based on the relationship (3.5.3) between the CTFS and DTFS. Note that $x_a(nT)$ can be written as

$$x_a[n] = x_a(nT) = \frac{1}{2}(e^{j2\pi n/10} + e^{-j2\pi n/10}) + \frac{1}{2}(e^{j8\pi n/10} + e^{-j8\pi n/10}) \tag{P5.1.2}$$

and it can somehow be matched with the IDFT formula

$$x\,[n] = \text{IDFT}_N\{X(k)\} \overset{(3.4.3)}{=} \frac{1}{N}\sum_{k=0}^{N-1} X(k)\,e^{\,j2\pi kn/N} \text{ for } n = 0 : N-1$$

to yield the DFT coefficients $X(k)$.

(b) For $x_b(t) = x_1(t) + x_3(t)$, do the same job as in (a).
(c) For $x_c(t) = x_1(t) + x_4(t)$, do the same job as in (a).
(d) For $x_d(t) = x_1(t) + x_5(t)$, do the same job as in (a).
(e) Complete the following MATLAB program "sig05p_01.m" and run it to find the $N = 10$ -point DFTs for $\{x_a(nT),\ n = 0 : 9\}$, $\{x_b(nT),\ n = 0 : 9\}$, $\{x_c(nT),\ n = 0 : 9\}$, and $\{x_d(nT),\ n = 0 : 9\}$. Do they agree with those obtained in (a)–(d)? For each of the four cases (a)–(d), tell whether the frequency folding or aliasing occurs or not by the sampling with period $T = 0.1$.

```
%sig05p_01.m
clear, clf
ws=[2*pi 8*pi 10*pi 18*pi 22*pi];
P=1; ts=0.001; tt=[0:ts:P]; T=0.1; N=round(P/T);
n=[0:T/ts:length(tt)]; n=n(1:N); nT=n*ts; k=[0:N-1];
x1t= cos(ws(1)*tt)+cos(ws(2)*tt); x1n= x1t(n+1); X1= fft(x1n);
subplot(521), plot(tt,x1t), hold on, stem(nT,x1n,'.')
subplot(522), stem(k,abs(X1),'.')
.. .. .. .. .. ..
```

(f) Among the sampling rates {15Hz, 20Hz, 25Hz, 40Hz, 50Hz}, choose the lowest one such that the frequency aliasing does not occur for any of (a)–(d).

5.2 Reconstruction
Consider the continuous-time waveform $x(t)$ depicted in Fig. P5.2.

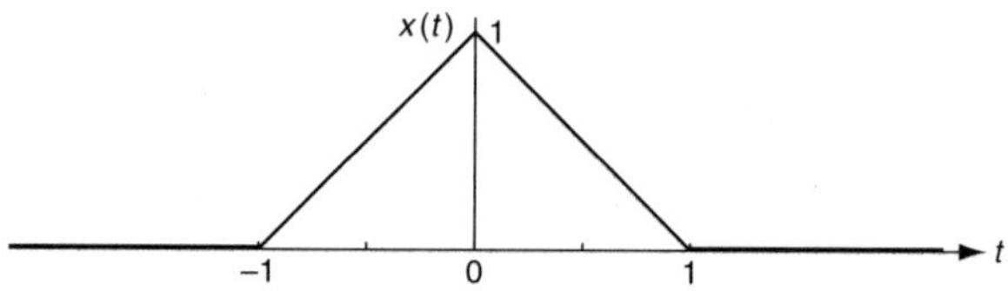

Fig. P5.2

(a) Use Eq. (E2.3.4) to find the CTFT of $x(t)$.

(b) Regarding $x(t)$ as a periodic waveform with period $P = 2$ [s], we can sample it with sampling period $T = 0.5$ [s] to get a (non-causal) discrete-time sequence

$$x_{NC}[n] = \{ \overset{n=-1}{0.5}, \ \overset{0}{1}, \overset{1}{0.5}, \overset{2}{0} \} \tag{P5.2.1}$$

and take the part of the periodic repetition of $x_{NC}[n]$ with period $N = 4$ for $n = 0 : 3$ to have a causal sequence

$$x[n] = \{ \overset{n=0}{1}, \ \overset{1}{0.5}, \ \overset{2}{0}, \ \overset{3}{0.5} \} \tag{P5.2.2}$$

Find the 4-point DFT of this sequence.

(c) Find the analytical expression of the DFS reconstruction of $x(t)$ using the DFT obtained in (b) and plot it on Fig. P5.2.

(d) Suppose we can apply an ideal LPF for reconstruction. Can the waveform depicted in P5.2 be reconstructed perfectly from its sampled version with some sampling period? Say 'yes' or 'no' and state the background of your answer.

5.3 Discrete-Time Interpolation
Consider the discrete-time signal $x[n]$ whose DTFT spectrum is depicted in Fig. P5.3.

(a) Sketch the spectrum of $y[n] = x_{(2)}[n]$ where

$$y[n] = x_{(2)}[n] = \begin{cases} x[n/2] & \text{for } n = 2m(m: \text{ integer}) \\ 0 & \text{elsewhere} \end{cases} \quad \text{(P5.3.1)}$$

(b) Suppose that we wish to form an interpolated version of $x[n]$ by passing $y[n]$ through a lowpass filter. Specify the requirements of the LPF on its gain and bandwidth.

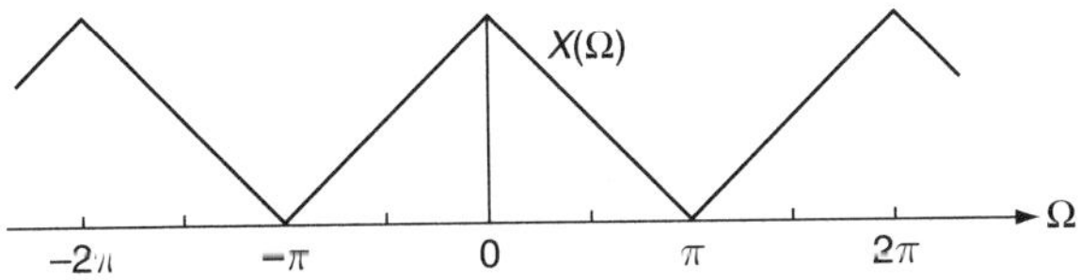

Fig. P5.3

5.4 Discrete-Time Interpolation
Consider the discrete-time signal $x[n]$ obtained by sampling the following continuous-time signal $x(t)$ at $t = nT(n = 0 : 50, \; T = 0.0001 \text{ sec})$ with sampling rate $F_s = 10\text{kHz}$.:

$$x(t) = \cos(2\pi \times 1000t) + \sin(2\pi \times 3000t) \quad \text{(P5.4.1)}$$

Using the discrete-time interpolation introduced in Sect. 5.4.4, insert 19 points between successive sample points of $x[n]$ to make a 1020-point interpolation $v[n]$ and plot it (continuously in a solid line) together with $x(nT)$ (discretely using the 'x' mark).

5.5 A/D- $G[z]$ (digital filter)-D/A Structure
Consider the A/D- $G[z]$ -D/A structure with $G[z] = 1 + z^{-1}$ where the input is to be sampled with sampling period $T = 0.1$ [s].

(a) Let the input be given by $x_1(t) = \sin(10\pi t/3)$. What is the digital frequency of $x_1[n] = x_1(nT)$?
(b) What is the highest analog frequency such that the input signal will not be aliased?
(c) Find the frequency response and DC gain of the digital filter $G[z]$.
(d) Find the steady-state output $y_{1,ss}[n]$ to $x_1[n] = x_1(nT) = \sin(10\pi Tn/3)$.
(e) Find the steady-state output $y_{2,ss}[n]$ to the unit step input $x_2[n] = u_s[n]$.
(f) Support your answers to (d) and (e) by the simulations for $n = 0 : 50$.

Chapter 6
Continuous-Time Systems and Discrete-Time Systems

Contents

In this chapter we are concerned how continuous-time and discrete-time systems are or can be related with each other. We will discuss the concept and criterion of discrete-time equivalent (digital simulator) and investigate various discretization methods to obtain the discrete-time equivalent for a given analog system, which can be used for the analysis and design of sampled-data systems.

6.1 Concept of Discrete-Time Equivalent

As the technology of digital processors becomes more advanced, it is often desirable to replace analog systems with digital ones. In some cases, we need to transform the analog systems into their "equivalent" digital systems in order to simulate their behaviors on digital computers. In other cases, rather than designing directly a

W.Y. Yang et al., *Signals and Systems with MATLAB®*,
DOI 10.1007/978-3-540-92954-3_6,

digital system, we make use of a variety of well-developed analog system design procedures to get an analog system with satisfactory performance and then convert it into a discrete-time equivalent that mimicks (emulates) the analog system. In either case, for a given continuous-time linear time-invariant (LTI) system $G_A(s)$ with input $x(t)$ and output $y_A(t)$, we wish to find a discrete-time system $G_D[z]$ with input $x[n]$ and output $y_D[n]$ such that

$$x[n] = x(nT) \Rightarrow y_D[n] = y_A(nT) \tag{6.1.1}$$

This means that $G_A(s)$ and $G_D[z]$ yield the equivalent outputs to equivalent inputs in the discrete-time domain. We will call such a system $G_D[z]$ the *discrete-time (z-domain) equivalent* or *digital simulator* for $G_A(s)$ where Eq. (6.1.1) is referred to as the (*discrete-time*) *equivalence criterion*. To establish necessary conditions for the validity of the above criterion, we shall first consider sinusoidal inputs, which will lead to the general case, because an arbitrary signal can be expressed as a linear combination of sine waves (the Fourier series representation). Suppose the input $x[n]$ of a discrete-time system $G_D[z]$ is the discrete-time version of $x(t) = e^{j\omega t}$ sampled with period T, i.e., $x[n] = x(nT) = e^{jn\omega T}$ where $x(t)$ is the input to the continuous-time system $G_A(s)$ (see Fig. 6.1):

$$x(t) = e^{j\omega t} \xrightarrow{\text{Sampling with period } T} x[n] = x(nT) = e^{jn\omega T} \tag{6.1.2}$$

Then we can use the convolutional input-output relationships (1.2.4)/(1.2.9) of continuous-time/discrete-time LTI systems and the definitions of CTFT and DTFT to write the outputs of $G_A(s)$ and $G_D[z]$ as

$$\begin{aligned} y_A(t) &= \int_{-\infty}^{\infty} g_A(\tau) e^{j\omega(t-\tau)} d\tau & \xrightarrow[\text{Sampling with period } T]{?} \quad y_D[n] &= \sum_{m=-\infty}^{\infty} g_D[m] e^{j\omega(n-m)T} \\ &\overset{(2.2.1a)}{=} G_A(j\omega) e^{j\omega t} & &\overset{(3.1.1)}{=} G_D[e^{j\omega T}] e^{jn\omega T} \end{aligned} \tag{6.1.3}$$

where

$$G_A(j\omega) \overset{(2.2.1a)}{=} \int_{-\infty}^{\infty} g_A(\tau) e^{-j\omega\tau} d\tau \underset{\substack{g_A(\tau)=0 \ \forall\ \tau<0 \\ \text{causal system}}}{\overset{(A.1)}{=}} G_A(s)|_{s=j\omega} \tag{6.1.4a}$$

$$G_D[e^{j\omega T}] \overset{(3.1.1)}{=} \sum_{m=-\infty}^{\infty} g_D[m]\, e^{-jm\,\omega T} \underset{\substack{g_D[n]=0 \ \forall\ n<0 \\ \text{causal system}}}{\overset{(4.1.1b)}{=}} G_D[z]|_{z=e^{j\omega T}} \tag{6.1.4b}$$

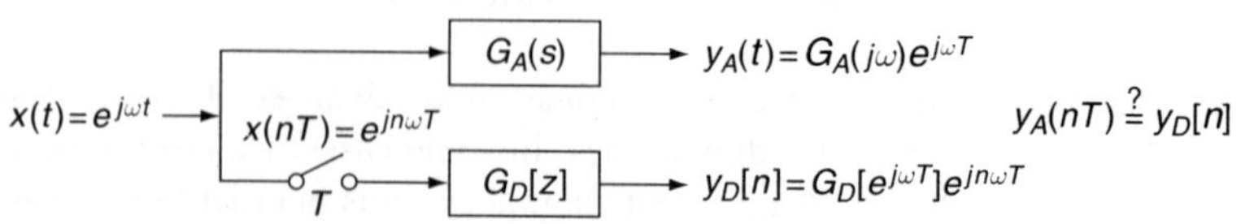

Fig. 6.1 The equivalence criterion between a continuous-time system and a discrete-time system

The above equivalence criterion (6.1.1) can be stated in terms of the frequency responses of the continuous-time and discrete-time systems as

$$G_D[e^{j\omega T}] = G_A(j\omega) \tag{6.1.5}$$

This is referred to as the frequency-domain *discrete-time equivalence criterion.*

However, this equivalence criterion cannot be satisfied perfectly because the digital frequency response $G_D[e^{j\omega T}]$ is periodic in ω (Sect. 3.2.1) while the frequency response $G_A(j\omega)$ is not. This implies that we cannot find the discrete-time equivalent of a continuous-time system that would work for all kinds of inputs. In principle, the criterion (6.1.5) can be satisfied only for the principal frequency range $(-\pi/T, +\pi/T)$ if we restrict the class of inputs into band-limited signals that do not contain frequencies higher than the folding frequency π/T, i.e., half the sampling frequency $\omega_s = 2\pi/T$. In this sense, the resulting system $G_D[z]$ satisfying the criterion (6.1.5) is a discrete-time equivalent of $G_A(s)$, which is valid only for such band-limited input signals.

Remark 6.1 Equivalence Criterion and Band-Limitedness Condition

(1) The band-limitedness condition is not so restrictive since we can increase π/T by decreasing the sampling period T. Furthermore, the condition is desirable in order for the sampled output $y(nT)$ to represent $y(t)$ faithfully.
(2) As a matter of fact, the equivalence criterion (6.1.5) cannot be met exactly even for the principal frequency range since $G_D[e^{j\omega T}]$ is a rational function of $e^{j\omega T}$, while $G_A(j\omega)$ is a rational function of $j\omega$. For engineering applications, however, approximate equality is good enough.

Now we will discuss the time-sampling method, which determines $G_D[z]$ so as to satisfy the criterion (6.1.5) with reasonable accuracy for almost every ω in the frequency band of the input signal. The basic idea is as follows. Suppose that the input $x(t)$ and the impulse response $g_A(t)$ of the analog (or continuous-time) system $G_A(s)$ are sufficiently smooth, i.e., nearly constant in any interval of length T. Then, letting

$$g_D[n] = T\ g_A(nT) \tag{6.1.6}$$

satisfies approximately the equivalence criterion (6.1.1), i.e.,

$$y_D[n] \cong y_A(nT) \quad \text{for } x[n] = x(nT)$$

since we have

$$y_A(nT) \overset{(1.2.4)}{=} \int_{-\infty}^{\infty} g_A(\tau)x(t-\tau)d\tau\Big|_{t=nT}$$

$$\cong \sum_{m=-\infty}^{\infty} T g_A(nT)x(nT-mT) \tag{6.1.7a}$$

$$y_D[n] \overset{(1.2.9)}{=} \sum_{m=-\infty}^{\infty} g_D[m]x[n-m] \tag{6.1.7b}$$

The condition (6.1.6) can be written in terms of the transforms as

$$G_D[z] = T\mathcal{Z}\{g_A(nT)\} = T\mathcal{Z}\left\{\mathcal{L}^{-1}\{G_A(s)\}|_{t=nT}\right\} \tag{6.1.8}$$

Remark 6.2 Time-Sampling Method – Impulse-Invariant Transformation

Note that sampling the continuous-time unit impulse $\delta(t) \cong (1/T)\text{sinc}(t/T)$ (Eq. (1.1.33a)) at $t = nT$ (i.e., every T s) yields $(1/T)\text{sinc}(n) \overset{(1.1.35)}{=} (1/T)\delta[n]$. Then, it is very natural that, in order for G_A and G_D to be similar, their responses to $\delta(t)$ and $(1/T)\delta[n]$ should be the same, i.e.,

$$g_A(nT) = G_A\{\delta(t)\}|_{t=nT} \equiv G_D\left\{\frac{1}{T}\delta[n]\right\} = \frac{1}{T}G_D\{\delta[n]\} = \frac{1}{T}g_D[n] \tag{6.1.9}$$

where $G\{x(t)\}$ denotes the output of a system G to an input $x(t)$. This is equivalent to the condition (6.1.6) for the *impulse-invariant transformation*.

Remark 6.3 Frequency Response Aspect of Impulse-Invariant Transformation

If $G_A(j\omega)$ is negligible for $|\omega| > \pi/T$ (in agreement with the smoothness conditions about $g_A(t)$), then we can take the DTFT of (6.1.6) and use Eq. (3.5.7) or (5.3.1b) to get

$$G_D(\Omega) = \mathcal{F}\{g_D[n]\}|_{\Omega=\omega T} \equiv T\mathcal{F}(g_A(nT))\} \overset{(5.3.1b)}{\cong} G_A(\omega) \tag{6.1.10}$$

6.2 Input-Invariant Transformation

We will consider the input-invariant transformations that are error-free for specific input signals. For example, the impulse-invariant/step-invariant transformation is accurate when the input is an impulse or a step function. If they were accurate only for the specified inputs, they would be of limited practical value. However, by superposition, an input-invariant transformation gives zero error in response to any linear combination of specified input functions.

6.2.1 Impulse-Invariant Transformation

As mentioned in Remark 6.2, this is identical to the time-sampling method discussed in the previous section. Note that Eq. (6.1.8) can be written as

$$\mathcal{Z}^{-1}\left\{G_D[z]\cdot\frac{1}{T}\right\} \equiv \mathcal{L}^{-1}\{G_A(s)\cdot 1\}|_{t=nT}$$

$$g_D[n]*\frac{1}{T}\delta[n] \overset{(1.1.28)}{=} \frac{1}{T}g_D[n] \equiv g_A(t)|_{t=nT} \overset{(1.1.21)}{=} g_A(t)*\delta(t)|_{t=nT} \qquad (6.2.1)$$

This implies that the (impulse) response of the continuous-time system $G_A(s)$ to the impulse input $\delta(t)$ is equal to the response of the discrete-time system $G_D[z]$ to the input $(1/T)\delta[n]$, which is the sampled version of $\delta(t)$.

The procedure to derive the *impulse-invariant equivalent* $G_{imp}[z]$ for a given analog system $G_A(s)$ is as follows:

1. Expand $G_A(s)$ into the partial fraction form:

$$G_A(s) = \sum_{i=1}^{N} \frac{K_i}{s-s_i} \qquad (6.2.2)$$

2. Replace each term $1/(s-s_i)$ by $T/(1-e^{s_iT}z^{-1})$:

$$G_{imp}[z] = \sum_{i=1}^{N} \frac{K_iT}{1-e^{s_iT}z^{-1}} \qquad (6.2.3)$$

Remark 6.4 Mapping of Stability Region by Impulse-Invariant Transformation

Comparing Eq. (6.2.2) with Eq. (6.2.3), we observe that a pole at $s = s_i$ in the s-plane is mapped to $z = e^{s_iT}$ in the z-plane. Consequently, if and only if s_i is in the left half plane (LHP), which is the stable region in the s-plane, then the corresponding pole is inside the unit circle, which is the stable region in the z-plane (see Fig. 6.2). However, the zeros will not in general be mapped in the same way as the poles are mapped.

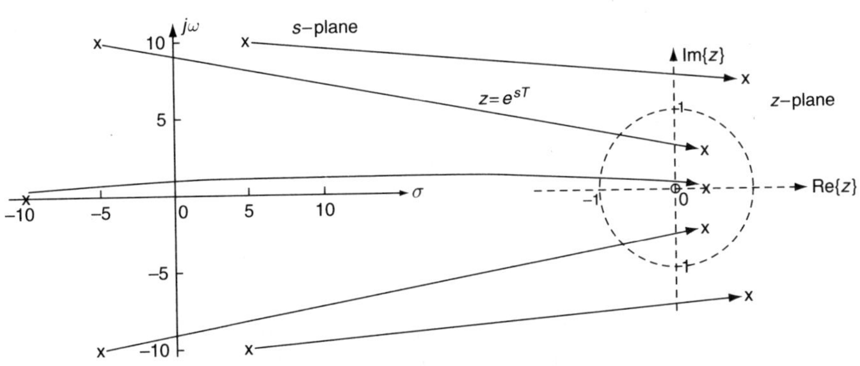

Fig. 6.2 Mapping of poles from the s-plane to the z-plane

Remark 6.5 Frequency Transformation by Impulse-Invariant Transformation

The relationship between the analog frequency ω_A and the corresponding digital frequency Ω is linear, that is, $\Omega = \omega_A T$ since

$$e^{j\Omega} \underset{\text{evaluation along the unit circle}}{\equiv} z = e^{sT} \underset{\text{evaluation along the } j\omega_A-\text{axis}}{\overset{s=j\omega_A}{\equiv}} e^{j\omega_A T} \tag{6.2.4}$$

Consequently, the shape of the frequency response is preserved. The negative aspect of this linear frequency relationship is that short sampling period T does not remove the frequency-aliasing problem caused by the impulse-invariant transformation (see Fig. 6.8).

Example 6.1 Impulse-Invariant Transformation – Time-Sampling Method

For a continuous-time system with the system function $G_A(s)$ and frequency response $G_A(j\omega)$ as

$$G_A(s) = \frac{a}{s+a}; \quad G_A(j\omega) = \frac{a}{j\omega + a} \tag{E6.1.1}$$

the impulse-invariant transformation yields the following discrete system function

$$G_{imp}[z] \overset{(6.2.3)}{=} \frac{a\,T}{1 - e^{-aT}z^{-1}}; \quad G_{imp}[e^{j\omega T}] = \frac{a\,T}{1 - e^{-aT}e^{-j\omega T}} \tag{E6.1.2}$$

6.2.2 Step-Invariant Transformation

If we let

$$G_{step}[z] = (1 - z^{-1})\mathcal{Z}\left\{ \mathcal{L}^{-1}\left\{ \frac{1}{s} G_A(s) \right\}\Bigg|_{t=nT} \right\} \tag{6.2.5}$$

then it will be an exact discrete-time equivalent of $G_A(s)$ for any input composed of step functions occurring at sample points. Note that Eq. (6.2.5) can be written as

$$\mathcal{Z}^{-1}\left\{ G_{step}[z] \cdot \frac{1}{1 - z^{-1}} \right\} = \mathcal{L}^{-1}\left\{ \frac{1}{s} G_A(s) \right\}\Bigg|_{t=nT};$$
$$g_{step}[n] * u_s[n] = g_A(t) * u_s(t)|_{t=nT} \tag{6.2.6}$$

This implies that the step response of the continuous-time system $G_A(s)$ is equal to the step response of the discrete-time system $G_{step}[z]$ on a sample-by-sample basis. That is why Eq. (6.2.5) is called the *step-invariant transformation*.

Let us consider the discrete-time error model for $G_A(s)$ and $G_{step}[z]$ in Fig. 6.3, in which the input to $G_A(s)$ is $\bar{x}(t)$, i.e., the zero-order-hold version of $x(t)$. The

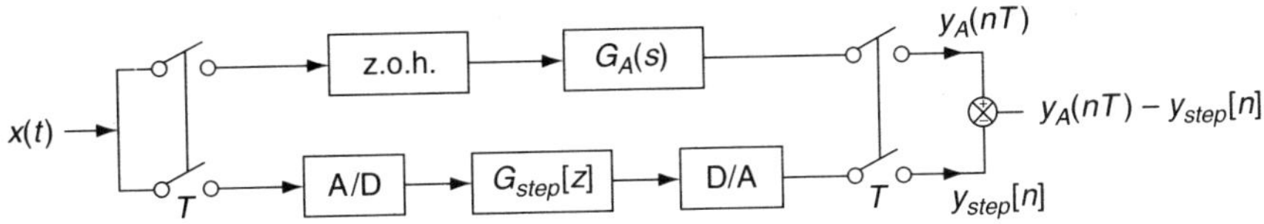

Fig. 6.3 The discrete–time error model for the step–invariant transformation

discrete-time transfer function of the system with a z.o.h. in the upper part of Fig. 6.3 is

$$\mathcal{Z}\left\{\mathcal{L}^{-1}\left\{G_{h0}(s)G_A(s)\right\}\Big|_{t=nT}\right\} \overset{(5.4.15)}{=} \mathcal{Z}\left\{\mathcal{L}^{-1}\left\{\frac{1-e^{-Ts}}{s}G_A(s)\right\}\Big|_{t=nT}\right\}$$

$$= \mathcal{Z}\left\{\mathcal{L}^{-1}\left\{\frac{1}{s}G_A(s)\right\}\Big|_{t=nT} - \mathcal{L}^{-1}\left\{\frac{e^{-Ts}}{s}G_A(s)\right\}\Big|_{t=nT}\right\}$$

$$= (1-z^{-1})\mathcal{Z}\left\{\mathcal{L}^{-1}\left\{\frac{1}{s}G_A(s)\right\}\Big|_{t=nT}\right\} \overset{(6.2.5)}{\equiv} G_{step}[z] \tag{6.2.7}$$

Therefore the step-invariant transformation is also called the *zero-order-hold equivalent mapping* and it is well suited to a digital computer implementation in the A/D-$G[z]$-D/A structure.

Example 6.2 Step-Invariant Transformation (Zero-Order-Hold Equivalent)

For a continuous-time system with the system function

$$G_A(s) = \frac{a}{s+a} \text{ with a pole at } s = s_p = -a \tag{E6.2.1}$$

the step-invariant transformation yields the following discrete system function:

$$\mathcal{Z}\left\{\frac{1}{s}G_A(s)\right\} = \mathcal{Z}\left\{\frac{a}{s(s+a)}\right\} = \mathcal{Z}\left\{\frac{1}{s} - \frac{1}{s+a}\right\}$$

$$\overset{\text{B.8(3),(6)}}{=} \frac{1}{1-z^{-1}} - \frac{1}{1-e^{-aT}z^{-1}} = \frac{(1-e^{-aT}z^{-1})-(1-z^{-1})}{(1-z^{-1})(1-e^{-aT}z^{-1})}$$

$$G_{step}[z] \overset{(6.2.5)}{=} (1-z^{-1})\mathcal{Z}\left\{\frac{1}{s}G_A(s)\right\} = \frac{1-e^{-aT}}{z-e^{-aT}} = G_{\text{zoh}}[z] \tag{E6.2.2}$$

This implies that the s-plane pole is mapped into the z-plane through the step-invariant transformation in the same way as through the impulse-invariant transformation.

$$s = s_p = -a \rightarrow z = z_p = e^{s_p T} = e^{-aT}$$

6.3 Various Discretization Methods [P-1]

A continuous-time LTI system can be described by the system or transfer function as

$$G_A(s) = \frac{Y(s)}{X(s)}$$

where $X(s)$ and $Y(s)$ are the Laplace transforms of the input and output. Its input-output relationship can be written as an integro-differential equation. We often use the numerical methods to convert the integro-differential equation into the corresponding difference equation, which can easily be solved on a digital computer. The difference equation may be represented by a discrete system function $G_D[z]$ (Sect. 4.4), which can be thought of representing a discrete-time equivalent to $G_A(s)$ (see Fig. 6.4).

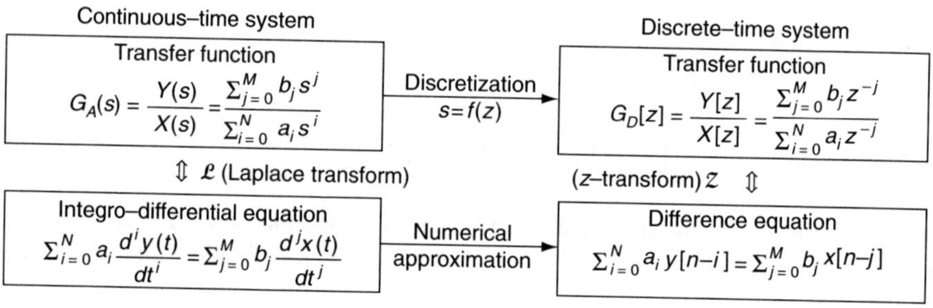

Fig. 6.4 Discretization and numerical approximation

Some numerical approximation techniques on differentiation/integration will now be presented with the corresponding discretization methods.

6.3.1 Backward Difference Rule on Numerical Differentiation

We could replace the derivative of a function $x(t)$

$$y(t) = \frac{d}{dt}x(t): \quad Y(s) = s\,X(s)$$

by

$$y(nT) = \frac{x(nT) - x(nT - T)}{T}: \; Y[z] = \frac{1 - z^{-1}}{T} X[z]$$

where the initial values have been neglected since we are focusing on describing the input-output relationship. This numerical differentiation method suggests the *backward difference mapping* rule (Fig. 6.5(a)):

$$s \to \frac{1 - z^{-1}}{T} \text{ or equivalently, } \frac{1}{1 - s\,T} \to z \tag{6.3.1}$$

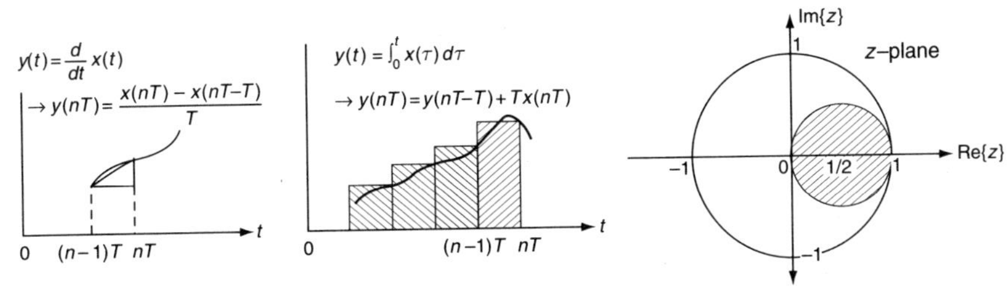

(a) Backward difference rule (b) Right–side integration rule (c) Mapping of stable region

Fig. 6.5 Backward difference or right–side integration rule

Example 6.3 Backward Difference Rule

For a continuous-time system with the system function $G_A(s) = a/(s + a)$, the backward difference rule yields the following discrete system function:

$$G_b[z] \overset{(6.3.1)}{=} \left.\frac{a}{s+a}\right|_{s=(1-z^{-1})/T} = \frac{a}{(1-z^{-1})/T + a}$$

$$= \frac{aT}{1 + aT - z^{-1}} = \frac{aTz/(1+aT)}{z - 1/(1+aT)} \tag{E6.3.1}$$

This implies that the s-plane pole is mapped into the z-plane through the backward difference or right-side integration rule as

$$s = s_p = -a \rightarrow z = z_p \overset{(6.3.1)}{=} \frac{1}{1 - s_pT} \underset{(D.25)}{\overset{\text{for } |s_pT|<<1}{\cong}} \frac{1}{e^{-s_pT}} = e^{s_pT} \tag{E6.3.2}$$

This implies that if the s-plane pole s_p is so close to the origin and/or the sampling period is so short that $|s_pT| << 1$, then the location of the pole mapped to the z-plane from the s-plane is

$$z_p \cong e^{s_pT}$$

This relationship between the s-plane poles and the corresponding z-plane poles is almost the same as those for other discretization methods.

Remark 6.6 Mapping of Stability Region and Frequency Transformation

From Eq. (6.3.1), it can be seen that the $j\omega$-axis (the stability boundary) in the s-plane is mapped to

$$z = \frac{1}{1 - j\omega T} = \frac{1 - j\omega T + 1 + j\omega T}{2(1 - j\omega T)} = \frac{1}{2}(1 + e^{j\theta}) \text{ with } \theta = 2\tan^{-1}(\omega T) \tag{6.3.2}$$

This describes the circle of radius 1/2 and with the center at $z = 1/2$, which is inside the unit circle in the z-plane (see Fig. 6.5(c)). It is implied that the backward difference rule always maps stable analog systems into stable discrete equivalents, but some unstable analog systems also yield stable discrete ones. Besides, since the $j\omega$-axis in the s-plane does not map to the unit circle in the z-plane, the digital frequency response will deviate from the analog frequency response as $\omega \to \pm\infty$ or $\Omega \to \pm\pi$ (farther from $s = 0$ or $z = 1$). Thus, in order to make the frequency response of the discrete-time equivalent close to that of the original analog system (for the principal frequency range), we must decrease the sampling T or equivalently, increase the sampling frequency ω_s so that significant pole/zeros are mapped to the neighborhood of $z = 1$.

6.3.2 Forward Difference Rule on Numerical Differentiation

We could replace the derivative of a function $x(t)$

$$y(t) = \frac{d}{dt}x(t) : \quad Y(s) = s\,X(s)$$

by

$$y(nT) = \frac{x(nT+T) - x(nT)}{T} : \; Y[z] = \frac{z-1}{T}X[z]$$

This numerical differentiation method suggests the *forward difference mapping* rule (Fig. 6.6(a)):

$$s \to \frac{z-1}{T} \text{ or equivalently, } 1 + s\,T \to z \tag{6.3.3}$$

Example 6.4 Forward Difference Rule

For a continuous-time system with the system function $G_A(s) = a/(s+a)$, the forward difference rule yields the following discrete system function:

$$G_f[z] = \left.\frac{a}{s+a}\right|_{s=(z-1)/T} = \frac{a}{(z-1)/T + a} = \frac{aT}{z - (1 - aT)} \tag{E6.4.1}$$

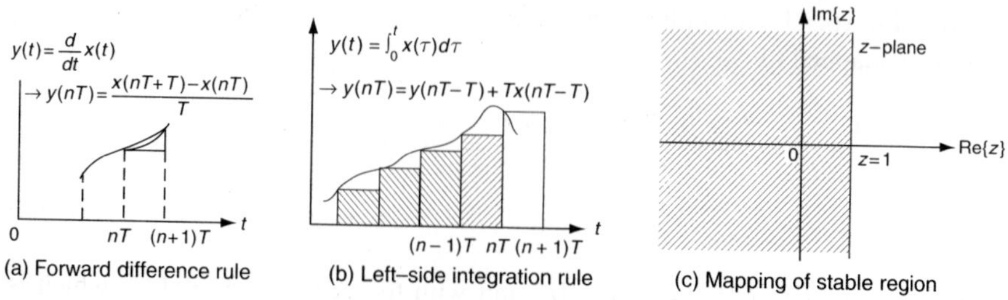

Fig. 6.6 Forward difference or left–side integration rule

This implies that the s-plane pole is mapped into the z-plane through the forward difference or left-side integration rule as

$$s = s_p = -a \;\rightarrow\; z = z_p \overset{(6.3.3)}{=} 1 + s_p T \underset{\text{for } |s_p T| << 1}{\overset{(D.25)}{\cong}} e^{s_p T} \tag{E6.4.2}$$

This implies that if the s-plane pole s_p is so close to the origin and/or the sampling period is so short that $|s_p T| << 1$, then the location of the poles mapped to the z-plane from the s-plane is

$$z_p \cong e^{s_p T}$$

Remark 6.7 Mapping of Stability Region By Forward Difference Rule

From Eq. (6.3.3), it can be seen that the $j\omega$-axis (the stability boundary) in the s-plane is mapped to

$$z = 1 + j\omega T \tag{6.3.4}$$

This describes the straight line parallel to the imaginary axis and crossing the real axis at $z = 1$. It is implied that the forward difference rule maps the left half plane (LHP) in the s-plane to the left of $z = 1$ in the z-plane with some portion outside the unit circle. Consequently, some stable analog systems may yield unstable discrete equivalents, while unstable analog systems always make unstable discrete ones. Hence, this is an undesirable mapping that cannot be used in practice.

6.3.3 Left-Side (Rectangular) Rule on Numerical Integration

We could replace the integral of a function $x(t)$

$$y(t) = \int_0^t x(\tau)d\tau \;:\; Y(s) = \frac{1}{s}X(s)$$

by

$$y(nT) = y(nT - T) + Tx(nT - T) \;:\; Y[z] = \frac{Tz^{-1}}{1 - z^{-1}}X[z]$$

This numerical integration method suggests the *left-side integration* rule (Fig. 6.6(b)):

$$s \rightarrow \frac{z-1}{T} \text{ or equivalently, } 1 + s\,T \rightarrow z \tag{6.3.5}$$

This is identical to the forward difference rule (6.3.3).

6.3.4 Right-Side (Rectangular) Rule on Numerical Integration

We could replace the integral of a function $x(t)$

$$y(t) = \int_0^t x(\tau)d\tau \;:\; Y(s) = \frac{1}{s}X(s)$$

by

$$y(nT) = y(nT - T) + T\,x(nT) : \; Y[z] = \frac{T}{1 - z^{-1}}X[z]$$

This numerical integration method suggests the *right-side integration* rule (Fig. 6.5(b)):

$$s \rightarrow \frac{1 - z^{-1}}{T} \text{ or equivalently, } \frac{1}{1 - sT} \rightarrow z \qquad (6.3.6)$$

This is identical to the backward difference rule (6.3.1).

6.3.5 Bilinear Transformation (BLT) – Trapezoidal Rule on Numerical Integration

By the trapezoidal rule (or Tustin's method), we could replace the integral of a function $x(t)$

$$y(t) = \int_0^t x(\tau)d\tau \;:\; Y(s) = \frac{1}{s}X(s)$$

by

$$y(nT) = y(nT - T) + \frac{T}{2}(x(nT) + x(nT - T)) : \; Y[z] = \frac{T}{2}\frac{1 + z^{-1}}{1 - z^{-1}}X[z]$$

which suggests the *bilinear transformation* rule (Fig. 6.7(a)):

$$s \rightarrow \frac{2}{T}\frac{1 - z^{-1}}{1 + z^{-1}} \text{ or equivalently, } \frac{1 + sT/2}{1 - sT/2} \rightarrow z \qquad (6.3.7)$$

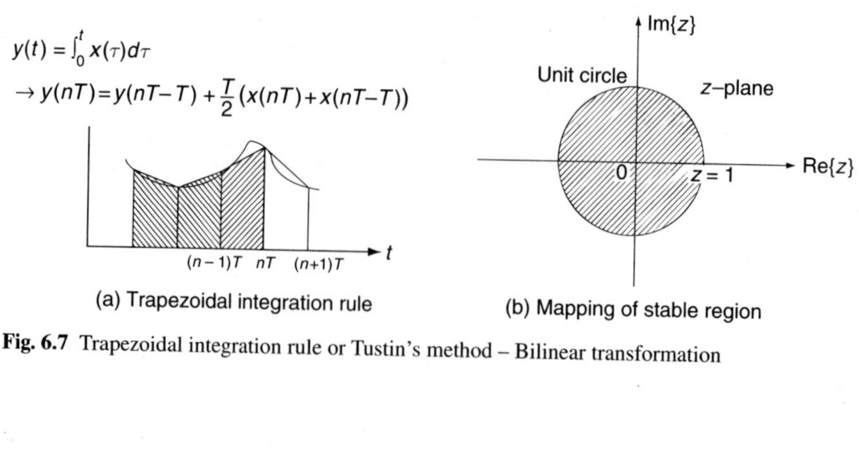

Fig. 6.7 Trapezoidal integration rule or Tustin's method – Bilinear transformation

Example 6.5 Bilinear Transformation (BLT)

For a continuous-time system with the system function $G_A(s) = a/(s+a)$ having a pole at $s = -a$ and cutoff frequency $\omega_{A,c} = a$, the bilinear transformation (BLT) yields the following discrete system function:

$$G_{bl}[z] = \left.\frac{a}{s+a}\right|_{s=\frac{2(1-z^{-1})}{T(1+z^{-1})}} = \frac{a}{\frac{2(1-z^{-1})}{T(1+z^{-1})} + a} = \frac{aT(z+1)/(2+aT)}{z - \frac{1-aT/2}{1+aT/2}} \qquad \text{(E6.5.1)}$$

This implies that the s-plane pole is mapped into the z-plane through the BLT as

$$s = s_p = -a \rightarrow z = z_p \overset{(6.3.7)}{=} \frac{1 + s_p T/2}{1 - s_p T/2} = \frac{1 - aT/2}{1 + aT/2} \qquad \text{(E6.5.2)}$$

$$\left(= \frac{e^{s_p T/2}}{e^{-s_p T/2}} \cong e^{s_p T} \text{ for } |s_p T| << 1 \right)$$

The cutoff frequency of this discrete-time equivalent can be found as

$$|G_{bl}[z]||_{z=e^{j\omega T}} = \left| \frac{a}{\frac{2(e^{j\omega T/2} - e^{-j\omega T/2})}{T(e^{j\omega T/2} + e^{-j\omega T/2})} + a} \right| = \left| \frac{a}{j\frac{2}{T}\tan\left(\frac{\omega T}{2}\right) + a} \right|$$

$$\equiv \frac{1}{\sqrt{2}};\ \omega_{D,c} = \frac{2}{T}\tan^{-1}\left(\frac{\omega_{A,c} T}{2}\right) \qquad \text{(E6.5.3)}$$

Remark 6.8 Mapping of Stability Region and Frequency Transformation by BLT

From Eq. (6.3.7), it can be seen that the $j\omega$-axis (the stability boundary) in the s-plane is mapped to

$$z \underset{s=j\omega}{\overset{(6.3.7)}{=}} \frac{1 + j\omega T/2}{1 - j\omega T/2} = \frac{\sqrt{1^2 + (\omega T/2)^2}}{\sqrt{1^2 + (\omega T/2)^2}} < \left(\tan^{-1}\left(\frac{\omega T}{2}\right) - \tan^{-1}\left(\frac{-\omega T}{2}\right) \right)$$

$$= e^{j2\tan(\omega T/2)} = e^{j\Omega} \qquad (6.3.8)$$

which describes the unit circle itself (see Fig. 6.7(b)). It is implied that the BLT always maps stable/unstable analog systems into stable/unstable discrete-time equivalents. However, since the entire $j\omega$-axis in the s-plane maps exactly once onto the unit circle in the z-plane, the digital frequency response will be distorted (warped) from the analog frequency response while no frequency aliasing occurs in the frequency response, i.e., the analog frequency response in the high frequency range is not wrapped up into the digital frequency response in the low frequency (see Fig. 6.8). We set $z = e^{j\Omega} = e^{j\omega_D T}$ with $\Omega = \omega_D T$ in Eq. (6.3.8) to get the

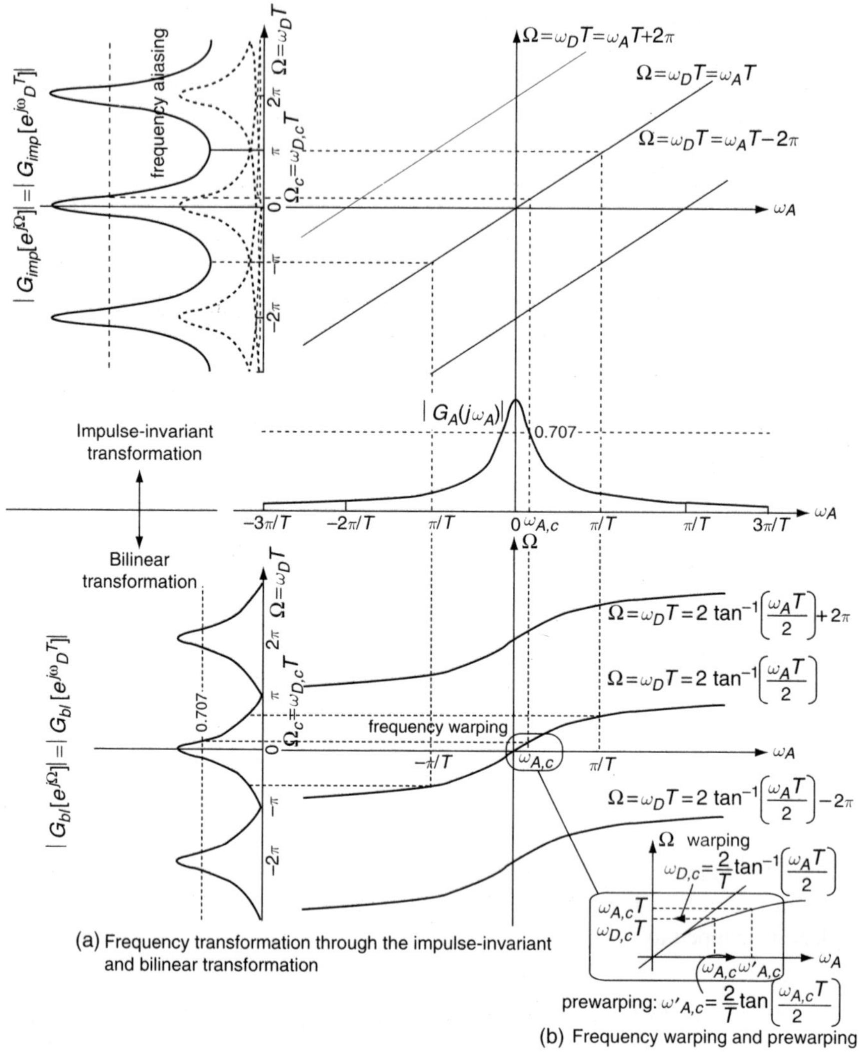

Fig. 6.8 Frequency transformation

relationship between the digital frequency (Ω or $\omega_D = \Omega/T$) in the z-plane and the analog frequency ω_A in the s-plane as

$$e^{j\Omega} \overset{(6.3.8)}{=} e^{j2\tan^{-1}(\omega_A T/2)};$$

$$\Omega \equiv \omega_D T = 2\tan^{-1}\left(\frac{\omega_A T}{2}\right); \ \omega_D = \frac{2}{T}\tan^{-1}\left(\frac{\omega_A T}{2}\right) : \textit{warping} \qquad (6.3.9)$$

For low frequencies such that $|\omega_A| << 1 \rightarrow \tan^{-1}(\omega_A T/2) \cong \omega_A T/2$, this relationship collapses down to

$$\omega_D \cong \omega_A \tag{6.3.10}$$

On the other hand, the nonlinear compression for high frequencies is more apparent, causing the frequency response at high frequencies to be highly distorted through the bilinear transformation.

Remark 6.9 Prewarping

(1) To compensate for the frequency warping so that frequency characteristics of the discrete-time equivalent in the frequency range of interest is reasonably similar to those of the continuous-time system, we should prewarp the critical frequencies before applying the BLT, by

$$\omega'_A \leftarrow \frac{2}{T}\tan\left(\frac{\omega_A T}{2}\right) : \textit{prewarping} \tag{6.3.11}$$

so that frequency transformation (6.3.9) with prewarping becomes

$$\omega'_D \overset{(6.3.9)}{=} \frac{2}{T}\tan^{-1}\left(\frac{\omega'_A T}{2}\right) \equiv \omega_A \tag{6.3.12}$$

(2) The BLT with prewarping at an analog frequency ω_c can be performed by substituting

$$s \rightarrow \frac{2}{T}\frac{1-z^{-1}}{1+z^{-1}} \times \frac{\omega_c}{(2/T)\tan(\omega_c T/2)} = \frac{\omega_c}{\tan(\omega_c T/2)}\frac{1-z^{-1}}{1+z^{-1}}$$

$$\text{instead of } s \rightarrow \frac{2}{T}\frac{1-z^{-1}}{1+z^{-1}} \tag{6.3.13}$$

for s in $G_A(s)$ where $\omega_c(< \pi/T)$ is a critical (prewarp) frequency.

(3) The warping and prewarping (against warping) by the frequency scale conversion is depicted in Fig. 6.8. It is implied that, if we design an analog filter, i.e., determine $G_A(j\omega'_A)$ to satisfy the specifications on frequency ω'_A, the original frequency response specifications on ω_A will be satisfied by the digital filter $G_{bl}[z]$ that is to be obtained by applying the BLT to $G_A(s')$.

Figure 6.8 show the frequency transformation through BLT (accompanied with frequency warping, but no frequency aliasing) and that through impulse-invariant transformation (accompanied with frequency aliasing, but no frequency warping).

Example 6.6 Bilinear Transformation with Prewarping

For a continuous-time system with the system function $G_A(s) = a/(s+a)$ having a pole at $s = -a$ and cutoff frequency $\omega_{A,c} = a$, the bilinear transformation (BLT) with prewarping yields the following discrete system function:

$$G_{bl,pre}[z] \overset{(6.3.13)}{=} \left.\frac{a}{s+a}\right|_{s=\frac{a(1-z^{-1})}{\tan(aT/2)(1+z^{-1})}} = \frac{(2/T)\tan(aT/2)}{\frac{2(1-z^{-1})}{T(1+z^{-1})}+\frac{2}{T}\tan(aT/2)} \tag{E6.6.1}$$

This implies that the s-plane pole is mapped into the z-plane through the BLT as

$$s = s_p = -a \rightarrow z = z_p \overset{(6.3.13)}{=} \frac{1+s_p\tan(aT/2)/a}{1-s_p\tan(aT/2)/a}$$
$$\left(\cong \frac{1-aT/2}{1+aT/2} \cong \frac{e^{s_pT/2}}{e^{-s_pT/2}} \cong e^{s_pT} \text{ for } |s_pT| << 1\right) \tag{E6.6.2}$$

Note that $G_A(s)$ and its discrete-time equivalent obtained by the BLT with prewarping have the same cutoff frequency (half-power point) $\omega_D = a$ at which

$$|G_A(j\omega_A)|\,|_{\omega_A=a} = \left|G_{bl,pre}[e^{j\omega_DT}]\right|\Big|_{\omega_D=a} = \frac{1}{\sqrt{2}} = -3\text{ [dB]} \tag{E6.6.3}$$

6.3.6 Pole-Zero Mapping – Matched z-Transform [F-1]

This technique consists of a set of heuristic rules given below.

(1) All poles and all finite zeros of $G_A(s)$ are mapped according to $z = e^{sT}$.
(2) All zeros of $G_A(s)$ at $s = \infty$ are mapped to the point $z = -1$ in $G[z]$. (Note that $s = j\infty$ and $z = -1 = e^{j\pi}$ represent the highest frequency in the s-plane and in the z-plane, respectively.)
(3) If a unit delay in the digital system is desired for any reason (e.g., because of computation time needed to process each sample), one zero of $G_A(s)$ at $s = \infty$ is mapped into $z = \infty$ so that $G[z]$ is proper, that is, the order of the numerator is less than that of the denominator.
(4) The DC gain of $G_{pz}[z]$ at $z = 1$ is chosen to match the DC gain of $G_A(s)$ at $s = 0$ so that

$$G_D[z]|_{z=e^{j\omega_pT}=1} \equiv G_A(s)|_{s=j\omega_A=0}\,;\; G_{pz}[1] \equiv G_A(0)$$

Example 6.7 Pole-Zero Mapping

For a continuous-time system with the system function $G_A(s) = a/(s+a)$, we can apply the pole-zero mapping procedure as

$$\frac{a}{s+a}\;\begin{matrix}\text{zero at } s=\infty\\ \text{pole at } s=-a\end{matrix}\;\overset{\text{Rule 1, 2}}{\rightarrow}\;\begin{matrix}\text{zero at } z=-1\\ \text{pole at } z=e^{-aT}\end{matrix}\;\frac{z+1}{z-e^{-aT}}$$
$$\overset{\text{Rule 3}}{\rightarrow}\;\begin{matrix}\text{zero at } z=\infty,\\ \text{pole at } z=e^{-aT}\end{matrix}\;\frac{1}{z-e^{-aT}}\;\overset{\text{Rule 4}}{\rightarrow}\;G_{pz}[z]=\frac{1-e^{-aT}}{z-e^{-aT}} \tag{E6.7.1}$$

At the last stage, $(1-e^{-aT})$ is multiplied so that the DC gain is the same as that of the analog system: $G_{pz}[1] \equiv G_A(0) = 1$. This happens to be identical with Eq. (E6.2.2), which is obtained through the step-invariant (or z.o.h. equivalent) transformation.

Remark 6.10 DC Gain Adjustment

Irrespective of which transformation method is used, we often adjust the DC gain of $G[z]$ by multiplying a scaling factor to match the DC steady-state response:

$$G_D[z]|_{z=e^{j\omega_p T}=1} \equiv \quad G_A(s)|_{s=j\omega_A=0}\,;\; G_{pz}[1] \equiv G_A(0) \tag{6.3.14}$$

6.3.7 Transport Delay – Dead Time

If an analog system contains a pure delay of d s in time, it can be represented by a continuous-time model of the form

$$G(s) = H(s)e^{-sd} \tag{6.3.15}$$

If the time delay is an integer multiple of the sampling period T, say, $d = MT$, then the delay factor $e^{-sd} = e^{-sMT}$ can be mapped to z^{-M} (with M poles at $z = 0$). More generally, if we have $d = MT + d_1$ with $0 \le d_1 < T$, then we can write

$$e^{-sd} = e^{-sMT} e^{-sd_1} \tag{6.3.16}$$

With sufficiently high sampling rate such that $d_1 < T << 1$, we can make a rational approximation of

$$e^{-d_1 s} \cong \frac{1 - d_1 s/2}{1 + d_1 s/2} \tag{6.3.17}$$

Now we can substitute Eq. (6.3.16) with Eq. (6.3.17) into Eq. (6.3.15) and apply a discretization method to obtain the discrete-time equivalent.

6.4 Time and Frequency Responses of Discrete-Time Equivalents

In the previous section we have obtained several discrete-time equivalents for the continuous-time system $G_A(s) = a/(s+a)$. Figure 6.9(a1, a2, b1, and b2) show the time and frequency responses of the two discrete-time equivalents obtained by the z.o.h. mapping and BLT for the sampling period of $T = 0.5$ and $T = 0.1$ [s] where $a = 1$. The following program "sig06f09.m" can be used to get Figs. 6.9(a1) and (b1).

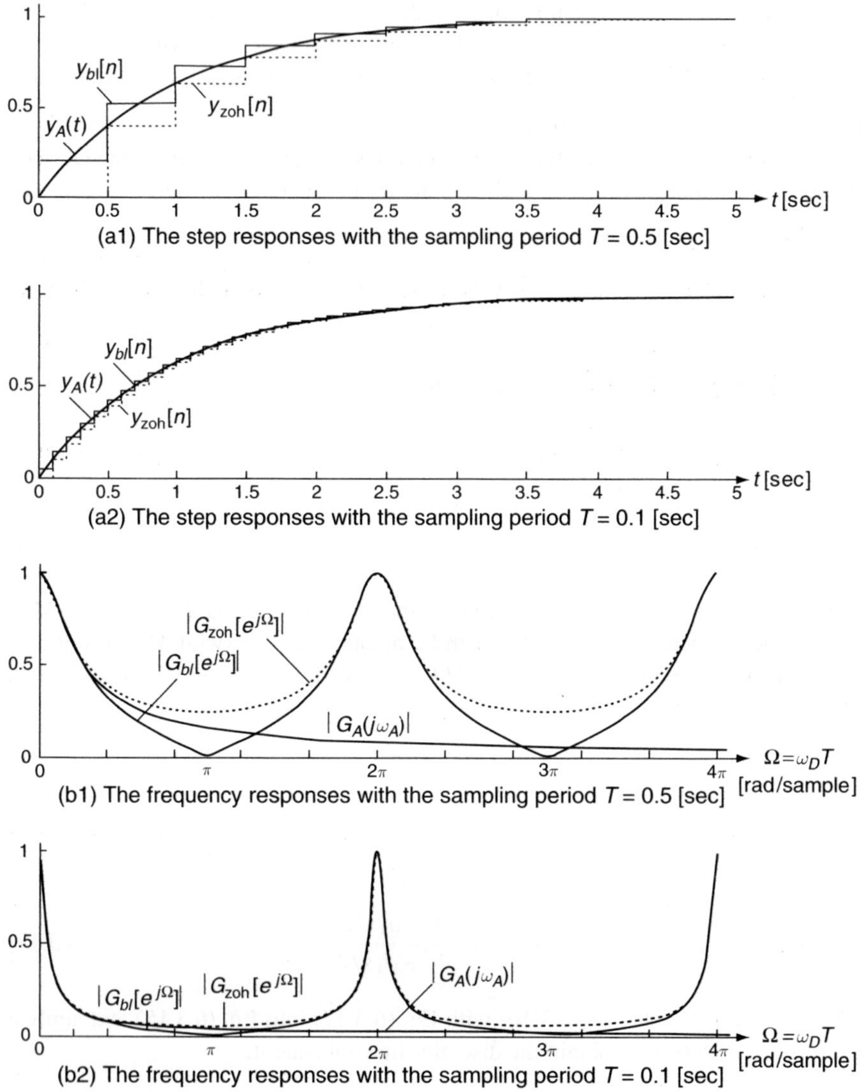

Fig. 6.9 The step and frequency responses of discrete–time equivalents

Note the following:

- All the discrete-time equivalents have the frequency characteristics that are periodic with period 2π in the digital frequency Ω or with period $2\pi/T$ in the analog frequency $\omega_D = \Omega/T$ and become similar to that of the continuous-time system for wider frequency range as the sampling period T gets shorter.
- The BLT gives a zero at $\omega_D = \pi/T$ corresponding to the zero at $z = -1 = e^{j\pi}$, assuring no aliasing problem in the frequency response.

- The BLT with prewarping gives the half-power (3dB) frequency at $\omega_D = a$ just like the continuous-time system.

```
%sig06f09.m
% Continuous-to-Discrete-time conversion of GA(s)=a/(s+a) with a=1
clear, clf
a=1; B=1; A=[1 a]; GAs=tf(B,A); % Analog transfer function
t_f=5; tt=[0:0.01:t_f]; % Time vector
fD=[0:0.001:2]; wD=2*pi*fD; % Digital/Analog frequency vector in Hz
T=0.5; % Sampling period
fA=fD/T; wA=wD/T; % Digital/Analog frequency vector in rad/s
GAw_mag= abs(freqs(B,A,wA)); % Analog frequency reponse |GA(jw)|
GD_zoh=c2d(GAs,T,'zoh'); [Bd_zoh,Ad_zoh]=tfdata(GD_zoh,'v')
GD_BLTp=c2d(GAs,T,'prewarp',a); [Bd_BLTp,Ad_BLTp]= tfdata(GD_BLTp,'v')
[yt,t]=step(GAs,t_f); % Step response
N=t_f/T; nT=[0:N-1]*T; yn_zoh=dstep(Bd_zoh,Ad_zoh,N);
subplot(411), plot(t,yt), hold on, stairs(nT,yn_zoh,':')
GDw_zoh_mag=abs(freqz(Bd_zoh,Ad_zoh,wD)); % Digital frequency response
subplot(413), plot(fD,GAw_mag, fD,GDw_zoh_mag,':')
```

6.5 Relationship Between s-Plane Poles and z-Plane Poles

In this section, we will derive the mathematical relationship between the s-plane and z-plane pole locations. We write the system function of a continuous-time second-order system in standard form as

$$G(s) = \frac{\omega_n^2}{s^2 + 2\zeta\omega_n s + \omega_n^2} = \frac{\omega_n^2}{(s+\sigma)^2 + \omega_d^2}, \quad 0 \le |\zeta| \le 1 \tag{6.5.1}$$

which has the poles at

$$s_{1,2} = -\zeta\omega_n \pm j\,\omega_n\sqrt{1-\zeta^2} = -\sigma \pm j\,\omega_d \text{ (Fig. 6.10(a))}$$

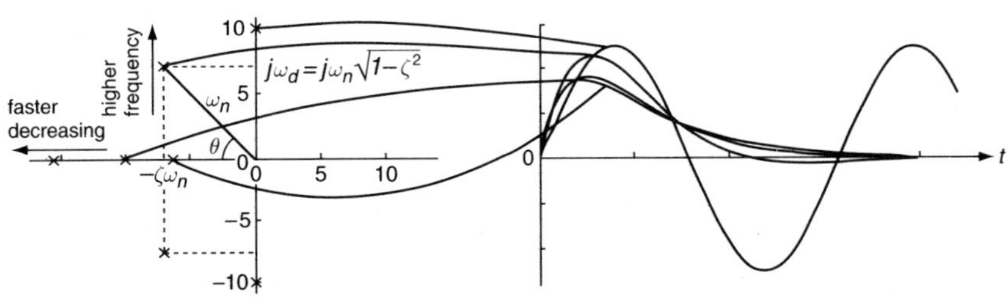

(a) The pole locations of an analog system (b) The impulse response corresponding to the pole locations

Fig. 6.10 Pole locations and the corresponding impulse responses where ζ, σ, ω_n, and ω_d are the *damping ratio*, *damping constant*, *natural frequency*, and *damped frequency*. Note that the impulse response of the system is

$$g(t) = \mathcal{L}^{-1}\{G(s)\} = \frac{\omega_n}{\sqrt{1-\zeta^2}} e^{-\sigma t} \sin(\omega_d t) \text{ (Fig. 6.10(b))} \tag{6.5.2}$$

The z-plane poles corresponding to the s-plane poles $s_{1,2}$ are located at

$$z_{1,2} = e^{sT}\Big|_{s_{1,2}} = e^{-\sigma T} \angle \pm \omega_d T = r\angle \pm \Omega \tag{6.5.3}$$

where

$$e^{-\zeta\omega_n T} = e^{-\sigma T} = r \quad \text{or} \quad \zeta\omega_n T = \sigma T = -\ln r \tag{6.5.4a}$$

$$\text{and} \quad \omega_n T\sqrt{1-\zeta^2} = \omega_d T = \Omega : \text{digital frequency} \tag{6.5.4b}$$

The discrete-time version of the impulse response can be written as

$$g[n] = g(nT) \overset{(6.5.2)}{=} \frac{\omega_n}{\sqrt{1-\zeta^2}} e^{-\sigma T n} \sin(\omega_d T n)$$

$$\overset{(6.5.4a,b)}{=} \frac{\omega_n}{\sqrt{1-\zeta^2}} r^n \sin(\Omega n) \tag{6.5.5}$$

Note that the larger the distance r of a pole from the origin is, the slower the output stemming from the pole converges; on the other hand, the larger the phase Ω (digital frequency) of a pole is, the shorter the discrete-time oscillation period of the output stemming from the pole Ω is.

Based on Eqs. (6.5.4a) and (6.5.4b), we can express the dampling ratio ζ and natural frequency ω_n in terms of the parameters of the pole location, i.e., the absolute value r and phase Ω of the pole(s) as

$$\frac{\zeta}{\sqrt{1-\zeta^2}} = \frac{-\ln r}{\Omega} \; ; \; \zeta = \frac{-\ln r}{\sqrt{\ln^2 r + \Omega^2}} \tag{6.5.6}$$

$$\omega_n = \frac{1}{T}\sqrt{\ln^2 r + \Omega^2} \tag{6.5.7}$$

The time constant of the pole(s) is

$$\tau = \frac{1}{\sigma} = \frac{1}{\zeta\omega_n} = \frac{-T}{\ln r} \tag{6.5.8}$$

Thus, given the complex pole locations $z = r\angle \pm \Omega$ in the z-plane, we can find the damping ratio ζ, natural frequency ω_n, and time constant τ. If we have sufficiently high sampling rate so that $T << \tau$, the poles of the discrete-time system are placed

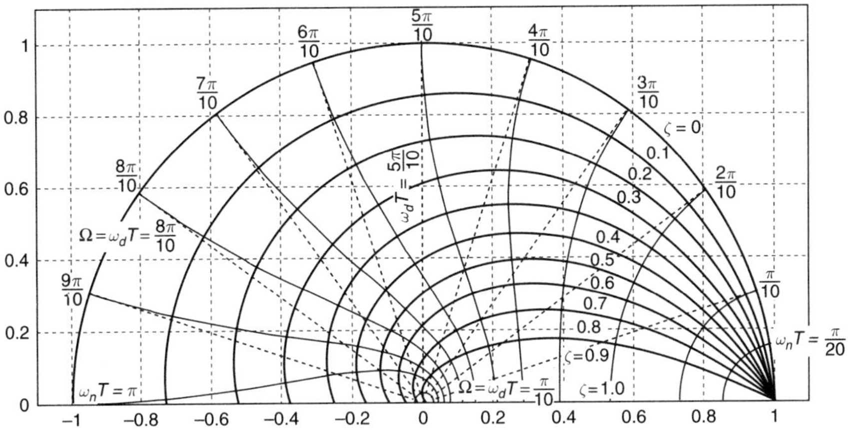

Fig. 6.11 The loci of the poles of a digital system for ζ = constant, $\omega_n T$ = constant, and $\omega_d T$ = constant

in the vicinity of $z = 1$. Note that high sampling rate does not make the response slower, but only makes the sampling interval shorter.

Figure 6.11 gives a picture of how the s-plane and z-plane poles are related where the loci of pole locations for constant damping ratio ζ are logarithmic spirals (with decreasing amplitude $r = e^{-\zeta\omega_n T}$ as $\omega_n T$ increases) and the loci of pole locations for constant $\omega_n T$ are drawn at right angles to the spirals for ζ = constant.

6.6 The Starred Transform and Pulse Transfer Function

6.6.1 The Starred Transform

We define the *starred transform* $X_*(s)$ of a signal $x(t)$ to be the Laplace transform of the impulse-modulated version $x_*(t) = x(t)\delta_T(t)$ representing the output signal of an ideal sampler with sampling period T. Using Eq. (2.1.10), we can express the ideal sampler output as

$$x_*(t) = x(t)\delta_T(t) = \frac{1}{T}\sum_{k=-\infty}^{\infty} x(t)\, e^{j\,k\omega_s t} \text{ with } \omega_s = \frac{2\pi}{T} \tag{6.6.1}$$

Taking the Laplace transform of the both sides and using the complex translation property (Table B.7(3)), we obtain the bilateral starred transform of $x(t)$ as

$$X_*(s) = \frac{1}{T}\sum_{k=-\infty}^{\infty} X(s - j\,k\omega_s) = \frac{1}{T}\sum_{k=-\infty}^{\infty} X(s + j\,k\omega_s) \tag{6.6.2}$$

Also, from Appendix III of [P-1], we borrow the following expressions for the unilateral starred transform:

$$X_*(s) = \sum_{m=0}^{\infty} x(mT)\, e^{-mTs} = X[z]|_{z=e^{Ts}} \tag{6.6.3a}$$

$$X_*(s) = \frac{1}{T}\sum_{k=-\infty}^{\infty} X(s+jk\omega_s) + \frac{1}{2}\sum_{n=0}^{\infty} \Delta\, x(nT)\, e^{-nTs} \tag{6.6.3b}$$

where $\Delta x(nT) = x(nT^+) - x(nT^-)$ is the amplitude of the discontinuity of $x(t)$ at $t = nT$.

The bilateral and unilateral starred transforms have the following properties:

<Properties of the starred transform>

1. They are periodic with period of $j\omega_s$ in s:

$$X_*(s+jm\omega_s) = \sum x(nT)\, e^{-nT(s+jm\omega_s)} = \sum x(nT)\, e^{-nTs} = X_*(s) \tag{6.6.4}$$

since $e^{-jn\,m\omega_s T} = e^{-jn\,m2\pi} = 1$ for any integer m.

2. If they have a pole at $s = s_1$, then they will also have poles at $s = s_1 + jm\omega_s,\ m = 0,\ \pm1,\ \pm2,\ \cdots$:

$$X_*(s) = \frac{1}{T}\{X(s)+X(s+j\omega_s)+X(s-j\omega_s)+X(s+j2\omega_s)+X(s-j2\omega_s)+\cdots \tag{6.6.5}$$

6.6.2 The Pulse Transfer Function

Consider the sampled-data systems depicted in Fig. 6.12 where the one has a sampled input and the other has a continuous-time input. Under the assumption of zero initial conditions, the output transform of system (a) can be written as

$$Y(s) = G(s)X_*(s) \tag{6.6.6}$$

If $x(t)$ and $y(t)$ are continuous at all sampling instants, we can use Eq. (6.6.2) to take the starred transform of Eq. (6.6.6) as

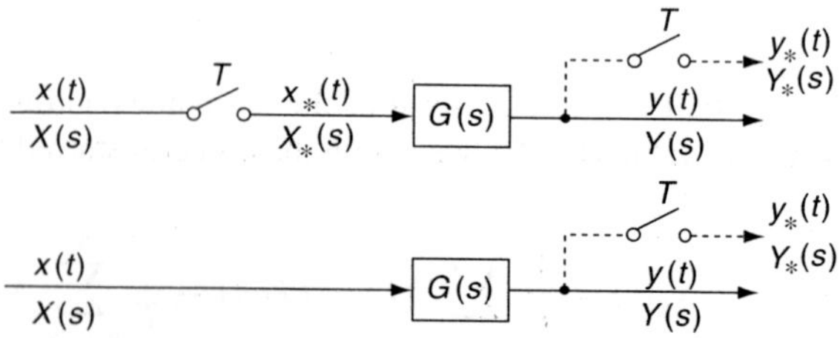

Fig. 6.12 Sampled–data systems

$$Y_*(s) \overset{(6.6.6)}{=} \{G(s)X_*(s)\}_* \overset{(6.6.2)}{=} \frac{1}{T}\sum_{k=-\infty}^{\infty} G(s+jk\omega_s)X_*(s+jk\omega_s)$$

$$\overset{(6.6.4)}{=} \frac{1}{T}\sum_{k=-\infty}^{\infty} G(s+jk\omega_s)X_*(s);\ Y_*(s) = G_*(s)X_*(s) \tag{6.6.7}$$

where $G_*(s)$ is called the *pulse transfer function*. This implies that, when taking the starred transform, we can factor out the existing starred transform. If we replace e^{Ts} by z in the starred transform, we will get the z-transform

$$Y[z] = G[z]\,X[z] \tag{6.6.8}$$

which describes the input-output relationship in the z-domain.

In contrast with this, the sampled output of system (b) in Fig. 6.12 can be expressed as

$$Y_*(s) = \{G(s)X(s)\}_* \neq G_*(s)X_*(s) \tag{6.6.9a}$$

$$Y[z] = \overline{GX}[z] \neq G[z]\,X[z] \tag{6.6.9b}$$

This implies that, if the input is applied to a continuous-time system before being sampled, the input transform can never be factored out to derive the transfer function.

6.6.3 Transfer Function of Cascaded Sampled-Data System

Consider the cascaded sampled-data systems depicted in Fig. 6.13. Each of the three sampled-data systems has the input-output relationship as below:

(a) $Y_*(s) = \{G_2(s)V_*(s)\}_* \overset{(6.6.7)}{=} G_{2*}(s)V_*(s) = G_{2*}(s)G_{1*}(s)X_*(s)$

$$Y[z] = G_2[z]G_1[z]X[z] \tag{6.6.10a}$$

(b) $Y_*(s) = \{G_2(s)G_1(s)X_*(s)\}_* \overset{(6.6.7)}{=} \{G_2(s)G_1(s)\}_*X_*(s) = \overline{G_2G_1}_*(s)X_*(s)$

$$Y[z] = \overline{G_2G_1}[z]\,X[z] \tag{6.6.10b}$$

(c) $Y_*(s) = \{G_2(s)V_*(s)\}_* \overset{(6.6.7)}{=} G_{2*}(s)V_*(s) = G_{2*}(s)\{G_1(s)X(s)\}_*$

$$= G_{2*}(s)\overline{G_1X}_*(s);\ Y[z] = G_2[z]\overline{G_1X}[z] \tag{6.6.10c}$$

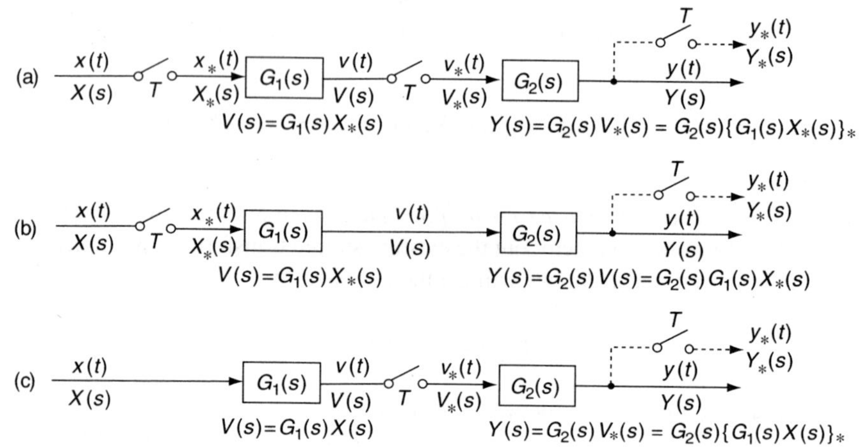

Fig. 6.13 Cascaded sampled–data systems

6.6.4 Transfer Function of System in A/D-$G[z]$-D/A Structure

Consider the sampled-data system containing an A/D-$G[z]$-D/A structure or a S/H device depicted in Fig. 6.14. Both of the two systems (a) and (b) have the same transfer function

$$\frac{Y[z]}{X[z]} = G_{ho}G_2[z]G_1[z] \tag{6.6.11}$$

since

$$V[z] = G_1[z]X[z] \;\;;\;\; V[e^{Ts}] = G_1[e^{Ts}]X[e^{Ts}] \;;\;\; V_*(s) = G_{1*}(s)X_*(s)$$

$$\bar{V}(s) = \frac{1-e^{-Ts}}{s}V_*(s) = \frac{1-e^{-Ts}}{s}G_{1*}(s)X_*(s)$$

$$Y(s) = G_2(s)\bar{V}(s) = \frac{1-e^{-Ts}}{s}G_2(s)G_{1*}(s)X_*(s)$$

$$Y[z] = Z\left\{\frac{1-e^{-Ts}}{s}G_2(s)\right\}G_1[z]X[z] = \overline{G_{ho}G_2}[z]\;G_1[z]\;X[z]$$

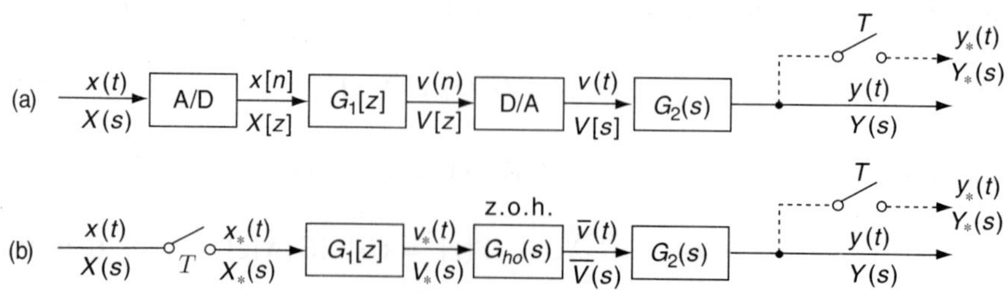

Fig. 6.14 Sampled–data system containing an A/D-$G[z]$-D/A structure or a S/H device

where

$$\overline{G_{ho}G_2}[z] = \mathcal{Z}\left\{(1 - e^{-Ts})\frac{G_2(s)}{s}\right\} = (1 - z^{-1})\,\mathcal{Z}\left\{\frac{1}{s}G_2(s)\right\}:$$

the z.o.h. equivalent of $G_2(s)$

(cf.) Note that the DAC usually has a data-hold device at its output, which allows us to model it as a zero-order-hold device.

Problems

6.1 Z.O.H. Equivalent

(a) Show that the zero-order-hold equivalent of $G(s) = 1/s(s+1)$ with sampling period T is

$$G(s) = \frac{1}{s(s+1)} \rightarrow G_{\text{zoh}}[z] = \frac{(T - 1 + e^{-T})z + 1 - e^{-T} - Te^{-T}}{(z-1)(z - e^{-T})} \tag{P6.1.1}$$

(b) Show that the zero-order-hold equivalent of $G(s) = 2/(s^2 + 2s + 2)$ with sampling period T is

$$G(s) = \frac{2}{(s+1)^2 + 1} \rightarrow$$

$$G_{\text{zoh}}[z] = \frac{(1 - e^{-T}(\cos T + \sin T))z + e^{-2T} - e^{-T}(\cos T - \sin T)}{z^2 - 2z\,e^{-T}\cos T + e^{-2T}} \tag{P6.1.2}$$

(c) Use the following MATLAB statements to find the discrete-time equivalent of analog system (b) through the BLT (bilinear transformation) with sampling period $T = 0.1$ [s] and critical frequency $\omega_p = \sqrt{2}$ [rad/s].

```
>>B=2; A=[1 2 2]; T=0.1; wp=sqrt(2);
>>GAs= tf(B,A);
>>Gz_BLT_prewarp= c2d(GAs,T,'prewarp',wp);
>>[Bd_BLT_p,Ad_BLT_p]= tfdata(Gz_BLT_prewarp,'v')
```

Also support the above results of (a) and (b) by completing and running the following MATLAB program "sig06p_01.m", which computes the z.o.h. equivalents of analog system (a) and (b) with sampling period $T = 0.1$ [s].

```
%sig06p_01.m
clear, clf
B=1; A=[1 1 0]; T=0.1; e_T= exp(-T); e_2T= exp(-2*T);
GAs= tf(B,A); % Analog transfer function
Gz_zoh= c2d(GAs,T); % z.o.h. equivalent
[Bd_zoh,Ad_zoh]= tfdata(Gz_zoh,'v')
Bd= [(T-1+e_T) 1-e_T-T*e_T], Ad= [1 -1-e_T e_T] % (P6.1.1)
```

(d) Referring to the Simulink block diagram of Fig. P6.1, perform the Simulink simulation to obtain the step responses of the analog system (b) and its two discrete-time equivalents, one through z.o.h. and one through BLT.

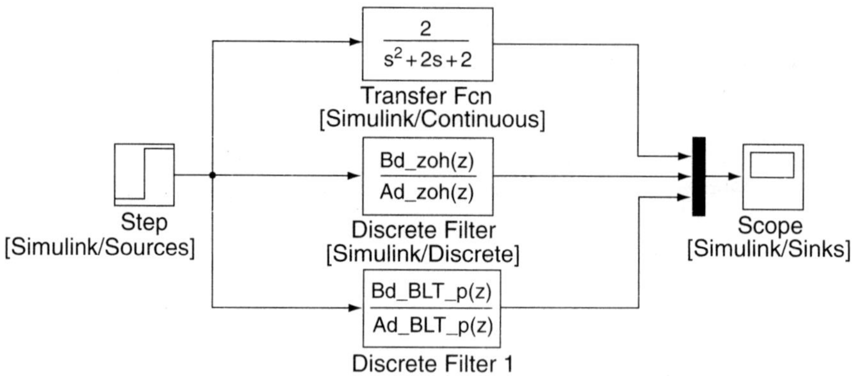

Fig. P6.1 Simulink block diagram for Problem 6.1

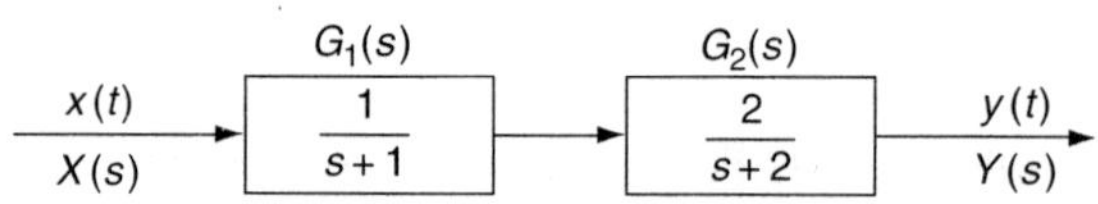

Fig. P6.2

6.2 Step-Invariant Transformation
Consider the system depicted in Fig. P6.2.

(a) Find the step-invariant equivalent $G_1[z]$ of $G_1(s)$.
(b) Find the step-invariant equivalent $G_2[z]$ of $G_2(s)$.
(c) Find the step-invariant equivalent $G[z]$ of $G_1(s)G_2(s)$.
(d) Is it true that $G[z] = G_1[z]G_2[z]$?

6.3 Bilinear Transformation without prewarping or with prewarping
Consider the second-order analog system whose transfer function is

$$G_A(s) = \frac{\omega_b s}{s^2 + \omega_b s + \omega_p^2} = \frac{2s}{s^2 + 2s + 1} \text{ with } \omega_b = 2, \quad \omega_p = 1 \quad \text{(P6.3.1)}$$

Note that the frequency response $G_A(j\omega)$, its peak frequency, and two 3dB-frequencies are

$$G_A(j\omega) = \frac{j\omega\omega_b}{(j\omega)^2 + j\omega\omega_b + \omega_p^2} = \frac{j2\omega}{(1-\omega^2) + j2\omega} \quad \text{(P6.3.2)}$$

$$\omega_p = 1, \omega_{3B,l} = \frac{-\omega_b + \sqrt{\omega_b^2 + 4\omega_p^2}}{2} = -1 + \sqrt{2},$$

$$\omega_{3B,u} = \frac{\omega_b + \sqrt{\omega_b^2 + 4\omega_p^2}}{2} = 1 + \sqrt{2} \quad \text{(P6.3.3)}$$

(a) Find the discrete-time equivalent $G_D[z]$ using the BLT with sampling period of $T = 1$ [s] and no prewarping. Also find the peak frequency Ω_p and lower/upper 3dB frequencies $\Omega_{3B,l}$ and $\Omega_{3B,u}$ of the digital frequency response $G_D[e^{j\Omega}]$. How are they related with ω_p, $\omega_{3B,l}$, and $\omega_{3B,u}$? You can modify and use the MATLAB program "sig06p_03.m" below.
(b) Find the discrete-time equivalent $G_D[z]$ using the BLT with sampling period of $T = 1$ [s] and prewarping three times, once at $\omega_p = 1$, once at $\omega_{3B,l}$, and once at $\omega_{3B,u}$. Also for each $G_D[z]$, find the peak frequency Ω_p and lower/upper 3dB frequencies $\Omega_{3B,l}$ and $\Omega_{3B,u}$ of the digital frequency response $G_D[e^{j\Omega}]$ and tell which frequency is closest to the corresponding analog frequency in terms of the basic realtionship $\Omega = \omega T$ between the analog and digital frequencies.

```
%sig06p_03.m
B=[2 0]; A=[1 2 1]; wp=1; w_3dB1= -1+sqrt(2); w_3dB2= 1+sqrt(2);
GAs= tf(B,A); T=1; % Analog transfer function and sampling period
Gz= c2d(GAs,T,'tustin'); % BLT without prewarping
W=[0:0.00001:1]*pi; GDW_mag= abs(freqz(BD,AD,W));
[GDW_max,i_peak]= max(GDW_mag); Wp= W(i_peak)
tmp= abs(GDW_mag-GDW_max/sqrt(2));
[tmp_3dB1,i_3dB1]= min(tmp(1:i_peak));
[tmp_3dB2,i_3dB2]= min(tmp(i_peak+1:end));
W_3dB1= W(i_3dB1); W_3dB2= W(i_peak+i_3dB2);
```

6.4 Pole Locations and Time Responses
Consider an analog system having the system function $G_a(s) = 1/(s^2 + 2s + 5)$ and its two z.o.h. equivalents $G_{zoh1}[z]$ and $G_{zoh2}[z]$, each with sampling period $T = 0.2$ and $T = 0.1$, respectively.

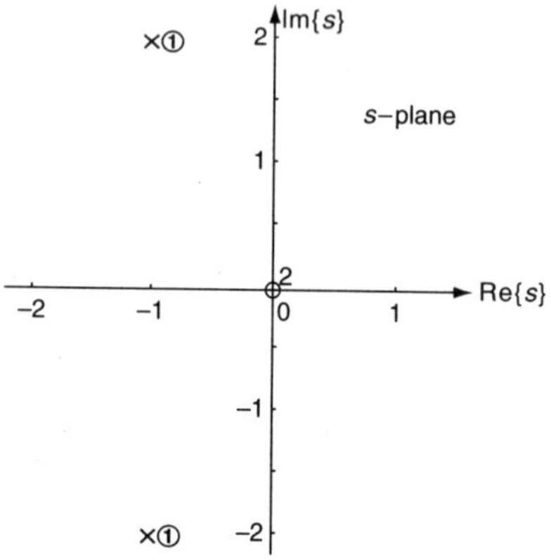

(a) The pole/zero plot of $G_a(s)$

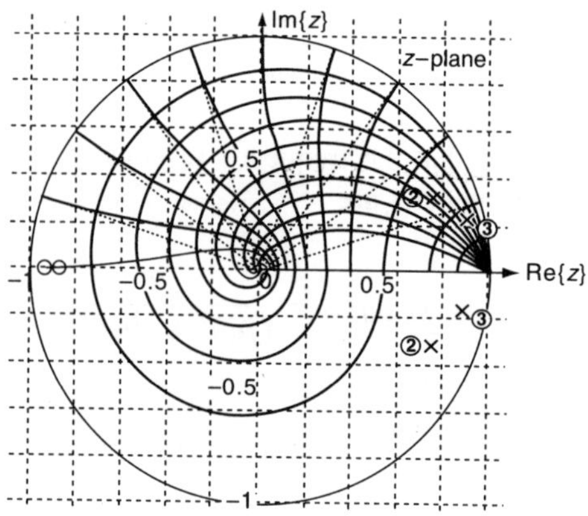

(b) The pole/zero plot of $G_{zoh1}[z]$ and $G_{zoh2}[z]$

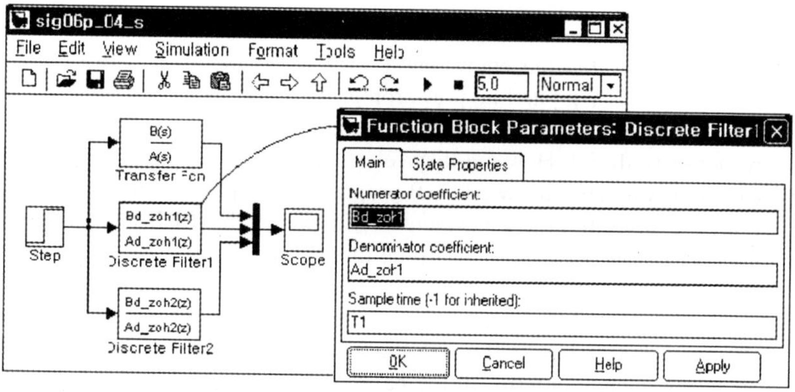

(c) The Simulink model window

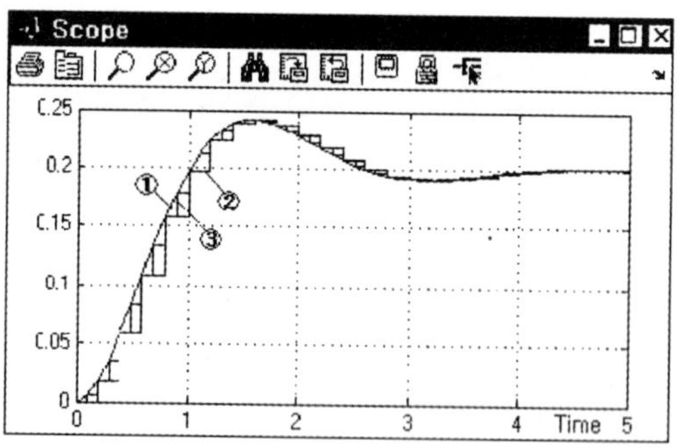

(d) The Scope window showing the simulation results

Fig. P6.4 Pole/zero plots and Simulink simulation

(a) Note that the pole-zero plot of $G_a(s)$ is depicted in Fig. P6.4(a). Referring to the pole/zero plots shown in Fig. P6.4(b), choose the pole locations $(r\angle\Omega)$ of $G_{\text{zoh1}}[z]$ and $G_{\text{zoh2}}[z]$ from ② and ③. The digital frequency $\omega_n T$ of the output corresponding to pole location ② is expected to be two times as high as that of the output corresponding to pole location ③ in proportion to the sampling period T where ω_n is determined from the pole location of the analog system $G_a(s)$. Does it mean that the output corresponding to pole location ② oscillates two times as fast as that corresponding to pole location ③?

(b) Referring to Fig. P6.4(c), perform the Simulink simulation for $G_a(s)$, $G_{\text{zoh1}}[z]$, and $G_{\text{zoh2}}[z]$ to get the simulation results as Fig. P6.4(d) and choose the output waveforms of $G_a(s)$, $G_{\text{zoh1}}[z]$ and $G_{\text{zoh2}}[z]$ from ①, ②, and ③.

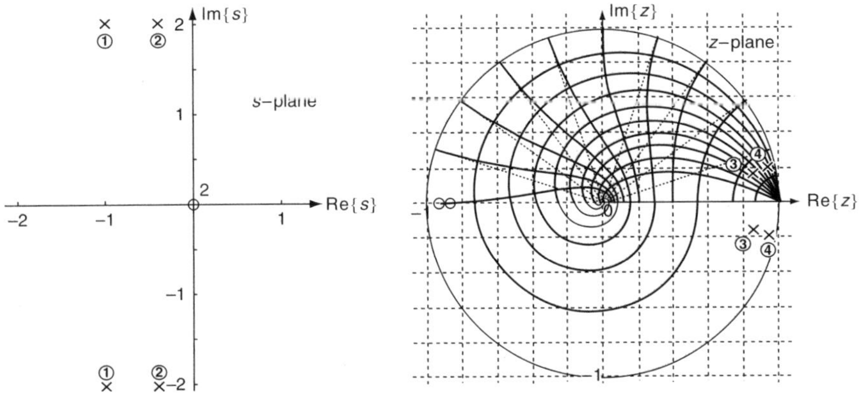

(a) The pole/zero plot of $G_{a1}(s)$ and $G_{a2}(s)$ (b) The pole/zero plot of $G_{\text{zoh1}}[z]$ and $G_{\text{zoh2}}[z]$

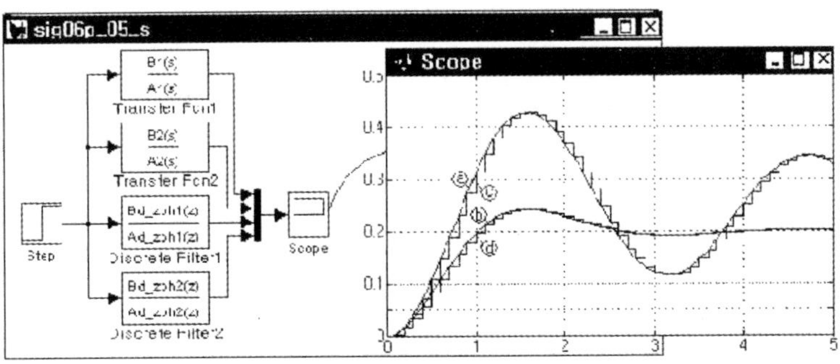

(c) The Simulink model window and the simulation results seen in the Scope window

Fig. P6.5 Pole/zero plots and Simulink simulation

6.5 Pole Locations and Time Responses

Consider two analog systems, each having the system function $G_1(s) = 1/(s^2 + 2s + 5)$ and $G_2(s) = 1/(s^2 + 0.4s + 4.04)$, respectively and their z.o.h. equivalents $G_{\text{zoh1}}[z]$ and $G_{\text{zoh2}}[z]$ with sampling period $T = 0.1$.

(a) Note that the pole-zero plots of $G_1(s)$ and $G_2(s)$ are depicted in Fig. P6.5(a). Referring to the pole/zero plots shown in Fig. P6.5(b), choose the pole locations of $G_{zoh1}[z]$ and $G_{zoh2}[z]$ from ③ and ④. Which converges faster of the outputs stemming from the poles ③ and ④? Note that the poles ③ are closer to the origin than the poles ④, i.e., $r_3 < r_4$.

(b) Referring to Fig. P6.5(c), perform the Simulink simulation for $G_1(s)$, $G_2(s)$, $G_{zoh1}[z]$, and $G_{zoh2}[z]$ and choose the output waveforms of the four systems from ⓐ, ⓑ, ⓒ, and ⓓ.

6.6 Pole-Zero Mapping (Matched z-Transform) of an Analog System with Delay
Using the pole-zero mapping (matched z-transform), find the discrete-time equivalent of an analog system

$$G(s) = \frac{2}{s+2} e^{-0.95s} \tag{P6.6.1}$$

with sampling period $T = 1/4$ [s].

Chapter 7
Analog and Digital Filters

Contents

This chapter introduces how to use MATLAB for designing analog and digital filters such that the given specification on the frequency response is satisfied. Especially for analog filters, the MATLAB routines to perform circuit-level design are also introduced.

7.1 Analog Filter Design

This section has been excerpted from Sect. 8.6 of [Y-2] (Yang, Won Y. and Seung C. Lee, *Circuit Systems with MATLAB and PSpice*, John Wiley & Sons, Inc., New Jersey, 2007.). Fig 7.1(a)/(b)/(c)/(d) show typical lowpass/bandpass/bandstop/ highpass filter specifications on their log-magnitude, $20\log_{10}|G(j\omega)|$ [dB], of frequency response. The filter specification can be described as follows:

$$20\log_{10}|G(j\omega_p)| \geq -R_p[\text{dB}] \text{ for the passband} \tag{7.1.1a}$$

$$20\log_{10}|G(j\omega_s)| \leq -A_s[\text{dB}] \text{ for the stopband} \tag{7.1.1b}$$

where ω_p, ω_s, R_p, and A_s are referred to as the *passband edge frequency*, the *stopband edge frequency*, the *passband ripple*, and the *stopband attenuation*, respectively. The most commonly used analog filter design techniques are the Butterworth, Chebyshev I, II, and elliptic ones ([K-2], Chap. 8). MATLAB has the built-in functions `butt()`, `cheby1()`, `cheby2()`, and `ellip()` for designing the four types of analog/digital filter. As summarized below, `butt()` needs the 3dB cutoff frequency while `cheby1()` and `ellip()` get the critical passband edge

W.Y. Yang et al., *Signals and Systems with MATLAB®*,
DOI 10.1007/978-3-540-92954-3_7, © Springer-Verlag Berlin Heidelberg 2009

frequency and `cheby2()` the critical stopband edge frequency as one of their input arguments. The parametric frequencies together with the filter order can be predetermined using `buttord()`, `cheb1ord()`, `cheb2ord()`, and `ellipord()`. The frequency input argument should be given in two-dimensional vector for designing BPF or BSF. Also for HPF/BSF, the string `'high'`/`'stop'` should be given as an optional input argument together with `'s'` for analog filter design.

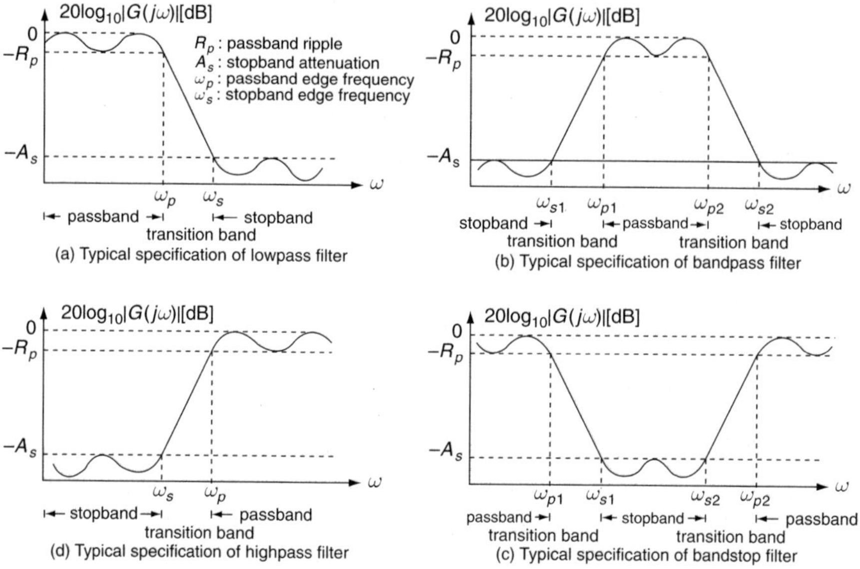

Fig. 7.1 Specification on the log-magnitude of the frequency response of an analog filter

```
function [N,wc] = buttord(wp,ws,Rp,As,opt)
% For opt='s', it selects the lowest order N and cutoff frequency wc of analog Butterworth filter.
%  that has the passband ripple<=Rp[dB] and stopband attenuation>=As[dB]
%  for the passband edge frequency wp and stopband edge frequency ws.
% Note that for the BPF/BSF, the passband edge frequency wp and stopband edge frequency ws should be
%  given as two-dimensional vectors like [wp1 wp2] and [ws1 ws2].

function [B,A]=butter(N,wc,opt)
% It designs a digital/analog Butterworth filter, returning the numerator/denominator of system function.
% [B,A]=butter(N,wc,'s') for the analog LPF of order N with the cutoff frequency wc[rad/s]
%  butter(N,[wc1 wc2],'s') for the analog BPF of order 2N with the passband wc1<w<wc2[rad/s]
%  butter(N,[wc1 wc2],'stop','s') for the analog BSF of order 2N with the stopband wc1<w<wc2[rad/s]
%  butter(N,wc,'high','s') for the analog HPF of order N with cutoff frequency wc[rad/s]
% Note that N and wc can be obtained from [N,wc]=buttord(wp,ws,Rp,As,opt).

function [B,A]=cheby1(N,Rp,wpc,opt)
% It designs a digital/analog Chebyshev type I filter with the passband ripple Rp[dB]
%   and the critical passband edge frequency wpc (Use Rp=0.5 as a starting point, if not sure).
% Note that N and wpc can be obtained from [N,wpc]=cheby1ord(wp,ws,Rp,As,opt).

function [B,A]=cheby2(N,As,wsc,opt)
% It designs a digital/analog Chebyshev type II filter with the stopband attenuation As[dB] down
%  and the critical stopband edge frequency wsc (Use As=20 as a starting point, if not sure).
% Note that N and wsc can be obtained from [N,wsc]=cheby2ord(wp,ws,Rp,As,opt).

function [B,A]=ellip(N,Rp,As,wpc,opt)
% It designs a digital/analog Elliptic filter with the passband ripple Rp, the stopband attenuation As,
%  and the critical passband edge frequency wpc (Use Rp=0.5[dB] & As=20[dB], if unsure).
% Note that N and wpc can be obtained from ellipord(wp,ws,Rp,As,opt).
```

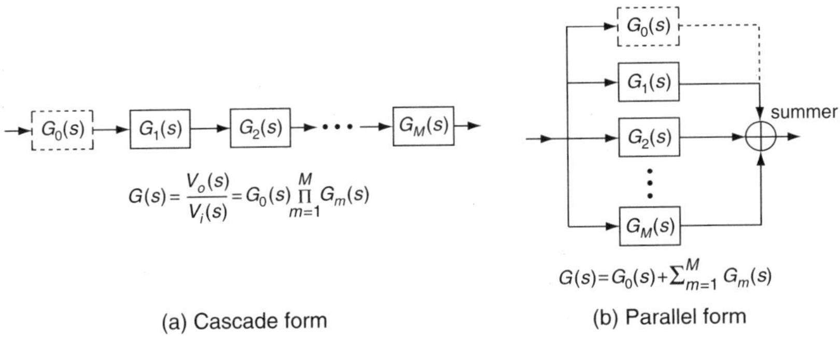

Fig. 7.2 Two realizations of an analog filter (system or transfer function)

The designed filter system functions are often factored into the sum or product of second-order sections called *biquads* (possibly with an additional first-order section in the case of an odd filter order) as

$$G(s) = K G_0(s) \prod_{m=1}^{M} G_m(s) = K \frac{b_{01}s + b_{02}}{s + a_{02}} \prod_{m=1}^{M} \frac{b_{m1}s^2 + b_{m2}s + b_{m3}}{s^2 + a_{m2}s + a_{m3}} \tag{7.1.2a}$$

with $\quad M = \text{floor}\left(\frac{N}{2}\right)$

$$G(s) = G_0(s) + \sum_{m=1}^{M} G_m(s) = \frac{b_{01}s + b_{02}}{s + a_{02}} + \sum_{m=1}^{M} \frac{b_{m1}s^2 + b_{m2}s + b_{m3}}{s^2 + a_{m2}s + a_{m3}} \tag{7.1.2b}$$

with $\quad M = \text{floor}\left(\frac{N}{2}\right)$

and then realized in *cascade* or *parallel* form, respectively as depicted in Fig. 7.2.

Rather than reviewing the design procedures, let us use the MATLAB functions to design a Butterworth lowpass filter, a Chebyshev I bandpass filter, a Chebyshev II bandstop filter, and elliptic highpass filter in the following example.

Example 7.1 Analog Filter Design Using the MATLAB Functions

Let us find the system functions of analog filters meeting the specifications given below.

(a) We are going to determine the system function of a *Butterworth* lowpass filter with

$$\omega_p = 2\pi \times 6000\,[\text{rad/s}],\ \omega_s = 2\pi \times 15000\,[\text{rad/s}],\ R_p = 2\,[\text{dB}],\ \text{and}\ A_s = 25\,[\text{dB}] \tag{E7.1.1}$$

First, we use the MATLAB function 'buttord()' to find the filter order N and the cutoff frequency ω_c at which $20\log_{10}|G(j\omega_c)| = -3$ [dB] by typing the following statements into the MATLAB command window:

```
>>wp=2*pi*6000; ws=2*pi*15000; Rp=2; As=25;
>>format short e, [N,wc]=buttord(wp,ws,Rp,As,'s')
   N = 4, wc = 4.5914e+004
```

We put these parameter values N and wc into the Butterworth filter design function 'butter()' as its first and second input arguments:

```
>>[Bb,Ab]=butter(N,wc,'s')
  Bb =       0             0             0             0             4.4440e+018
  Ab = 1.0000e+000   1.1998e+005   7.1974e+009   2.5292e+014   4.4440e+018
```

This means that the system function of the designed Butterworth LPF of order N=4 is

$$G(s) = \frac{4.444 \times 10^{18}}{s^4 + 1.1998 \times 10^5 s^3 + 7.1974 \times 10^9 s^2 + 2.5292 \times 10^{14} s + 4.444 \times 10^{18}} \tag{E7.1.2}$$

We can find the cascade and parallel realizations of this system function by typing the following statements into the MATLAB command window:

```
>>[SOS,K]=tf2sos(Bb,Ab); % cascade realization
>>Ns=size(SOS,1); >>Gm=K^(1/Ns), BBc=SOS(:,1:3), AAc=SOS(:,4:6)
  Gm = 2.1081e+009
  BBc = 0     0     1      AAc = 1.0000e+000   3.5141e+004   2.1081e+009
        0     0     1            1.0000e+000   8.4838e+004   2.1081e+009
>>[BBp,AAp]=tf2par_s(Bb,Ab)   % parallel realization (see Sect. E.15)
  BBp = 0    4.2419e+004    3.5987e+009
        0   -4.2419e+004   -1.4906e+009
  AAp = 1.0000e+000   8.4838e+004   2.1081e+009
        1.0000e+000   3.5141e+004   2.1081e+009
```

This means that the designed system function can be realized in cascade and parallel form as

$$G(s) = \frac{2.108 \times 10^9}{s^2 + 3.514 \times 10^4 s + 2.108 \times 10^9} \times \frac{2.108 \times 10^9}{s^2 + 8.484 \times 10^4 s + 2.108 \times 10^9} \tag{E7.1.3a}$$

$$G(s) = \frac{4.242 \times 10^4 s + 3.599 \times 10^9}{s^2 + 8.484 \times 10^4 s + 2.108 \times 10^9} - \frac{4.242 \times 10^4 s + 1.491 \times 10^9}{s^2 + 3.514 \times 10^4 s + 2.108 \times 10^9} \tag{E7.1.3b}$$

(b) We are going to determine the system function of a *Chebyshev* I bandpass filter with

$$\omega_{s1} = 2\pi \times 6000,\ \omega_{p1} = 2\pi \times 10000,\ \omega_{p2} = 2\pi \times 12000,$$
$$\omega_{s2} = 2\pi \times 15000\ [\text{rad/s}],\ R_p = 2\ [\text{dB}],\ \text{and}\ A_s = 25\ [\text{dB}] \tag{E7.1.4}$$

First, we use the MATLAB function 'cheb1ord()' to find the filter order N and the critical passband edge frequencies ω_{pc1} and ω_{pc2} at which the passband ripple condition is closely met, i.e., $20\log_{10}|G(j\omega_{pc})| = -R_p$ [dB] by typing the following statements:

```
>>ws1=2*pi*6e3; wp1=2*pi*1e4; wp2=2*pi*12e3; ws2=2*pi*15e3; Rp=2; As=25;
>>[N,wpc]=cheb1ord([wp1 wp2],[ws1 ws2],Rp,As,'s')
   N = 2, wpc = 6.2832e+004   7.5398e+004
```

We put the (half) filter order N, the passband ripple Rp, and the critical passband edge frequency vector wpc $= [\omega_{pc1}\ \omega_{pc2}]$ into the Chebyshev I filter design function 'cheby1()' as

```
>>[Bc1,Ac1]=cheby1(N,Rp,wpc,'s')
   Bc1 =          0          0  1.0324e+008           0           0
   Ac1 = 1.0000e+000  1.0101e+004  9.6048e+009 4.7853e+013  2.2443e+019
```

This means that the system function of the designed Chebyshev I BPF of order 2N=4 is

$$G(s) = \frac{1.0324 \times 10^8 s^2}{s^4 + 10101s^3 + 9.6048 \times 10^9 s^2 + 4.7853 \times 10^{13} s + 2.2443 \times 10^{19}} \tag{E7.1.5}$$

We can find the cascade and parallel realizations of this system function by typing the following statements into the MATLAB command window:

```
>>[SOS,K]=tf2sos(Bc1,Ac1); % cascade realization
>>Ns=size(SOS,1); Gm=K^(1/Ns), BBc=SOS(:,1:3), AAc=SOS(:,4:6)
   Gm = 1.0161e+004
   BBc = 0     0     1      AAc = 1.0000e+000  5.4247e+003  5.4956e+009
         1     0     1            1.0000e+000  4.6763e+003  4.0838e+009
>>[BBp,AAp]=tf2par_s(Bc1,Ac1)  % parallel realization
   BBp = 0   1.8390e+002  4.0242e+008
         0  -1.8390e+002 -2.9904e+008
   AAp = 1.0000e+000  5.4247e+003  5.4956e+009
         1.0000e+000  4.6763e+003  4.0838e+009
```

This means that the designed system function can be realized in cascade and parallel form as

$$G(s) = \frac{1.0161 \times 10^4 s}{s^2 + 5.425 \times 10^3 s + 5.496 \times 10^9} \times \frac{1.0161 \times 10^4 s}{s^2 + 4.676 \times 10^3 s + 4.084 \times 10^9} \tag{E7.1.6a}$$

$$G(s) = \frac{1.839 \times 10^2 s + 4.024 \times 10^8}{s^2 + 5.425 \times 10^3 s + 5.496 \times 10^9} - \frac{1.839 \times 10^2 s + 2.990 \times 10^8}{s^2 + 4.676 \times 10^3 s + 4.084 \times 10^9} \tag{E7.1.6b}$$

(c) We are going to determine the system function of a *Chebyshev* II bandstop filter with

$$\omega_{p1} = 2\pi \times 6000,\ \omega_{s1} = 2\pi \times 10000,\ \omega_{s2} = 2\pi \times 12000,$$
$$\omega_{p2} = 2\pi \times 15000\,[\text{rad/s}],\ R_p = 2\,[\text{dB}],\ \text{and}\ A_s = 25\,[\text{dB}] \tag{E7.1.7}$$

First, we use the MATLAB function 'cheb2ord()' to find the filter order N and the critical stopband edge frequencies ω_{sc1} and ω_{sc2} at which the stopband attenuation condition is closely met, i.e., $20\log_{10}|G(j\omega_{sc})| = -A_s$ [dB] by typing the following statements:

```
>>wp1=2*pi*6000; ws1=2*pi*10000; ws2=2*pi*12000; wp2=2*pi*15000;
>>Rp=2; As=25;
>>[N,wsc]=cheb2ord([wp1 wp2],[ws1 ws2],Rp,As,'s')
   N = 2, wsc = 6.2798e+004  7.5438e+004
```

We put the (half) filter order N, the stopband attenuation As, and the critical stopband edge frequency vector wsc $= [\omega_{sc1}\ \omega_{sc2}]$ into the Chebyshev II filter design function 'cheby2()' as

```
>>[Bc2,Ac2]=cheby2(N,As,wsc,'stop','s')
   Bc2 = 1.0000e+000  1.0979e-010  9.5547e+009  4.9629e-001  2.2443e+019
   Ac2 = 1.0000e+000  5.1782e+004  1.0895e+010  2.4531e+014  2.2443e+019
```

This means that the system function of the designed Chebyshev II BSF of order 2N = 4 is

$$G(s) = \frac{s^4 + 9.5547\times 10^9 s^2 + 4.9629\times 10^{-1} s + 2.2443\times 10^{19}}{s^4 + 51782 s^3 + 1.0895\times 10^{10} s^2 + 2.4531\times 10^{14} s + 2.2443\times 10^{19}} \tag{E7.1.8}$$

We can find the cascade and parallel realizations of this system function by typing the following statements into the MATLAB command window:

```
>>[SOS,K]=tf2sos(Bc2,Ac2); % cascade realization [BBc,AAc]=tf2cas(B,A)
>>Ns=size(SOS,1); Gm=K^(1/Ns), BBc=SOS(:,1:3), AAc=SOS(:,4:6)
   Gm = 1
   BBc = 1.0000e+000  7.7795e-011  5.3938e+009
         1.0000e+000  2.9104e-011  4.1609e+009
   AAc = 1.0000e+000  3.1028e+004  7.0828e+009
         1.0000e+000  2.0754e+004  3.1687e+009
>>[BBp,AAp]=tf2par_s(Bc2,Ac2)  % parallel realization
   BBp = 5.0000e-001 -1.5688e+004  3.4426e+009
         5.0000e-001 -1.0204e+004  1.6285e+009
   AAp = 1.0000e+000  3.1028e+004  7.0828e+009
         1.0000e+000  2.0754e+004  3.1687e+009
```

This means that the designed system function can be realized in cascade and parallel form as

$$G(s) = \frac{s^2 + 5.394\times 10^9}{s^2 + 3.103\times 10^4 s + 7.083\times 10^9} \times \frac{s^2 + 4.161\times 10^9}{s^2 + 2.075\times 10^4 s + 3.169\times 10^9} \tag{E7.1.9a}$$

$$G(s) = \frac{0.5s^2 - 1.569\times 10^4 s + 3.443\times 10^9}{s^2 + 3.103\times 10^4 s + 7.083\times 10^9} + \frac{0.5s^2 - 1.020\times 10^4 s + 1.6285\times 10^9}{s^2 + 2.075\times 10^4 s + 3.169\times 10^9} \tag{E7.1.9b}$$

(d) We are going to determine the system function of an *elliptic* highpass filter with

$$\omega_s = 2\pi\times 6000\,[\text{rad/s}],\ \omega_p = 2\pi\times 15000\,[\text{rad/s}],\ R_p = 2\,[\text{dB}],\ \text{and}\ A_s = 25\,[\text{dB}] \tag{E7.1.10}$$

First, we use the MATLAB function 'ellipord()' to find the filter order N and the critical passband edge frequency ω_{pc} at which $20\log_{10}|G(j\omega_{pc})| = -R_p$ [dB] by typing the following statements into the MATLAB command window:

```
>>ws=2*pi*6000; wp=2*pi*15000; Rp=2; As=25;
>>format short e, [N,wc]=ellipord(wp,ws,Rp,As,'s')
   N = 3, wc = 9.4248e+004
```

We put the parameter values N, Rp, As, and wc into the elliptic filter design function 'ellip()' as

```
>>[Be,Ae]=ellip(N,Rp,As,wc,'high','s')
  Be = 1.0000e+000  8.9574e-009  3.9429e+009 -5.6429e+002
  Ae = 1.0000e+000  2.3303e+005  1.4972e+010  1.9511e+015
```

This means that the system function of the designed elliptic HPF of order N = 3 is

$$G(s) = \frac{s^3 + 3.9429 \times 10^9 s - 5.6429 \times 10^2}{s^3 + 2.3303 \times 10^5 s^2 + 1.4972 \times 10^{10} s + 1.9511 \times 10^{15}} \quad \text{(E7.1.11)}$$

We can find the cascade and parallel realizations of this system function by typing the following statements into the MATLAB command window:

```
>>[SOS,K]=tf2sos(Be,Ae); % cascade realization
>>Ns=size(SOS,1); Gm=K^(1/Ns), BBc=SOS(:,1:3), AAc=SOS(:,4:6)
   Gm = 1.0000e+000
   BBc = 1.0000e+000 -1.4311e-007            0
         1.0000e+000  1.5207e-007  3.9429e+009
   AAc = 1.0000e+000  2.0630e+005            0
         1.0000e+000  2.6731e+004  9.4575e+009
>>[BBp,AAp]=tf2par_s(Be,Ae)   % parallel realization
   BBp = 5.0000e-001   -1.3365e+004    4.7287e+009
         0              5.0000e-001   -1.0315e+005
   AAp = 1.0000e+000    2.6731e+004    9.4575e+009
                   0    1.0000e+000    2.0630e+005
```

This means that the designed system function can be realized in cascade and parallel form as

$$G(s) = \frac{s}{s + 2.063 \times 10^5} \times \frac{s^2 + 3.943 \times 10^9}{s^2 + 2.673 \times 10^4 s + 9.458 \times 10^9} \quad \text{(E7.1.12a)}$$

$$G(s) = \frac{0.5s^2 - 1.337 \times 10^4 s + 4.729 \times 10^9}{s^2 + 2.673 \times 10^4 s + 9.458 \times 10^9} + \frac{0.5s - 1.032 \times 10^5}{s + 2.063 \times 10^5} \quad \text{(E7.1.12b)}$$

(e) All the above filter design works are put into the M-file named "sig07e01.m", which plots the frequency responses of the designed filters so that one can check if the design specifications are satisfied. Figure 7.3, obtained by running the program "sig07e01.m", shows the following points:

- Figure 7.3(a) shows that the cutoff frequency ω_c given as an input argument of 'butter()' is the frequency at which $20\log_{10}|G(j\omega_c)| = -3$ [dB]. Note that the frequency response magnitude of a Butterworth filter is monotonic, i.e., has no ripple.

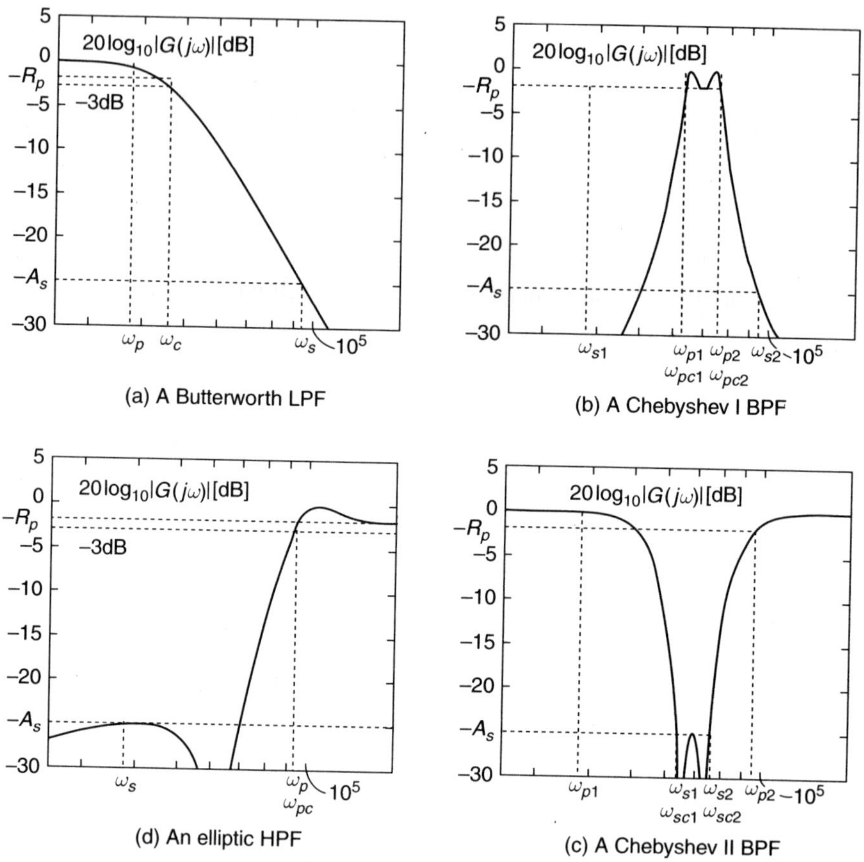

Fig. 7.3 Frequency responses of the filters designed in Example 7.1

- Figure 7.3(b) shows that the critical passband edge frequencies ω_{pc1} and ω_{pc2} given as an input argument wpc = [wpc1 wpc2] of 'cheby1()' are the frequencies at which the passband ripple condition is closely met, i.e., $20\log_{10}|G(j\omega_{pc})| = -R_p$ [dB]. Note that the frequency response magnitude of a Chebyshev I filter satisfying the passband ripple condition closely has a ripple in the passband, which is traded off for a narrower transition band than the Butterworth filter (with the same filter order).
- Figure 7.3(c) shows that the critical stopband edge frequencies ω_{sc1} and ω_{sc2} given as an input argument wsc = [wsc1 wps2] of 'cheby2()' are the frequencies at which the stopband attenuation condition is closely met, i.e., $20\log_{10}|G(j\omega_{sc})| = -A_s$ [dB]. Note that the frequency response magnitude of a Chebyshev II filter satisfying the stopband attenuation condition closely has a ripple in the stopband.
- Figure 7.3(d) shows that the critical passband edge frequency ω_{pc} given as an input argument wpc of 'ellip()' is the frequency at which the passband ripple

condition is closely met, i.e., $20\log_{10}|G(j\omega_{pc})| = -R_p$ [dB]. Note that the frequency response magnitude of an elliptic filter has ripples in both the passband and the stopband, yielding a relatively narrow transition band with the smallest filter order N = 3 among the four filters.

```
%sig07e01.m for analog filter design and frequency response plot
clear, clf, format short e
disp('(a) Butterworth LPF')
wp=2*pi*6000; ws=2*pi*15000; Rp=2; As=25;
[Nb,wcb]= buttord(wp,ws,Rp,As,'s') % Order of analog BW LPF
[Bb,Ab]= butter(Nb,wcb,'s') % num/den of analog BW LPF system ftn
[SOS,K]= tf2sos(Bb,Ab); % cascade realization [BBc,AAc]=tf2cas(B,A)
Ns=size(SOS,1); Gm=K^(1/Ns), BBc=SOS(:,1:3), AAc=SOS(:,4:6)
[BBp,AAp]= tf2par_s(Bb,Ab) % parallel realization -- see Sect. E.15
ww= logspace(4,6,1000); % log frequency vector from 1e4 to 1e6[rad/s]
subplot(221), semilogx(ww,20*log10(abs(freqs(Bb,Ab,ww))))
title('Butterworth LPF')

disp('(b) Chebyshev I BPF')
ws1=2*pi*6e3; wp1=2*pi*1e4; wp2=2*pi*12e3; ws2=2*pi*15e3; Rp=2; As=25;
[Nc1,wpc]= cheb1ord([wp1 wp2],[ws1 ws2],Rp,As,'s')
[Bc1,Ac1]= cheby1(Nc1,Rp,wpc,'s')
[SOS,K]= tf2sos(Bc1,Ac1); % cascade realization
Ns=size(SOS,1); Gm=K^(1/Ns), BBc=SOS(:,1:3), AAc=SOS(:,4:6)
[BBp,AAp]= tf2par_s(Bc1,Ac1) % parallel realization
subplot(222), semilogx(ww,20*log10(abs(freqs(Bc1,Ac1,ww))))
title('Chebyshev I BPF')

disp('(c) Chebyshev II BSF')
wp1=2*pi*6e3; ws1=2*pi*1e4; ws2=2*pi*12e3; wp2=2*pi*15e3; Rp=2; As=25;
[Nc2,wsc]= cheb2ord([wp1 wp2],[ws1 ws2],Rp,As,'s')
[Bc2,Ac2]= cheby2(Nc2,As,wsc,'stop','s')
[SOS,K]= tf2sos(Bc2,Ac2); % cascade realization
Ns=size(SOS,1); Gm=K^(1/Ns), BBc=SOS(:,1:3), AAc=SOS(:,4:6)
[BBp,AAp]= tf2par_s(Bc2,Ac2) % parallel realization
subplot(224), semilogx(ww,20*log10(abs(freqs(Bc2,Ac2,ww))))
title('Chebyshev II BSF')
disp('(d) Elliptic HPF')

ws=2*pi*6000; wp=2*pi*15000; Rp=2; As=25;
[Ne,wpc]= ellipord(wp,ws,Rp,As,'s')
[Be,Ae]= ellip(Ne,Rp,As,wpc,'high','s')
[SOS,K]= tf2sos(Be,Ae); % cascade realization
Ns=size(SOS,1); Gm=K^(1/Ns), BBc=SOS(:,1:3), AAc=SOS(:,4:6)
[BBp,AAp]= tf2par_s(Be,Ae) % parallel realization
subplot(223), semilogx(ww,20*log10(abs(freqs(Be,Ae,ww))))
```

Now we are going to conclude this section with some MATLAB routines that can be used to determine the parameters of the circuits depicted in Figs. 7.4, 7.5, and 7.6 so that they can realize the designed (second-order) system functions. See Chap. 8 of [Y-2] for more details.

```
function [CR1,CR2,Gs]= filter_LPF_7_4a(A2,A3,K,RC1,RC2,KC)
% Design an LPF with the circuit in Fig. 7.4(a)
%                    KG1G2/C1C2                              B3=K*A3
% G(s) = ----------------------------------------- = ---------------
%        s^2 +((G1+G2)/C1+(1-K)G2/C2)*s +G1G2/C1C2    s^2 + A2*s + A3
if K<1, error('We must have K=(R3+R4)/R3 >= 1!'); end
if nargin<6, KC=1; end
if KC==1 % Find C1 and C2 for given K, R1, and R2.
  R1= RC1; R2= RC2; G1= 1/R1; G2= 1/R2;
  a= G1+G2; b= -(K-1)*G2; c= A2; d= A3/G1/G2;  tmp = c^2-4*a*b*d;
  C1= 2*a/(c + sqrt(tmp)); C2= 1/d/C1; CR1= C1; CR2= C2;
 else  % Find R1 and R2 for given K, C1, and C2.
  C1= RC1; C2= RC2;
  a= 1/C1; b= 1/C1 - (K-1)/C2; c= A2; d= A3*C1*C2;  tmp = c^2-4*a*b*d;
  if tmp<0, error('Increase C1 and K, or decrease C2'); end
  G1= (c + sqrt(tmp))/2/a; G2= d/G1; R1= 1/G1; R2= 1/G2; CR1= R1; CR2= R2;
end
B3= K*A3; A2= (G1+G2)/C1 + (1-K)*G2/C2; A3= G1*G2/C1/C2;
syms s; Gs = B3/(s^2+A2*s+A3);
```

```
function [CR1,CR2,Gs]= filter_HPF_7_4b(A2,A3,K,RC1,RC2,KC)
% Design a HPF with the circuit in Fig. 7.4(b)
%                        K*s^2                                  K*s^2
% G(s) = ------------------------------------------- = ---------------
%        s^2 +(G2(1/C1+1/C2)-(K-1)G1/C1)s +G1G2/C1C2    s^2 + A2*s + A3
if K<1, error('We must have K=(R3+R4)/R3 >= 1!'); end
if nargin<6, KC=1; end
if KC==1 % Find C1 and C2 for given K, R1, and R2.
  R1= RC1; R2= RC2; G1= 1/R1; G2= 1/R2;
  a= G2+(1-K)*G1; b= G2; c= A2; d= A3/G1/G2; tmp= c^2-4*a*b*d;
  if tmp<0, error('Try with smaller/greater values of R1/K'); end
  C1= 2*a/(c + sign(a)*sqrt(tmp)); C2= 1/d/C1; CR1= C1; CR2= C2;
 else  % Find R1 and R2 for given K, C1, and C2.
  C1=RC1; C2=RC2;
  a=(1-K)/C1; b=1/C1+1/C2; c=A2; d=A3*C1*C2; tmp=c^2-4*a*b*d;
  if tmp<0, error('Try with smaller/greater values of C2/K'); end
  if abs(a)<eps, G2= A2/b; G1= d/G2;
    else G1= (c + sign(a)*sqrt(tmp))/2/a; G2= d/G1;
  end
  R1= 1/G1; R2= 1/G2; CR1= R1; CR2= R2;
end
B1= K; A2= G2*(1/C1+1/C2) - (K-1)*G1/C1; A3= G1*G2/C1/C2;
syms s; Gs = B1*s^2/(s^2+A2*s+A3);
```

```
function [R1,C2R3,C5R4,Gs]= filter_LPF_7_5a(B3,A2,A3,R3C2,R4C5,KC)
% Design an LPF with the circuit in Fig. 7.5(a)
%                 -G1G4/C2C5                         -B3
% G(s) = -------------------------------- = ---------------
%        s^2 + (G1+G3+G4)/C2*s + G3G4/C2C5   s^2 + A2*s + A3
if nargin<6, KC=1; end
if KC==1 % Find R1, C2 and C5 for given R3 and R4.
  R3= R3C2; R4= R4C5; G3= 1/R3; G4= 1/R4;
  G1=G3*B3/A3; C2=(G1+G3+G4)/A2; C5=G3*G4/C2/A3; R1=1/G1; C2R3=C2; C5R4=C5;
 else  % Find R1, R3 and R4 for given C2 and C5.
  C2=R3C2; C5=R4C5; a=1+B3/A3; b=1; c=A2*C2; d=A3*C2*C5; tmp = c^2-4*a*b*d;
  if tmp<0, error('Try with greater/smaller values of C2/C5'); end
  G3= (c + sign(a)*sqrt(tmp))/2/a; G4= d/G3;
  G1= B3/A3*G3; R3= 1/G3; R4= 1/G4; R1=1/G1; C2R3= R3; C5R4= R4;
end
B3= G1*G4/C2/C5; A2= (G1+G3+G4)/C2; A3= G3*G4/C2/C5;
syms s; Gs = -B3/(s^2+A2*s+A3);
```

```
function [C1,C3R2,C4R5,Gs]= filter_HPF_7_5b(B1,A2,A3,R2C3,R5C4,KC)
% Design a HPF with the circuit in Fig. 7.5(b)
%                     -(C1/C3)*s^2                          -B1*s^2
% G(s) = -------------------------------------- = ---------------
%         s^2 + G5(C1+C3+C4)/C3/C4*s + G2G5/C3C4   s^2 + A2*s + A3
if nargin<6, KC=1; end
if KC==1 % Find C1, C3 and C4 for given R2 and R5.
  R2= R2C3; R5= R5C4; G2= 1/R2; G5= 1/R5;
  a= 1; b= 1+B1; c= A2/G5; d= A3/G2/G5;  tmp = c^2-4*a*b*d;
  if tmp<0, error('Try with smaller/greater values of R2/R5'); end
  C3= 2*a/(c + sqrt(tmp)); C4= 1/d/C3; C1= B1*C3; C3R2= C3; C4R5= C4;
 else  % Find C1, R2 and R5 for given C3 and C4.
  C3= R2C3; C4= R5C4;
  C1 = B1*C3; G5= A2/(C1+C3+C4)*C3*C4; G2= A3*C3*C4/G5;
  R2= 1/G2; R5= 1/G5; C3R2= R2; C4R5= R5;
end
B1= C1/C3; A2= G5*(C1+C3+C4)/C3/C4; A3= G2*G5/C3/C4;
syms s; Gs = -B1*s^2/(s^2+A2*s+A3);
% Examples of Usage
%>>B1=2; A2=100; A3=10000;
%>> R2=1e4; R5=2e5; [C1,C3,C4,Gs]=filter_HPF_7_5b(B1,A2,A3,R2,R5,1)
%>> C3=1e-7; C4=2e-6; [C1,R2,R5,Gs]=filter_HPF_7_5b(B1,A2,A3,C3,C4,2)
```

```
function [C3R1,C4R2,R5,Gs]= filter_BPF_7_6a(B2,A2,A3,R1C3,R2C4,KC)
% Design a BPF with the circuit in Fig. 7.6(a)
%                     -(G1/C3)*s                          -B2*s
% G(s) = ------------------------------------ = ---------------
%         s^2 + G5(1/C3+1/C4)*s + (G1+G2)G5/C3C4   s^2 + A2*s + A3
if nargin<6, KC=1; end
if KC==1 % Find C3, C4, and R5 for given R1 and R2.
  R1= R1C3; R2= R2C4; G1= 1/R1; G2= 1/R2; C3= G1/B2;
  G5= (A2 - A3*C3/(G1+G2))*C3;
  if G5<0, error('Try with smaller values of R2'); end
  C4= G5*(G1+G2)/C3/A3;
  R5= 1/G5; C3R1= C3; C4R2= C4;
  fprintf('C3=%10.4e, C4=%10.4e, R5=%10.4e\n', C3,C4,R5)
 elseif KC==2  % Find R1, R2 and R5 for given C3 and C4.
  C3= R1C3; C4= R2C4;
  G1 = B2*C3; G5= A2/(1/C3+1/C4); G2= A3*C3*C4/G5-G1;
  R5= 1/G5; R1=1/G1; R2=1/G2; C3R1= R1; C4R2= R2;
  fprintf('R1=%10.4e, R2=%10.4e, R5=%10.4e\n', R1,R2,R5)
 else  % Find R1, R5, and C3=C4=C for given R2 and C3=C4.
  R2= R1C3; G2=1/R2;
 nonlinear_eq= inline('[2*x(1)-A2*x(2); x(1).*(B2*x(2)+G2)-A3*x(2).^2]',...
      'x','G2','B2','A2','A3');
 G50=0.1; C0=0.1; x0=[G50 C0]; % Initial guesses of G5 and C
 x= fsolve(nonlinear_eq,x0,optimset('TolFun',1e-8),G2,B2,A2,A3)
 %tol=1e-5; MaxIter=100; x=newtons(nonlinear_eq,x0,tol,MaxIter,G2,B2,A2,A3)
 G5= x(1); C=x(2); C3=C; C4=C; G1=B2*C3;
 R1=1/G1; R5=1/G5; C3R1=C3; C4R2=R1;
 fprintf('C3=C4=%10.4e, R1=%10.4e, R5=%10.4e\n', C,R1,R5)
end
B1= G1/C3; A2= G5*(1/C3+1/C4); A3= (G1+G2)*G5/C3/C4;
syms s; Gs = -B2*s/(s^2+A2*s+A3);
% Examples of Usage
%>>B2=100; A2=100; A3=10000;
%>> R1=1e2; R2=1e2; [C3,C4,R5,Gs]=filter_BPF_7_6a(B2,A2,A3,R1,R2,1)
%>> C3=1e-4; C4=1e-4; [R1,R2,R5,Gs]=filter_BPF_7_6a(B2,A2,A3,C3,C4,2)
%>> R2=1e2; [C3,R1,R5,Gs]=filter_BPF_7_6a(B2,A2,A3,R2,0,3)
```

```
function [C1,C2R3,C5R4,Gs]= filter_BPF_7_6b(B2,A2,A3,R3C5,R4C5,KC)
% Design a BPF with the circuit in Fig. 7.6(b)
%                  -(C1G4/(C1+C2)C5)*s                          -B2*s
% G(s) = ------------------------------------------ = ---------------
%          s^2 + ((G3+G4)/(C1+C2))*s + G3G4/(C1+C2)C5    s^2 + A2*s + A3
if nargin<6, KC=1; end
if KC==1 % Find C1, C2 and C5 for given R3 and R4.
  R3= R3C5; R4=R4C5; G3= 1/R3; G4=1/R4;
  C1pC2= (G3+G4)/A2; C5= G3*G4/A3/C1pC2; C1= B2*C1pC2*C5/G4 %=B2*G3/A3
  C2= C1pC2 - C1; C2R3= C2; C5R4= C5;
  if C2<0, error('Try with greater/smaller values of R3/R4'); end
  fprintf('C1=%10.4e, C2=%10.4e, C5=%10.4e\n', C1,C2,C5)
 else  % Find C1, R3 and R4 for given C5 and C1=C2.
  C5=R3C5; G4= 2*C5*B2; G3_2C= A3/G4*C5; %=A3/2/B2: not adjustable
  C= G4/2/(A2-G3_2C); C1=C; C2=C;
  if C<0, error('How about increasing B2 & A2 and/or decreasing A3'); end
  G3= G3_2C*2*C; R3= 1/G3; R4= 1/G4;  C2R3= R3; C5R4= R4;
  fprintf('C1=C2=%10.4e, R3=%10.4e, R4=%10.4e\n', C,R3,R4)
end
B3= C1*G4/(C1+C2)/C5; A2= (G3+G4)/(C1+C2); A3= G3*G4/(C1+C2)/C5;
syms s; Gs = -B2*s/(s^2+A2*s+A3);
```

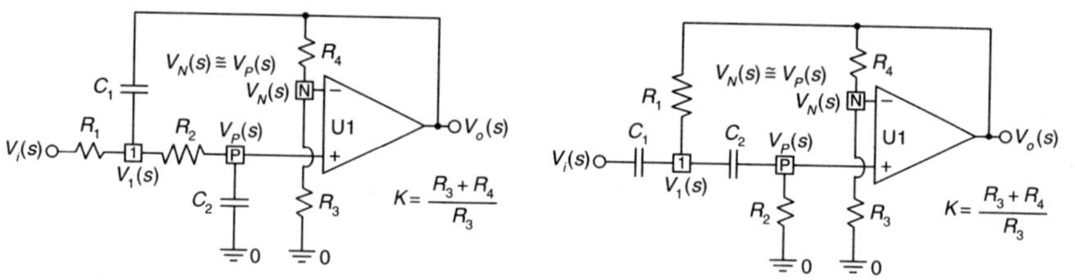

$$G_a(s) = \frac{V_o(s)}{V_i(s)} = \frac{KG_1G_2/C_1C_2}{s^2 + ((G_1+G_2)/C_1 + (1-K)G_2/C_2)s + G_1G_2/C_1C_2}$$

$$G_b(s) = \frac{V_o(s)}{V_i(s)} = \frac{Ks^2}{s^2 + G_2(1/C_1 + 1/C_2) + (1-K)G_1/C_1)s + G_1G_2/C_1C_2}$$

(a) A second-order Sallen-Key lowpass filter (b) A second-order Sallen-Key highpass filter

Fig. 7.4 Second-order active filters

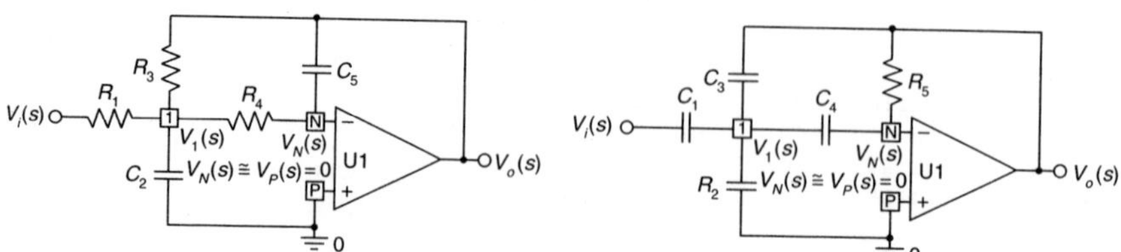

$$G_a(s) = \frac{V_o(s)}{V_i(s)} = \frac{-G_1G_4/C_2C_5}{s^2 + sC_5(G_1+G_3+G_4)/C_2C_5 + G_3G_4/C_2C_5}$$

$$G_b(s) = \frac{V_o(s)}{V_i(s)} = \frac{-C_1C_4s^2/C_3C_4}{s^2 + sG_5(C_1+C_3+C_4)/C_3C_4 + G_2G_5/C_3C_4}$$

(a) A second-order MFB (multi-feedback) LPF (b) A second-order MFB (multi-feedback) HPF

Fig. 7.5 Second-order active filters

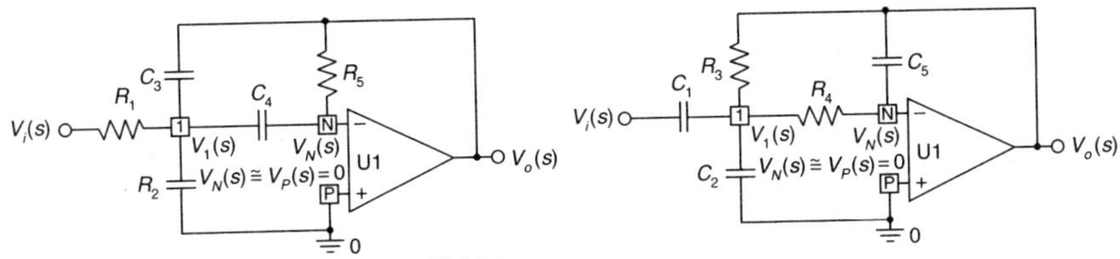

$$G_a(s) = \frac{V_o(s)}{V_i(s)} = \frac{-(G_1/G_3)s}{s^2 + (G_5(C_3+C_4)/C_3C_4)s + (G_1+G_2)G_5/C_3C_4}$$

$$G_b(s) = \frac{V_o(s)}{V_i(s)} = \frac{-(C_1G_4/(C_1+C_2)C_5)s}{s^2 + ((G_3+G_4)/(C_1+C_2)s + G_3G_4/(C_1+C_2)C_5}$$

(a) A second-order MFB (multi-feedback) BPF (b) A second-order MFB (multi-feedback) BPF

Fig. 7.6 Second-order active filters

For example, we can use the MATLAB routine 'filter_BPF_7_6a()' to tune the parameters of the MFB (multiple feedback) circuit of Fig. 7.6(a) so that the circuit realizes the following BPF system function

$$G(s) = \frac{-(G_1/C_3)s}{s^2 + (G_5(C_3+C_4)/C_3C_4)s + (G_1+G_2)G_5/C_3C_4} = \frac{-100s}{s^2 + 100s + 100^2} \tag{7.1.3}$$

To this end, we have only to type the following statements into the MATLAB command window:

```
>>B2=100; A2=100; A3=10000; % The desired system function B2*s/(s^2+A2*s+A3)
>>R2=1e2; KC=3; % With the given value of R2=100 and the assumption that C3=C4
>>[C3,R1,R5,Gs]=filter_BPF_7_6a(B2,A2,A3,R2,0,KC)
   C3=C4=1.0000e-004, R1=9.9999e+001, R5=2.0000e+002
   Gs = -100*s/(s^2+100*s+5497540047884755/549755813888)
>>5497540047884755/549755813888 % To see the weird constant term
   1.0000e+004
```

For another example, we can use the MATLAB routine 'filter_LPF_7_4a()' to tune the parameters of the Sallen-Key circuit of Fig. 7.4(a) so that the circuit realizes the following LPF system function:

$$G(s) = \frac{KG_1G_2/C_1C_2}{s^2 + ((G_1+G_2)/C_1 + (1-K)G_2/C_2)s + G_1G_2/C_1C_2} = \frac{K\omega_r^2}{s^2 + \omega_b s + \omega_r^2} \tag{7.1.4}$$

More specifically, suppose we need to determine the values of R_1 and R_2 of the Sallen-Key circuit of Fig. 7.4(a) with the pre-determined values of capacitances $C_1 = C_2 = 100\text{pF}$ so that it realizes a second-order LPF with the DC gain $K = 1.5$, the corner frequency $\omega_r = 2\pi \times 10^7$ [rad/s], and the quality factor $Q = 0.707$ (for $\omega_b = \omega_r/Q$). To this end, we have only to type the following statements into the MATLAB command window:

```
>>K=1.5; C1=1e-10; C2=1e-10; wr=2*pi*1e7; Q=0.707;
   % The coefficients of denominator of desired system ftn
>>A2=wr/Q; A3=wr^2; % G(s)=K*A3/(s^2+A2*s+A3)
>>KC=2; [R1,R2,Gs]= filter_LPF_7_4a(A2,A3,K,C1,C2,KC)
```

```
R1 = 221.2010  % tuned resistance
R2 = 114.5126
Gs = 5921762640653615/(s^2+5964037174912491/67108864*s+7895683520871487/2)
```

(cf) For reference, you can visit the web site <http://www.national.com/pf/LM/LMH6628.html> to see the application note OA-26 for Designing Active High Speed Filters.

To illustrate the filter design and realization procedure collectively, let us find the cascade realization of a fourth-order Butterworth LPF with cutoff frequency $\omega_c = 10\,\text{kHz}$ using the Sallen-Key circuit of Fig. 7.4(a). For this job, we compose the following program and run it:

```
%sig07_01_1.m
N=4; fc=1e4; wc=2*pi*fc; % the order and cutoff frequency of the LPF
format short e
[B,A]= butter(N,wc,'s') % Butterworth LPF system function G(s)=B(s)/A(s)
f= logspace(3,5,400); % frequency vector of 400 points between 1e3~1e5[Hz]
Gw= freqs(B,A,2*pi*f); % frequency response G(jw)
semilogx(f,20*log10(abs(Gw))) % plot |G(jw)| in dB versus frequency[Hz]
[SOS,K0]= tf2sos(B,A); % cascade realization
BBc=SOS(:,1:3); AAc=SOS(:,4:6); % numerator/denominator of each SOS
K=1; R1= 1e4; R2= 1e4; KC=1; % predetermined values of R1 and R2
for n=1:floor(N/2)
   A2 = AAc(n,2); A3 = AAc(n,3);
   [C1,C2,Gs]= filter_LPF_7_4a(A2,A3,K,R1,R2,KC) % filter tuning
end
% Check the LPF design results obtained at the web site
%  <http://www.daycounter.com/Filters/Sallen-Key-LP-Calculator.phtml>
```

(cf) Note that multiplying all the resistances/capacitances by the same constant does not change the system function and frequency response. This implies that if you want to scale up/down the tuned capacitances/resistances without affecting the system function and frequency response, you can scale up/down the predetermined values of resistances/capacitances in the same ratio.

7.2 Digital Filter Design

Digital filters can be classified into IIR (infinite-duration impulse response) or FIR (finite-duration impulse response) filter depending on whether the duration of the impulse response is infinite or finite. If the system function of a filter has a polynomial in z or z^{-1} of degree one or higher (in addition to a single z^N or z^{-N} term) in its denominator, its impulse response has an infinite duration. For example, consider a filter whose system function is

$$G[z] = \frac{Y[z]}{X[z]} = \frac{1}{z - 0.5} \tag{7.2.1}$$

so that the z-domian and time-domain input-output relationships are

$$(z - 0.5)Y[z] = X[z];\ zY[z] - 0.5Y[z] = X[z];\ zY[z] = 0.5Y[z] + X[z];$$
$$y[n+1] = 0.5y[n] + x[n] \tag{7.2.2}$$

This means that the output $y[n+1]$ of this *recursive* difference equation is affected by not only the input $x[n]$ but also the previous output $y[n]$. That is why the impulse response of the filter has indefinitely long duration. In contrast with this, the duration of the impulse response of a FIR filter is equal to one plus the degree of the (numerator) polynomial in z or z^{-1} of its system function (having no denominator). For example, consider a filter whose system function is

$$G[z] = \frac{Y[z]}{X[z]} = 1 - 2z^{-1} + 3z^{-3} = \frac{z^3 - 2z^2 + 0z + 3}{z^3} \tag{7.2.3}$$

so that the z-domian and time-domain input-output relationships are

$$Y[z] = (1 - 2z^{-1} + 3z^{-3})X[z];$$
$$y[n] = x[n] - 2x[n-1] + 3x[n-3] \tag{7.2.4}$$

The output $y[n]$ of this *nonrecursive* difference equation is affected not by any previous output, but only by the input terms. That is why the impulse response of the filter has a finite duration. In fact, the impulse response of a FIR filter is identical to the filter coefficients, say, $[1\ \ -2\ 0\ 3]$ in the case of this filter having the system function (7.2.3).

In this section we will see how to design the IIR and FIR filters using the MATLAB software.

7.2.1 IIR Filter Design

The methods of designing IIR filters introduced here are basically the discretizations of analog filters dealt with in Sect. 7.1. We will use the same MATLAB functions that are used for analog filter design.

Example 7.2 IIR Filter Design

Let us find the system functions of digital filters (with sampling frequency $F_s = 50$ [kHz]) meeting the specifications given below.

(a) We are going to determine the system function of a digital *Butterworth* lowpass filter with the passband/stopband edge frequencies, passband ripple, and stopband attenuation as

$$\omega_p = 2\pi \times 6000\,[\text{rad/s}],\ \omega_s = 2\pi \times 15000\,[\text{rad/s}],\ R_p = 2\,[\text{dB}],\ \text{and}\ A_s = 25\,[\text{dB}] \tag{E7.2.1}$$

First, we prewarp the edge frequencies, design an analog Butterworth LPF satisfying the given specifications on the passband ripple and stopband attenuation at the prewarped passband and stopband edge frequencies, and then discretize the LPF through bilinear transformation:

```
>>Fs=5e4; T=1/Fs; format short e %Sampling frequency and sampling period
>>wp=2*pi*6000; ws=2*pi*15000; Rp=2; As=25;
>>wp_p=prewarp(wp,T); ws_p=prewarp(ws,T); % Prewarp the edge frequencies
>>[Nb,wcb]=buttord(wp_p,ws_p,Rp,As,'s');%Order, cutoff freq of analog BW LPF
>>[Bb,Ab]= butter(Nb,wcb,'s'); % num/den of analog BW LPF system ftn
>>[Bb_d0,Ab_d0]= bilinear(Bb,Ab,Fs) % Bilinear transformation
```

We can also use the MATLAB function `butter()` to design a digital Butterworth filter directly:

```
>>fp=wp*T/pi; fs=ws*T/pi; %Normalize edge freq into [0,1](1:pi[rad/sample])
>>[Nb,fcb]= buttord(fp,fs,Rp,As) % Order, Cutoff freq of digital BW LPF
   Nb =    3, fcb =   3.0907e-001
>>[Bb_d,Ab_d]= butter(Nb,fcb) % num/den of digital BW LPF system ftn
   Bb_d =  5.3234e-002  1.5970e-001  1.5970e-001  5.3234e-002
   Ab_d =  1.0000e+000 -1.1084e+000  6.6286e-001 -1.2856e-001
>>[SOS,Kc]= tf2sos(Bb_d,Ab_d) % Cascade form realization
   SOS =  1    1    0       1.0000e+00   -3.0925e-01           0
          1    2    1       1.0000e+00   -7.9918e-01  4.1571e-01
   Kc =  5.3234e-002
>>[BBp,AAp,Kp]= tf2par_z(Bb_d,Ab_d) % Parallel form realization
   BBp =  -9.9489e-001   5.6162e-001             0
                     0   1.4622e+000             0
   AAp =   1.0000e+000  -7.9918e-001   4.1571e-001
                     0   1.0000e+000  -3.0925e-001
   Kp =  -4.1408e-001
```

```
%sig07e02a.m for digital Butterworth LPF design
clear, clf, format short e
Fs=5e4; T=1/Fs; % Sampling frequency and sampling period
wp=2*pi*6000; ws=2*pi*15000; Rp=2; As=25;
% analog filter design and discretization through bilinear transformation
wp_p=prewarp(wp,T); ws_p=prewarp(ws,T); % Prewarp the edge frequencies
[Nb,wcb]=buttord(wp_p,ws_p,Rp,As,'s'); % Order, cutoff freq of analog BW LPF
[Bb,Ab]= butter(Nb,wcb,'s'); % num/den of analog BW LPF system ftn
[Bb_d0,Ab_d0]= bilinear(Bb,Ab,Fs) % Bilinear transformation
% direct digital filter design
fp=wp*T/pi; fs=ws*T/pi; % Normalize edge freq into [0,1] (1: pi[rad/sample])
[Nb,fcb]= buttord(fp,fs,Rp,As) % Order of digital BW LPF
[Bb_d,Ab_d]= butter(Nb,fcb) % num/den of digital BW LPF system ftn
% Plot the frequency response magnitude
fn=[0:512]/512; W=pi*fn; % Normalized and digital frequency range
subplot(221), plot(fn,20*log10(abs(freqz(Bb_d,Ab_d,W))+eps))
% To check if the filter specifications are satisfied
hold on, plot(fp,-Rp,'o', fcb,-3, '+', fs,-As,'x')
axis([0 1 -80 10]), title('Butterworth LPF')
[SOS,Kc]= tf2sos(Bb_d,Ab_d) % Cascade form realization
[BBp,AAp,Kp]= tf2par_z(Bb_d,Ab_d) % Cascade form realization
```

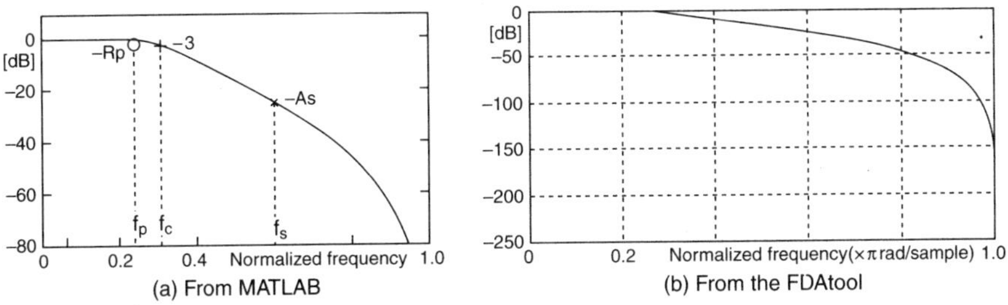

Fig. 7.7 The magnitude curves of the frequency response of the digital Butterworth LPF

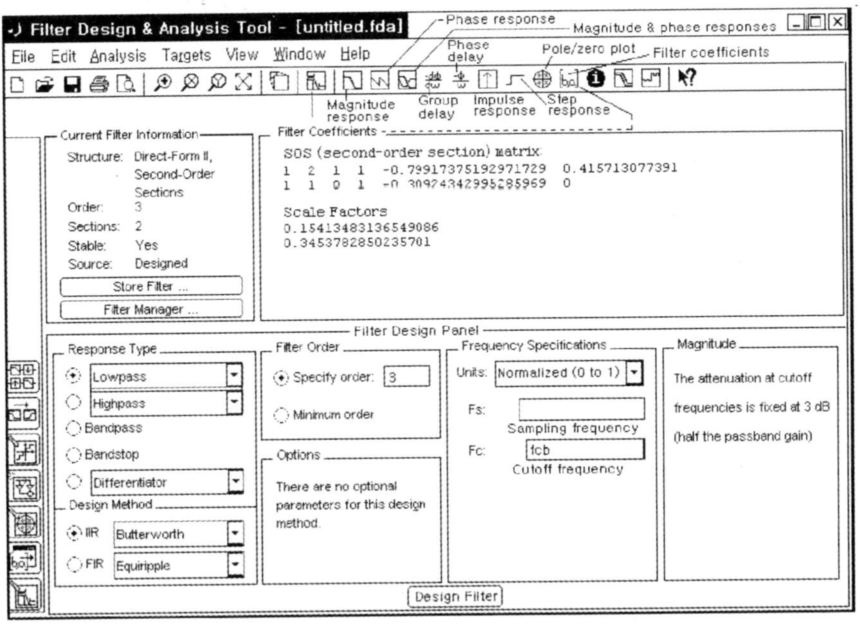

Fig. 7.8 Design of a digital Butterworth LPF with order 3 and cutoff frequency 0.309 using FDAtool

This result means that the system function of the designed Butterworth LPF of order N = 3 is

$$G[z] = \frac{0.05323z^3 + 0.1597z^2 + 0.1597z + 0.05323}{z^3 - 1.1084z^2 + 0.6629z - 0.1286} \text{ : Direct form} \quad \text{(E7.2.2a)}$$

$$= \frac{0.05323(z^2 + 2z + 1)(z + 1)}{(z^2 - 0.7992z + 0.4157)(z - 0.3092)} \text{ : Cascade form} \quad \text{(E7.2.2b)}$$

$$= \frac{-0.9949z^2 + 0.5616z}{z^2 - 0.7992z + 0.4157} + \frac{1.4622z}{z - 0.3092} - 0.4141 \text{ : Parallel form} \quad \text{(E7.2.2c)}$$

The magnitude curve of the frequency response of the designed filter is shown in Fig. 7.7. If the order and cutoff frequency are determined, then we can use the FDATool to design a digital filter as depicted in Fig. 7.8 where the same cascade realization is obtained as with `butter()`. Note that the cutoff frequency fcb determined by `buttord()` and used by `butter()` and FDATool is normalized to the range of [0,1] with 1 corresponding to Fs/2[Hz] (half the sampling frequency).

(b) We are going to determine the system function of a *Chebyshev* I BPF with

$$\omega_{s1} = 2\pi \times 6000,\ \omega_{p1} = 2\pi \times 10000,\ \omega_{p2} = 2\pi \times 12000,$$
$$\omega_{s2} = 2\pi \times 15000\,[\text{rad/s}],\ R_p = 2\,[\text{dB}],\ \text{and}\ A_s = 25\,[\text{dB}] \quad (\text{E7.2.3})$$

First, we prewarp the edge frequencies, design an analog chebyshev I BPF satisfying the given specifications on the passband ripple and stopband attenuation at the prewarped passband and stopband edge frequencies, and then discretize the BPF through bilinear transformation:

```
>>Fs=5e4; T=1/Fs; % Sampling frequency and sampling period
>>ws1=2*pi*6e3; wp1=2*pi*1e4; wp2=2*pi*12e3; ws2=2*pi*15e3; Rp=2; As=25;
>>wp_p=prewarp([wp1 wp2],T); ws_p=prewarp([ws1 ws2],T); % Prewarp
>>[N,wpc]=cheb1ord(wp_p,ws_p,Rp,As,'s') %Order & cutoff freq of A-C1 BPF
   N = 2, wpc = 7.2654e+004   9.3906e+004
>>[Bc1,Ac1]=cheby1(N,Rp,wpc,'s') % num/den of analog C1 BPF system ftn
>>[Bc1_d0,Ac1_d0]= bilinear(Bc1,Ac1,Fs) % Bilinear transformation
```

The MATLAB function `cheby1()` can also be used to design a digital Chebyshev I filter directly as follows:

```
>>fp=[wp1 wp2]*T/pi; fs=[ws1 ws2]*T/pi; %Normalize edge freq into [0,1]
>>[Nc1,fcc1]=cheb1ord(fp,fs,Rp,As) % Order & Cutoff freq of D-C1 BPF
   Nc1 =    2,      fcc1 =  0.4    0.48
>>[Bc1_d,Ac1_d]= cheby1(Nc1,Rp,fcc1) % num/den of D-C1 BPF system ftn
   Bc1_d = 9.3603e-03                0 -1.8721e-02              0  9.3603e-03
   Ac1_d = 1.0000e+00 -7.1207e-01  1.8987e+00 -6.4335e-01  8.1780e-01
>>[SOS,Kc]= tf2sos(Bc1_d,Ac1_d) % Cascade form realization
   SOS =  1    2    1        1.0000e+000 -1.6430e-001  9.0250e-001
          1   -2    1        1.0000e+000 -5.4780e-001  9.0610e-001
   Kc =  9.3603e-003
>>[BBp,AAp,Kp]= tf2par_z(Bc1_d,Ac1_d) % Parallel form realization
   BBp =   -1.9910e-003 -8.9464e-002             0
           -9.4316e-005  9.7236e-002             0
   AAp =    1.0000e+000 -5.4785e-001  9.0612e-001
            1.0000e+000 -1.6432e-001  9.0253e-001
   Kp =  1.1446e-002
```

This means that the system function of the designed Chebyshev I BPF of order $2\text{Nc1} = 4$ is

$$G[z] = \frac{9.3603 \times 10^{-3} z^4 - 1.8721 \times 10^{-2} z^2 + 9.3603 \times 10^{-3}}{z^4 - 0.7121 z^3 + 1.8987 z^2 - 0.6434 z + 0.8178} : \text{Direct form} \tag{E7.2.4a}$$

$$= \frac{9.3603 \times 10^{-3} (z^2 + 2z + 1)(z^2 - 2z + 1)}{(z^2 - 0.1643z + 0.9025)(z^2 - 0.5478z + 0.9061)} : \text{Cascade form} \tag{E7.2.4b1}$$

$$= \frac{0.10855 \times 0.10855 \times 0.7943 (z^2 - 1)(z^2 - 1)}{(z^2 - 0.1643z + 0.9025)(z^2 - 0.5478z + 0.9061)} : \text{Cascade form} \tag{E7.2.4b2}$$

$$= 0.011446 + \frac{-9.4316 \times 10^{-5} z^2 + 9.7236 \times 10^{-2} z}{z^2 - 0.1643z + 0.9025} + \frac{-1.991 \times 10^{-3} z^2 - 8.9464 \times 10^{-2} z}{z^2 - 0.5478z + 0.9061} : \text{Parallel form} \tag{E7.2.4c}$$

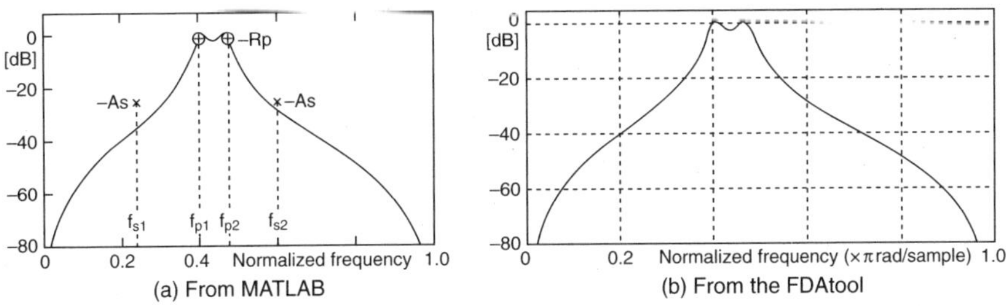

Fig. 7.9 The magnitude curves of the frequency response of the digital Chebyshev I BPF

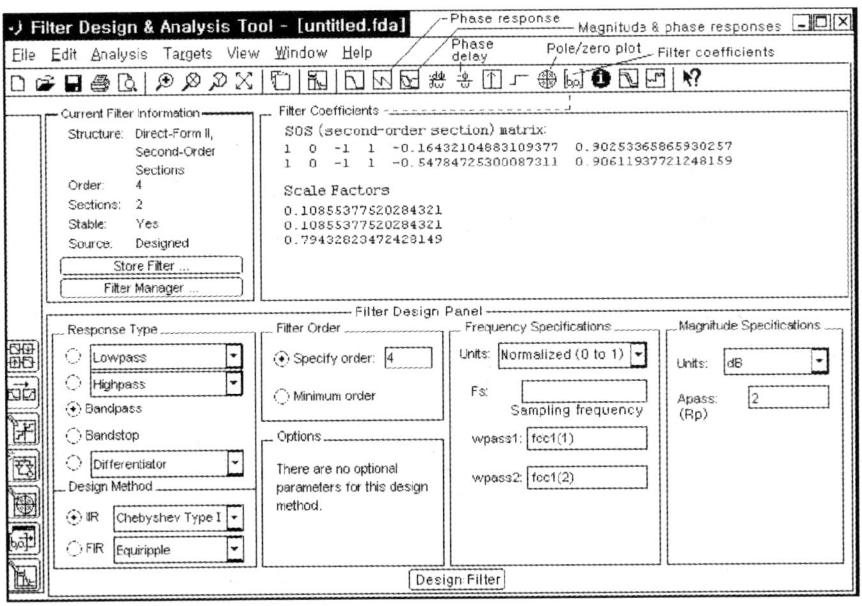

Fig. 7.10 Design of a digital Chebyshev I BPF with order 4 and passband edge frequencies 0.4 & 0.48 using FDAtool

The magnitude curve of the frequency response of the designed filter is shown in Fig. 7.9. If the order and passband edge frequencies are determined, then we use the FDATool to design a digital filter as depicted in Fig. 7.10 where the same cascade realization is obtained as with `cheby1()`. Note that the critical passband edge frequencies fcc1(1) and fcc1(2) determined by `cheb1ord()` and used by `cheby1()` and FDATool are normalized to the range of [0,1] with 1 corresponding to Fs/2[Hz] (half the sampling frequency, i.e., the Nyquist frequency).

(c) We are going to determine the system function of a *Chebyshev* II BSF with

$$\omega_{p1} = 2\pi \times 6000,\ \omega_{s1} = 2\pi \times 10000,\ \omega_{s2} = 2\pi \times 12000,$$
$$\omega_{p2} = 2\pi \times 15000\,[\text{rad/s}],\ R_p = 2\,[\text{dB}],\ \text{and}\ A_s = 25\,[\text{dB}] \qquad \text{(E7.2.5)}$$

Let us use the MATLAB function `cheby2()` to design a digital Chebyshev II filter directly as follows:

```
>>Fs=5e4; T=1/Fs; % Sampling frequency and sampling period
>>wp1=2*pi*6e3; ws1=2*pi*1e4; ws2=2*pi*12e3; wp2=2*pi*15e3; Rp=2; As=25;
>>fp=[wp1 wp2]*T/pi; fs=[ws1 ws2]*T/pi; %Normalize edge freq into [0,1]
>>[Nc2,fcc2]=cheb2ord(fp,fs,Rp,As) % Order & Cutoff freq of D-C2 BSF
    Nc2 =    2,    fcc2 =  0.4    0.48
>>[Bc2_d,Ac2_d]=cheby2(Nc2,As,fcc2,'stop') %D-C2 BSF system ftn
    Bc2_d = 6.0743e-01 -4.5527e-01  1.2816e+00 -4.5527e-01  6.0743e-01
    Ac2_d = 1.0000e+00 -5.7307e-01  1.1202e+00 -3.3746e-01  3.7625e-01
>>[SOS,Kc]= tf2sos(Bc2_d,Ac2_d) % Cascade form realization
    SOS =  1   -0.2000   1         1.0000e+000  1.4835e-001  5.9730e-001
           1   -0.5495   1         1.0000e+000 -7.2143e-001  6.2992e-001
    Kc =  6.0743e-001
>>[BBp,AAp,Kp]= tf2par_z(Bc2_d,Ac2_d) % Parallel form realization
    BBp =   -4.7229e-001  2.3377e-001            0
            -5.3469e-001 -7.9541e-002            0
    AAp =    1.0000e+000 -7.2143e-001  6.2992e-001
             1.0000e+000  1.4835e-001  5.9730e-001
    Kp =  1.6144e+000
```

This means that the system function of the designed Chebyshev II BSF of order 2Nc2 = 4 is

$$G[z] = \frac{0.6074z^4 - 0.4553z^3 + 1.2816z^2 - 0.4553z + 0.6074}{z^4 - 0.5731z^3 + 1.1202z^2 - 0.3375z + 0.3763} : \text{Direct form} \qquad \text{(E7.2.6a)}$$

$$= \frac{0.6074(z^2 - 0.2z + 1)(z^2 - 0.5495z + 1)}{(z^2 - 0.1484z + 0.5973)(z^2 - 0.7214z + 0.6299)} : \text{Cascade form} \qquad \text{(E7.2.6b)}$$

$$= \frac{-0.4723z^2 + 0.2338z}{z^2 - 0.7214z + 0.6299} + \frac{-0.5347z^2 - 0.07954z}{z^2 + 0.1484z + 0.5973} + 1.6144 \quad : \text{Parallel form} \qquad \text{(E7.2.6c)}$$

The magnitude curve of the frequency response of the designed filter is shown in Fig. 7.11. If the order and stopband edge frequencies are determined, then we can use the FDATool to design a digital filter as depicted in Fig. 7.12 where the same cascade realization is obtained as with `cheby2()`. Note that the critical stopband edge frequencies fcc2(1) and fcc2(2) determined by `cheb2ord()` and used by `cheby2()` and FDATool are normalized to the range of [0,1] with 1 corresponding to Fs/2[Hz]. Note also that as can be seen from the magnitude curves in Fig. 7.11, the Chebyshev II type filter closely meets the specification on the stopband attenuation ($A_s \geq 25$[dB]), while it satisfies that on the passband ripple ($R_p \leq 2$[dB]) with some margin.

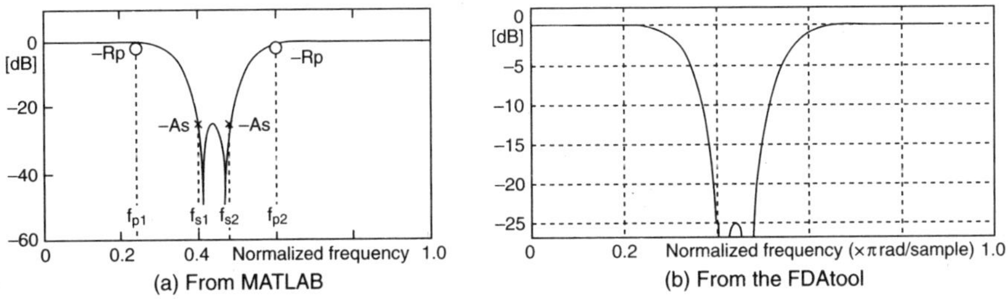

Fig. 7.11 The magnitude curves of the frequency response of the digital Chebyshev II BSF

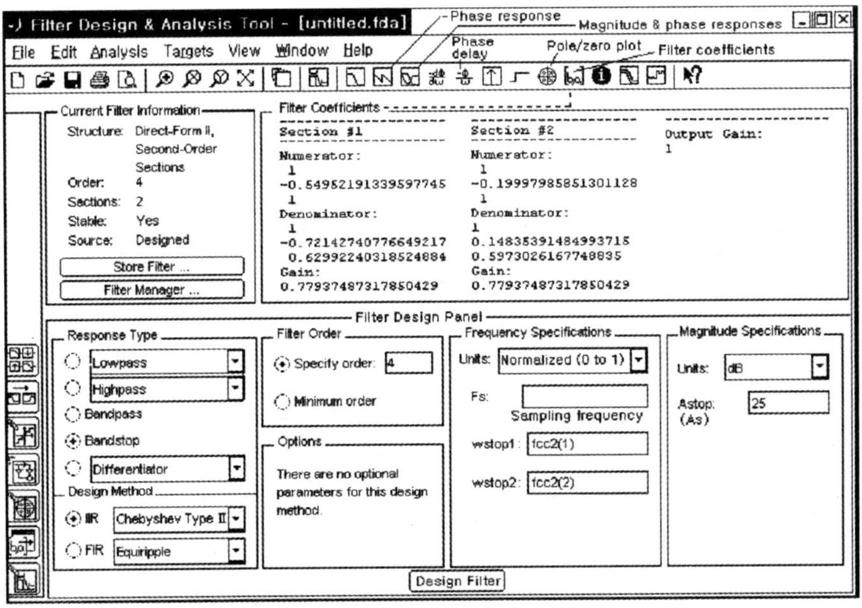

Fig. 7.12 Design of a digital Chebyshev II BSF with order 4 and stopband edge frequencies 0.4 & 0.48 using FDAtool (version 7.2-R2006a)

(d) We are going to determine the system function of an *elliptic* HPF with

$$\omega_s = 2\pi \times 6000\,[\text{rad/s}],\ \omega_p = 2\pi \times 15000\,[\text{rad/s}],\ R_p = 2\,[\text{dB}],\ \text{and}\ A_s = 25\,[\text{dB}] \tag{E7.2.7}$$

Let us use the MATLAB function `ellip()` to design a digital elliptic filter directly as follows:

```
>>Fs=5e4; T=1/Fs; % Sampling frequency and sampling period
>>ws=2*pi*6e3; wp=2*pi*15e3; Rp=2; As=25;
>>fp=wp*T/pi; fs=ws*T/pi; %Normalize edge freq into [0,1]
>>[Ne,fce]=ellipord(fp,fs,Rp,As) % Order & Cutoff freq of D-elliptic HPF
   Ne = 2, fce = 0.6
>>[Be_d,Ae_d]=ellip(Ne,Rp,As,fce,'high') %D-elliptic HPF system ftn
   Be_d = 2.0635e-001 -3.0101e-001  2.0635e-001
   Ae_d = 1.0000e+000  5.4365e-001  4.4217e-001

>>[SOS,Kc]= tf2sos(Be_d,Ae_d) % Cascade form realization
   SOS =  1  -1.4587e+000  1   1.0000e+000  5.4365e-001  4.4217e-001
   Kc =  2.0635e-001
>>[BBp,AAp,Kp]= tf2par_z(Be_d,Ae_d) % Parallel form realization
   BBp =  -2.6034e-001 -5.5472e-001            0
   AAp =   1.0000e+000  5.4365e-001  4.4217e-001
   Kp =   4.6669e-001
```

This means that the system function of the designed elliptic HPF of order Ne = 2 is

$$G[z] = \frac{0.2064z^2 - 0.3010z + 0.2064}{z^2 + 0.5437z + 0.4422} = \frac{0.2064(z^2 - 1.4587z + 1)}{z^2 + 0.5437z + 0.4422} : \text{Cascade form} \tag{E7.2.8a}$$

$$= \frac{-0.2603z^2 - 0.5547z}{z^2 + 0.5437z + 0.4422} + 0.4667 : \text{Parallel form} \tag{E7.2.8b}$$

```
%sig07e02.m for digital filter design and frequency response plot
clear, clf, format short e
Fs=5e4; T=1/Fs; % Sampling frequency and sampling period
disp('(a) Digital Butterworth LPF')
wp=2*pi*6000; ws=2*pi*15000; Rp=2; As=25;
fp=wp*T/pi; fs=ws*T/pi;
[Nb,fcb]= buttord(fp,fs,Rp,As) % Order of analog BW LPF
[Bb_d,Ab_d]= butter(Nb,fcb) % num/den of digital BW LPF system ftn
fn=[0:512]/512; W=pi*fn;
% Plot the frequency response magnitude curve
subplot(221), plot(fn,20*log10(abs(freqz(Bb_d,Ab_d,W))+eps))
hold on, plot(fp,-Rp,'o', fcb,-3,'+', fs,-As,'x')
[SOS,Kc]= tf2sos(Bb_d,Ab_d)  % Cascade form realization
[BBp,AAp,Kp]= tf2par_z(Bb_d,Ab_d) % Parallel form realization
disp('(b) Digital Chebyshev I BPF')
ws1=2*pi*6e3; wp1=2*pi*1e4; wp2=2*pi*12e3; ws2=2*pi*15e3; Rp=2; As=25;
```

```
fp=[wp1 wp2]*T/pi; fs=[ws1 ws2]*T/pi; %Normalize edge freq into [0,1]
[Nc1,fcc1]=cheb1ord(fp,fs,Rp,As) % Order & critical passband edge freq
[Bc1_d,Ac1_d]= cheby1(Nc1,Rp,fcc1) % num/den of D-C1 BPF system ftn
subplot(222), plot(fn,20*log10(abs(freqz(Bc1_d,Ac1_d,W))+eps))
[SOS,Kc]= tf2sos(Bc1_d,Ac1_d)  % Cascade form realization
[BBp,AAp,Kp]= tf2par_z(Bc1_d,Ac1_d) % Parallel form realization
disp('(c) Digital Chebyshev II BSF')
wp1=2*pi*6e3; ws1=2*pi*1e4; ws2=2*pi*12e3; wp2=2*pi*15e3; Rp=2; As=25;
fp=[wp1 wp2]*T/pi; fs=[ws1 ws2]*T/pi; %Normalize edge freq into [0,1]
[Nc2,fcc2]=cheb2ord(fp,fs,Rp,As) % Order & critical edge frequencies
[Bc2_d,Ac2_d]=cheby2(Nc2,As,fcc2,'stop') %num/den of D-C2 BSF system ftn
subplot(223), plot(fn,20*log10(abs(freqz(Bc2_d,Ac2_d,W))+eps))
[SOS,Kc]= tf2sos(Bc2_d,Ac2_d)  % Cascade form realization
[BBp,AAp,Kp]= tf2par_z(Bc2_d,Ac2_d) % Parallel form realization
disp('(d) Digital elliptic HPF')
ws=2*pi*6000; wp=2*pi*15000; Rp=2; As=25;
fp=wp*T/pi; fs=ws*T/pi; %Normalize edge freq into [0,1]
[Ne,fce]=ellipord(fp,fs,Rp,As) % Order & Cutoff freq of D-elliptic HPF
[Be_d,Ae_d]=ellip(Ne,Rp,As,fce,'high') % D-elliptic HPF system ftn
subplot(224), plot(fn,20*log10(abs(freqz(Be_d,Ae_d,W))+eps))
[SOS,Kc]= tf2sos(Be_d,Ae_d) % Cascade form realization
[BBp,AAp,Kp]= tf2par_z(Be_d,Ae_d) % Parallel form realization:
```

The magnitude curve of the frequency response of the designed filter is shown in Fig. 7.13. If the order and cutoff frequency are determined, then we use the FDATool to design a digital filter as depicted in Fig. 7.14, yielding the cascade realization close to that obtained with `ellip()`. Note that the cutoff frequency fce determined by `ellipord()` and used by `ellip()` and FDATool is normalized to the range of [0,1] with 1 corresponding to Fs/2[Hz] (half the sampling frequency).

(e) All the above filter design works are put into the M-file named "sig07e02.m", which plots the frequency responses of the designed filters so that one can check if the design specifications are satisfied. Figs. 7.7(a), 7.9(a), 7.11(a), and 7.13(a), obtained by running the program "sig07e02.m", show the following points:

- Figure 7.7(a) shows the monotone frequency response magnitude of a Butterworth filter.
- Figure 7.9(a) shows that the critical passband edge frequencies f_{p1} and f_{p2} given as an input argument fcc1 = [fp1 fp2] of 'cheby1()' are the frequencies at which the passband ripple condition is closely met. Note that the frequency response magnitude of a Chebyshev I filter satisfying the passband ripple condition closely has a ripple in the passband, which is traded off for a narrower transition band than the Butterworth filter (with the same filter order).
- Figure 7.11(a) shows that the critical stopband edge frequencies f_{s1} and f_{s2} given as an input argument fcc2 = [fs1 fs2] of 'cheby2()' are the frequencies at which the stopband attenuation condition is closely met. Note that the frequency response magnitude of a Chebyshev II filter satisfying the stopband attenuation condition closely has a ripple in the stopband.

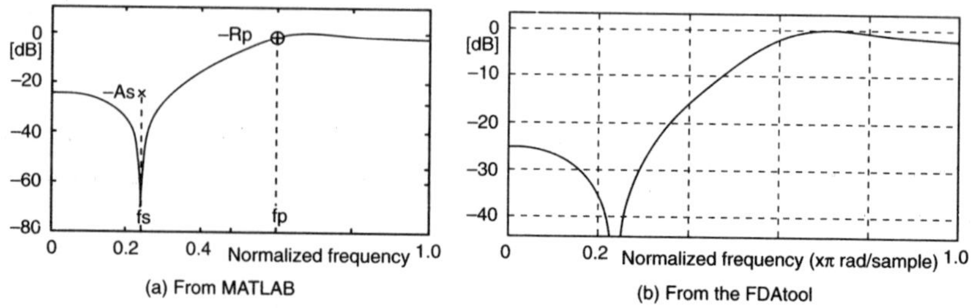

Fig. 7.13 The magnitude curves of the frequency response of the digital elliptic HPF

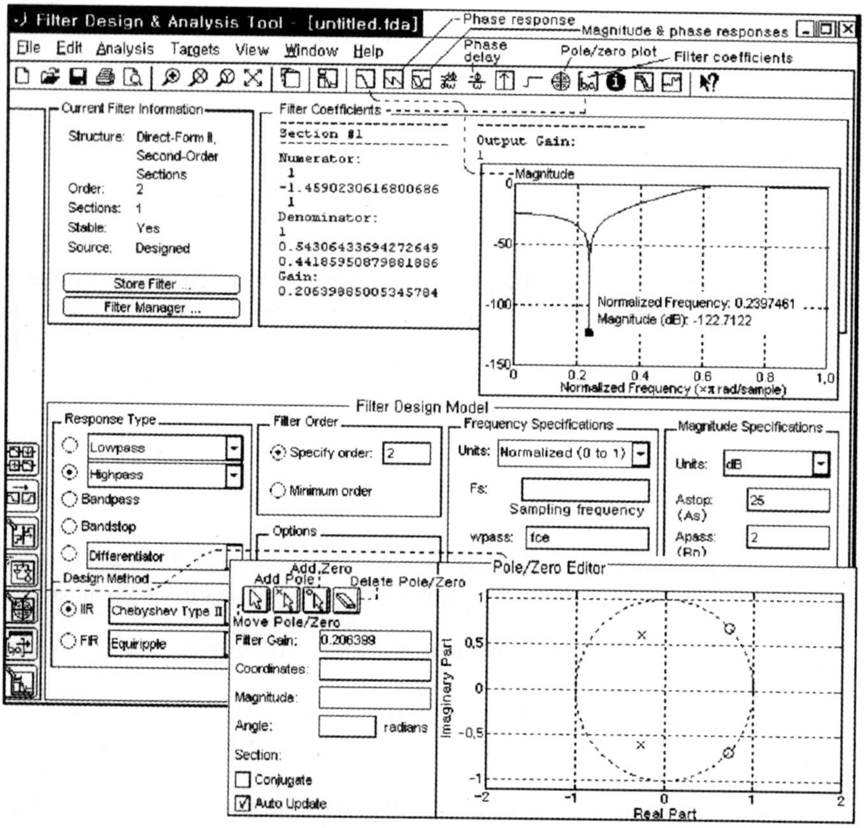

Fig. 7.14 Design of a digital elliptic HPF with order 2 and stopband/passband edge frequencies 0.24/0.6 using FDAtool (version 7.2-R2006a)

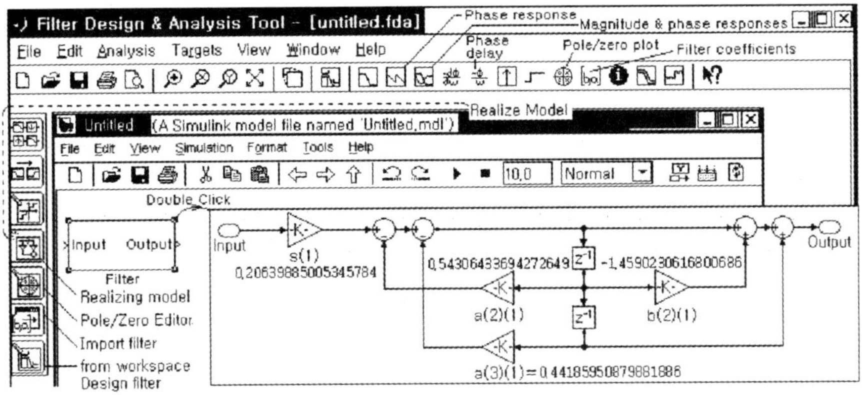

Fig. 7.15 The state diagram of a filter drawn by the 'Realize Model' function of FDAtool

- Figure 7.13(a) shows that the critical passband edge frequency f_p given as an input argument fce of 'ellip()' is the frequency at which the passband ripple condition is closely met. Note that the frequency response magnitude of an elliptic filter has ripples in both the passband and the stopband, yielding a relatively narrow transition band with the smallest filter order N = 2 among the four filters.

Note that we can click the Realize Model button or the Pole/Zero Editor button on the left-hand side of the FDATool window to create a Simulink model file having the designed filter block as depicted in Fig. 7.15 or to see the pole/zero plot of the designed filter as illustrated in Fig. 7.14.

7.2.2 FIR Filter Design

FIR filters are compared with IIR filters as follows:

Advantages:	Disadvantages:
- FIR filters can have exactly linear phase.	- FIR filters require a much higher filter order than IIR filters to achieve a given level of performance.
- FIR filters are stable.	- The delay of FIR filters is often much greater than that of an equal performance IIR filter.

According to the signal processing toolbox user's guide (page 2–16), FIR filter design methods cast into MATLAB functions are listed in Table 7.1.

Table 7.1 FIR filter design methods

Windowing	Apply windowing to truncated IDFT of desired "brick wall" filter	`fir1(), fir2()`
Multibands with transition bands	Equiripple or least squares approach over subbands of the frequency range	`firpm(), firls()`
Constrained least squares	Minimize squared integral error over entire frequency range subject to maximum error constraints	`fircls(), fircls1()`
Arbitrary responses	Arbitrary responses, including nonlinear phase and complex filters	`cfirpm()`
Raised cosine	Lowpass response with smooth, sinusoidal transition	`firrcos()`

As discussed in Sect. 4.6, the phases of FIR filters having symmetric or anti-symmetric coefficients are linear so that the phase delay τ_p and group delay τ_g are equal and constant over the frequency band, where the *phase delay* and *group delay* are related to the phase shift $\theta(\Omega) = \angle G(\Omega)$ of the frequency response as

$$\tau_p = -\frac{\theta(\Omega)}{\Omega} \text{ and } \tau_g = -\frac{d\theta(\Omega)}{d\Omega} \text{ with } \theta(\Omega) = \angle G(\Omega)(\text{phase response}) \quad (7.2.5)$$

Table 7.2 shows some restrictions on the frequency response depending on the symmetry of the FIR filter (tap) coefficients $g[n]$ and whether the filter order is even or odd. Note that fir1(N,fc,'high') and fir1(N,[fc1 fc2], 'stop') with an odd filter order N design an (N + 1)th-order HPF and BSF (of type I), respectively since no linear phase HPF or BSF of odd order (type II) can be made.

Table 7.2 FIR filter design methods (see Sect. 4.6)

Linear phase filter type	Filter order N	Symmetry of filter (tap) coefficients $g[n]$	Restriction on frequency response
Type I	Even	Symmetric $g[n] = g[N-n]$ for $n = 0 : N$	No restriction
Type II	Odd	Symmetric $g[n] = g[N-n]$ for $n = 0 : N$	$G(\Omega)\|_{\Omega=\pi} = 0$ (No HPF or BSF)
Type III	Even	Anti-symmetric $g[n] = -g[N-n]$ for $n = 0 : N$	$G(\Omega)\|_{\Omega=0} = 0$, $G(\Omega)\|_{\Omega=\pi} = 0$ (No LPF or HPF or BSF)
Type IV	Odd	Anti-symmetric $g[n] = -g[N-n]$ for $n = 0 : N$	$G(\Omega)\|_{\Omega=0} = 0$ (No LPF)

7.2.2.1 Windowing Method

Consider the frequency response of an ideal ("brick wall") LPF with cutoff frequency Ω_c as

$$G_I(\Omega) = G_I[e^{j\Omega}] = \begin{cases} 1 & \text{for } |\Omega - 2m\pi| \le \Omega_c \le \pi (m : \text{an integer}) \\ 0 & \text{elsewhere} \end{cases} \quad (7.2.6)$$

Referring to Eq. (E3.5.2), we can take the inverse DTFT of this frequency response to write the impulse response of an ideal LPF as

$$g_I[n] = \mathcal{F}^{-1}\{G_I(\Omega)\} \overset{(3.1.3)}{=} \frac{1}{2\pi}\int_{-\Omega_c}^{\Omega_c} G(\Omega)e^{j\Omega n}d\Omega = \frac{\Omega_c}{\pi}\text{sinc}\left(\frac{\Omega_c n}{\pi}\right) \quad (7.2.7)$$

Since this infinite-duration impulse response cannot be realized by a FIR filter, we apply a (rectangular) window of, say, length $N+1$ ($N = 2M$: an even number) to truncate it to make the tap coefficients of an N th-order FIR LPF as

$$g_{LP}[n] = g_I[n]r_{2M+1}[n+M] = \frac{\Omega_c}{\pi}\text{sinc}\left(\frac{\Omega_c n}{\pi}\right) \text{ for } -M \le n \le M (M = \frac{N}{2}) \quad (7.2.8)$$

We can multiply this with $2\cos(\Omega_p n)$ (for modulation) to obtain the tap coefficients of an Nth-order FIR BPF with passband center frequency Ω_p and bandwidth $2\Omega_c$ as

$$g_{BP}[n] = 2g_{LP}[n]\cos(\Omega_p n) \text{ for } -M \le n \le M (M = \frac{N}{2}) \quad (7.2.9)$$

We can also get the tap coefficients of Nth-order FIR HPF (with cutoff frequency Ω_c) and BSF (with stopband center frequency Ω_s and bandwidth $2\Omega_c$) by subtracting the tap coefficients of LPF and BPF from the unit impulse sequence $\delta[n]$ as

$$g_{HP}[n] = \delta[n] - g_{LP}[n] = \delta[n] - \frac{\Omega_c}{\pi}\text{sinc}\left(\frac{\Omega_c n}{\pi}\right) \text{ for } -M \le n \le M (M = N/2) \quad (7.2.10)$$

$$g_{BS}[n] = \delta[n] - g_{BP}[n] = \delta[n] - 2g_{LP}[n]\cos(\Omega_s n) \text{ for } -M \le n \le M (M = N/2) \quad (7.2.11)$$

where $\delta[n]$ is the impulse response of an all-pass filter with a flat frequency response (see Example 3.4). Note the following:

- The tap coefficients or impulse responses of FIR filters (7.2.8), (7.2.9), (7.2.10), and (7.2.11) should be delayed by $M = N/2$ to make the filters causal for physical realizability (see Sect. 1.2.9).

- Cutting off the impulse response of an ideal filter abruptly for truncation results in ripples near the band edge of the frequency response. Thus a non-rectangular window such as a Hamming window is often used to reduce the Gibbs effect.

Given a filter order N and description of the desired frequency response including 6dB band edge frequency (vector), the MATLAB functions `fir1()` and `fir2()` return the N + 1 tap coefficients of a FIR filter designed by the windowing method. Let us try using the MATLAB functions for FIR filter design in the following examples:

Example 7.3 Standard Band FIR Filter Design

Let us design several standard band FIR filters with sampling frequency $F_s = 50$ [kHz]. The following program "sig07e03.m" finds the tap coefficients of an LPF with cutoff frequency 5kHz, a BPF with passband 7.5 $\sim$ 17.5kHz, a HPF with cutoff frequency 20kHz, and a BSF with stopband 7.5 $\sim$ 17.5kHz by using Eqs. (7.2.8), (7.2.9), (7.2.10), and (7.2.11) and the MATLAB function `fir1()`. Fig. 7.16 shows the impulse responses (tap coefficients) and frequency responses of the filters obtained by running the program.

```
%sig07e03.m to design standard band FIR filters
clear, clf
Fs= 5e4; T=1/Fs; % Sampling frequency and sampling period
N=30; M=N/2; % Filter order and its half
nn=[-M:M]; nn1=nn+M; % Duration of filter impulse response
fn=[0:512]/512; W= fn*pi; % normalized and digital frequency ranges
% LPF design
fc_LP= 5000*T*2; % Normalized 6dB cutoff frequency corresponding to 5kHz
gLP= fc_LP*sinc(fc_LP*nn); % Eq. (7.2.8)
gLP1= fir1(N,fc_LP); % filter impulse response or tap coefficients
GLP= abs(freqz(gLP,1,W))+eps; GLP1= abs(freqz(gLP1,1,W))+eps;
subplot(421), stem(nn1,gLP), hold on, stem(nn1,gLP1,'r.')
subplot(422), plot(fn,20*log10(GLP), fn,20*log10(GLP1),'r:')
% BPF design
fc_BP=[7500 17500]*T*2; % Normalized 6dB band edge frequencies
fp= sum(fc_BP)/2*pi; % Passband center frequency[rad/sample]
gBP= 2*gLP.*cos(fp*nn); % Eq. (7.2.9)
gBP1= fir1(N,fc_BP); % filter impulse response or tap coefficients
GBP= abs(freqz(gBP,1,W))+eps; GBP1= abs(freqz(gBP1,1,W))+eps;
subplot(423), stem(nn1,gBP), hold on, stem(nn1,gBP1,'r.')
subplot(424), plot(fn,20*log10(GBP), fn,20*log10(GBP1),'r:')
% HPF design
impulse_delayed= zeros(1,N+1); impulse_delayed(M+1)= 1; % Impulse
fc_HP= 20000*T*2; % Normalized 6dB cutoff frequency 20kHz
gHP= impulse_delayed-fc_HP*sinc(fc_HP*nn); % Eq. (7.2.10)
gHP1=fir1(N,fc_HP,'high'); %filter impulse response/tap coefficients
GHP= abs(freqz(gHP,1,W))+eps; GHP1= abs(freqz(gHP1,1,W))+eps;
subplot(425), stem(nn1,gHP), hold on, stem(nn1,gHP1,'r.')
subplot(426), plot(fn,20*log10(GHP), fn,20*log10(GHP1),'r:')
% BSF design
fc_BS=[7500 17500]*T*2; % Normalized 6dB band edge frequencies
```

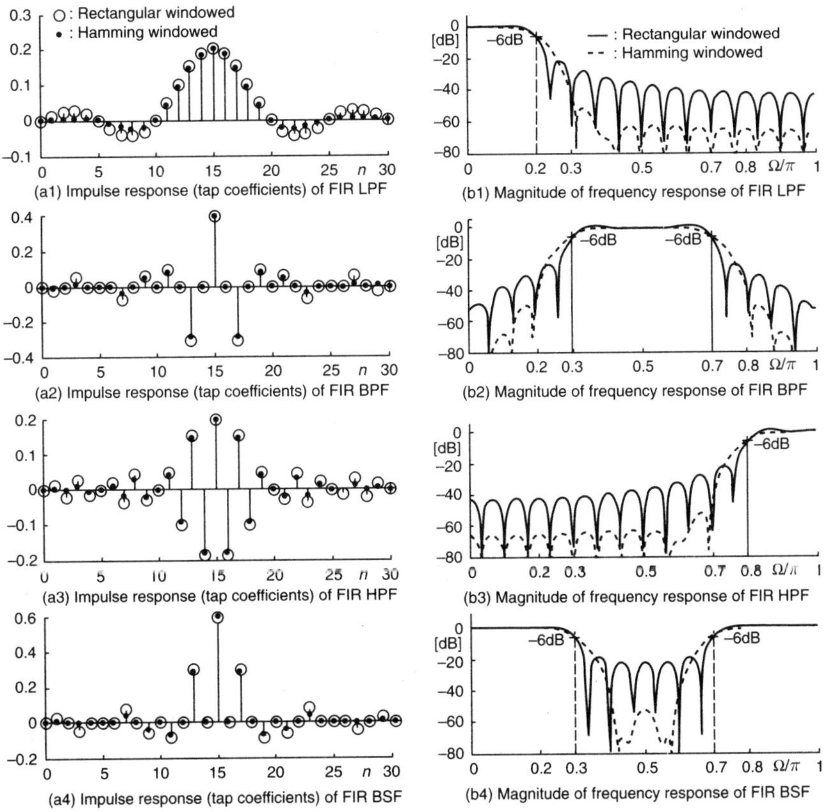

Fig. 7.16 Impulse responses (tap coefficients) and frequency responses of FIR filters

```
Ws= sum(fc_BS)/2*pi; % Stopband center frequency[rad/sample]
fc_LP= (fc_BS(2)-fc_BS(1))/2; gLP= fc_LP*sinc(fc_LP*nn);
gBS= impulse_delayed-2*gLP.*cos(Ws*nn); % Eq. (7.2.11)
gBS1=fir1(N,fc_BS,'stop'); % filter impulse response/tap coefficients
GBS= abs(freqz(gBS,1,W))+eps; GBS1= abs(freqz(gBS1,1,W))+eps;
subplot(427), stem(nn1,gBS), hold on, stem(nn1,gBS1,'r.')
subplot(428), plot(fn,20*log10(GBS), fn,20*log10(GBS1),'r:')
```

7.2.2.2 Multi-Band FIR Filter Design

We can use the MATLAB function `fir1(N,f,'DC-1')` or `fir1(N,f,'DC-0')` to design FIR filters whose frequency response magnitude is close to [1 0 1 0 ...] or [0 1 0 1 ...] for each band [f(1) ∼ f(2) f(2) ∼ f(3) f(3) ∼ f(4) f(4) ∼ f(5)...] between the neighboring edge frequencies. We can also use the MATLAB function `fir2(N,f,A)` or `firpm(N,f,A)` or `firls(N,f,A)` to design FIR filters whose frequency response magnitude is close to an arbitrary piecewise linear shape specified by the second and third input arguments f and A. Note the following:

- The firpm() function implements the Parks-McClellan algorithm, which uses the Remez exchange algorithm and Chebyshev approximation theory to design *minimax* or *equiripple* filters minimizing the maximum error between the desired and actual frequency responses [W-3]. It yields the frequency response exhibiting an equiripple behavior.
- The firls() function minimizes the integral of the squared error between the desired and actual frequency responses.
- The number of desired frequency-amplitude points must be even for firpm() and firls().
- While the desired frequency response for fir1() is supposed to be the brick wall type, firpm and firls() allow us to describe the desired frequency response so that we can insert the transition band(s) between the passband(s) and stopband(s) to reduce the ripple in the magnitude of the frequency response.

Example 7.4 Multi-Band FIR Filter Design

Figure 7.17 shows the impulse responses (tap coefficients) and frequency responses of the filters obtained by running the following program "sig07e04.m". Especially, Fig. 7.17(b4) illustrates the difference between the magnitude responses of the equiripple and LSE filters that are designed using firpm() and firls(), respectively.

```
%sig07e04.m
% to design multi-band FIR filters using fir1()/fir2()/firpm()
clear, clf
fn=[0:512]/512; W= fn*pi; % normalized and digital frequency ranges
N=30; M=N/2; % Filter order and its half
nn=[-M:M]; nn1=nn+M; % Duration of filter impulse response
%Multi pass/stop-band filter using fir1()
ffe=[0.2 0.4 0.6 0.8]; % Band edge frequency vector
B_fir1_DC1=fir1(N,ffe,'DC-1'); % Tap coefficient vector
G_fir1_DC1= freqz(B_fir1_DC1,1,W); % Frequency response magnitude
subplot(421), stem(nn1,B_fir1_DC1)
subplot(422), plot(fn,abs(G_fir1_DC1))
%Multi stop/pass-band filter using fir1()
B_fir1_DC0=fir1(N,ffe,'DC-0'); % Tap coefficient vector
G_fir1_DC0= freqz(B_fir1_DC0,1,W); % Frequency response magnitude
subplot(423), stem(nn1,B_fir1_DC0)
subplot(424), plot(fn,abs(G_fir1_DC0))
%Multi pass/stop-band filter using fir2()
% Desired piecewise linear frequency response
ffd=[0 0.18 0.22 0.38 0.42 0.58 0.62 0.78 0.82 1]; % Band edges in pairs
GGd=[1 1    0    0    1    1    0    0    1    1];
B_fir2=fir2(N,ffd,GGd);
G_fir2= freqz(B_fir2,1,W);
subplot(425), stem(nn1,B_fir2)
subplot(426), plot(fn,abs(G_fir2), ffd,GGd,'k:')
```

```
%Multi-band pass/stop filter using firpm()
B_firpm=firpm(N,ffd,GGd); % The number of frequency points must be even
B_firls=firls(N,ffd,GGd); % The number of frequency points must be even
G_firpm= freqz(B_firpm,1,W);
G_firls= freqz(B_firls,1,W);
subplot(427), stem(nn1,B_firpm), hold on, stem(nn1,B_firls,'.')
subplot(428), plot(fn,abs(G_firpm), fn,abs(G_firls),'r-.', ffd,GGd,'k:')
```

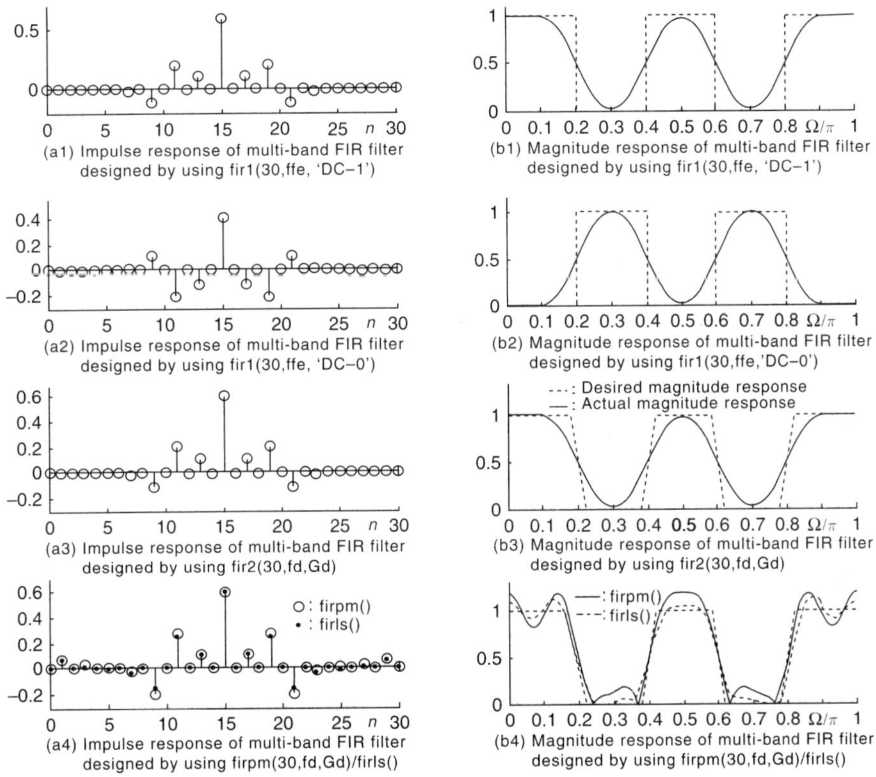

Fig. 7.17 Multi-band FIR filters designed using fir1(), fir2(), and firpm()/firls()

7.2.2.3 Anti-Symmetric FIR Filter Design

We can use the MATLAB functions `firpm()` or `firls()` with the fourth input argument 'h' or 'd' to design anti-symmetric FIR filters. Note that in the 'd' mode, `firpm()` and `firls()` weight the error by $1/\Omega$ and $1/\Omega^2$, respectively in nonzero amplitude bands to minimize the maximum relative error or sum of relative squared error. Note from Table 7.2 that the desired frequency response of an anti-symmetric FIR filter of an even order should be allowed to be zero at $\Omega = 0$ and π and that of an anti-symmetric FIR filter of an odd order should be allowed to be zero at $\Omega = 0$.

Example 7.5 Anti-Symmetric Filters – Hilbert Transformer and Differentiator

(a) Hilbert Transformer

The impulse response of the Hilbert transformer dealt with in Problems 2.10 and 3.2 has an odd symmetry so that it can easily be designed using `firpm()` or `firls()` with the fourth input argument `'h'` as in the following program "sig07e05a.m". We can run it to get Fig. 7.18, where the frequency response of the ideal Hilbert transformer is

$$H(\Omega) \overset{(P3.2.1)}{=} \begin{cases} -j & \text{for } 0 < \Omega < \pi \\ +j & \text{for } -\pi < \Omega < 0 \end{cases} \tag{7.2.12}$$

```
%sig07e05a.m
% to design Hilbert transformers as anti-symmetric FIR filters
% by using firpm() or firls().
clear, clf
fn=[0:512]/512; W= fn*pi; % normalized and digital frequency ranges
% Hilbert transformer using firpm() or firls()
%  Type III (anti-symmetric and even order) with G(0)=0 and G(pi)=0
B20_H_firpm= firpm(20,[0.05 0.95],[1 1],'h'); % Type III
G20_H_firpm= freqz(B20_H_firpm,1,W);
B20_H_firls= firls(20,[0.05 0.95],[1 1],'h'); % Type III
G20_H_firls= freqz(B20_H_firls,1,W);
% Type IV (anti-symmetric and odd order) with G(0)=0
B21_H_firpm= firpm(21,[0.05 1],[1 1],'h'); % Type IV
G21_H_firpm= freqz(B21_H_firpm,1,W);
B21_H_firls= firls(21,[0.05 1],[1 1],'h'); % Type IV
G21_H_firls= freqz(B21_H_firls,1,W);
subplot(421), nn=[0:20]; % Duration of filter impulse response
stem(nn,B20_H_firpm), hold on, stem(nn,B20_H_firls,'.')
subplot(422), plot(fn,abs(G20_H_firpm), fn,abs(G20_H_firls),':')
subplot(423), nn=[0:21]; % Duration of filter impulse response
stem(nn,B21_H_firpm), hold on, stem(nn,B21_H_firls,'.')
subplot(424), plot(fn,abs(G21_H_firpm), fn,abs(G21_H_firls),':')
% Use filter visualization tool to see the filter
fvtool(B20_H_firpm,1,B20_H_firls,1)
fvtool(B21_H_firpm,1,B21_H_firls,1)
```

(b) Differentiator

The impulse response of the differentiator dealt with in Problem 3.3 has an odd symmetry so that it can easily be designed by using `firpm()` or `firls()` with the fourth input argument `'d'` as in the following program "sig07e05b.m" where `firpm()` and `firls()` in the 'd' mode weight the error by $1/\Omega$ and $1/\Omega^2$, respectively. We can run it to get Fig. 7.19.

(Q)Why is the rightmost edge frequency in the second input argument of firpm()/firls() set to 0.95 instead of 1 for the type III FIR filter design despite the nonzero desired magnitude response at $\Omega = \pi$?

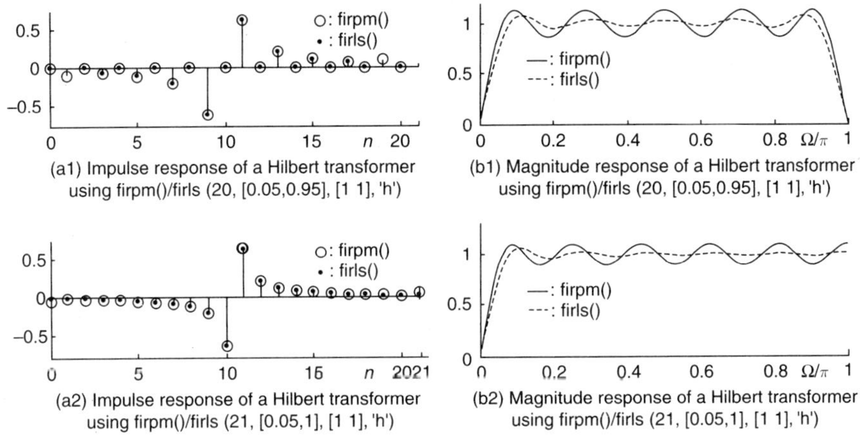

(a1) Impulse response of a Hilbert transformer using firpm()/firls (20, [0.05,0.95], [1 1], 'h')

(b1) Magnitude response of a Hilbert transformer using firpm()/firls (20, [0.05,0.95], [1 1], 'h')

(a2) Impulse response of a Hilbert transformer using firpm()/firls (21, [0.05,1], [1 1], 'h')

(b2) Magnitude response of a Hilbert transformer using firpm()/firls (21, [0.05,1], [1 1], 'h')

Fig. 7.18 Hilbert transformers designed using firpm() and firls()

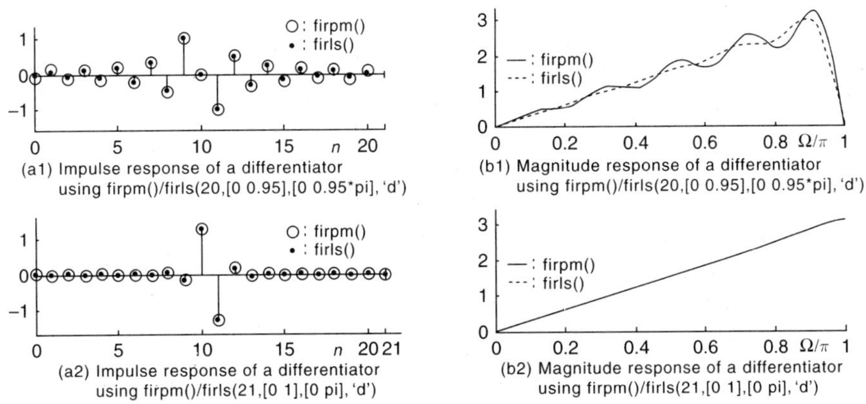

(a1) Impulse response of a differentiator using firpm()/firls(20,[0 0.95],[0 0.95*pi], 'd')

(b1) Magnitude response of a differentiator using firpm()/firls(20,[0 0.95],[0 0.95*pi], 'd')

(a2) Impulse response of a differentiator using firpm()/firls(21,[0 1],[0 pi], 'd')

(b2) Magnitude response of a differentiator using firpm()/firls(21,[0 1],[0 pi], 'd')

Fig. 7.19 Differentiators designed using firpm() and firls()

```
%sig07e05b.m
% to design differentiators as anti-symmetric FIR filters
% by using firpm() or firls().
clear, clf
fn=[0:512]/512; W= fn*pi; % normalized and digital frequency ranges
% Differentiator using firpm()or firls()
B20_d_firpm= firpm(20,[0 0.95],[0 0.95*pi],'d'); % Type III
G20_d_firpm= freqz(B20_d_firpm,1,W);
B20_d_firls= firls(20,[0 0.95],[0 0.95*pi],'d'); % Type III
G20_d_firls= freqz(B20_d_firls,1,W);
B21_d_firpm= firpm(21,[0 1],[0 pi],'d'); % Type IV with G(0)=0
G21_d_firpm= freqz(B21_d_firpm,1,W);
B21_d_firls= firls(21,[0 1],[0 pi],'d'); % Type IV with G(0)=0
G21_d_firls= freqz(B21_d_firls,1,W);
subplot(421), nn=[0:20]; % Duration of filter impulse response
stem(nn,B20_d_firpm), hold on, stem(nn,B20_d_firls,'.')
subplot(422), plot(fn,abs(G20_d_firpm), fn,abs(G20_d_firls),':')
subplot(423), nn=[0:21]; % Duration of filter impulse response
stem(nn,B21_d_firpm), hold on, stem(nn,B21_d_firls,'.')
subplot(424), plot(fn,abs(G21_d_firpm), fn,abs(G21_d_firls),':')
% Use filter visualization tool to see the filter
fvtool(B20_d_firpm,1,B20_d_firls,1)
fvtool(B21_d_firpm,1,B21_d_firls,1)
```

7.2.2.4 Constrained Least Squares (CLS) FIR Filter Design

There are two CLS filter design functions, `fircls()` and `fircls1()`. The `fircls()` function designs FIR filters whose magnitude responses are close to the desired piecewise constant one specified by a vector of band edges and the corresponding vector of band amplitudes. The upperbound and lowerbound of the desired magnitude response can also be given. The `fircls1()` function designs a FIR LPF or HPF whose magnitude response is close to the brick wall type for the passband/stopband specified by the cutoff frequency within the given tolerance on the passband and stopband ripples. Given the passband and stopband edge frequencies together with the weighting factor on stopband error relative to passband error, it applies the weighted least squares algorithm. In the case of HPF design, the filter order N given as the first input argument should be even; otherwise, it will be incremented by one to make an even filter order.

Example 7.6 Multi-Band CLS FIR Filter Design

The MATLAB function `fircls()` is used to design a two-band FIR filter with the desired magnitude response vector A = [0 0.5 0 1 0] for the bands specified by the band edge vector f = [0 0.3 0.5 0.7 0.9 1] in the following program "sig07e06.m". We run it to get Fig. 7.20, which shows the magnitude responses of the two filters where filter 1 and filter 2 are designed with the upperbound and lowerbound stricter on the stopband and the passband, respectively.

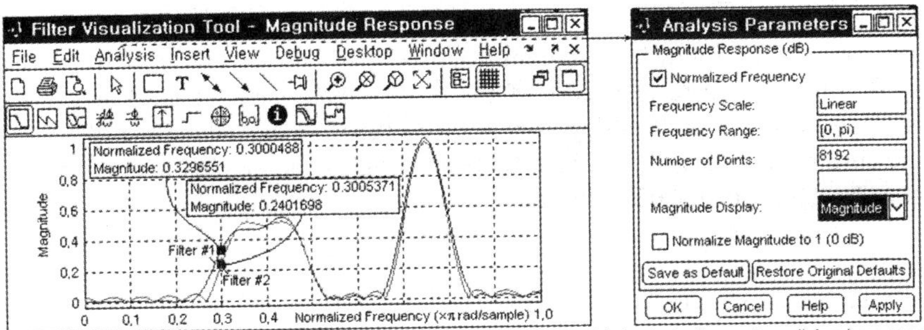

(a) Filter Visualization window opened by using fvtool() or clicking View/Filter_Visualization_Tool on the top menu bar

(b) Analysis Parameters dialog box opened by clicking Analysis/Analysis Parameters

Fig. 7.20 Multiband FIR filters designed using fircls(30,f,A,ub,lb) with f = [0 0.3 0.5 0.7 0.8 1] and A = [0 0.5 0 1 0]

(cf) We can change the unit of magnitude and frequency scale in the Analysis Parameters dialog box opened by clicking Analysis/Analysis_Parameters on the menu bar of the Filter Visualization window.

```
%sig07e06.m
% CLS design multi-band filters using fircls().
clear, refresh(1)
N=50; % Filter order
f= [0 0.3 0.5 0.7 0.8 1]; % The band edge frequency vector
A= [0 0.5 0 1 0]; % The magnitude response vector
ub1= [0.01 0.54 0.02 1.05 0.02]; lb1= [-0.01 0.46 -0.02 0.95 -0.02];
B1_fircls= fircls(N,f,A,ub1,lb1); % stricter on stopband ripple condition
ub2= [0.05 0.51 0.05 1.02 0.05]; lb2= [-0.05 0.49 -0.05 0.98 -0.05];
B2_fircls= fircls(N,f,A,ub2,lb2); % stricter on passband ripple condition
fvtool(B1_fircls,1, B2_fircls,1) %filter visualization tool to see filter
% Click any point on the frequency curve to add data markers
```

Example 7.7 CLS FIR LPF/HPF Design

(a)FIR LPF Dsign

We can run the following program "sig07e07a.m" to get Fig. 7.21(a), which shows the magnitude responses of the two filters where filter 2 is designed with more weighting (Kw > 1) on the stopband.

```
%sig07e07a.m
% to CLS design LPF filters using fircls1().
clear, refresh(1)
N=30; fc= 0.3; % Filter order and Cutoff frequency
fp=0.28; fs=0.32; % Passband and stopband edge frequencies
rp= 0.05; rs= 0.02; % Tolerances on passband and stopband ripple
% FIR LPF design using fircls1()
B1_LPF_fircls1= fircls1(N,fc,rp,rs);
Kw=10; % For more weighting on the stopband ripple condition
B2_LPF_fircls1= fircls1(N,fc,rp,rs,fp,fs,Kw);
% Use filter visualization tool to see the filter
fvtool(B1_LPF_fircls1,1, B2_LPF_fircls1,1)
```

(b)FIR HPF Dsign

We can run the following program "sig07e07b.m" to get Fig. 7.21(b), which shows the magnitude responses of the three filters where filters 2 and 3 are designed with more weighting ($Kw < 1$) on the passband. Note that filters 2/3 are ensured to satisfy the passband/stopband ripple condition at the frequency of $ft = fp + 0.02/fs - 0.02$, respectively.

```
%sig07e07b.m
% to CLS design HPF filters using fircls1().
clear, refresh(1)
N=30; fc= 0.3; % Filter order and Cutoff frequency
fp=0.28; fs=0.32; % Passband and stopband edge frequencies
rp= 0.05; rs= 0.02; % tolerances on passband and stopband ripple
% FIR HPF design using fircls1()
B_HPF_fircls1= fircls1(N,fc,rp,rs,'high');
Kw=0.1; %more weighting on passband ripple condition
% To ensure error(ft)<rp with ft within the passband
ft=fp+0.02;
B1_HPF_fircls1= fircls1(N,fc,rp,rs,fp,fs,Kw,ft,'high');
% To ensure error(ft)<rs with ft within the stopband
ft=fs-0.02;
B2_HPF_fircls1= fircls1(N,fc,rp,rs,fp,fs,Kw,ft,'high');
% Use filter visualization tool to see the filter
fvtool(B_HPF_fircls1,1, B1_HPF_fircls1,1, B2_HPF_fircls1,1)
```

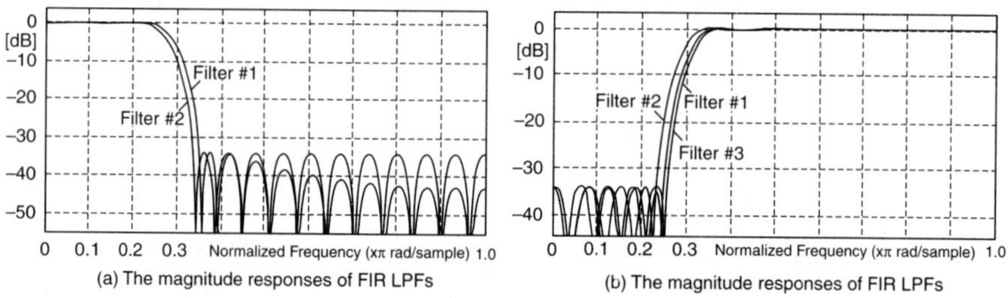

Fig. 7.21 FIR LPF/HPFs designed using fircls1()

7.2.2.5 Arbitrary-Response FIR Filter Design

The `cfirpm()` function designs a possibly nonlinear phase, asymmetric, complex-coefficient, and equiripple frequency response FIR filter minimizing the Chebyshev (or minimax) error between the actual magnitude response and desired one. Note the following:

- $B = \text{cfirpm}(N, f, A, w)$ returns a length $(N + 1)$ FIR filter whose magnitude response is the best approximation to the desired frequency response described by f and A where

  N: the filter order.

f: the vector of (an even number of) band edge frequencies arranged in ascending order between -1 and $+1$ where 1 is half the sampling frequency, i.e., the Nyquist frequency. The frequency bands span f(k) to f(k + 1) for k odd; the intervals f(k + 1) to f(k + 2) for k odd are "transition bands" or "don't care regions" during optimization.

A: a real vector (of the same size as f) which specifies the desired magnitude response as the line connecting the points (F(k),A(k)) and (F(k+1), A(k+1)) for odd k.

W: a vector of real, positive weights, one per band, for use during optimization. If not specified, it is set to unity.

- For filters with a gain other than zero at Fs/2, e.g., highpass and bandstop filters, N must be even. Otherwise, N will be incremented by one.
- B = cfirpm(N, f, {@fresp, p1, p2, ⋯}, w) returns a length (N + 1) FIR filter whose magnitude is the best approximation to the desired frequency response as returned by function @fresp with optional arguments p1,p2, ⋯. The function is called from within cfirpm() using the syntax [fg, wg] = fresp(N, f, fg, w, p1, p2, ⋯) where

fg: a vector of interpolated grid frequencies over each specified frequency band at which the response function will be evaluated.

Ag and wg: the desired complex frequency response and optimization weight vectors, respectively, evaluated at each frequency in grid fg.

- Predefined frequency response functions for @fresp include:

'lowpass', 'bandpass', 'hilbfilt', 'allpass', 'highpass', 'bandstop', 'differentiator', 'invsinc'

- B = cfirpm(⋯ , Sym) imposes a symmetry constraint on the impulse response, where Sym may be one of the following:

'none': Default if any negative band edge frequencies are passed, or if @fresp does not supply a default value of Sym.

'even': Impulse response will be real and even. This is the default for highpass, lowpass, bandpass, bandstop, and multiband designs.

'odd': Impulse response will be real and odd. This is the default for Hilbert and differentiator designs. The desired magnitude response at $\Omega = 0$ must be zero.

'real': Impose conjugate symmetry on the frequency response.

(cf) If any Sym option other than 'none' is specified, the band edges should only be specified over positive frequencies; the negative frequency region will be filled in from symmetry.

Example 7.8 Complex-Coefficient, Arbitrary Magnitude Response FIR Filter Design

The MATLAB function `cfirpm()` is used to design two FIR filters with different filter order N = 30 and 40, with the desired piecewise linear magnitude response connecting the frequency-amplitude points $\{(-1, 5), (-0.5, 1), (-0.4, 2), (0.3, 2), (0.4, 2), (0.9, 1)\}$ in the following program "sig07e08.m". We run it to get Fig. 7.22, which shows the impulse and magnitude responses of the two filters.

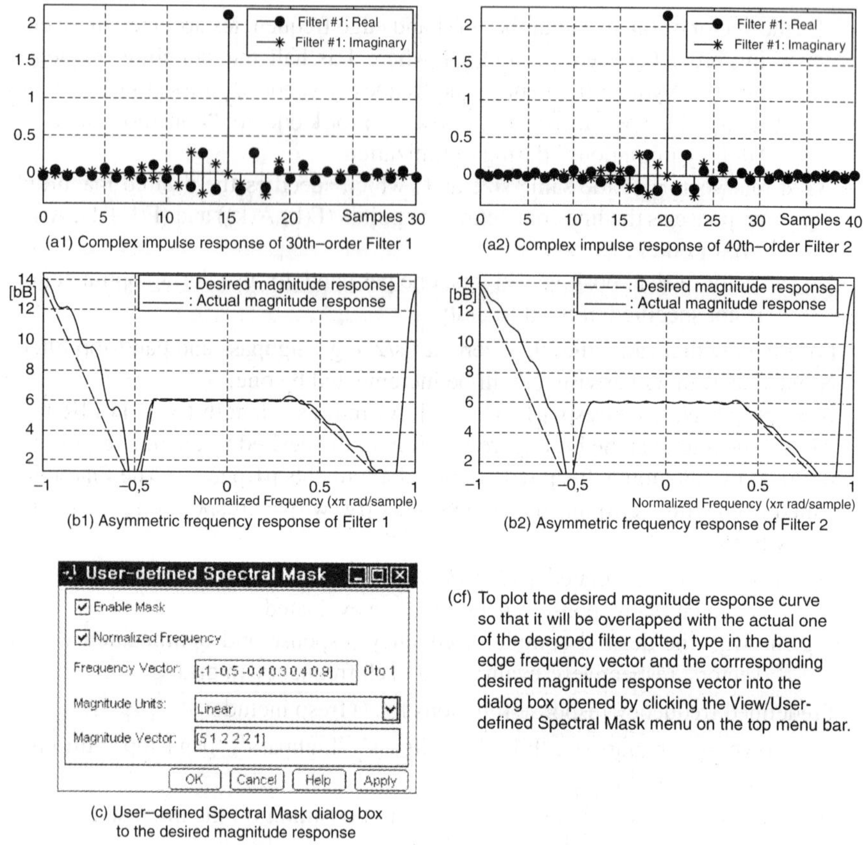

Fig. 7.22 Complex impulse response and frequency response of (nonlinear phase) FIR filters designed using cfirpm()

```
%sig07e08.m
% use cfirpm() to design FIR filters
%  having an arbitrary complex frequency response (with nonlinear phase)
clear, refresh
for N=[30 40]
    % Frequency/magnitude vectors describing desired frequency response
    f=[-1 -0.5 -0.4 0.3 0.4 0.9];
    A=[5 1 2 2 2 1]; %[14 0 6 6 6 0](dB)
    Kw=[1 10 5]; % A vector describing the weighting factor for each band
    B_cfirpm= cfirpm(N,f,A,Kw);
    % Use filter visualization tool to see the filter
    fvtool(B_cfirpm,1)
end
```

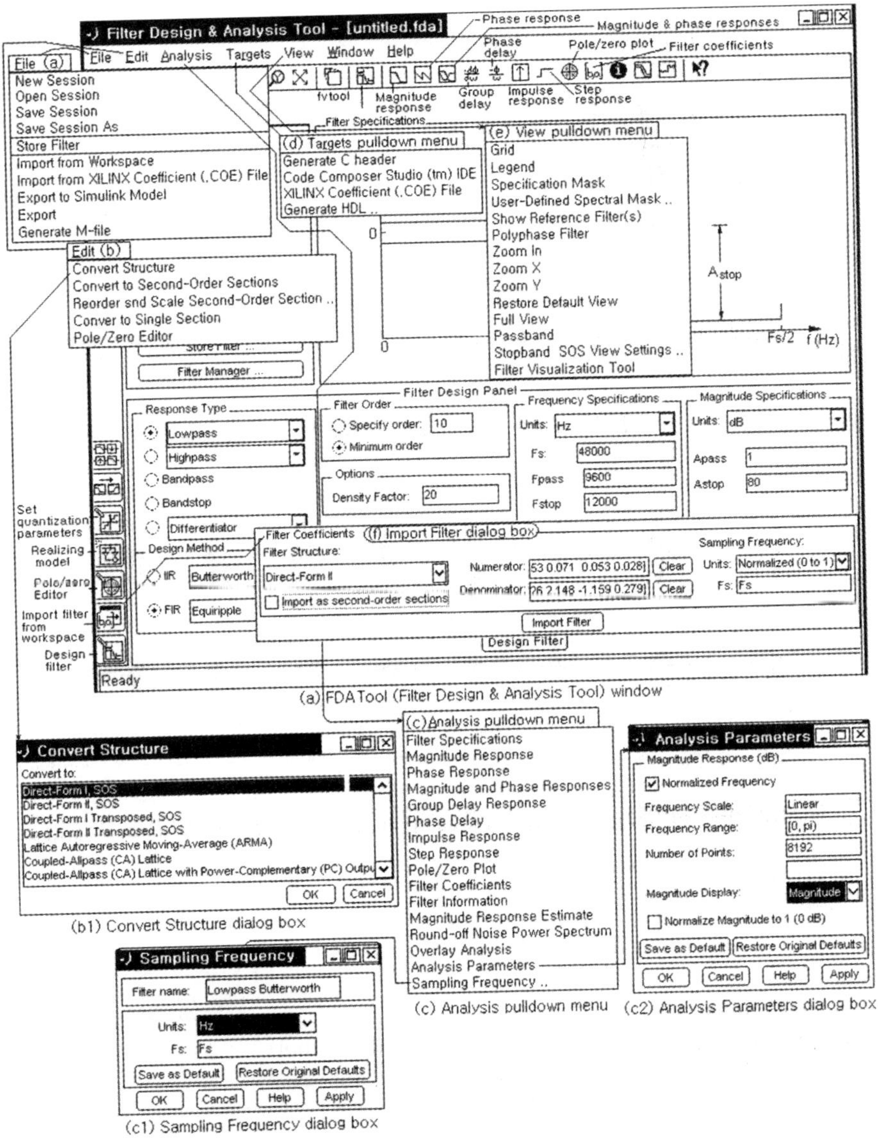

Fig. 7.24 FDA Tool window and its menu

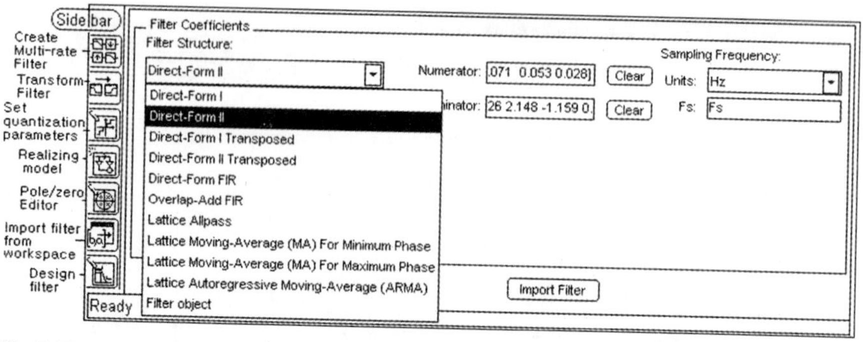

Fig. 7.25 Import Filter panel opened by clicking the Import Filter button in the side bar

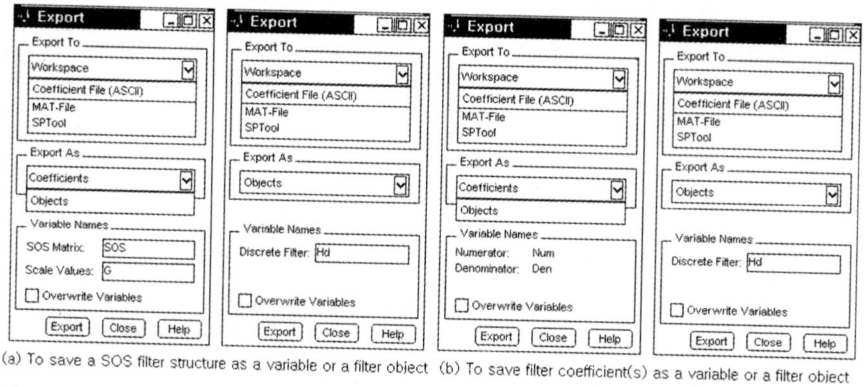

Fig. 7.26 Exporting a filter design result

7.3 How to Use SPTool

SPTool is an interactive GUI for digital signal processing that can be used to

- Analyze signals (Signal Browser)
- Design filters (FDATool: Filter Design & Analysis Tool)
- Analyze (view) filters (FVTool: Filter Visualization Tool)
- Filter signals
- Analyze signal spectra (Spectrum Viewer)

Signals, filters, and spectra can be brought from the MATLAB workspace into the SPTool workspace using File/Import. Signals, filters, and spectra created/modified in or imported into the SPTool workspace can be saved (as MATLAB structures) using File/Export.

Figure 7.27(a) shows the SPTool window opened by typing 'sptool' into the MATLAB command window where you can access the three GUIs, i.e., Signal

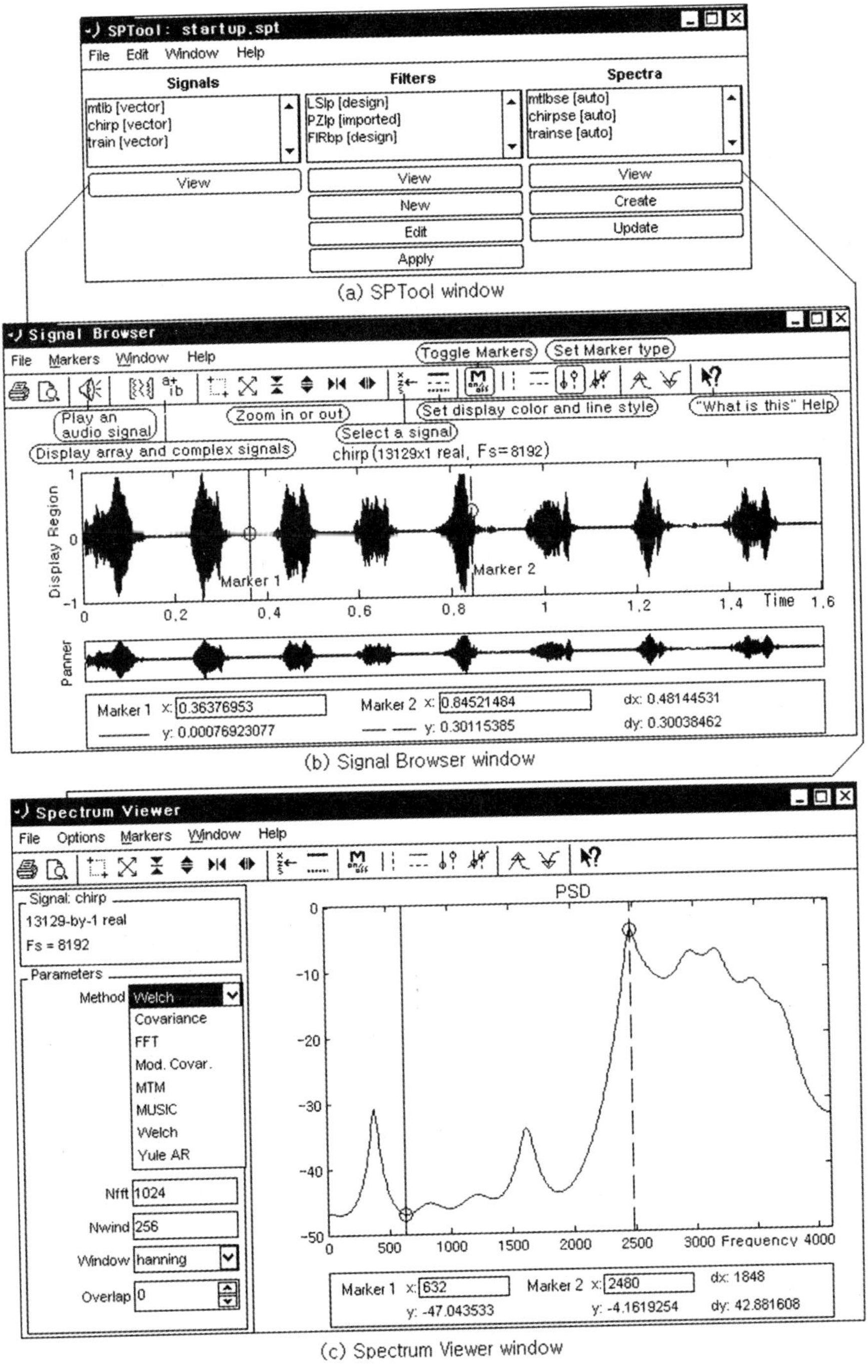

Fig. 7.27 SPTool window, Signal Browser window, and Spectrum Viewer window

Browser (Fig. 7.27(b)), Filter Visualization Tool (Fig. 7.20(a)), and Spectrum Viewer (Fig. 7.27(c)) by selecting a signal, filter, or spectrum and clicking the appropriate View button. Note that if you start FVTool by clicking the SPTool Filter View button, FVTool is linked to SPTool so that any changes made in SPTool are immediately reflected in FVTool where the FVTool title bar includes "SPTool" to indicate the link. (Every time you click the Filter View button, a new, linked FVTool starts, which allows you to view multiple analyses simultaneously. Any parameter except the sampling frequency can be changed in a linked FVTool. The sampling frequency can be changed through the Sampling Frequency dialog box opened by selecting the Edit/Sampling_Frequency menu in the SPTool window or by selecting the Analysis/Sampling_Frequency menu in the FDATool window, which will be opened by clicking the Filters/Edit button.) If you start an FVTool by clicking the New button or by selecting File/New from within FVTool, that FVTool is a stand-alone version that is not linked to SPTool. You can also access a reduced version of FDATool (Fig. 7.24(a)) by clicking the New button to create a new filter or the Edit button to edit a selected filter. You can apply a selected filter to a selected signal by clicking the Apply button. Clicking the Create button opens the Spectrum Viewer and shows the PSD (power spectral density) of the selected signal. Clicking the Update button opens the Spectrum Viewer for the selected spectrum.

Let us try using the SPTool in the following steps:

1. Create a noisy two-tone signal in the MATLAB workspace and import it into SPTool.
2. Design a BPF using FDATool.
3. Apply the designed filter to the signal to create a bandlimited noisy signal.
4. Analyze the input and output signals. For example, you can compare the original and filtered signals in the time domain using the Signal Browser.
5. Compare the original and filtered signals in the frequency domain using the Spectrum Viewer.
6. Save the filter design and signal analysis results by exporting to disk (MAT-file) or workspace.

<Step 1: Creating a noisy signal in the MATLAB workspace and importing it into SPTool>

You can type the following statements into the MATLAB command window:

```
>>Fs=1e4; Ts=1/Fs; tt=[0:Ts:0.5]; randn('state',0);
>>x=sin(2*pi*941*tt)+cos(2*pi*1209*tt)+0.1*randn(size(tt));
>>sptool % Open a SPTool window to start a new SPTool session
```

Then in the SPTool window, select the File/Import menu to open the Import-to-SPTool dialog box in which you can import x and Fs into the SPTool in the following way (see Fig. 7.28(a)):

- Select x from the Workspace Contents list, click the upper Right-Arrow button, and name the signal by typing, say, 'sig1' (default name) in the Name field to import x as a signal data.

7.2.3 Filter Structure and System Model Available in MATLAB

Figure 7.23 shows various filter structures that can be constructed, realized, and cast into a Simulink block by using FDATool or the dfilt()/realizemdl() command (see Sect. E.15). Note that all of the (transposed) direct I and II forms in Fig. 7.23(a)–(d) commonly have the following system function

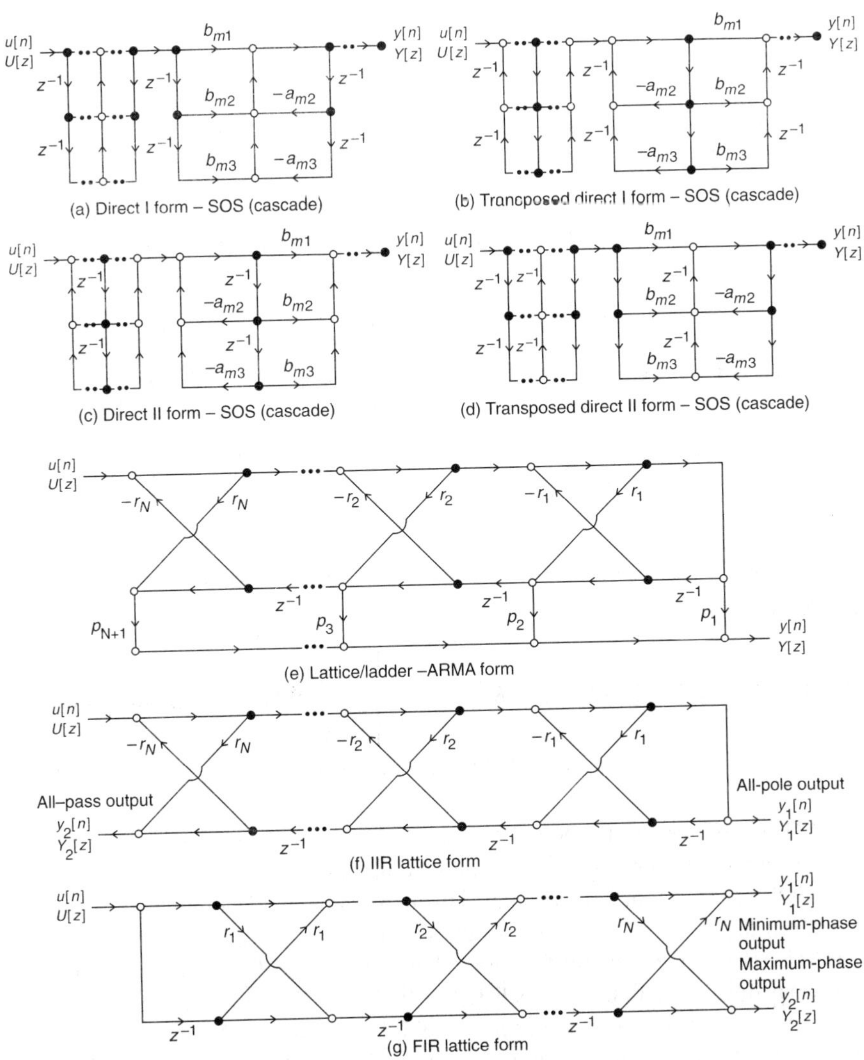

Fig. 7.23 Various filter structures

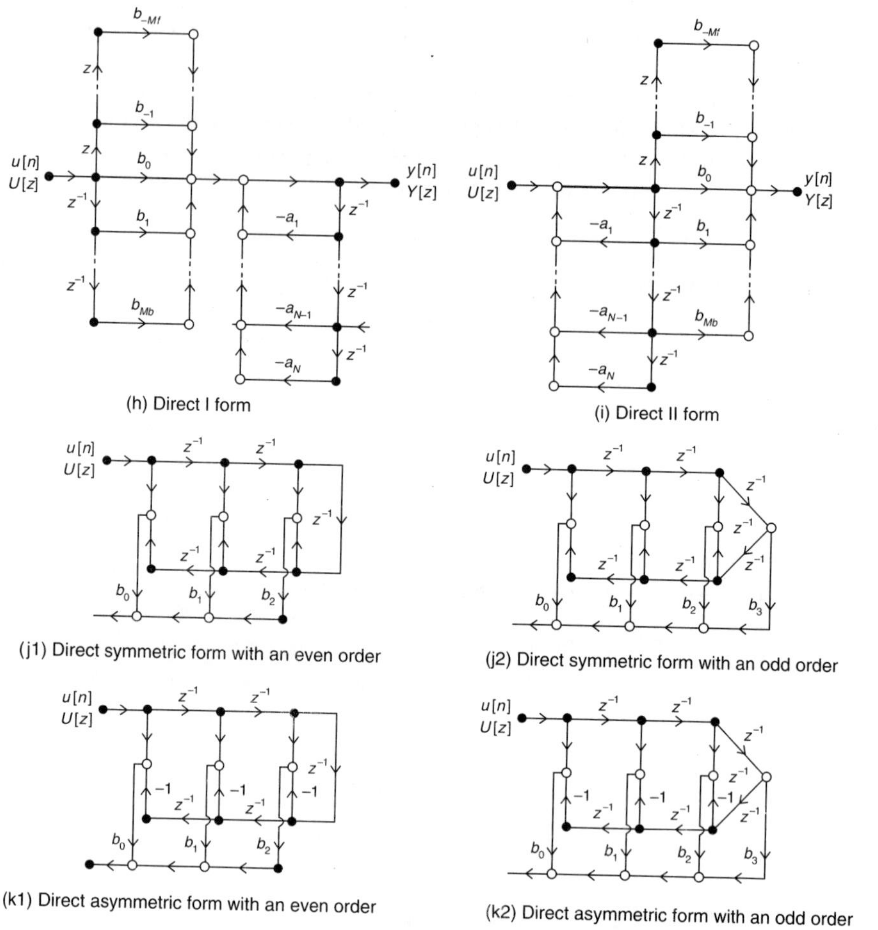

(h) Direct I form

(i) Direct II form

(j1) Direct symmetric form with an even order

(j2) Direct symmetric form with an odd order

(k1) Direct asymmetric form with an even order

(k2) Direct asymmetric form with an odd order

Fig. 7.23 (continued)

$$G[z] = KG_{M+1}[z] \prod_{m=1}^{M} G_m[z] = K \frac{b_{M+1,1}z + b_{M+1,2}}{z + a_{M+1,2}} \prod_{m=1}^{M} \frac{b_{m1}z^2 + b_{m2}z + b_{m3}}{z^2 + a_{m2}z + a_{m3}}$$

$$\text{with } M = \text{floor}(\frac{N}{2}) \tag{7.2.13}$$

MATLAB represents the SOSs (second-order sections) connected in cascade (with the output of each stage applied into the input of the next stage) as an $M \times 6$ array 'SOS', each row of which contains the numerator and denominator coefficient vector $[b_{m1}\ b_{m2}\ b_{m3}\ 1\ a_{m2}\ a_{m3}]$ of a single section.

MATLAB provides several functions that can be used for conversion between the various linear system models as listed in Table 7.3. Converting from one system model or filter structure to another may yield a result with different characteristic than the original due to the computer's finite-precision arithmetic operations and the variations in the conversion's round-off computations.

Table 7.3 MATLAB functions for conversion between the various linear system models

	System Function	State-Space	Zero-Pole-Gain	SOS (cascade)	SOS (parallel)	Lattice
System Function	—	`tf2ss()`	`tf2zp()`	`tf2sos()`	`tf2par_z()`*	`tf2latc()`
State-Space	`ss2tf()`	—	`ss2zp()`	`ss2sos()`	None	None
Zero-Pole-Gain	`zp2tf()`	`zp2ss()`	—	`zp2sos()`	None	None
SOS (cascade)	`sos2tf()`	`sos2ss()`	`sos2zp()`	—	None	None
SOS (parallel)	`par2tf()`*	None	None	None	—	None
Lattice	`latc2tf()`	None	None	None	None	—

(cf) * means that the MATLAB function is fabricated in this book.

In the Filter Design & Analysis window (Fig. 7.24(a)) opened by typing 'fdatool' in the MATLAB command window, we can convert the structure of a designed filter through the Convert Structure dialog box (Fig. 7.24(b1)) opened by clicking Edit/Convert_Structure (Fig. 7.24(b)) on the top menu bar. The following structures are available for conversion in MATLAB:

- Direct form I, Direct form II, Direct form I transposed, or Direct form II transposed
- Second-Order Sections
- Lattice minimum/maximum phase from minimum/maximum-phase FIR filter
- Lattice allpass from Allpass filter
- Lattice ARMA from IIR filter
- State-Space model

The MATLAB function dfilt() can also be used to convert the filter structure. For example,

```
>>B=firls(30,[0 .5 .6 1],[0 0 1 1]); Gd1=dfilt.dffir(B); %direct-form FIR
>>[B,A]=butter(7,0.2); Gd2=dfilt.df2tsos(tf2sos(B,A)); %direct IIt-SOS IIR
>>Gd_par=dfilt.parallel(Gd1,Gd2); %parallel structure of two or more objects
>>realizemdl(Gd_par) % To create Simulink model
>>fvtool(Gd1,Gd2,Gd_par) % To analyze filter
```

7.2.4 Importing/Exporting a Filter Design

In the FDATool window (Fig. 7.24), we can click the Import Filter button in the side bar to open the Import Filter panel (Fig. 7.25) and import a filter in any of the representations in the Filter Structure or a SOS by clicking on the check box. In the fields of Numerator and Denominator, type the filter coefficient vectors explicitly or the names of the variables whose values are in the MATLAB workspace. Then select the frequency unit from the options in the Units menu and if the selected frequency unit is not 'Normalized', specify the value or MATLAB workspace variable of the sampling frequency in the Fs field. Lastly click the Import Filter button at the bottom of the Import Filter panel to import the filter that you have specified. You can edit the imported filter using the Pole/Zero Editor panel in which you can move poles, add, or delete poles/zeros.

On the other hand, we can select the File/Export menu in the FDATool window to open the Export dialog box (Fig. 7.26) and save your filter design result by

- exporting the filter coefficients or objects to the workspace, a MAT-file, or a C header file,
- exporting the filter coefficients to an ASCII file,
- or exporting to SPTool or Simulink.

(cf) FDATool GUI is opened through the Digital Filter Design block in the Signal Processing Blockset.

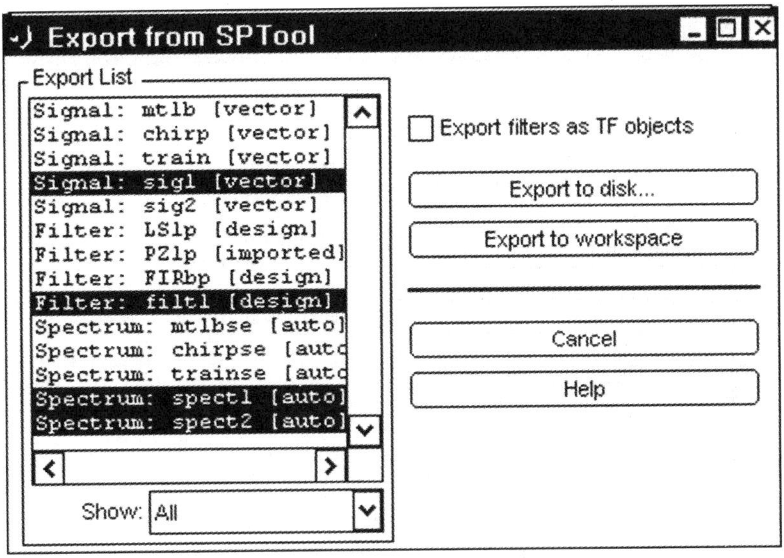

Fig. 7.30 Export_from_SPTool dialog box

Problems

7.1 Design and Use of a Filter to Remove/Reduce a Noise from a Music Signal

(a) Using any recording software, create a wave file of sampling frequency 8kHz, duration 16 s, and any melody that you like in the name of, say, "melody.wav". Then take the following steps:

(1) Use the MATLAB command 'wavread()' to extract the signal vector x and sampling frequency Fs from the wave file. Use 'soundsc()' to listen to the melody signal. Use 'fft()' & 'plot()' to plot x together with the magnitude of its DFT spectrum in dB.

(2) Add a noise of amplitude 0.5 and frequency 3.9kHz to the signal, listen to the noise-contaminated signal xn, and plot xn together with its DFT spectrum magnitude in dB.

(3) Design a Butterworth LPF with passband/stopband edge frequencies of 3.6kHZ/3.9kHz and use the LPF to filter the noise-contaminated signal to obtain a filtered signal xf. Then listen to the filtered signal xf and plot xf together with the magnitude of its DFT spectrum in dB.

```
%test_filtering.m
clear, clf
[x,Fs]=wavread('melody.wav');
N=2^17; x=x(end-N+1:end,1).'; % convert into row vector
soundsc(x,Fs);
Ts=1/Fs; t=(0:N-1)*Ts;  nn=1:N/32; tt=t(nn); % time vector
subplot(4,2,1), plot(tt,x(nn)), axis([tt([1 end]) -2 2])
xlabel('time[s]'), ylabel('signal x[n]')
X=fftshift(fft(x,N)); X_mag=20*log10(abs([X X(1)]));
f=(-N/2:N/2)*(Fs/N); fkHz=f/1000; % frequency vector
subplot(4,2,2), plot(fkHz,X_mag), axis([-5 5 -40 100])
xlabel('f[kHz]'), ylabel('20*log10|X(k)|[dB]')
% Add a High Frequency Noise
omega=2*pi*3900*Ts; % convert 3.9 kHz into digital (DT) frequency
%omega=2*pi*5000*Ts; % convert 5 kHz into digital (DT) frequency
n=0:N-1; noise=0.5*cos(omega*n); xn = x + noise; soundsc(xn,Fs);
subplot(4,2,3), plot(tt,xn(nn)), axis([tt([1 end]) -3 3])
xlabel('time[s]'), ylabel('noise-contaminated signal xn[n]')
Xn=fftshift(fft(xn,N)); Xn_mag=20*log10(abs([Xn Xn(1)]));
subplot(4,2,4), plot(fkHz,Xn_mag), axis([-5 5 -40 100])
xlabel('f[kHz]'), ylabel('20*log10|Xn(k)|[dB]'),
% Butterworth LPF Design
Rp=3; As=40; % Passband Ripple and Stopband Attenuation in dB
fp=3600*Ts*2; fs=3900*Ts*2; % passband/stopband edge frequency
[Nb,fcb]=buttord(fp,fs,Rp,As);
[Bb,Ab]=butter(Nb,fcb);
H=fftshift(freqz(Bb,Ab,N,'whole')); H_mag=20*log10(abs([H; H(1)]));
subplot(4,2,6), plot(fkHz,H_mag), axis([-5 5 -100 5])
xlabel('f[kHz]'), ylabel('20*log10|H(k)|[dB]'),
% Filtering to remove the 10kHz noise
xf=filter(Bb,Ab,xn);  soundsc(xf,Fs);
subplot(4,2,7), plot(tt,xf(nn)), axis([tt([1 end]) -2 2])
xlabel('time[s]'), ylabel('filetred signal xf[n]')
Xf=fftshift(fft(xf,N)); Xf_mag=20*log10(abs([Xf Xf(1)]));
subplot(4,2,8), plot(fkHz,Xf_mag); axis([-5 5 -40 100])
xlabel('f[kHz]'), ylabel('20*log10|Xf(k)|[dB]'),
```

(b) Referring to Fig. P7.1, make a Simulink model file to perform the filtering operation as done by the above MATLAB program "test_filtering.m". You can relocate the To_Wave_Device block to listen to the original signal, the noise-contaminated one, and the filtered one.

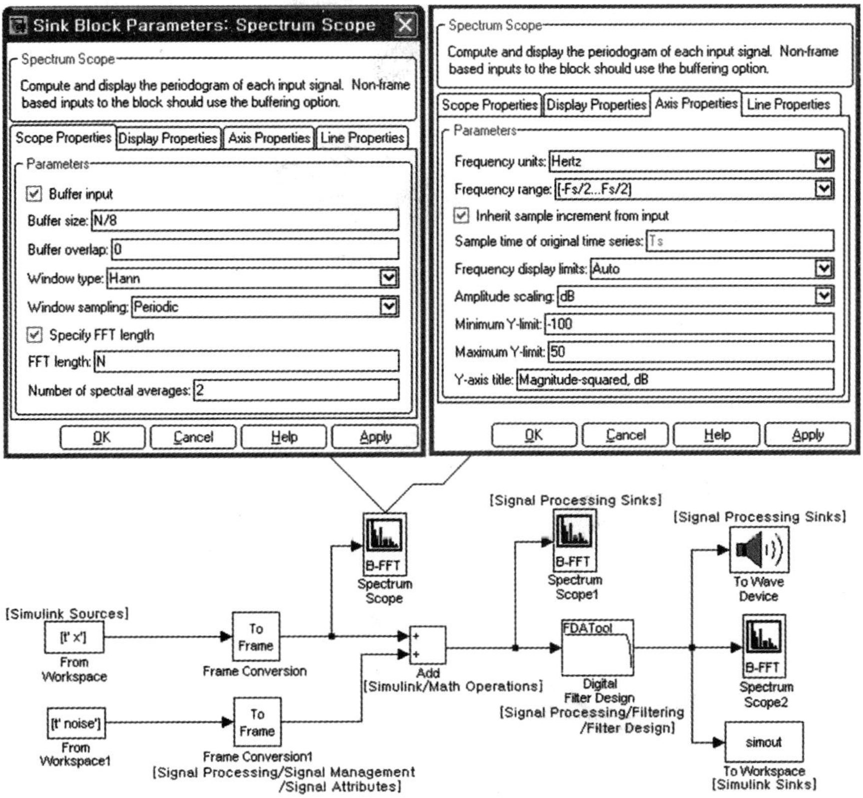

Fig. P7.1 Simulink block diagram for a signal filtering and the parameter setting dialog boxes for spectrum scope block

Problem

[illegible]

[illegible]

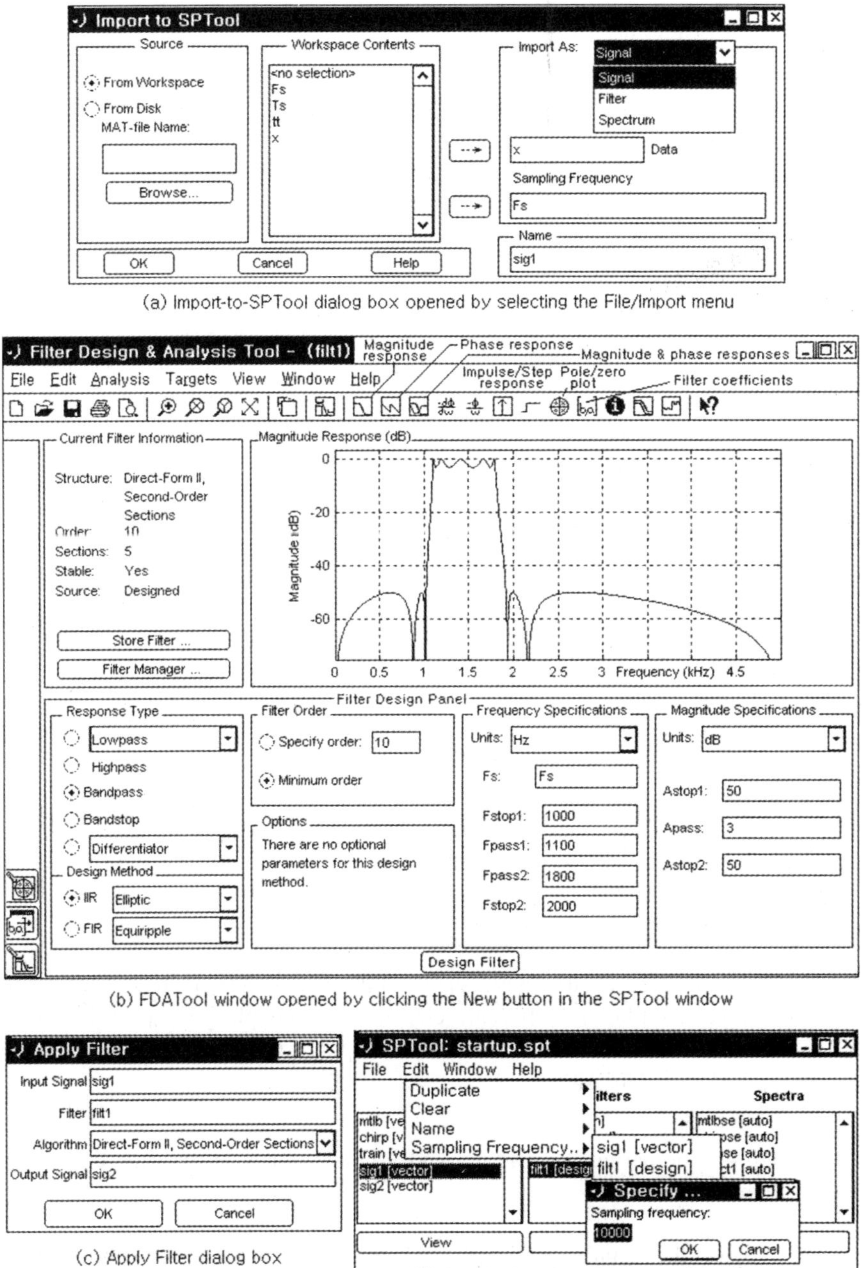

(a) Import-to-SPTool dialog box opened by selecting the File/Import menu

(b) FDATool window opened by clicking the New button in the SPTool window

(c) Apply Filter dialog box

(d) Specify_Sampling_Frequency dialog box

Fig. 7.28 Importing signal, designing filter, applying designed filter, and editing sampling frequency

- Select Fs from the workspace contents list and click the lower Right-Arrow button to import Fs as the sampling frequency.
- Click the OK button to close the Import-to-SPTool dialog box. Then you will see sig1[vector] in the SPTool's Signals list (see Fig. 7.29(a)).

(cf) You can also import signals from MAT-files on your disk.
(cf) You can import filters and spectra into SPTool in the same way as you import signals.

<Step 2: Using FDATool to design a BPF>
You might import an existing filter or design/edit a new filter using FDATool. Here, click the New button to open FDATool, type the following specification into the appropriate fields, and then click the Design Filter button to design a filter named 'filt1' by default (Fig. 7.28(b)):
Sampling frequency: Fs = 10kHz, Stopband attenuation: $A_s = 50$dB, and Passband ripple: $R_p = 3$dB

Passband edge frequencies: $f_{p1} = 1.1\,\text{kHz},\ f_{p2} = 1.8\,\text{kHz}$
Stopband edge frequencies: $f_{s1} = 1\,\text{kHz},\ f_{s2} = 2\text{kHz}$

<Step 3: Applying the designed filter to the signal to create a bandlimited signal>

- Select the signal 'sig1[vector]' from the Signals list by clicking on it (Fig. 7.27(a)).
- Select the filter 'filt1[design]' from the Filters list by clicking on it.
- Click the Apply button under the Filters list to open the Apply Filter dialog box (Fig. 7.28(c)), in which you can select the filtering algorithm, name the output signal, say, 'sig2', and click OK to close the dialog box. Then you will see sig2[vector] in the SPTool's Signals list.

(cf) If you see a warning message that the sampling frequencies for the signal and filter are different, select the Edit/Sampling_Frequency menu to make them equal and then apply the filter to the signal (Fig. 7.28(d)).

<Step 4: Comparing the original and filtered signals in the time domain in Signal Browser>

- Select the signals 'sig1' and 'sig2' from the Signals list by (Shift/Ctrl+) clicking on them.
- Click the View button under the Signals list to open the Signal Browser window and see the two signals in the time domain.
- If needed or desired, you can click the Select Trace button to select one of the displayed signals and then click the Line Properties button to change the color and/or line style to display the selected signal.

- You can also click the Play_Selected_Signal button to play the selected signal. You can click the Vertical_Markers button to use the vertical Markers to select a portion of the signal you want to play. If you want to print the signals (or their spectra), click the Print button.

<Step 5: Compare the original and filtered signals in the frequency domain in Spectrum Viewer>

- In the SPTool window (Fig. 7.29(a)), select the signal 'sig1[vector]' from the Signals list by clicking on it.
- Click the Create button under the Spectra list to create a PSD 'spect1' corresponding to the selected signal 'sig1' and open the Spectrum Viewer window (Fig. 7.29(b1)). Note that the PSD is not yet computed and displayed.
- Through the Parameters region, set the parameters such as the spectral analysis method (Welch), FFT size (Nfft = 1024), window size (Nwind = 1024), Window type (hanning), and Overlap length (512).
- Click the Apply button in the Spectrum Viewer window to compute and display the PSD spect1.
- Follow the above three steps for the filter output signal 'sig2' to create another PSD, say, spect2. In the Parameters region, you had better select 'spect1' from the PSD list in the Inherit_from field (Fig. 7.29(b2)) so that every spectral analysis parameter can be inherited from the existent PSD 'spect1'.
- You can Shift + click on 'spect1' and 'spect2' in the Spectra list to select them and click the View button under the Spectra list to reactivate the Spectrum Viewer and display the two spectra together.
- You can export the signals, filters, and spectra through the Export_from_SPTool dialog box opened by selecting the File/Export menu.

<Step 6: Save the filter design and signal analysis results>

You can save the designed filter(s) and signal analysis result(s) by exporting on your disk (as a MAT-file) or the MATLAB workspace in the following way:

- In the SPTool window, select the File/Export menu to open the Export_from_SPTool dialog box in which you can select the signals, filters, and spectra you want to save from the Export list and save them on your disk (as a MAT-file) or the MATLAB workspace by clicking the Export_to_Disk or Export_to_workspace button, respectively (see Fig. 7.30).
- If you have exported the filter structure 'filt1' into the workspace, you can refer to its numerator and denominator as 'filt1.tf.num' and 'filt1.tf.den'.

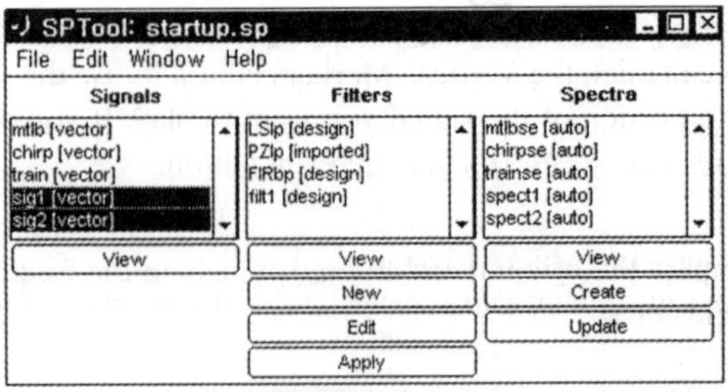

(a) SPTool window

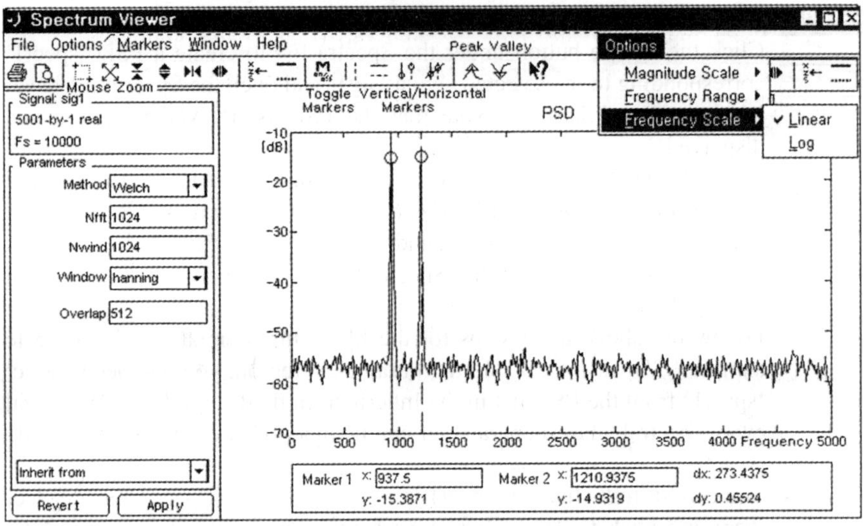

(b1) Spectrum Viewer window for the original signal 'sig1'

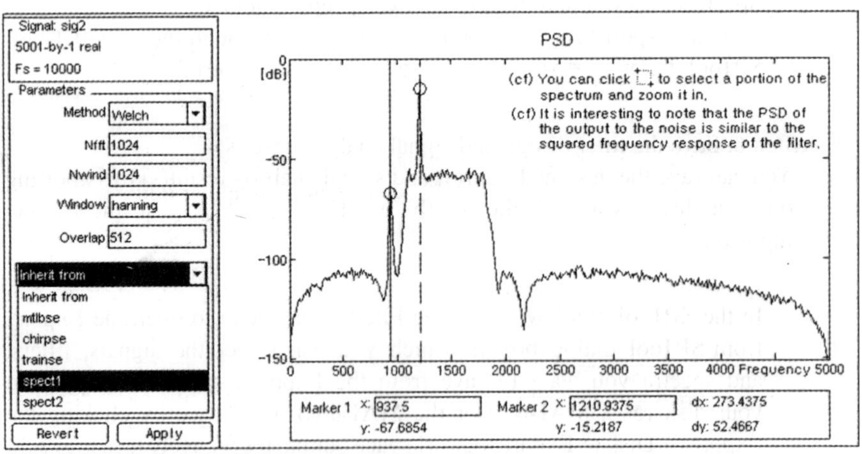

(b2) Spectrum Viewer window for the filtered signal 'sig2'

Fig. 7.29 SPTool window and Spectrum Viewer window

Chapter 8
State Space Analysis of LTI Systems

Contents

In this chapter we will introduce the state space description of a system, which consists of the state and output equations. It has several distinct features compared with the transfer function approach:

- It allows us to deal with multi-input multi-output (MIMO) systems in a systematic way.
- It describes a system more completely than the transfer function does. It describes not only the input-output relationship, but also what is happening under any initial condition, while the transfer function covers only systems with zero initial condition.
- It can be applied to certain types of nonlinear and/or time-varying systems.
- It is not easy to determine the state equation through experiments, while the transfer function of an LTI system can be measured by employing signal generators and spectrum analyzer.

W.Y. Yang et al., *Signals and Systems with MATLAB®*,
DOI 10.1007/978-3-540-92954-3_8,

8.1 State Space Description – State and Output Equations

In this section we introduce the state space description of an N th order LTI system, which consists of a set of equations describing the relations among the input, output, and state:

State equation: $\mathbf{x}'(t) = f(\mathbf{x}(t), \mathbf{u}(t), t)$ (8.1.1a) | $\mathbf{x}[n+1] = f(\mathbf{x}[n], \mathbf{u}[n], n)$ (8.1.1b)

Output equation: $\mathbf{y}(t) = g(\mathbf{x}(t), \mathbf{u}(t), t)$ (8.1.2a) | $\mathbf{y}[n] = g(\mathbf{x}[n], \mathbf{u}[n], n)$ (8.1.2b)

where

State vector: $\mathbf{x}(t) = [x_1(t), \cdots, x_N(t)]^T$ | $\mathbf{x}[n] = [x_1[n], \cdots, x_N[n]]^T$

Input vector: $\mathbf{u}(t) = [u_1(t), \cdots, u_K(t)]^T$ | $\mathbf{u}[n] = [u_1[n], \cdots, u_K[n]]^T$

Output vector: $\mathbf{y}(t) = [y_1(t), \cdots, y_M(t)]^T$ | $\mathbf{y}[n] = [y_1[n], \cdots, y_M[n]]^T$

(cf.) Note that, in this chapter, the notation $u(t)/u[n]$ denotes the general input function, while the unit step function/sequence is denoted by $u_s(t)/u_s[n]$.

Here, we have the definitions of the state and the state variable:

Definition 8.1 *State, State Variables, and State Vector*
The state of a system at time t_0 is the amount of information at t_0 that, together with the input from t_0, determines uniquely the behavior of the system for all $t > t_0$. Note that the 'behavior' means all the responses, including the state, of the system.

The state variables of a dynamic system are the variables forming the smallest set of variables which determine the state of the system. The state vector is composed of the state variables.

(Ex) For an *RLC* circuit driven by a source $e(t)$, the inductor current $i_L(t)$ and capacitor voltage $v_C(t)$ can form the state. The charge $q_C(t)$ and inductor current $i_L(t)$ can also make the state. It is because $\{i_L(t),\ v_C(t)\}$ or $\{q_C(t),\ i_L(t)\}$ can be determined for any $t > t_0$ if the value of input $e(t)$ is known for $t > t_0$ together with the initial condition $\{i_L(t_0),\ v_C(t_0)\}$ or $\{q_C(t_0),\ i_L(t_0)\}$.

(Ex) For a moving body, the set of the position $\mathbf{x}(t)$ and velocity $\mathbf{x}'(t)$ qualifies the state of the mechanical system since the information about $\mathbf{x}(t_0)$, $\mathbf{x}'(t_0)$, and $\mathbf{u}(t)$ (force) for $t > t_0$ is necessary and sufficient for determining $\mathbf{x}(t)$ and $\mathbf{x}'(t)$ for any time $t > t_0$.

Especially, the state space descriptions of continuous-time/discrete-time LTI systems are

State equation: $\mathbf{x}'(t) = A\mathbf{x}(t) + B\mathbf{u}(t)$ (8.1.3a) | $\mathbf{x}[n+1] = A\mathbf{x}[n] + B\mathbf{u}[n]$ (8.1.3b)

Output equation: $\mathbf{y}(t) = C\mathbf{x}(t) + D\mathbf{u}(t)$ (8.1.4a) | $\mathbf{y}[n] = C\mathbf{x}[n] + D\mathbf{u}[n]$ (8.1.4b)

In Sect. 1.3.4, we illustrated how a continuous-time/discrete-time state diagram can be constructed for a given differential/difference equation. Once a state diagram is constructed, the corresponding state equation can easily be obtained by the following procedure:

1. Assign a state variable $x_i(t)/x_i[n]$ to the output of each integrator s^{-1}/delay z^{-1}.
2. Write an equation for the input $x_i'(t)/x_i[n+1]$ of each integrator/delay.
3. Write an equation for each system output in terms of state variables and input(s).

Applying this procedure to the state diagrams in Figs. 1.19(a)/1.20(c) yields

$$\begin{bmatrix} x_1'(t) \\ x_2'(t) \end{bmatrix} = \begin{bmatrix} 0 & 1 \\ -a_0 & -a_1 \end{bmatrix} \begin{bmatrix} x_1(t) \\ x_2(t) \end{bmatrix} + \begin{bmatrix} 0 \\ 1 \end{bmatrix} u(t) \quad (8.1.5a)$$

$$\begin{bmatrix} x_1[n+1] \\ x_2[n+1] \end{bmatrix} = \begin{bmatrix} 0 & 1 \\ -a_0 & -a_1 \end{bmatrix} \begin{bmatrix} x_1[n] \\ x_2[n] \end{bmatrix} + \begin{bmatrix} 0 \\ 1 \end{bmatrix} u[n] \quad (8.1.5b)$$

$$y(t) = \begin{bmatrix} b_0 & b_1 \end{bmatrix} \begin{bmatrix} x_1(t) \\ x_2(t) \end{bmatrix} \quad (8.1.6a)$$

$$y[n] = \begin{bmatrix} b_0 & b_1 \end{bmatrix} \begin{bmatrix} x_1[n] \\ x_2[n] \end{bmatrix} \quad (8.1.6b)$$

which is referred to as the *controllable canonical form*. Also, for Figs. 1.19(b)/1.20(d), we obtain

$$\begin{bmatrix} x_1'(t) \\ x_2'(t) \end{bmatrix} = \begin{bmatrix} 0 & -a_0 \\ 1 & -a_1 \end{bmatrix} \begin{bmatrix} x_1(t) \\ x_2(t) \end{bmatrix} + \begin{bmatrix} b_0 \\ b_1 \end{bmatrix} u(t) \quad (8.1.7a)$$

$$\begin{bmatrix} x_1[n+1] \\ x_2[n+1] \end{bmatrix} = \begin{bmatrix} 0 & -a_0 \\ 1 & -a_1 \end{bmatrix} \begin{bmatrix} x_1[n] \\ x_2[n] \end{bmatrix} + \begin{bmatrix} b_0 \\ b_1 \end{bmatrix} u[n] \quad (8.1.7b)$$

$$y(t) = \begin{bmatrix} 0 & 1 \end{bmatrix} \begin{bmatrix} x_1(t) \\ x_2(t) \end{bmatrix} \quad (8.1.8a)$$

$$y[n] = \begin{bmatrix} 0 & 1 \end{bmatrix} \begin{bmatrix} x_1[n] \\ x_2[n] \end{bmatrix} \quad (8.1.8b)$$

which is referred to as the *observable canonical form*.

(cf.) Note that the controllable and observable canonical forms of state/output equations are convenient for designing a controller and an observer, respectively. [F-1]

8.2 Solution of LTI State Equation

8.2.1 State Transition Matrix

For later use, we will define the continuous-time/discrete-time LTI state transition matrices and examine their properties.

Definition 8.2 *LTI State Transition Matrix – Fundamental Matrix*
For an LTI system described by the state equations (8.1.3a)/(8.1.3b), the LTI state transition matrix or fundamental matrix $\phi(t)/\phi[n]$ is an $N \times N$ matrix, which is multiplied with the initial state $\mathbf{x}(0)/\mathbf{x}[0]$ to make the state $\mathbf{x}(t)/\mathbf{x}[n]$ at any time $t/n \geq 0$ as

$$\mathbf{x}(t) = \phi(t)\mathbf{x}(0) \qquad \Big| \qquad \mathbf{x}[n] = \phi[n]\mathbf{x}[0]$$

and satisfies the homogeneous state equation with zero input

$$\begin{aligned} &\mathbf{x}'(t) = A\mathbf{x}(t) \\ ;\ &\phi'(t)\mathbf{x}(0) = A\phi(t)\mathbf{x}(0) \\ ;\ &\phi'(t) = A\phi(t) \qquad (8.2.1a) \end{aligned} \qquad \Big| \qquad \begin{aligned} &\mathbf{x}[n+1] = A\mathbf{x}[n] \\ ;\ &\phi[n+1]\mathbf{x}[0] = A\phi[n]\mathbf{x}[0] \\ ;\ &\phi[n+1] = A\phi[n] \qquad (8.2.1b) \end{aligned}$$

where the initial condition is $\phi(0)/\phi[0] = I$ (an $N \times N$ identity matrix).

To find $\phi(t)$, we make use of Tables A.2(5)/B.7(2) (with $n_1 = 1$) to take the (unilateral) Laplace/z -transform of both sides of Eqs. (8.2.1a)/(8.2.1b) as

$$s\Phi(s) - \phi(0) = A\Phi(s) \qquad \Big| \qquad z\Phi[z] - z\phi[0] = A\Phi[z]$$

and solve this for $\Phi(s)/\Phi[z]$ as

$$\begin{aligned} \Phi(s) &= [sI - A]^{-1}\phi(0) = [sI - A]^{-1} \\ &= [I - As^{-1}]^{-1}s^{-1} \\ &= Is^{-1} + As^{-2} + A^2s^{-3} + \cdots \end{aligned} \qquad \Big| \qquad \begin{aligned} \Phi[z] &= [zI - A]^{-1}z\phi[0] = [zI - A]^{-1}z \\ &= [I - z^{-1}A]^{-1} \\ &= I + Az^{-1} + A^2z^{-2} + \cdots \end{aligned}$$

Now we use Table A.1(4)/Eq. (4.1.1) to take the inverse transform of $\Phi(s)/\Phi[z]$ to get

$$\begin{aligned} \phi(t) &= \mathcal{L}^{-1}\{\Phi(s)\} = \mathcal{L}^{-1}\{[sI - A]^{-1}\} \\ &= I + At + \frac{A^2}{2!}t^2 + \cdots = e^{At} \qquad (8.2.2a) \end{aligned} \qquad \Big| \qquad \begin{aligned} \phi[n] &= \mathcal{Z}^{-1}\{\Phi[z]\} = \mathcal{Z}^{-1}\{[zI - A]^{-1}z\} \\ &= A^n \qquad (8.2.2b) \end{aligned}$$

which is the continuous-time/discrete-time *LTI state transition* or *fundamental matrix*. This result can be verified by substituting Eqs. (8.2.2a)/(8.2.2b) into Eqs. (8.2.1a)/(8.2.1b), respectively.

The LTI state transition matrices possess the following properties:
<Properties of the LTI state transition matrix>

1) $\phi(t_1)\phi(t_2) = \phi(t_1 + t_2)\ \forall\ t_1, t_2$
2) $\phi(-t) = \phi^{-1}(t);\ \phi(0) = I$
3) $\phi(t) = e^{At}$ is nonsingular $\forall\ t < \infty$

1) $\phi[n_1]\phi[n_2] = \phi[n_1 + n_2]\ \forall\ n_1, n_2$
2) $\phi[-n] = \phi^{-1}[n];\ \phi[0] = I$ if $\phi[n]$ is nonsingular.

8.2.2 Transformed Solution

To solve the LTI state equations (8.1.3a)/(8.1.3b), we make use of Tables A.2(5)/B.7(2) (with $n_1 = 1$) to take the (unilateral) Laplace/z -transform of both sides and write

$$sX(s) - \mathbf{x}(0) = AX(s) + BU(s)$$
$$;\ [sI - A]X(s) = \mathbf{x}(0) + BU(s)$$

$$zX[z] - z\mathbf{x}[0] = AX[z] + BU[z]$$
$$;\ [zI - A]X[z] = z\mathbf{x}[0] + BU[z]$$

which can be solved for $X(s)/X[z]$ as

$$X(s) = [sI - A]^{-1}\mathbf{x}(0) + [sI - A]^{-1}BU(s) \tag{8.2.3a}$$

$$X[z] = [zI - A]^{-1}z\mathbf{x}[0] + [zI - A]^{-1}BU[z] \tag{8.2.3b}$$

Now, we will find the inverse transform of this transformed solution:

$$x(t) = \mathcal{L}^{-1}\{[sI - A]^{-1}\}\mathbf{x}(0) + \mathcal{L}^{-1}\{[sI - A]^{-1}BU(s)\} \tag{8.2.4a}$$

$$\mathbf{x}[n] = \mathcal{Z}^{-1}\{[zI - A]^{-1}z\}\mathbf{x}[0] + \mathcal{Z}^{-1}\{[zI - A]^{-1}BU[z]\} \tag{8.2.4b}$$

We can use Eq. (8.2.2), the convolution property, and the causality assumption to write

$$\mathcal{L}^{-1}\{[sI - A]^{-1}\} \overset{(8.2.2a)}{=} e^{At} \tag{8.2.5a}$$

$$\mathcal{L}^{-1}\{[sI - A]^{-1}BU(s)\} \overset{B.7(4)}{=} \mathcal{L}^{-1}\{[sI - A]^{-1}\} * \mathcal{L}^{-1}\{BU(s)\} \overset{(8.2.5a)}{=} e^{At} * \mathbf{Bu}(t) \overset{(A.17)}{=} \int_0^t e^{A(t-\tau)}\mathbf{Bu}(\tau)d\tau \tag{8.2.6a}$$

$$\mathcal{Z}^{-1}\{[zI - A]^{-1}z\} \overset{(8.2.2b)}{=} A^n \tag{8.2.5b}$$

$$\mathcal{Z}^{-1}\{[zI - A]^{-1}BU[z]\} \overset{B.7(4)}{=} \mathcal{Z}^{-1}\{[zI - A]^{-1}\} * \mathcal{Z}^{-1}\{BU[z]\} \overset{(8.2.5b)}{=} A^{n-1} * \mathbf{Bu}[n] = \sum_{m=0}^{n-1} A^{n-1-m}\mathbf{Bu}[m] \tag{8.2.6b}$$

Substituting Eqs. (8.2.5) and (8.2.6) into Eq. (8.2.4) yields the solution of the LTI state equation as

$$\mathbf{x}(t) = e^{At}\mathbf{x}(0) + \int_0^t e^{A(t-\tau)}B\mathbf{u}(\tau)d\tau = \phi(t)\mathbf{x}(0) + \int_0^t \phi(t-\tau)B\mathbf{u}(\tau)d\tau \tag{8.2.7a}$$

$$\mathbf{x}[n] = A^n\mathbf{x}[0] + \sum_{m=0}^{n-1} A^{n-1-m}B\mathbf{u}[m] = \phi[n]\mathbf{x}[0] + \sum_{m=0}^{n-1}\phi[n-1-m]B\mathbf{u}[m] \tag{8.2.7b}$$

which is referred to as the *state transition equation*. Note that, if the initial time is t_0/n_0, then we will have

$$\mathbf{x}(t) = e^{A(t-t_0)}\mathbf{x}(t_0) + \int_{t_0}^t e^{A(t-\tau)}B\mathbf{u}(\tau)d\tau = \phi(t-t_0)\mathbf{x}(t_0) + \int_{t_0}^t \phi(t-\tau)B\mathbf{u}(\tau)d\tau \tag{8.2.8a}$$

$$\mathbf{x}[n] = A^{n-n_0}\mathbf{x}[n_0] + \sum_{m=n_0}^{n-1} A^{n-1-m}B\mathbf{u}[m] = \phi[n-n_0]\mathbf{x}[n_0] + \sum_{m=n_0}^{n-1}\phi[n-1-m]B\mathbf{u}[m] \tag{8.2.8b}$$

Example 8.1 Solving a State Equation

(a) Consider a continuous-time LTI system described by the following differential equation:

$$y''(t) + y'(t) = u(t) \tag{E8.1.1}$$

Applying the procedure illustrated in Sects. 1.3.4 and 8.1, we can write the state equation as

$$\begin{bmatrix} x_1'(t) \\ x_2'(t) \end{bmatrix} \underset{a_0=0,a_1=1}{\overset{(8.1.5a)}{=}} \begin{bmatrix} 0 & 1 \\ 0 & -1 \end{bmatrix}\begin{bmatrix} x_1(t) \\ x_2(t) \end{bmatrix} + \begin{bmatrix} 0 \\ 1 \end{bmatrix} u(t) \tag{E8.1.2}$$

$$y(t) \underset{b_0=1,b_1=0}{\overset{(8.1.6a)}{=}} \begin{bmatrix} 1 & 0 \end{bmatrix}\begin{bmatrix} x_1(t) \\ x_2(t) \end{bmatrix} \tag{E8.1.3}$$

where

$$A = \begin{bmatrix} 0 & 1 \\ 0 & -1 \end{bmatrix},\; B = \begin{bmatrix} 0 \\ 1 \end{bmatrix},\; C = \begin{bmatrix} 1 & 0 \end{bmatrix},\; D = 0$$

Thus we have

$$[sI - A]^{-1} = \left[\begin{bmatrix} s & 0 \\ 0 & s \end{bmatrix} - \begin{bmatrix} 0 & 1 \\ 0 & -1 \end{bmatrix}\right]^{-1} = \begin{bmatrix} s & -1 \\ 0 & s+1 \end{bmatrix}^{-1} = \frac{1}{s(s+1)}\begin{bmatrix} s+1 & 1 \\ 0 & s \end{bmatrix} = \begin{bmatrix} s^{-1} & s^{-1}-(s+1)^{-1} \\ 0 & (s+1)^{-1} \end{bmatrix} \tag{E8.1.4}$$

so that the state transition matrix is

$$\phi(t) \overset{(8.2.2a)}{=} \mathcal{L}^{-1}\{[sI - A]^{-1}\} \underset{\text{B.8(3),(6)}}{\overset{\text{(E8.1.4)}}{=}} \begin{bmatrix} 1 & 1 - e^{-t} \\ 0 & e^{-t} \end{bmatrix} \text{ for } t \geq 0 \tag{E8.1.5}$$

Therefore, from Eq. (8.2.7a), we can write the solution of the state equation as

$$\begin{bmatrix} x_1(t) \\ x_2(t) \end{bmatrix} \underset{\text{(E8.1.5)}}{\overset{(8.2.7a)}{=}} \begin{bmatrix} 1 & 1 - e^{-t} \\ 0 & e^{-t} \end{bmatrix} \begin{bmatrix} x_1(0) \\ x_2(0) \end{bmatrix} + \int_0^t \begin{bmatrix} 1 - e^{-(t-\tau)} \\ e^{-(t-\tau)} \end{bmatrix} u(\tau) d\tau \text{ for } t \geq 0 \tag{E8.1.6}$$

(b) Consider a discrete-time LTI system described by the following state equation:

$$\begin{bmatrix} x_1[n+1] \\ x_2[n+1] \end{bmatrix} = \begin{bmatrix} 1 & 1 - e^{-T} \\ 0 & e^{-T} \end{bmatrix} \begin{bmatrix} x_1[n] \\ x_2[n] \end{bmatrix} + \begin{bmatrix} T - 1 + e^{-T} \\ 1 - e^{-T} \end{bmatrix} u[n] \tag{E8.1.7}$$

$$y[n] = \begin{bmatrix} 1 & 0 \end{bmatrix} \begin{bmatrix} x_1[n] \\ x_2[n] \end{bmatrix} \tag{E8.1.8}$$

where

$$A = \begin{bmatrix} 1 & 1 - e^{-T} \\ 0 & e^{-T} \end{bmatrix}, \quad B = \begin{bmatrix} T - 1 + e^{-T} \\ 1 - e^{-T} \end{bmatrix}, \quad C = \begin{bmatrix} 1 & 0 \end{bmatrix}, \quad D = 0$$

For this system, we have

$$\begin{aligned} [zI - A]^{-1} z &= \left[\begin{bmatrix} z & 0 \\ 0 & z \end{bmatrix} - \begin{bmatrix} 1 & 1 - e^{-T} \\ 0 & e^{-T} \end{bmatrix} \right]^{-1} z = \begin{bmatrix} z - 1 & -1 + e^{-T} \\ 0 & z - e^{-T} \end{bmatrix}^{-1} z \\ &= \frac{z}{(z-1)(z - e^{-T})} \begin{bmatrix} z - e^{-T} & 1 - e^{-T} \\ 0 & z - 1 \end{bmatrix} \\ &= \begin{bmatrix} \frac{z}{z-1} & \frac{(1-e^{-T})z}{(z-1)(z-e^{-T})} \\ 0 & \frac{z}{z-e^{-T}} \end{bmatrix} = \begin{bmatrix} \frac{z}{z-1} & \frac{z}{z-1} - \frac{z}{z-e^{-T}} \\ 0 & \frac{z}{z-e^{-T}} \end{bmatrix} \end{aligned} \tag{E8.1.9}$$

so that the state transition matrix is

$$\phi[n] \overset{(8.2.2b)}{=} \mathcal{Z}^{-1}\{[zI - A]^{-1} z\} \underset{\text{B.8(3),(6)}}{\overset{\text{(E8.1.9)}}{=}} \begin{bmatrix} 1 & 1 - e^{-nT} \\ 0 & e^{-nT} \end{bmatrix} \text{ for } n \geq 0 \tag{E8.1.10}$$

as would be obtained from $\phi[n] \overset{(8.2.2b)}{=} A^n$. Therefore, from Eq. (8.2.7b), we can write the solution of the state equation as

$$\begin{bmatrix} x_1[n] \\ x_2[n] \end{bmatrix} \underset{\text{(E8.1.10)}}{\overset{(8.2.7b)}{=}} \begin{bmatrix} 1 & 1 - e^{-nT} \\ 0 & e^{-nT} \end{bmatrix} \begin{bmatrix} x_1[0] \\ x_2[0] \end{bmatrix} + \sum_{m=0}^{n-1} \begin{bmatrix} T - (1 - e^{-T}) e^{-(n-1-m)T} \\ (1 - e^{-T}) e^{-(n-1-m)T} \end{bmatrix} u[m]$$

$$\text{for } n \geq 0 \tag{E8.1.11}$$

8.2.3 Recursive Solution

The discrete-time state equation has a recursive form and therefore is well-suited for a digital computer, which can be programmed to perform the following computation:

$$\begin{aligned}
\mathbf{x}[1] &= A\mathbf{x}[0] + B\mathbf{u}[0] \\
\mathbf{x}[2] &= A\mathbf{x}[1] + B\mathbf{u}[1] = A^2\mathbf{x}[0] + AB\mathbf{u}[0] + B\mathbf{u}[1] \\
\mathbf{x}[3] &= A\mathbf{x}[2] + B\mathbf{u}[2] = A^3\mathbf{x}[0] + A^2B\mathbf{u}[0] + AB\mathbf{u}[1] + B\mathbf{u}[2] \\
&\cdots\cdots \\
\mathbf{x}[n] &= A^n\mathbf{x}[0] + \sum_{m=0}^{n-1} A^{n-1-m}B\mathbf{u}[m]
\end{aligned} \tag{8.2.9}$$

If the continuous-time state equation is somehow discretized, it can be solved recursively, too.

8.3 Transfer Function and Characteristic Equation

8.3.1 Transfer Function

Using the transformed solution (8.2.3) of the state equation, we can write the transform of the output equation (8.1.4) with zero initial condition as

$$Y(s) = CX(s) + DU(s) \underset{\mathbf{x}(t)=0}{\overset{(8.2.3a)}{=}} C[sI - A]^{-1}BU(s) + DU(s) \tag{8.3.1a}$$

$$Y[z] = CX[z] + DU[z] \underset{\mathbf{x}[n]=0}{\overset{(8.2.3b)}{=}} C[zI - A]^{-1}BU[z] + DU[z] \tag{8.3.1b}$$

Thus the *transfer function matrix* describing the input-output relationship turns out to be

$$G(s) = C[sI - A]^{-1}B + D \tag{8.3.2a}$$

$$G[z] = C[zI - A]^{-1}B + D \tag{8.3.2b}$$

Taking the inverse transform of this transfer function matrix yields the *impulse response matrix*:

$$g(t) = C\phi(t)B + D\delta(t) \tag{8.3.3a}$$

$$g[n] = C\phi[n-1]B + D\delta[n] \tag{8.3.3b}$$

8.3.2 Characteristic Equation and Roots

Note that the transfer function matrix (8.3.2) can be written as

$$G(s) = C\frac{\text{Adj}\,[sI - A]}{|sI - A|}B + D \quad (8.3.4\text{a}) \qquad G[z] = C\frac{\text{Adj}\,[zI - A]}{|zI - A|}B + D \quad (8.3.4\text{b})$$

where Adj$[sI - A]$ denotes the adjoint, i.e., the transpose of the cofactor matrix.

The *characteristic equation* of an LTI system is obtained by equating the denominator of the transfer function to zero:

$$|sI - A| = 0 \quad (8.3.5\text{a}) \qquad |zI - A| = 0 \quad (8.3.5\text{a})$$

Therefore, the roots of the characteristic equation are the eigenvalues of the matrix A and are often called the *characteristic roots* of the system. The physical implication of a characteristic root s_0/z_0 is that there will appear such terms as $Ke^{s_0t}/K(z_0)^n$ in the natural response of the system.

(cf.) The *natural response* means the output stemming from the initial condition, while the *forced response* means the output caused by the input.

Example 8.2 Transfer Function

(a) For the continuous-time LTI system described by Eq. (E8.1.2) in Example 8.1(a), we can use Eq. (8.3.2a) to find the transfer function

$$G(s) \overset{(8.3.2\text{a})}{=} C[sI - A]^{-1}B + D = [1 \;\; 0]\begin{bmatrix} s & -1 \\ 0 & s+1 \end{bmatrix}^{-1}\begin{bmatrix} 0 \\ 1 \end{bmatrix}$$

$$= \frac{[1 \;\; 0]}{s(s+1)}\begin{bmatrix} s+1 & 1 \\ 0 & s \end{bmatrix}\begin{bmatrix} 0 \\ 1 \end{bmatrix} = \frac{1}{s(s+1)} \quad \text{(E8.2.1)}$$

This could be obtained directly by taking the transform of the differential equation (E8.1.1). We can equate the denominator of this transfer function with zero to write the characteristic equation

$$|sI - A| = s(s+1) = 0 \quad \text{(E8.2.2)}$$

which has the roots $s = 0$ and $s = -1$ as the eigenvalues.

(b) For the discrete-time LTI system described by Eq. (E8.1.7) in Example 8.1(b), we can use Eq. (8.3.2b) to find the transfer function

$$G[z] \overset{(8.3.2b)}{=} C[zI - A]^{-1}B + D = [1 \quad 0]\begin{bmatrix} z-1 & -1+e^{-T} \\ 0 & z-e^{-T} \end{bmatrix}^{-1}\begin{bmatrix} T-1+e^{-T} \\ 1-e^{-T} \end{bmatrix}$$
$$= \frac{[1 \quad 0]}{(z-1)(z-e^{-T})}\begin{bmatrix} z-e^{-T} & 1-e^{-T} \\ 0 & z-1 \end{bmatrix}\begin{bmatrix} T-1+e^{-T} \\ 1-e^{-T} \end{bmatrix}$$
$$= \frac{(T-1+e^{-T})z + 1 - e^{-T} - Te^{-T}}{(z-1)(z-e^{-T})} \tag{E8.2.3}$$

We can equate the denominator with zero to write the characteristic equation

$$|zI - A| = (z-1)(z-e^{-T}) = 0 \tag{E8.2.4}$$

which has the roots $z = 1$ and $z = e^{-T}$ as the eigenvalues.

8.4 Discretization of Continuous-Time State Equation

In Chap. 6, we studied various discretzation methods that can be used for converting a given s -transfer function or a differential equation into an 'equivalent' z -transfer function or a difference equation. In this section, we will develop a technique to discretize a continuous-time state equation into an 'equivalent' discrete-time state equation.

8.4.1 State Equation Without Time Delay

Let us consider a continuous-time state equation

$$\mathbf{x}'(t) = A\mathbf{x}(t) + Bu(t) \tag{8.4.1}$$
$$y(t) = C\mathbf{x}(t) + Du(t) \tag{8.4.2}$$

As shown in Sect. 8.2.2, the solution to these equations is given by Eq. (8.2.8a):

$$\mathbf{x}(t) = \phi(t-t_0)\mathbf{x}(t_0) + \int_{t_0}^{t} \phi(t-\tau)Bu(\tau)d\tau \tag{8.4.3}$$

To obtain an 'equivalent' discrete-time state equation, we write the state transition equation for an interval of sampling period T from $t_0 = nT$ to $t = (n+1)T$:

$$\mathbf{x}((n+1)T) = \phi(T)\mathbf{x}(nT) + \int_{nT}^{(n+1)T} \phi(nT+T-\tau)Bu(\tau)d\tau \tag{8.4.4}$$

Assuming that the input is constant during each sampling interval, i.e., $u(t) = u(nT) = u[n]$ for $nT \le t < (n+1)T$, we can rewrite this equation as

$$\mathbf{x}[n+1] = \phi(T)\mathbf{x}[n] + \int_{nT}^{(n+1)T} \phi(nT+T-\tau)d\tau\, Bu[n]$$
$$; \qquad \mathbf{x}[n+1] = A_D\mathbf{x}[n] + B_D u[n] \tag{8.4.5}$$

where we have let $\mathbf{x}(nT) = \mathbf{x}[n],\ u(nT) = u[n]$, and

$$A_D = \phi(T) = e^{AT} \tag{8.4.6a}$$

$$B_D = \int_{nT}^{(n+1)T} \phi(nT+T-\tau)d\tau\, B \overset{nT+T-\tau\to\sigma}{=} -\int_T^0 \phi(\sigma)d\sigma\, B = \int_0^T \phi(\tau)d\tau\, B \tag{8.4.6b}$$

We can get the discrete-time system matrices A_D and B_D by substituting the state transition matrix (8.2.2a) into these Eqs. (8.4.6a) and (8.4.6b), which is cumbersome in general. As an alternative, it may be better to use a digital computer for evaluating them in the following way [F-1]:

$$A_D \overset{(8.4.6a)}{=} e^{AT} \overset{(D.25)}{=} \sum_{m=0}^{\infty} \frac{A^m T^m}{m!} = I + AT\sum_{m=0}^{\infty} \frac{A^m T^m}{(m+1)!} = I + AT\Psi \tag{8.4.7a}$$

$$B_D = \int_0^T \sum\nolimits_{m=0}^{\infty} \frac{A^m \tau^m}{m!} d\tau\, B = \sum\nolimits_{m=0}^{\infty} \frac{A^m T^{m+1}}{(m+1)!} B = \Psi T B \tag{8.4.7b}$$

with

$$\Psi = \sum_{m=0}^{\infty} \frac{A^m T^m}{(m+1)!} \simeq I + \frac{AT}{2}\left(I + \frac{AT}{3}\left(I + \cdots \left(I + \frac{AT}{N-1}\left(I + \frac{AT}{N}\right)\right)\cdots\right)\right) \tag{8.4.8}$$

Here, the infinite number of summations in Eq. (8.4.8) is truncated to some finite number N, which will be chosen in consideration of the desired accuracy and computation time.

Example 8.3 Discretization of a Continuous-Time State Equation

Consider a continuous-time LTI system described by Eqs. (E8.1.2) and (E8.1.3) in Example 8.1(a), where the state transition matrix is given by Eq. (E8.1.5) as

$$\phi(t) \overset{(8.2.2a)}{=} \mathcal{L}^{-1}\{[sI-A]^{-1}\} \overset{(E8.1.5)}{=} \begin{bmatrix} 1 & 1-e^{-t} \\ 0 & e^{-t} \end{bmatrix} \text{ for } t \ge 0 \tag{E8.3.1}$$

(a) Find the discretized state equation.
We can use Eqs. (8.4.6a) and (8.4.6b) to get

$$A_D \overset{(8.4.6a)}{=} \phi(T) = \begin{bmatrix} 1 & 1-e^{-T} \\ 0 & e^{-T} \end{bmatrix} \tag{E8.3.2}$$

$$B_D \overset{(8.4.6b)}{=} \int_0^T \phi(\tau)d\tau B = \begin{bmatrix} \tau & \tau+e^{-\tau} \\ 0 & -e^{-\tau} \end{bmatrix}\Bigg|_0^T \begin{bmatrix} 0 \\ 1 \end{bmatrix} = \begin{bmatrix} T & T+e^{-T}-1 \\ 0 & -e^{-T}+1 \end{bmatrix}\begin{bmatrix} 0 \\ 1 \end{bmatrix}$$
$$= \begin{bmatrix} T+e^{-T}-1 \\ 1-e^{-T} \end{bmatrix} \tag{E8.3.3}$$

so that the discretized state equation can be obtained as

$$\begin{bmatrix} x_1[n+1] \\ x_2[n+1] \end{bmatrix} \overset{(8.4.5)}{=} \begin{bmatrix} 1 & 1-e^{-T} \\ 0 & e^{-T} \end{bmatrix}\begin{bmatrix} x_1[n] \\ x_2[n] \end{bmatrix} + \begin{bmatrix} T+e^{-T}-1 \\ 1-e^{-T} \end{bmatrix} u[n] \tag{E8.3.4}$$

```
%sig08e03.m
clear, clf
syms s z
Ns=1; Ds=[1 1 0]; Gs=tf(Ns,Ds); % Analog transfer function
[A,B,C,D]=tf2ss(Ns,Ds) % Transfer function to state equation
N=size(A,2); % the dimension of the system
% The numerator/denominator of transfer function (8.3.2a)
[Ns,Ds]=ss2tf(A,B,C,D)
Gs1= C*(s*eye(N)-A)^-1*B + D; pretty(Gs1) % Eq.(8.3.2a)
T=0.1; NT=101; t=[0:NT-1]*T;
% To find the response to step input applied to the 1st input terminal
[y,x,tt]=step(A,B,C,D,1,t);
syms s t %for symbolic solution of the state eq.
A=[0 1;0 -1]; B=[0 1]'; % Eq.(E8.1.2)
x0=[0 0]'; % zero initial condition
% Laplace transform solution
Xs=(s*eye(N)-A)^-1*(x0+B/s) % Eq.(8.2.3a)
% Inverse Laplace transform
for n=1:N, xt(n)=ilaplace(Xs(n)), end
for n=1:length(tt)
   t=tt(n); y1(n)=eval(xt(1)); % Eq.(E8.1.3) with C=[1 0]
end
% To solve the differential equation directly
x=dsolve('Dx1=x2,Dx2=-x2+1','x1(0)=0,x2(0)=0'); % Eq.(E8.1.2)
t=tt; y2= eval(x.x1); % Eq.(E8.1.3) with C=[1 0]
plot(tt,y,'k', tt,y1,'b', tt,y2,'r')
% Discretization
A=[0 1;0 -1]; B=[0 1]'; C=[1 0]; D=0; % Eq.(E8.1.2)
[Ad,Bd,Cd,Dd]=c2dm(A,B,C,D,T,'zoh') % Discretized state equation
[Ad1,Bd1]=c2d_steq(A,B,T) % Eq.(8.4.7a,b)
e_T=exp(-T); Ad2=[1 1-e_T; 0 e_T], Bd2=[T+e_T-1; 1-e_T] % Eq.(E8.3.4)
% The numerator/denominator of transfer function (8.3.2b)
[Nz,Dz]=ss2tf(Ad,Bd,Cd,Dd) % Eq.(8.3.2b) to (E8.2.3) or (E8.3.5)
Nz1=[T-1+e_T 1-e_T-T*e_T], Dz1=[1 -1-e_T e_T]
%Gz1= Cd*(z*eye(N)-Ad)^-1*Bd + Dd, pretty(Gz1)
% The z.o.h. equivalent of the analog transfer function
```

```
Gz_zoh=c2d(Gs,T,'zoh');
[Bd_zoh,Ad_zoh]=tfdata(Gz_zoh,'v') % Eq.(E8.3.7)
% To find the response to step input applied to the 1st input terminal
yd=dstep(Ad,Bd,Cd,Dd,1,NT);
hold on, stairs(tt,yd)

function [Ad,Bd]=c2d_steq(A,B,T,N)
if nargin<4, N=100; end
I= eye(size(A,2));  PSI= I;
for m=N:-1:1, PSI= I +A*PSI*T/(m+1); end  % Eq.(8.4.8)
Ad= I +A*T*PSI;  Bd= PSI*T*B; % Eq.(8.4.7a,b)
```

Note that this is identical to Eq. (E8.1.7) in Example 8.1(b) and the transfer function of the system was computed via Eq. (8.3.2b) as Eq. (E8.2.3) in Example 8.2:

$$G_D[z] = \frac{(T - 1 + e^{-T})z + 1 - e^{-T} - Te^{-T}}{(z-1)(z-e^{-T})} \tag{E8.3.5}$$

(b) Find the transfer function for the z.o.h. (step-invariant) equivalent of the continuous-time system.
Noting from Eq. (E8.2.1) that the transfer function of the continuous-time system is

$$G(s) \overset{\text{(E8.2.1)}}{=} \frac{1}{s(s+1)}, \tag{E8.3.6}$$

we can use Eq. (6.2.5) to get

$$\begin{aligned}
G_{step}[z] &\overset{(6.2.5)}{=} (1-z^{-1})\mathcal{Z}\left\{\mathcal{L}^{-1}\left\{\frac{1}{s}G(s)\right\}\Big|_{t=nT}\right\} \\
&\overset{\text{(E8.3.6)}}{=} \frac{z-1}{z}\mathcal{Z}\left\{\mathcal{L}^{-1}\left\{\frac{1}{s^2(s+1)}\right\}\right\} \overset{\text{PFE}}{=} \frac{z-1}{z}\mathcal{Z}\left\{\mathcal{L}^{-1}\left\{\frac{1}{s^2}+\frac{-1}{s}+\frac{1}{s+1}\right\}\right\} \\
&\overset{\text{B.8(3),(4)\&(6)}}{=} \frac{z-1}{z}\left(\frac{Tz}{(z-1)^2}-\frac{z}{z-1}+\frac{z}{z-e^{-T}}\right) = \frac{T}{z-1}-1+\frac{z-1}{z-e^{-T}} \\
&= \frac{(T-1+e^{-T})z+1-e^{-T}-Te^{-T}}{(z-1)(z-e^{-T})}
\end{aligned} \tag{E8.3.7}$$

This is the same as the transfer function of the system described by the discretized state equation (E8.3.4) and output equation (E8.1.8). It is because we have assumed that the input is constant during each sampling period in discretizing the state equation as if a S/H (sampler and zero-order-hold) device were installed at the input stage.
Interested readers are invited to run the above program "sig08e03.m", check the validity of the above results, and see how the MATLAB commands can be used to discretize a continuous-time state equation.

Remark 8.1 Discretized State Equation and Zero-Order-Hold Equivalent

The z.o.h. (step-invariant) equivalent of a continuous-time system has the same transfer function with the discrete-time system obtained by discretizing the state and output equations.

Example 8.4 Discretization of a Double Integrator

Consider the continuous-time LTI system described by the state and output equations as

$$\begin{bmatrix} x_1'(t) \\ x_2'(t) \end{bmatrix} = \begin{bmatrix} 0 & 1 \\ 0 & 0 \end{bmatrix} \begin{bmatrix} x_1(t) \\ x_2(t) \end{bmatrix} + \begin{bmatrix} 0 \\ 1 \end{bmatrix} u(t) \tag{E8.4.1}$$

$$y(t) = \begin{bmatrix} 1 & 0 \end{bmatrix} \begin{bmatrix} x_1(t) \\ x_2(t) \end{bmatrix} \tag{E8.4.2}$$

For this system, we have

$$[sI - A]^{-1} = \left[\begin{bmatrix} s & 0 \\ 0 & s \end{bmatrix} - \begin{bmatrix} 0 & 1 \\ 0 & 0 \end{bmatrix} \right]^{-1} = \begin{bmatrix} s & -1 \\ 0 & s \end{bmatrix}^{-1} = \frac{1}{s^2} \begin{bmatrix} s & 1 \\ 0 & s \end{bmatrix} = \begin{bmatrix} s^{-1} & s^{-2} \\ 0 & s^{-1} \end{bmatrix} \tag{E8.4.3}$$

so that the state transition matrix is

$$\phi(t) \overset{(8.2.2a)}{=} \mathcal{L}^{-1}\{[sI - A]^{-1}\} \underset{B.8(3),(4)}{\overset{(E8.4.3)}{=}} \begin{bmatrix} 1 & t \\ 0 & 1 \end{bmatrix} \text{ for } t \geq 0 \tag{E8.4.4}$$

Thus we can use Eqs. (8.4.6a) and (8.4.6b) to get the system and input coefficient matrices for the discretized state equation as

$$A_D \overset{(8.4.6a)}{=} \phi(T) = \begin{bmatrix} 1 & T \\ 0 & 1 \end{bmatrix} \text{ and } B_D \overset{(8.4.6b)}{=} \int_0^T \phi(\tau) d\tau B = \int_0^T \begin{bmatrix} \tau \\ 1 \end{bmatrix} d\tau = \begin{bmatrix} T^2/2 \\ T \end{bmatrix} \tag{E8.4.5}$$

8.4.2 State Equation with Time Delay

A continuous-time LTI system with the input $\mathbf{u}(t)$ delayed by d [s] can be described by

$$\mathbf{x}'(t) = A\mathbf{x}(t) + B\mathbf{u}(t - d) \tag{8.4.9}$$

There are two cases: (i) $0 < d \leq T$ (ii) $T < d = MT + d_1$ with $0 < d_1 < T$ and $M \geq 1$.

<**Case I**> $0 \leq d \leq T$

When the time delay d is not longer than the sampling period T, i.e., $0 < d \leq T$, the second term of the RHS of Eqs. (8.4.4) or (8.4.5) becomes

$$\int_{nT}^{(n+1)T} \phi(nT + T - \tau)B\mathbf{u}(\tau - d)d\tau$$
$$= \int_{nT}^{nT+d} \phi(nT + T - \sigma)d\sigma\, B\mathbf{u}(nT - T)$$
$$+ \int_{nT+d}^{nT+T} \phi(nT + T - \sigma)d\sigma\, B\mathbf{u}(nT) \tag{8.4.10}$$

where the input signal $\mathbf{u}(t)$ is assumed to be constant over each sampling interval. Thus Eq. (8.4.5) becomes

$$\mathbf{x}[n+1] = A_D\mathbf{x}[n] + B_{D1}\mathbf{u}[n-1] + B_{D0}\mathbf{u}[n] \tag{8.4.11}$$

where

$$A_D \overset{(8.4.6a)}{=} \phi(T) = e^{AT} \tag{8.4.12a}$$

$$B_{D1} \overset{(8.4.6b)}{=} \int_{nT}^{nT+d} \phi(nT + T - \sigma)d\sigma\, B$$
$$\overset{nT+T-\sigma\to\tau}{=} -\int_{d}^{0} \phi(T - d + \tau)d\tau\, B = \phi(T - d)\int_{0}^{d} \phi(\tau)d\tau\, B \tag{8.4.12b}$$

$$B_{D0} \overset{(8.4.6b)}{=} \int_{nT+d}^{nT+T} \phi(nT + T - \sigma)d\sigma\, B$$
$$\overset{nT+T-\sigma\to\tau}{=} -\int_{T-d}^{0} \phi(\tau)d\tau\, B = \int_{0}^{T-d} \phi(\tau)d\tau\, B \tag{8.4.12c}$$

This can be written in the form of state equation as

$$\begin{bmatrix} \mathbf{x}[n+1] \\ \mathbf{u}[n] \end{bmatrix} = \begin{bmatrix} A_D & B_{D1} \\ O & O \end{bmatrix} \begin{bmatrix} \mathbf{x}[n] \\ \mathbf{u}[n-1] \end{bmatrix} + \begin{bmatrix} B_{D0} \\ I \end{bmatrix} \mathbf{u}[n] \tag{8.4.13}$$

where K extra state variables $\mathbf{u}[n-1]$ representing the past input values are introduced to augment the state vector.

<**Case II**> $T < d = MT + d_1$ with $0 < d_1 < T$ and $M \ge 1$
When the time delay d is longer than the sampling period T so that $d = MT + d_1$ with $0 < d_1 < T$ and some integer $M \ge 1$, Eq. (8.4.11) becomes

$$\mathbf{x}[n+1] = A_D\mathbf{x}[n] + B_{D1}\mathbf{u}[n - M - 1] + B_{D0}\mathbf{u}[n - M] \tag{8.4.14}$$

where the matrices A_D, B_{D1}, and B_{D0} are as defined by Eqs. (8.4.12a,b,c). This can be written in the form of state equation as

$$\begin{bmatrix} \mathbf{x}[n+1] \\ \mathbf{u}[n-M] \\ \bullet \\ \bullet \\ \mathbf{u}[n-1] \\ \mathbf{u}[n] \end{bmatrix} = \begin{bmatrix} A_D & B_{D1} & B_{D0} & O \cdots O \\ O & O & I & O \cdots O \\ \bullet & \bullet & \bullet & \bullet \cdots \bullet \\ \bullet & \bullet & \bullet & \bullet \cdots \bullet \\ O & O & O & O \cdots I \\ O & O & O & O \cdots O \end{bmatrix} \begin{bmatrix} \mathbf{x}[n] \\ \mathbf{u}[n-M-1] \\ \bullet \\ \bullet \\ \mathbf{u}[n-2] \\ \mathbf{u}[n-1] \end{bmatrix} + \begin{bmatrix} O \\ O \\ \bullet \\ \bullet \\ O \\ I \end{bmatrix} \mathbf{u}[n] \tag{8.4.15}$$

where $(M+1)K$ extra state variables $\{\mathbf{u}[n-1], \mathbf{u}[n-2], \cdots, \mathbf{u}[n-M-1]\}$ representing the past input values are introduced.

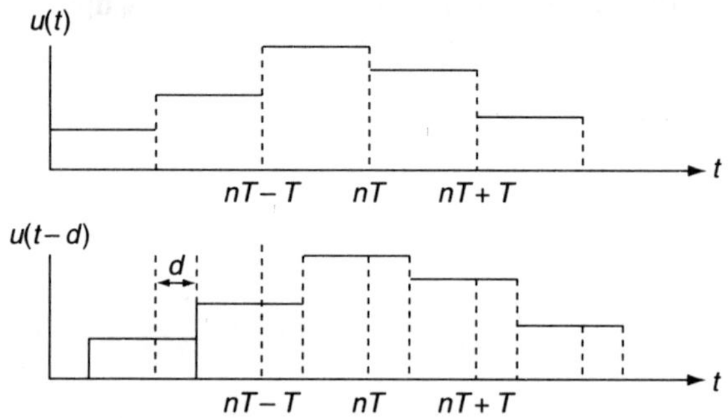

Fig. 8.1 Delayed input signal by d

Example 8.5 Discretization of a Double Integrator with Time Delay

Consider the double integrator dealt with in Example 8.4, where the input is delayed by d $(0 < d < T)$. With the state transition matrix $\phi(t)$ given by Eq. (E8.4.4), we use Eq. (8.4.12) to get the discrete-time system matrices as

$$A_D = \phi(T) = \begin{bmatrix} 1 & T \\ 0 & 1 \end{bmatrix} \tag{E8.5.1}$$

$$B_{D1} = \phi(T-d)\int_0^d \phi(\tau)d\tau B = \begin{bmatrix} 1 & T-d \\ 0 & 1 \end{bmatrix}\begin{bmatrix} d^2/2 \\ d \end{bmatrix} = \begin{bmatrix} d(T-d/2) \\ d \end{bmatrix} \tag{E8.5.2}$$

$$B_{D0} = \int_0^{T-d} \phi(\tau)d\tau B = \begin{bmatrix} \tau^2/2 \\ \tau \end{bmatrix}\Bigg|_0^{T-d} = \begin{bmatrix} (T-d)^2/2 \\ T-d \end{bmatrix} \tag{E8.5.3}$$

8.5 Various State Space Description – Similarity Transformation

As can be seen in Sect. 1.3.4, we could construct many state diagrams for a given transfer function and each of them can be represented by a state equation. This implies that there is no unique state space model for a certain system.

As a matter of fact, we can use *similarity transformation* to derive any number of different state equations, all of which are equivalent in terms of the input-output relationship. Consider the state and output equation of an LTI system as

State equation:

$$\mathbf{x}'(t) = A\mathbf{x}(t) + B\mathbf{u}(t) \quad (8.5.1a) \qquad \mathbf{x}[n+1] = A\mathbf{x}[n] + B\mathbf{u}[n] \quad (8.5.1b)$$

Output equation:

$$\mathbf{y}(t) = C\mathbf{x}(t) + D\mathbf{u}(t) \quad (8.5.2a) \qquad \mathbf{y}[n] = C\mathbf{x}[n] + D\mathbf{u}[n] \quad (8.5.2b)$$

With a nonsingular transformation matrix P and a new state vector $\mathbf{w}(t)/\mathbf{w}[n]$, we substitute

$$\mathbf{x}(t) = P\,\mathbf{w}(t) \quad (8.5.3a) \qquad \mathbf{x}[n] = P\,\mathbf{w}[n] \quad (8.5.3b)$$

into Eqs. (8.5.1) and (8.5.2) to write

$$\mathbf{w}'(t) = P^{-1}AP\mathbf{w}(t) + P^{-1}B\mathbf{u}(t) \qquad \mathbf{w}[n+1] = P^{-1}AP\mathbf{w}[n] + P^{-1}B\mathbf{u}[n]$$
$$\mathbf{y}(t) = CP\mathbf{w}(t) + D\mathbf{u}(t) \qquad \mathbf{y}[n] = CP\mathbf{w}[n] + D\mathbf{u}[n]$$

This can be written as another set of state and output equations:

$$\mathbf{w}'(t) = A_p\mathbf{w}(t) + B_p\mathbf{u}(t) \quad (8.5.4a) \qquad \mathbf{w}[n+1] = A_p\mathbf{w}[n] + B_p\mathbf{u}[n] \quad (8.5.4b)$$

$$\mathbf{y}(t) = C_p\mathbf{w}(t) + D_p\mathbf{u}(t) \quad (8.5.5a) \qquad \mathbf{y}[n] = C_p\mathbf{w}[n] + D_p\mathbf{u}[n] \quad (8.5.5b)$$

where

$$A_p = P^{-1}AP,\; B_p = P^{-1}B,\; C_p = CP,\; \text{and } D_p = D \tag{8.5.6}$$

Remark 8.2 Similarity Transformation – Equivalence Transformation

(1) Note that, with different nonsingular matrices P, we could obtain different set of state and output equations as many as we want.
(2) Transfer function, characteristic equation, and eigenvalues are not changed under similarity transformation:

$$|sI - A_p| = |sI - P^{-1}AP| = |P^{-1}sIP - P^{-1}AP| = |P^{-1}||sI - A||P| = |sI - A| \tag{8.5.7}$$

(3) The determinant/trace of A_p is the same as that of A, since the determinant/trace of a matrix is equal to the product/sum of the eigenvalues. Especially if a system has distinct characteristic values, we may derive a state equation with a diagonal system matrix having the characteristic values on its diagonal, which is referred to as the *Jordan canonical* or *normal form.*

Suppose we have an $N \times N$ system matrix A with N distinct eigenvalues λ_i's for $i = 1 : N$ and the corresponding eigenvectors $\mathbf{m}_i$'s. Then we can write

$$A\mathbf{m}_i = \lambda_i \mathbf{m}_i \quad \text{for} \quad i = 1, 2, \cdots, N \tag{8.5.8}$$

or in a more compact form,

$$A[\mathbf{m}_1\ \mathbf{m}_2\ \cdots\ \mathbf{m}_N] = [\mathbf{m}_1\ \mathbf{m}_2\ \cdots\ \mathbf{m}_N]\begin{bmatrix} \lambda_1 & 0 & \cdots & 0 \\ 0 & \lambda_2 & \cdots & 0 \\ \cdot & \cdot & \cdots & \cdot \\ \cdot & \cdot & \cdots & \cdot \\ \cdot & \cdot & \cdots & \cdot \\ 0 & 0 & \cdots & \lambda_N \end{bmatrix}$$

$$; \qquad AM = M\Lambda \tag{8.5.9}$$

where M, called a *modal matrix*, is composed of the eigenvectors as columns, and $M^{-1}AM = \Lambda$ is a diagonal matrix with the eigenvalues on its diagonal. The modal matrix M is nonsingular if the eigenvalues are distinct and consequently, the eigenvectors are linearly independent. This implies that, with the similarity transformation $P = M$, the resulting system matrix A_p will appear to be diagonal in the new state equation.

Example 8.6 Diagonal/Jordan Canonical Form

Consider a system described by the following discrete-time state and output equations:

$$\begin{bmatrix} x_1[n+1] \\ x_2[n+1] \end{bmatrix} = \begin{bmatrix} 1 & 3 \\ 0 & 2 \end{bmatrix}\begin{bmatrix} x_1[n] \\ x_2[n] \end{bmatrix} + \begin{bmatrix} 0 \\ 1 \end{bmatrix} u[n] \tag{E8.6.1}$$

$$y[n] = \begin{bmatrix} 1 & 0 \end{bmatrix}\begin{bmatrix} x_1[n] \\ x_2[n] \end{bmatrix} \tag{E8.6.2}$$

We can get its eigenvalues and the corresponding eigenvectors as below:

$$|sI - A| = (s-1)(s-2) = 0; \ s_1 = 1, \ s_2 = 2$$

$s_1 = 1$

$$\begin{bmatrix} 1 & 3 \\ 0 & 2 \end{bmatrix}\begin{bmatrix} m_{11} \\ m_{21} \end{bmatrix} = 1\begin{bmatrix} m_{11} \\ m_{21} \end{bmatrix}$$

$; m_{21} = 0, \ m_{11}$: arbitrary

$s_2 = 2$

$$\begin{bmatrix} 1 & 3 \\ 0 & 2 \end{bmatrix}\begin{bmatrix} m_{12} \\ m_{22} \end{bmatrix} = 2\begin{bmatrix} m_{12} \\ m_{22} \end{bmatrix}$$

$; m_{22}$: arbitrary, $m_{12} = 3m_{22}$

Thus we have a modal matrix and its inverse as

$$M = \begin{bmatrix} m_{11} & m_{12} \\ m_{21} & m_{22} \end{bmatrix} = \begin{bmatrix} 1 & 3 \\ 0 & 1 \end{bmatrix}, \ M^{-1} = \begin{bmatrix} 1 & -3 \\ 0 & 1 \end{bmatrix} \tag{E8.6.3}$$

Now, with the similarity transformation $P = M$, we use Eq. (8.5.6) to obtain another set of state and output equations with

$$A_p = M^{-1}AM = \begin{bmatrix}1 & -3\\0 & 1\end{bmatrix}\begin{bmatrix}1 & 3\\0 & 2\end{bmatrix}\begin{bmatrix}1 & 3\\0 & 1\end{bmatrix} = \begin{bmatrix}1 & -3\\0 & 2\end{bmatrix}\begin{bmatrix}1 & 3\\0 & 1\end{bmatrix} = \begin{bmatrix}1 & 0\\0 & 2\end{bmatrix} : \text{diagonal} \tag{E8.6.4a}$$

$$B_p = M^{-1}B = \begin{bmatrix}1 & -3\\0 & 1\end{bmatrix}\begin{bmatrix}0\\1\end{bmatrix} = \begin{bmatrix}-3\\1\end{bmatrix} \tag{E8.6.4b}$$

$$C_p = CM = \begin{bmatrix}1 & 0\end{bmatrix}\begin{bmatrix}1 & 3\\0 & 1\end{bmatrix} = \begin{bmatrix}1 & 3\end{bmatrix} \tag{E8.6.4c}$$

We can use the MATLAB function '`jordan()`' to get a modal matrix and the corresponding Jordan canonical form of state equation:

```
>>A=[1 3; 0 2]; [M,Ap]=jordan(A) % [M,L]=eig(A)
    M =   -3        3           Ap =   1       0
           0        1                  0       2
>>B=[0; 1]; C=[1  0]; Bp=M\B, Cp=C*M
    Bp =   1                    Cp =  -3     3
           1
```

8.6 Summary

In this chapter we have studied the state space description of LTI systems in the form of the state and output equations. We have also defined the state transition matrix and derived the solution of continuous-time/discrete-time LTI state equations. Then we have discussed the techniques of discretizing the continuous-time state equations and the similarity transformation.

Problems

8.1 Controllable/Observable Canonical Form of State Equations

Complete the controllable/observable form of state diagrams in Fig. P8.1 for a system having the following transfer function:

$$G(s) = \frac{b_2 s^2 + b_1 s + a_0}{s^3 + a_2 s^2 + a_1 s + a_0} \tag{P8.1.1}$$

and show that the corresponding state and output equations are

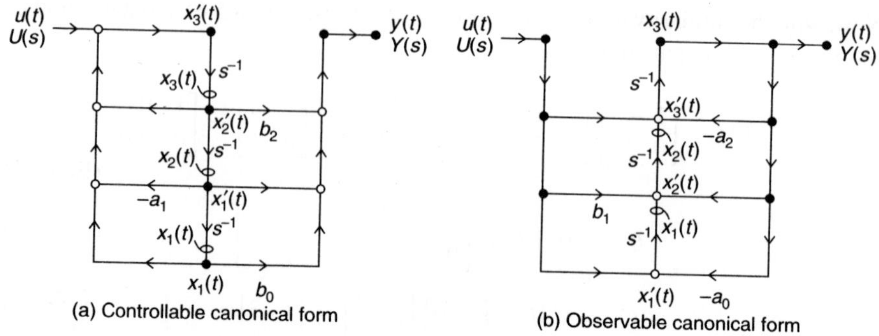

Fig. P8.1 Controllable/observable canonical state diagrams for the transfer function (P8.1.1)

$$\begin{bmatrix} x_1'(t) \\ x_2'(t) \\ x_3'(t) \end{bmatrix} = \begin{bmatrix} 0 & 1 & 0 \\ 0 & 0 & 1 \\ -a_0 & -a_1 & -a_2 \end{bmatrix} \begin{bmatrix} x_1(t) \\ x_2(t) \\ x_3(t) \end{bmatrix} + \begin{bmatrix} 0 \\ 0 \\ 1 \end{bmatrix} u(t) \quad \text{(P8.1.2a)}$$

$$\begin{bmatrix} x_1'(t) \\ x_2'(t) \\ x_3'(t) \end{bmatrix} = \begin{bmatrix} 0 & 0 & -a_0 \\ 1 & 0 & -a_1 \\ 0 & 1 & -a_2 \end{bmatrix} \begin{bmatrix} x_1(t) \\ x_2(t) \\ x_3(t) \end{bmatrix} + \begin{bmatrix} b_0 \\ b_1 \\ b_2 \end{bmatrix} u(t) \quad \text{(P8.1.2b)}$$

$$y(t) = \begin{bmatrix} b_0 & b_1 & b_2 \end{bmatrix} \begin{bmatrix} x_1(t) \\ x_2(t) \\ x_3(t) \end{bmatrix} \quad \text{(P8.1.3a)}$$

$$y(t) = \begin{bmatrix} 0 & 0 & 1 \end{bmatrix} \begin{bmatrix} x_1(t) \\ x_2(t) \\ x_3(t) \end{bmatrix} \quad \text{(P8.1.3b)}$$

8.2 Controllable/Observable Canonical Form of State Equations
Consider a system having the following transfer function:

$$G(s) = \frac{Y(s)}{U(s)} = \frac{4(s+3)}{2s^2+6s+4} \quad \text{(P8.2.1)}$$

(a) Write the controllable/observable form of state and output equations and find the state transition matrices.
(b) Show that the step response is

$$y(t) = (3 - 4e^{-t} + e^{-2t})u_s(t) \quad \text{(P8.2.2)}$$

8.3 State Transition Matrix
Find the state transition matrix $\phi(t) = e^{At}$ for a state equation having the following system matrix:

$$A = \begin{bmatrix} \lambda & 1 \\ 0 & \lambda \end{bmatrix} \quad \text{(P8.3.1)}$$

8.4 State Diagram and State Equation

(a) Show that the state and output equations for the state diagram of Fig. 1.20(a) are

$$\begin{bmatrix} x_1[n+1] \\ x_2[n+1] \\ x_3[n+1] \\ x_4[n+1] \end{bmatrix} = \begin{bmatrix} 0 & 1 & 0 & 0 \\ -a_0 & -a_1 & b_0 & b_1 \\ 0 & 0 & 0 & 1 \\ 0 & 0 & 0 & 0 \end{bmatrix} \begin{bmatrix} x_1[n] \\ x_2[n] \\ x_3[n] \\ x_4[n] \end{bmatrix} + \begin{bmatrix} 0 \\ 0 \\ 0 \\ 1 \end{bmatrix} u[n] \quad \text{(P8.4.1)}$$

$$y[n] = \begin{bmatrix} -a_0 & -a_1 & b_0 & b_1 \end{bmatrix} \begin{bmatrix} x_1[n] \\ x_2[n] \\ x_3[n] \\ x_4[n] \end{bmatrix} \quad \text{(P8.4.2)}$$

(b) Use Eq. (8.3.2b) to find the transfer function of the system described by the above state equation.

8.5 Discretization of a Continuous-Time State Equation
Consider a system described by the following state and output equations:

$$\begin{bmatrix} x_1'(t) \\ x_2'(t) \end{bmatrix} = \begin{bmatrix} 0 & 1 \\ 0 & -2 \end{bmatrix} \begin{bmatrix} x_1(t) \\ x_2(t) \end{bmatrix} + \begin{bmatrix} 0 \\ 1 \end{bmatrix} u(t) \quad \text{(P8.5.1)}$$

$$y(t) = \begin{bmatrix} 1 & 0 \end{bmatrix} \begin{bmatrix} x_1(t) \\ x_2(t) \end{bmatrix} \quad \text{(P8.5.2)}$$

(a) Use Eq. (8.3.2a) to find the transfer function $G(s)$ of this continuous-time system.
(b) Discretize the above state equation and then use Eq. (8.3.2b) to find the transfer function $G_D[z]$ of the discretized system. Compare $G_D[z]$ with the z.o.h. equivalent of $G(s)$.

8.6 Discretization of a Continuous-Time State Equation
Note that for an oscillator having the transfer function

$$G(s) = \frac{\omega^2}{s^2 + \omega^2} \quad \text{(P8.6.1)}$$

we can write the following state and output equations:

$$\begin{bmatrix} x_1'(t) \\ x_2'(t) \end{bmatrix} = \begin{bmatrix} 0 & \omega \\ \omega & 0 \end{bmatrix} \begin{bmatrix} x_1(t) \\ x_2(t) \end{bmatrix} + \begin{bmatrix} 0 \\ \omega \end{bmatrix} u(t) \quad \text{(P8.6.2)}$$

$$y(t) = \begin{bmatrix} 1 & 0 \end{bmatrix} \begin{bmatrix} x_1(t) \\ x_2(t) \end{bmatrix} \quad \text{(P8.6.3)}$$

(a) Use Eq. (8.3.2a) to find the transfer function $G(s)$ of this continuous-time system.

(b) Discretize the above state equation and then use Eq. (8.3.2b) to find the transfer function $G_D[z]$ of the discretized system. Compare $G_D[z]$ with the z.o.h. equivalent $G_{step}[z]$ of $G(s)$.

8.7 PWM (Pulse-Width Modulated) Input
Consider a system described by the following state equation

$$\begin{bmatrix} x_1'(t) \\ x_2'(t) \end{bmatrix} = \begin{bmatrix} 0 & 1 \\ 0 & -1 \end{bmatrix} \begin{bmatrix} x_1(t) \\ x_2(t) \end{bmatrix} + \begin{bmatrix} 0 \\ 1 \end{bmatrix} u(t) \tag{P8.7.1}$$

where the input $u(t)$ is a PWM signal as depicted in Fig. P8.7. Show that the discretized state equation is

$$\begin{bmatrix} x_1[n+1] \\ x_2[n+1] \end{bmatrix} = \begin{bmatrix} 1 & 1-e^{-T} \\ 0 & e^{-T} \end{bmatrix} \begin{bmatrix} x_1[n] \\ x_2[n] \end{bmatrix} + \begin{bmatrix} d_n - e^{-(T-d_n)} + e^{-T} \\ e^{-(T-d_n)} - e^{-T} \end{bmatrix} \tag{P8.7.2}$$

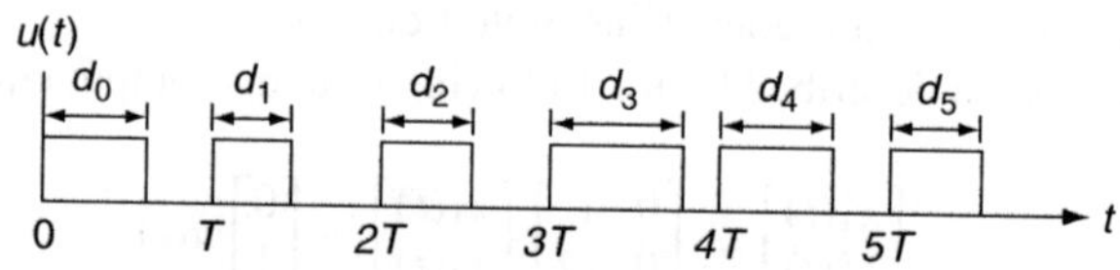

Fig. P8.7 A PWM (pulse-width modulated) signal

8.8 Modal Matrix and Diagonalization of a Circulant Matrix
Consider the following circulant matrix:

$$C = \begin{bmatrix} c(0) & c(1) \cdot\cdot & c(N-1) \\ c(N-1) & c(0) \cdot\cdot & c(N-2) \\ \cdot & \cdot \;\cdot\cdot & \cdot \\ \cdot & \cdot \;\cdot\cdot & \cdot \\ c(1) & c(2) \cdot\cdot & c(0) \end{bmatrix} \tag{P8.8.1}$$

(a) Show that the eigenvalues and the corresponding eigenvectors of the circulant matrix are

$$\lambda_i = \sum_{n=0}^{N-1} c(n)W^{in} = c(0) + c(1)W^i + \cdots + c(N-1)W^{(N-1)i}, \quad W = e^{j2\pi/N} \tag{P8.8.2}$$

$$\mathbf{m}_i = [1,\ W^i,\ \cdots,\ W^{(N-1)i}]^T,\ i = 0, 1, \cdots, N-1 \tag{P8.8.3}$$

so that

$$C\mathbf{m}_i = \lambda_i \mathbf{m}_i \tag{P8.8.4}$$

(b) Note that a modal matrix consisting of the eigenvectors can be written as

$$M = \frac{1}{\sqrt{N}}[\mathbf{m}_0,\ \mathbf{m}_1,\ \cdots,\ \mathbf{m}_{N-1}] = \frac{1}{\sqrt{N}}\begin{bmatrix} 1 & 1 & \cdot\cdot & 1 \\ 1 & W^1 & \cdot\cdot & W^{N-1} \\ \cdot & \cdot & \cdot\cdot & \cdot \\ \cdot & \cdot & \cdot\cdot & \cdot \\ 1 & W^{N-1} & \cdot\cdot & W^{(N-1)^2} \end{bmatrix} \tag{P8.8.5}$$

Show that the inverse of the modal matrix is

$$\boldsymbol{M}^{-1} = \begin{bmatrix} 1 & 1 & \cdot\cdot & 1 \\ 1 & W^{-1} & \cdot\cdot & W^{-(N-1)} \\ \cdot & \cdot & \cdot\cdot & \cdot \\ \cdot & \cdot & \cdot\cdot & \cdot \\ 1 & W^{-(N-1)} & \cdot\cdot & W^{-(N-1)^2} \end{bmatrix} = \boldsymbol{M}^* \text{ (Conjugate transpose)} \tag{P8.8.6}$$

(c) With the matrix dimension $N = 4$, diagonalize the circulant matrix C by the similarity transformation. For reference, you can run the following program "sig08p_08.m".

```
%sig08p_08.m
clear, clf
syms c0 c1 c2 c3
C=[c0 c1 c2 c3; c3 c0 c1 c2; c2 c3 c0 c1; c1 c2 c3 c0];
[M,L]=jordan(C) % [M,L]=eig(C)
M*M'
```

Appendix A
The Laplace Transform

The Laplace transform was discovered originally by Leonhard Euler (1707–1783), the great 18th-century Swiss mathematician and physicist, but is named in honor of a French mathematician and astronomer Pierre-Simon Laplace (1749–1827), who used the transform in his work on probability theory. He was such a genius not only in mathematics that the great mathematician Simeon Poisson (1781–1840) labeled him the Isaac Newton of France, but also in politics that he could serve three regimes in revolutionary France – the republic, the empire of Napoleon, and the Bourbon restoration, having been bestowed a count from Napoleon and a marquis from Louis XVIII.

The Laplace transform is a very useful tool in solving differential equations and much further, it plays an important role in dealing with linear time-invariant systems.

A.1 Definition of the Laplace Transform

The (unilateral or one-sided) Laplace transform is defined for a function $x(t)$ of a real variable t (often meaning the time) as

$$X(s) = \mathcal{L}\{x(t)\} = \int_{0-}^{\infty} x(t)e^{-st}dt \tag{A.1}$$

where s is a complex variable, the lower limit, t_-, of the integration interval is the instant just before $t = 0$, and $x(t)$ is often assumed to be causal in the sense that it is zero for all $t < 0$.

A.2 Examples of the Laplace Transform

A.2.1 Laplace Transform of the Unit Step Function

The unit step function is defined as

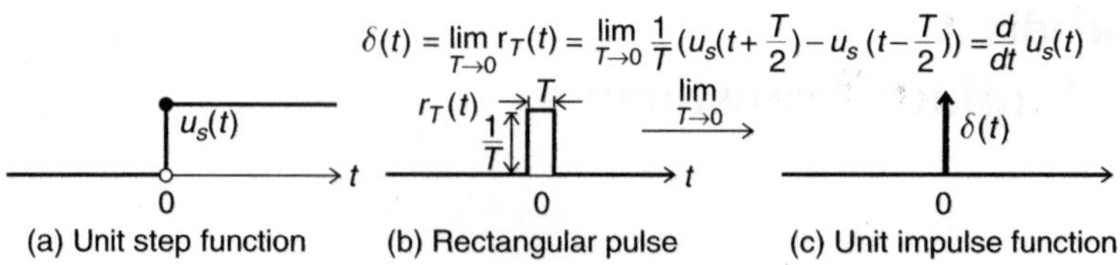

Fig. A.1 Unit step, rectangular pulse, and unit impulse functions

$$u_s(t) = \begin{cases} 1 & \text{for } t \geq 0 \\ 0 & \text{for } t < 0 \end{cases} \tag{A.2}$$

which is depicted in Fig. A.1(a). We can use Eq. (A.1) to obtain the Laplace transform of the unit step function as

$$\mathcal{L}\{u_s(t)\} \overset{(A.1)}{=} \int_0^\infty u_s(t)e^{-st}dt \overset{(A.2)}{=} \int_0^\infty e^{-st}dt \overset{(D.33)}{=} \frac{1}{-s}\, e^{-st}\Big|_0^\infty = \frac{1}{-s}(0-1) = \frac{1}{s}$$

This Laplace transform pair is denoted by

$$u_s(t) \overset{\mathcal{L}}{\longleftrightarrow} \frac{1}{s} \tag{A.3}$$

A.2.2 Laplace Transform of the Unit Impulse Function

The unit impulse function can be defined to be the limit of a rectangular pulse function $r_T(t)$ with the pulsewidth $T \to 0$ (converging to zero) or simply, the time derivative of the unit step function as

$$\delta(t) = \lim_{T\to 0} r_T(t) = \lim_{T\to 0} \frac{1}{T}\left(u_s\left(t+\frac{T}{2}\right) - u_s\left(t-\frac{T}{2}\right)\right) = \frac{d}{dt}u_s(t) \tag{A.4}$$

which is depicted in Fig. A.1(c). We can use Eq. (A.1) to obtain the Laplace transform of the unit impulse function as

$$\mathcal{L}\{\delta(t)\} = \int_{0-}^\infty \delta(t)e^{-st}dt = 1$$

This Laplace transform pair is denoted by

$$\delta(t) \overset{\mathcal{L}}{\longleftrightarrow} 1 \tag{A.5}$$

A.2.3 Laplace Transform of the Ramp Function

The Laplace transform of the unit ramp function $tu_s(t)$ is obtained as

$$\mathcal{L}\{tu_s(t)\} \overset{(A.1)}{=} \int_0^\infty tu_s(t)e^{-st}dt \overset{(A.2)}{=} \int_0^\infty te^{-st}dt \overset{(D.36)}{=} \frac{t}{-s}e^{-st}\Big|_0^\infty + \frac{1}{s}\int_0^\infty e^{-st}dt \overset{(D.33)}{=} \frac{1}{s^2}$$

This Laplace transform pair is denoted by

$$tu_s(t) \overset{\mathcal{L}}{\longleftrightarrow} \frac{1}{s^2} \tag{A.6}$$

A.2.4 Laplace Transform of the Exponential Function

The Laplace transform of the exponential function $e^{-at}u_s(t)$ is obtained as

$$\mathcal{L}\{e^{-at}u_s(t)\} \overset{(A.1)}{=} \int_0^\infty e^{-at}u_s(t)e^{-st}dt \overset{(A.2)}{=} \int_0^\infty e^{-at}e^{-st}dt = \int_0^\infty e^{-(s+a)t}dt \overset{(D.33)}{=} \frac{1}{s+a}$$

This Laplace transform pair is denoted by

$$e^{-at}u_s(t) \overset{\mathcal{L}}{\longleftrightarrow} \frac{1}{s+a} \tag{A.7}$$

A.2.5 Laplace Transform of the Complex Exponential Function

Substituting $\sigma + j\omega$ for a into (A.7) yields the Laplace transform of the complex exponential function as

$$e^{-(\sigma+j\omega)t}u_s(t) \overset{(D.20)}{=} e^{-\sigma t}(\cos\omega t - j\sin\omega t)u_s(t) \overset{\mathcal{L}}{\longleftrightarrow}$$

$$\frac{1}{s+\sigma+j\omega} = \frac{s+\sigma}{(s+\sigma)^2+\omega^2} - j\frac{\omega}{(s+\sigma)^2+\omega^2}$$

$$e^{-\sigma t}\cos\omega t\; u_s(t) \overset{\mathcal{L}}{\longleftrightarrow} \frac{s+\sigma}{(s+\sigma)^2+\omega^2} \tag{A.8}$$

$$e^{-\sigma t}\sin\omega t\; u_s(t) \overset{\mathcal{L}}{\longleftrightarrow} \frac{\omega}{(s+\sigma)^2+\omega^2} \tag{A.9}$$

A.3 Properties of the Laplace Transform

Let the Laplace transforms of two functions $x(t)$ and $y(t)$ be $X(s)$ and $Y(s)$, respectively.

A.3.1 Linearity

The Laplace transform of a linear combination of $x(t)$ and $y(t)$ can be written as

$$\alpha\, x(t) + \beta y(t) \overset{\mathcal{L}}{\longleftrightarrow} \alpha X(s) + \beta Y(s) \tag{A.10}$$

A.3.2 Time Differentiation

The Laplace transform of the derivative of $x(t)$ w.r.t. t can be written as

$$x'(t) \overset{\mathcal{L}}{\longleftrightarrow} s\, X(s) - x(0) \tag{A.11}$$

Proof

$$\mathcal{L}\{x'(t)\} \overset{\text{(A.1)}}{=} \int_0^{\infty} \frac{dx}{dt} e^{-st}\, dt \overset{\text{(D.36)}}{=} x(t)e^{-st}\big|_0^{\infty} - (-s)\int_0^{\infty} x(t)e^{-st}dt \overset{\text{(A.1)}}{=} s\, X(s) - x(0)$$

Repetitive application of this time differentiation property yields the Laplace transform of n th-order derivative of $x(t)$ w.r.t. t as

$$x^{(n)}(t) \overset{\mathcal{L}}{\longleftrightarrow} s^n X(s) - s^{n-1}x(0) - s^{n-2}x'(0) - \cdots - x^{(n-1)}(0) \tag{A.12}$$

A.3.3 Time Integration

The Laplace transform of the integral of $x(t)$ w.r.t. t can be written as

$$\int_{-\infty}^{t} x(\tau)d\tau \overset{\mathcal{L}}{\longleftrightarrow} \frac{1}{s}X(s) + \frac{1}{s}\int_{-\infty}^{0} x(\tau)d\tau \tag{A.13}$$

This can be derived by substituting $\int_{-\infty}^{t} x(\tau)d\tau$ and $x(t)$ for $x(t)$ and $x'(t)$ into Eq. (A.11) as

$$x(t) \overset{\mathcal{L}}{\longleftrightarrow} X(s) = s\mathcal{L}\left\{\int_{\infty}^{t} x(\tau)d\tau\right\} - \int_{-\infty}^{0} x(\tau)d\tau$$

Repetitive application of this time integration property yields the Laplace transform of n th-order integral of $x(t)$ w.r.t. t as

$$\int_{-\infty}^{t}\int_{-\infty}^{t}\cdots\int_{-\infty}^{t} x(\tau)d\tau^n \overset{\mathcal{L}}{\longleftrightarrow} s^{-n}X(s) + s^{-n}\int_{-\infty}^{0} x(\tau)d\tau + \cdots + \int_{-\infty}^{0}\int_{-\infty}^{0}\cdots\int_{-\infty}^{0} x(\tau)d\tau^n \tag{A.14}$$

A.3.4 Time Shifting – Real Translation

The Laplace transform of a delayed function $x(t)$ can be obtained as follows.

$$\mathcal{L}\{x(t-t_1)\} \overset{(A.1)}{=} \int_{-\infty}^{t_1} x(t-t_1)e^{-st}d\tau + \int_{t_1}^{\infty} x(t-t_1)e^{-st}d\tau$$

$$\overset{t-t_1=\tau}{=} e^{-st_1}\int_{-t_1}^{0} x(\tau)e^{-s\tau}d\tau + e^{-st_1}\int_{0}^{\infty} x(\tau)e^{-s\tau}d\tau;$$

$$x(t-t_1),\ t_1>0 \overset{\mathcal{L}}{\longleftrightarrow} e^{-st_1}\left\{X(s)+\int_{-t_1}^{0} x(\tau)e^{-s\tau}d\tau\right\} \overset{\text{if } x(t)=0\ \forall\ t<0}{\rightarrow} e^{-st_1}X(s) \tag{A.15}$$

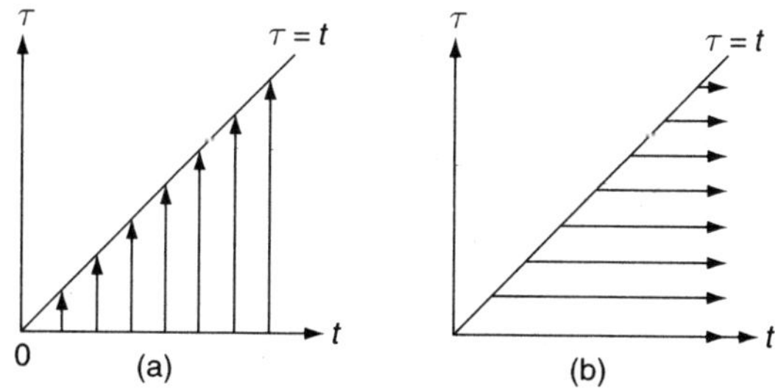

Fig. A.2 Switching the order of two integrations for double integral

A.3.5 Frequency Shifting – Complex Translation

$$e^{s_1 t}x(t) \overset{\mathcal{L}}{\longleftrightarrow} X(s-s_1) \tag{A.16}$$

A.3.6 Real Convolution

The (real) convolution of two (causal) functions $g(t)$ and $x(t)$ is defined as

$$g(t) * x(t) = \int_{-\infty}^{\infty} g(\tau)x(t-\tau)d\tau \underset{\text{causality}}{\overset{\text{if } g(t)=0 \text{ and } x(t)=0\ \forall\ t<0}{=}} \int_{0}^{t} g(\tau)x(t-\tau)d\tau \tag{A.17}$$

The Laplace transform of the convolution $y(t) = g(t) * x(t)$ turns out to be the product of the Laplace transforms of the two functions as

$$y(t) = g(t) * x(t) \overset{\mathcal{L}}{\longleftrightarrow} Y(s) = G(s)X(s) \tag{A.18}$$

Proof

$$
\begin{aligned}
\mathcal{L}\{g(t)*x(t)\} &\overset{(A.1)}{=} \int_0^\infty g(t)*x(t)e^{-st}dt \overset{(A.17)}{=} \int_0^\infty \left\{\int_0^t g(\tau)x(t-\tau)d\tau\right\}e^{-st}dt \\
&\overset{\text{Fig.A.2}}{=} \int_0^\infty g(\tau)e^{-s\tau}\int_\tau^\infty x(t-\tau)e^{-s(t-\tau)}dt\,d\tau \\
&\overset{(t-\tau=v)}{=} \int_0^\infty g(\tau)e^{-s\tau}\left\{\int_0^\infty x(v)e^{-sv}dv\right\}d\tau \\
&\overset{(A.1)}{=} \int_0^\infty g(\tau)e^{-s\tau}d\tau\, X(s) \overset{(A.1)}{=} G(s)\,X(s)
\end{aligned}
$$

This property is used to describe the input-output relationship of a linear time-invariant (LTI) system which has the input $x(t)$, the impulse function $g(t)$, and the output $y(t)$, where the Laplace transform of the impulse function, i.e., $G(s) = \mathcal{L}\{g(t)\}$, is referred to as the *system function* or *transfer function* of the system.

A.3.7 Partial Differentiation

Given the Laplace transform of a function having a parameter a, that is $\mathcal{L}\{x(t,a)\} = X(s,a)$, the Laplace transform of the derivative of the parameterized function $x(t,a)$ w.r.t. a equals the derivative of its Laplace transform $X(s,a)$ w.r.t. a.

$$\frac{\partial}{\partial a}x(t,a) \overset{\mathcal{L}}{\longleftrightarrow} \frac{\partial}{\partial a}X(s,a) \tag{A.19}$$

A.3.8 Complex Differentiation

The Laplace transform of $-tx(t)$ equals the derivative of $\mathcal{L}\{x(t)\} = X(s)$ w.r.t. s.

$$tx(t) \overset{\mathcal{L}}{\longleftrightarrow} -\frac{d}{ds}X(s) \tag{A.20}$$

This can be derived by differentiating Eq. (A.1) w.r.t. s.

Example A.1 Applying Partial/Complex Differentiation Property

To find the Laplace transform of $t\,e^{-at}u_s(t)$, we differentiate Eq. (A.7) w.r.t. a and multiply the result by -1 to get

$$t\,e^{-at}u_s(t) \overset{\mathcal{L}}{\longleftrightarrow} \frac{1}{(s+a)^2} \tag{A.21}$$

Multiplying t by the left-hand side (LHS) and applying the complex differentiation property (A.20) repetitively, we get

$$t^2 e^{-at} u_s(t) \overset{\mathcal{L}}{\longleftrightarrow} \frac{2!}{(s+a)^3} \tag{A.22}$$

$$\cdots\cdots\cdots\cdots\cdots\cdots\cdots\cdots$$

$$t^m e^{-at} u_s(t) \overset{\mathcal{L}}{\longleftrightarrow} \frac{m!}{(s+a)^{m+1}} \tag{A.23}$$

A.3.9 Initial Value Theorem

We can get the initial value of $x(t)$ from its Laplace transform $\mathcal{L}\{x(t)\} = X(s)$ as follows.

$$x(0) = \lim_{s\to\infty} s\, X(s) \tag{A.24}$$

This can be derived by substituting $s = \infty$ into the time differentiation property (A.11) as

$$\lim_{s\to\infty} s\, X(s) - x(0) \overset{\text{(A.11)}}{=} \lim_{s\to\infty} \mathcal{L}\left\{x'(t)\right\} \overset{\text{(A.1)}}{=} \lim_{s\to\infty} \int_0^\infty x'(t) e^{-st} dt = 0$$

A.3.10 Final Value Theorem

We can get the final value of $x(t)$ from its Laplace transform $\mathcal{L}\{x(t)\} = X(s)$ as follows.

$$x(\infty) = \lim_{s\to 0} s\, X(s) \tag{A.25}$$

on the premise that $x(t)$ is convergent, or equivalently, all poles of $X(s)$ are in the left-half plane (LHP) except for a simple pole at $s = 0$. This can be derived by substituting $s = 0$ into the time differentiation property (A.11) as

$$\lim_{s\to 0} s\, X(s) - x(0) \overset{\text{(A.11)}}{=} \lim_{s\to 0} \mathcal{L}\left\{x'(t)\right\} \overset{\text{(A.1)}}{=} \lim_{s\to 0} \int_0^\infty x'(t) e^{-st} dt = x(\infty) - x(0)$$

(Q) Can we apply this final value theorem to have $\sin\omega t|_{t=\infty} = \lim_{s\to 0} s\, X(s) \overset{\text{Table A.1(7)}}{=} \lim_{s\to 0} s \frac{\omega}{s^2+\omega^2} = 0$?

A.4 Inverse Laplace Transform

Suppose a s-function $X(s)$ is given in the form of a rational function, i.e., a ratio of an M th-degree polynomial $Q(s)$ to an N th-degree polynomial $P(s)$ in s and it is expanded into the partial fractions as

$$X(s) = \frac{Q(s)}{P(s)} = \frac{b_M s^M + \ldots + b_1 s + b_0}{a_N s^N + \ldots + a_1 s + a_0} \; (M \le N) \tag{A.26}$$

$$= \left(\sum_{n=1}^{N-L} \frac{r_n}{s - p_n} \right) + \frac{r_{N-L+1}}{s-p} + \ldots + \frac{r_N}{(s-p)^L} + K \tag{A.27}$$

where

$$r_n = (s - p_n) \left. \frac{Q(s)}{P(s)} \right|_{s = p_n}, \quad n = 1, 2, \ldots, N - L \tag{A.28.1}$$

$$r_{N-l} = \frac{1}{l!} \frac{d^l}{ds^l} \left\{ (s-p)^L \frac{Q(s)}{P(s)} \right\} \Bigg|_{s = p}, \quad l = 0, 1, \ldots, L - 1 \tag{A.28.2}$$

Then the inverse Laplace transform of $X(s)$ can be obtained as

$$x(t) = \left\{ \sum_{n=1}^{N-L} r_n e^{p_n t} + r_{N-L+1} e^{pt} + r_{N-L+2} t\, e^{pt} + \ldots + \frac{r_N}{(L-1)!} t^{L-1} e^{pt} \right\} u_s(t) + K\delta(t) \tag{A.29}$$

Example A.2 Inverse Laplace Transform
Let us find the inverse Laplace transform of the following s-functions

$$\text{(a)} \quad X(s) = \frac{3s^2 + 11s + 11}{s^3 + 4s^2 + 5s + 2} = \frac{3s^2 + 11s + 11}{(s+1)^2(s+2)}$$

$$= \frac{r_1}{s+2} + \frac{r_2}{s+1} + \frac{r_3}{(s+1)^2} \text{ (with a double pole at } s = -1) \tag{1}$$

We can use the formula (A.28.1) to find the simple-pole coefficient r_1 as

$$r_1 = (s+2) X(s) |_{s=-2} \overset{(1)}{=} (s+2) \left. \frac{3s^2 + 11s + 11}{(s+1)^2(s+2)} \right|_{s=-2} = 1 \tag{2}$$

and can also use the formula (A.28.2) to find the multiple-pole coefficient r_2 and r_3 as

$$r_3 = (s+1)^2 X(s)|_{s=-1} \overset{(1)}{=} \left. \frac{3s^2+11s+11}{s+2} \right|_{s=-1} = 3 \tag{3}$$

$$r_2 = \frac{d}{ds}(s+1)^2 X(s)\bigg|_{s=-1} = \frac{d}{ds}\left. \frac{3s^2+11s+11}{s+2} \right|_{s=-1}$$

$$\overset{(D.31)}{=} \left. \frac{(6s+11)(s+2)-(3s^2+11s+11)}{(s+2)^2} \right|_{s=-1} = 2 \tag{4}$$

Thus the inverse Laplace transform can be written as

$$x(t) \underset{(A.29)}{\overset{(1)}{=}} (r_1 e^{-2t} + r_2 e^{-t} + r_3 t e^{-t}) u_s(t) = (e^{-2t} + 2e^{-t} + 3te^{-t}) u_s(t) \tag{5}$$

```
>>Ns=[3 11 11]; Ds=[1 4 5 2]; [r,p,k]=residue(Ns,Ds), [A,B]=residue(r,p,k)
```

(b) $$\begin{aligned} X(s) &= \frac{s}{s^2+2s+2} \\ &= \frac{s}{(s+1)^2+1^2} \text{ (with complex conjugate poles at } s=-1\pm j) \end{aligned} \tag{6}$$

We may use the formula (A.28.1) to find the coefficients of the partial fraction expansion form

$$X(s) = \frac{s}{s^2+2s+2} = \frac{r_1}{s+1-j} + \frac{r_2}{s+1+j} \tag{7}$$

as

$$\begin{aligned} r_1 &\overset{(A.28.1)}{=} (s+1-j)X(s)|_{s=-1+j} \overset{(6)}{=} \left. \frac{s}{s+1+j} \right|_{s=-1+j} = 0.5(1+j) \\ r_2 &\overset{(A.28.1)}{=} (s+1+j)X(s)|_{s=-1-j} \overset{(6)}{=} \left. \frac{s}{s+1-j} \right|_{s=-1-j} = 0.5(1-j) \end{aligned} \tag{8}$$

Thus we can write the inverse Laplace transform as

$$x(t) \underset{(A.29)}{\overset{(7)}{=}} (0.5(1+j)e^{(-1+j)t} + 0.5(1-j)e^{(-1-j)t}) u_s(t) \overset{(F.20)}{=} e^{-t}(\cos t - \sin t) u_s(t) \tag{9}$$

In the case of complex conjugate poles, it may be simpler to equate $X(s)$ with the following form

$$X(s) = \frac{s}{s^2+2s+2} = \frac{A(s+1)}{(s+1)^2+1^2} + \frac{B\times 1}{(s+1)^2+1^2} \overset{\text{common denominator}}{=} \frac{As+(A+B)}{(s+1)^2+1^2} \tag{10}$$

and get the coefficients as $A = 1$ and $B = -1$. Then we can directly find each term of the inverse Laplace transform from the Laplace transform Table A.1 and write the inverse Laplace transform as

Table A.1 Laplace transforms of basic functions

$x(t)$	$X(s)$	$x(t)$	$X(s)$	$x(t)$	$X(s)$
(1) $\delta(t)$	1	(5) $e^{-at}u_s(t)$	$\frac{1}{s+a}$	(9) $e^{-at}\sin\omega t\, u_s(t)$	$\frac{\omega}{(s+a)^2+\omega^2}$
(2) $\delta(t-t_1)$	$e^{-t_1 s}$	(6) $t^m e^{-at}u_s(t)$	$\frac{m!}{(s+a)^{m+1}}$	(10) $e^{-at}\cos\omega t\, u_s(t)$	$\frac{s+a}{(s+a)^2+\omega^2}$
(3) $u_s(t)$	$\frac{1}{s}$	(7) $\sin\omega t\, u_s(t)$	$\frac{\omega}{s^2+\omega^2}$		
(4) $t^m u_s(t)$	$\frac{m!}{s^{m+1}}$	(8) $\cos\omega t\, u_s(t)$	$\frac{s}{s^2+\omega^2}$		

$$X(s) \overset{(10)}{=} \frac{s+1}{(s+1)^2+1^2} + \frac{(-1)\times 1}{(s+1)^2+1^2} \overset{\text{Table A.1(9),(10)}}{\rightarrow} x(t) = (e^{-t}\cos t - e^{-t}\sin t)u_s(t) \tag{11}$$

(c) Use of MATLAB for Partial Fraction Expansion and Inverse Laplace Transform
We can use the MATLAB command "residue ()" to get the partial fraction expansion and "ilaplace ()" to obtain the whole inverse Laplace transform. It should, however, be noted that "ilaplace ()" might not work properly for high-degree rational functions.

```
>>Ns=[3 11 11]; Ds=[1 4 5 2]; [r,p,k]=residue(Ns,Ds); [r p],k % (1)
    r = 1.0000     p = -2.0000     % (2) 1/(s-(-2))
        2.0000         -1.0000     % (4) 2/(s-(-1))
        3.0000         -1.0000     % (3) 3/(s-(-1))^2
    k = []
>>syms s, x= ilaplace((3*s^2+11*s+11)/(s^3+4*s^2+5*s+2))
    x = exp(-2*t)+3*t*exp(-t)+2*exp(-t)      % (5)
>>Ns=[1 0]; Ds=[1 2 2]; [r,p,k]=residue(Ns,Ds); [r p],k  % (6)
    r = 0.5000 + 0.5000i     p = -1.0000 + 1.0000i  % (8) (0.5+0.5i)/(s+1-i)
        0.5000 - 0.5000i     p = -1.0000 - 1.0000i  % (8) (0.5-0.5i)/(s+1+i)
    k = []
>>syms s, x= ilaplace(s/(s^2+2*s+2))
    x = exp(-t)*cos(t)-exp(-t)*sin(t)     % (9) or (11)
>>ilaplace(s/(s^4+10*s^3+9*s^2+6*s+1)) %?? Ns=[1 0]; Ds=[1 10 9 6 1];
```

A.5 Using the Laplace Transform to Solve Differential Equations

One can realize how useful the Laplace transform is for dealing with linear ordinary differential equations.

Example A.3 Solving a Differential Equation
Let us solve the following differential equation.

$$\frac{d^2}{dt^2}y(t) + 3\frac{d}{dt}y(t) + 2y(t) = 10u_s(t),\ t \geq 0 \tag{1}$$

with the initial conditions $y(0) = 1$ and $y'(0) = -2$.

(Solution)
We make use of the time differentiation properties (A.11) and (A.12) of Laplace transform to write the Laplace transform of both sides as

$$s^2Y(s) - sy(0) - y'(0) + 3(sY(s) - y(0)) + 2Y(s) = \frac{10}{s};$$
$$(s^2 + 3s + 2)Y(s) = sy(0) + y'(0) + 3y(0) + \frac{10}{s} = \frac{10}{s} + s + 1 \tag{2}$$

This algebraic equation can be solved for $Y(s)$ as

$$Y(s) = \frac{s(s+1)+10}{s(s^2+3s+2)} \tag{3}$$

We expand this s-domain solution into partial fractions as

$$Y(s) = \frac{s(s+1)+10}{s(s^2+3s+2)} = \frac{r_1}{s} + \frac{r_2}{s+1} + \frac{r_3}{s+2} = \frac{5}{s} + \frac{-10}{s+1} + \frac{6}{s+2} \tag{4}$$

where the coefficients are found by using the formula (A.28.1) as

$$r_1 = sY(s)|_{s=0} \overset{(3)}{=} \left.\frac{s^2+s+10}{s^2+3s+2}\right|_{s=0} = 5$$
$$r_2 = (s+1)Y(s)|_{s=-1} \overset{(3)}{=} \left.\frac{s^2+s+10}{s(s+2)}\right|_{s=-1} = -10 \tag{5}$$
$$r_3 = (s+2)Y(s)|_{s=-2} \overset{(3)}{=} \left.\frac{s^2+s+10}{s(s+1)}\right|_{s=-2} = 6$$

Thus we can write the time-domain solution $y(t)$ as

$$y(t) = (r_1e^{-0t} + r_2e^{-t} + r_3e^{-2t})u_s(t) = (5 - 10e^{-t} + 6e^{-2t})u_s(t) \tag{6}$$

```
>>syms s, Ns=[1 1 10]; Ds=[1 3 2 0];
>>y=ilaplace(poly2sym(Ns,s)/poly2sym(Ds,s)) % Inverse Laplace transform
>>y1=ilaplace((s^2+s+10)/(s^3+3*s^2+2*s)) % Or, directly from Eq.(3)
>>t=0:0.01:10; yt=eval(y); plot(t,yt), hold on % plot y(t) for t=[0,10]
>>[r,p,k]=residue(Ns,Ds), % Partial fraction expansion
>>yt1=real(r.'*exp(p*t)); plot(t,yt1,'r') % Another way to get y(t)
```

Another alternative way to solve differential equations using MATLAB is to use the symbolic differential solver "`dsolve()`". To solve Eq. (1), we need the following MATLAB statement:

```
>>y=dsolve('D2y+3*Dy+2*y=0','y(0)=1,Dy(0)=-2')
   y = 5+6*exp(-2*t)-10*exp(-t)
```

Table A.2 Properties of Laplace transform

(0) Definition	$X(s) = \mathcal{L}\{x(t)\} = \int_0^\infty x(t)e^{-st}\,dt;\ x(t) \overset{\mathcal{L}}{\leftrightarrow} X(s)$
(1) Linearity	$\alpha x(t) + \beta y(t) \leftrightarrow \alpha X(s) + \beta Y(s)$
(2) Time shifting (Real translation)	$x(t - t_1)u_s(t - t_1),\ t_1 > 0 \leftrightarrow$ $e^{-st_1}\left\{X(s) + \int_{-t_1}^{0} x(\tau)e^{-s\tau}d\tau\right\} = e^{-st_1}X(s)$ for $x(t) = 0\ \forall t < 0$
(3) Frequency shifting (Complex translation)	$e^{s_1 t}x(t) \leftrightarrow X(s - s_1)$
(4) Real convolution	$g(t) * x(t) \leftrightarrow G(s)X(s)$
(5) Time derivative (Differentiation property)	$x'(t) = \frac{d}{dt}x(t) \leftrightarrow sX(s) - x(0)$
(6) Time integral (Integration property)	$\int_{-\infty}^{t} x(\tau)d\tau \leftrightarrow \frac{1}{s}X(s) + \frac{1}{s}\int_{-\infty}^{0} x(\tau)d\tau$
(7) Complex derivative	$t\,x(t) \leftrightarrow -\frac{d}{ds}X(s)$
(8) Complex convolution	$x(t)y(t) \leftrightarrow \frac{1}{2\pi j}\int_{\sigma_0-\infty}^{\sigma_0+\infty} X(v)Y(s - v)dv$
(9) Initial value theorem	$x(0) = \lim_{s\to\infty} s\,X(s)$
(10) Final value theorem	$x(\infty) = \lim_{s\to 0} s\,X(s)$

Theorem A.1 *Stability Condition of an LTI system on its Pole Location*
A linear time-invariant (LTI) system having the system or transfer function $G(s)/G[z]$ is stable iff (if and only if) all the poles of $G(s)/G[z]$ are strictly within the left-half s-plane/the unit circle in the z-plane (see Remarks 2.5, 3.2, and 4.5).

Proof Noting that the system or transfer function of a continuous-time/discrete-time system is the Laplace/ z-transform of its impulse response $g(t)/g[n]$, let the system function of a causal LTI (linear time-invariant) system be

$$G(s) = \mathcal{L}\{g(t)\} = \int_0^\infty g(t)e^{-st}dt \quad \text{(1a)}$$

$$G[z] = \mathcal{Z}\{g[n]\} = \sum_{n=0}^{\infty} g[n]z^{-n} \quad \text{(1b)}$$

Taking the absolute value yields

$$|G(s)| = \left|\int_0^\infty g(t)e^{-st}dt\right| \le \int_0^\infty |g(t)||e^{-st}|dt \quad \text{(2a)}$$

$$|G[z]| = \left|\sum_{n=0}^{\infty} g[n]z^{-n}\right| \le \sum_{n=0}^{\infty} |g[n]||z^{-n}| \quad \text{(2b)}$$

Suppose that

$G(s)$ has a pole at $s = s_p = \sigma_p + j\omega_p$ with $\sigma_p > 0$ on the RHP (right-half s-plane)

$G[z]$ has a pole at $z = z_p = r_p e^{j\Omega_p}$ with $r_p > 1$ outside the unit circle in the z-plane so that

$$|G(s_p)| = \infty \quad (3a)$$

$$|G[z_p]| = \infty \quad (3b)$$

We can substitute this pole into Eq. (2a)/(2b) to get

$$\infty \overset{(3a)}{=} |G(s_p)| \overset{(2a)}{\le} \int_0^\infty |g(t)||e^{-\sigma_p t}||e^{-j\omega_p t}|dt;$$
$$\infty \le \int_0^\infty |g(t)|dt \quad (4a)$$

$$\infty \overset{(3b)}{=} |G[z_p]| \overset{(2b)}{\le} \sum_{n=0}^{\infty} |g[n]||r_p^{-n}||e^{-j\Omega_p n}|;$$
$$\infty \le \sum_{n=0}^{\infty} |g[n]| \quad (4b)$$

which contradicts the stability condition (1.2.27a)/(1.2.27b). This implies that if even a single pole of a system exists in the unstable region, i.e., on the RHP in the s-plane or outside the unit circle in the z-plane, then the system is unstable. This completes the proof of the theorem.

The *stability condition* can be understood more clearly if we expand the system function into the partial fractions and take its inverse transform to write the impuse response as

$$G(s) = \sum_{k=0}^{K} \frac{A_k}{s - s_k};$$
$$g(t) = \sum_{k=0}^{k} a_k e^{s_k t} \quad (5a)$$

$$G[z] = \sum_{k=0}^{k} \frac{A_k}{z - z_k};$$
$$g[n] = \sum_{k=0}^{K} A_k z_k^n \quad (5b)$$

This shows that even a single pole with $\text{Re}\{s_k\} > 0$ or $|z_k| > 1$ causes the impulse response to diverge, which violates the BIBO stability condition (1.2.27).

Appendix B
Tables of Various Transforms

Table B.1 Definition and properties of the CTFS (continuous-time Fourier series)

(0) Definition	Synthesis: $\tilde{x}_P(t) \overset{(2.1.5a)}{=} \frac{1}{P}\sum_{k=-\infty}^{\infty} X_k e^{jk\omega_0 t}$, P: period of $\tilde{x}_P(t)$
	Analysis: $X_k \overset{(2.1.5b)}{=} \int_P \tilde{x}_P(t) e^{-jk\omega_0 t}dt,\ \omega_0 = \frac{2\pi}{P}$
(1) Linearty	$a\tilde{x}_P(t) + b\tilde{y}_P(t) \overset{\mathcal{F}}{\leftrightarrow} a\ X_k + bY_k$
(2) Time reversal	$\tilde{x}_P(-t) \overset{\mathcal{F}}{\leftrightarrow} X_{-k}$
(3) Symmetry for real-valued functions	Real-valued $\tilde{x}_P(t) = \tilde{x}_{P.e}(t) + \tilde{x}_{P.o}(t) \overset{\mathcal{F}}{\leftrightarrow} X_k = X^*_{-k}$
	Real-valued and even $\tilde{x}_{P.e}(t) \overset{\mathcal{F}}{\leftrightarrow} X_{k.e} = \mathrm{Re}\{X_k\}$
	Real-valued and odd $\tilde{x}_{P.o}(t) \overset{\mathcal{F}}{\leftrightarrow} X_{k.o} = j\mathrm{Im}\{X_k\}$
(4) Conjugate	$\tilde{x}^*_P(t) \overset{\text{CTFS}}{\leftrightarrow} X^*_{-k}$
(5) Time shifting (Real translation)	$\tilde{x}_P(t - t_1) \leftrightarrow X_k e^{-jk\omega_0 t_1} = \lvert X_k\rvert \angle(\phi_k - k\omega_0 t_1)$
(6) Frequency shifting (Complex translation)	$\tilde{x}_P(t)e^{jM\omega_0 t} \overset{\text{CTFS}}{\leftrightarrow} X_{k-M}$
(7) Real convolution (periodic/circular)	$\tilde{x}_P(t) * \tilde{y}_P(t) = \int_P \tilde{x}(\tau)\tilde{y}(t-\tau)d\tau \overset{\text{CTFS}}{\leftrightarrow} X_k Y_k$
(8) Complex convolution	$\tilde{x}_P(t)\tilde{y}_P(t) \overset{\text{CTFS}}{\leftrightarrow} \frac{1}{P}\sum_{m=-\infty}^{\infty} X_m Y_{k-m}$
(9) Time differentiation	$\frac{dx(t)}{dt} \overset{\text{CTFS}}{\leftrightarrow} jk\omega_0 X_k$
(10) Time integration	$\int_{-\infty}^{t} x(\tau)d\tau \begin{pmatrix}\text{finite-valued and periodic} \\ \text{only if } X_0 = 0\end{pmatrix} \overset{\text{CTFS}}{\leftrightarrow} \frac{1}{jk\omega_0} X_k$
(11) Scaling	$x(at),\ a > 0$ (periodic with period P/a) $\overset{\text{CTFS}}{\leftrightarrow}$ X_k (with fundamental frequency $a\omega_0$)
(12) Parseval's relation	$\int_{-\infty}^{\infty} \lvert\tilde{x}_P(t)\rvert^2 dt = \frac{1}{P}\sum_{k=-\infty}^{\infty} \lvert X_k\rvert^2$

Table B.2 Definition and properties of the CTFT (continuous-time Fourier transform)

Property	Relation
(0) Definition	$X(j\omega)=\mathcal{F}\{x(t)\}\overset{(2.2.1a)}{=}\int_{-\infty}^{\infty}x(t)e^{-j\omega t}dt$
(1) Linearity	$ax(t)+by(t)\overset{\mathcal{F}}{\leftrightarrow}aX(\omega)+bY(\omega)$
(2) Time reversal	$x(-t)\underset{(2.5.2)}{\overset{\mathcal{F}}{\leftrightarrow}}X(-\omega)$
(3) Symmetry for real-valued functions	Real-valued $x(t)=x_e(t)+x_o(t)\overset{\mathcal{F}}{\leftrightarrow}X(\omega)=X^*(-\omega)$
	Real-valued and even $x_e(t)\underset{(2.5.5a)}{\overset{\mathcal{F}}{\leftrightarrow}}X_e(\omega)=\text{Re}\{X(\omega)\}$
	Real-valued and odd $x_o(t)\underset{(2.5.5a)}{\overset{\mathcal{F}}{\leftrightarrow}}X_o(\omega)=j\,\text{Im}\{X(\omega)\}$
(4) Conjugate	$x^*(t)\overset{\mathcal{F}}{\leftrightarrow}X^*(-\omega)$
(5) Time shifting (Real translation)	$x(t-t_1)\underset{(2.5.6)}{\overset{\mathcal{F}}{\leftrightarrow}}X(\omega)e^{-j\omega t_1}=X(\omega)\angle-t_1\omega$
(6) Frequency shifting (Complex translation)	$x(t)e^{j\omega_1 t}\underset{(2.5.7)}{\overset{\mathcal{F}}{\leftrightarrow}}X(\omega-\omega_1)$
(7) Duality	$g(t)\overset{\mathcal{F}}{\leftrightarrow}f(\omega)\underset{(2.5.9)}{\Leftrightarrow}f(t)\overset{\mathcal{F}}{\leftrightarrow}2\pi g(-\omega)$
(8) Real convolution	$y(t)=x(t)*g(t)\underset{(2.5.11)}{\overset{\mathcal{F}}{\leftrightarrow}}Y(\omega)=X(\omega)G(\omega)$
(9) Complex convolution	$y(t)=x(t)m(t)\underset{(2.5.14)}{\overset{\mathcal{F}}{\leftrightarrow}}Y(\omega)=\frac{1}{2\pi}X(\omega)*M(\omega)$
(10) Time differentiation	$\frac{dx(t)}{dt}\underset{(2.5.17)}{\overset{\mathcal{F}}{\leftrightarrow}}j\omega X(\omega)$
(11) Time integration	$\int_{-\infty}^{t}x(\tau)d\tau\underset{(2.5.18)}{\overset{\mathcal{F}}{\leftrightarrow}}\pi X(0)\delta(\omega)+\frac{1}{j\omega}X(\omega)$
(12) Scaling	$x(at)\underset{(2.5.21)}{\overset{\mathcal{F}}{\leftrightarrow}}\frac{1}{\lvert a\rvert}X\left(\frac{\omega}{a}\right)$
(13) Time multiplication-Frequency differentiation	$tx(t)\underset{(2.5.20)}{\overset{\mathcal{F}}{\leftrightarrow}}j\frac{dX(\omega)}{d\omega}$
(14) Parseval's relation	$\int_{-\infty}^{\infty}\lvert x(t)\rvert^2dt\underset{(2.5.22)}{=}\frac{1}{2\pi}\int_{-\infty}^{\infty}\lvert X(\omega)\rvert^2d\omega$

Table B.3 CTFS/CTFT for basic functions

function $x(t)$	CTFS X_k (for periodic function)	CTFT $X(\omega)$ (for periodic/aperiodic function)
(1) $\frac{1}{P}\sum_{k=-\infty}^{\infty} X_k e^{jk\omega_0 t}$, $\omega_0 = \frac{2\pi}{P}$	X_k	$\overset{(2.33)}{\rightarrow} \frac{2\pi}{P}\sum_{k=-\infty}^{\infty} X_k \delta(\omega - k\omega_0)$
(2) $e^{j\omega_0 t}$, $\omega_0 = \frac{2\pi}{P}$	$X_k = \begin{cases} P & \text{for } k=1 \\ 0 & \text{elsewhere} \end{cases}$	$2\pi\delta(\omega - \omega_0)$
(3) $\cos\omega_0 t$, $\omega_0 = \frac{2\pi}{P}$	$X_k = \begin{cases} P/2 & \text{for } k=\pm 1, \cdots \\ 0 & \text{elsewhere} \end{cases}$	$\overset{(E2.11.2)}{\rightarrow} \pi(\delta(\omega+\omega_0) + \delta(\omega-\omega_0))$
(4) $\sin\omega_0 t$, $\omega_0 = \frac{2\pi}{P}$	$X_k = \begin{cases} -jP/2 & \text{for } k=1 \\ jP/2 & \text{for} k=-1 \\ 0 & \text{elsewhere} \end{cases}$	$\overset{(E2.11.1)}{\rightarrow} j\pi(\delta(\omega+\omega_0) - \delta(\omega-\omega_0))$
(5) Constant function $x(t) = 1$	$X_k = \begin{cases} P & \text{for } k=0 \\ 0 & \text{elsewhere} \end{cases}$	$\overset{(E2.7.1)}{\rightarrow} 2\pi\delta(\omega)$
(6) Impulse train $\sum_{i=-\infty}^{\infty} \delta(t - iP)$	$X_k = 1 \forall k$	$\frac{2\pi}{P}\sum_{k=-\infty}^{\infty} \delta(\omega - k\omega_0)$, $\omega_0 = \frac{2\pi}{P}$
(7) Rectangular wave $\sum_{i=-\infty}^{\infty} r_D(t - iP)$, $D \le P$	$X_k = D\ \text{sinc}\left(\frac{k\omega_0 D}{2\pi}\right)$, $\omega_0 = \frac{2\pi}{P}$	$\frac{2\pi}{P}\sum_{k=-\infty}^{\infty} X_k \delta(\omega - k\omega_0)$
(8) Rectangular pulse $r_D(t) = u_s(t + \frac{D}{2}) - u_s(t - \frac{D}{2})$	N/A for an aperiodic function	$D\ \text{sinc}(\frac{D}{2\pi}\omega) \overset{(E2.3.2)}{=} \frac{\sin(D\omega/2)}{\omega/2}$
(9) $\frac{\sin(Bt)}{\pi t} = \frac{B}{\pi}\text{sinc}\left(\frac{Bt}{\pi}\right)$	N/A for an aperiodic function	$\overset{(E2.9.1)}{\rightarrow} \begin{cases} 1 & \text{for } \|\omega\| \le B \\ 0 & \text{elsewhere} \end{cases}$
(10) $\delta(t)$	N/A for an aperiodic function	1
(11) $u_s(t)$	N/A for an aperiodic function	$\overset{(E2.8.6)}{\rightarrow} \frac{1}{j\omega} + \pi\delta(\omega)$
(12) $e^{-at}u_s(t)$, $\text{Re}\{a\} > 0$	N/A for an aperiodic function	$\frac{1}{a + j\omega}$
(13) $te^{-at}u_s(t)$, $\text{Re}\{a\} > 0$	N/A for an aperiodic function	$\frac{1}{(a + j\omega)^2}$
(14) $\frac{t^{M-1}}{(M-1)!}e^{-at}u_s(t)$, $\text{Re}\{a\} > 0$	N/A for an aperiodic function	$\frac{1}{(a + j\omega)^M}$

Table B.4 Definition and properties of the DTFT (discrete-time Fourier transform)

(0) Definition	DTFT: $X(\Omega) = X[e^{j\Omega}] = \mathcal{F}\{x[n]\} \overset{(3.1.1)}{=} \sum_{n=-\infty}^{\infty} x[n]e^{-j\Omega n}$
	IDTFT: $x[n] = \mathcal{F}^{-1}\{X(\Omega)\} \overset{(3.1.3)}{=} \frac{1}{2\pi}\int_{2\pi} X(\Omega)e^{j\Omega n}d\Omega$
(1) Periodicity	$X(\Omega) = X[e^{j\Omega}] = X(\Omega + 2m\pi)$ for any integer m
(2) Linearity	$ax[n] + by[n] \overset{\mathcal{F}}{\leftrightarrow} aX(\Omega) + bY(\Omega)$
(3) Time reversal	$x[-n] \underset{(3.2.3)}{\overset{\mathcal{F}}{\leftrightarrow}} X(-\Omega)$
(4) Symmetry for real-valued sequences	Real-valued $x[n] = x_e[n] + x_o[n] \overset{\mathcal{F}}{\leftrightarrow} X(\Omega) = X^*(-\Omega)$
	Real-valued and even $x_e[n] \underset{(3.2.5a)}{\overset{\mathcal{F}}{\leftrightarrow}} X_e(\Omega) = \mathrm{Re}\{X(\Omega)\}$
	Real-valued and odd $x_o[n] \underset{(3.2.5b)}{\overset{\mathcal{F}}{\leftrightarrow}} X_o(\Omega) = j\ \mathrm{Im}\{X(\Omega)\}$
(5) Conjugate	$x^*[n] \overset{\mathcal{F}}{\leftrightarrow} X^*(-\Omega)$
(6) Time shifting (Real translation)	$x[n-n_1] \overset{\mathcal{F}}{\leftrightarrow} X(\Omega)e^{-j\Omega n_1} = X(\Omega)\angle - n_1\Omega$
(7) Frequency shifting (Complex translation)	$x[n]e^{j\Omega_1 n} \overset{\mathcal{F}}{\leftrightarrow} X(\Omega - \Omega_1)$
(8) Real convolution	$x[n] * g[n] \overset{\mathcal{F}}{\leftrightarrow} X(\Omega)G(\Omega)$
(9) Complex convolution (circular)	$x[n]m[n] \overset{\mathcal{F}}{\leftrightarrow} \frac{1}{2\pi}X(\Omega) * M(\Omega)$
(10) Differencing in time	$x[n] - x[n-1] \overset{\mathcal{F}}{\leftrightarrow} (1 - e^{-j\Omega})X(\Omega)$
(11) Summation in time	$\sum_{m=-\infty}^{n} x[m] = x[n] * u_s[n] \overset{\mathcal{F}}{\leftrightarrow}$
	$\frac{1}{1-e^{-j\Omega}}X(\Omega) + \pi X(0)\sum_{i=-\infty}^{\infty}\delta(\Omega - 2\pi i)$
(12) Scaling	$x_{(K)}[n] = \begin{cases} x[n/K] = x[r] & \text{for } n = rK \\ 0 & \text{elsewhere} \end{cases} \underset{(3.2.14)}{\overset{\mathcal{F}}{\leftrightarrow}} X(K\Omega)$
(13) Time multiplication-Frequency differentiation	$nx[n] \overset{\mathcal{F}}{\leftrightarrow} j\frac{dX(\Omega)}{d\Omega}$
(14) Parseval's relation	$\sum_{n=-\infty}^{\infty} \lvert x[n]\rvert^2 = \frac{1}{2\pi}\int_{2\pi}\lvert X(\Omega)\rvert^2 d\Omega$

Table B.5 Definition and properties of the DFT/DFS (discrete-time Fourier series)

(0) Definition	$X(k) = \text{DFT}_N\{x[n]\} \overset{(3.4.2)}{=} \sum_{n=0}^{N-1} x[n]e^{-j2\pi kn/N} = \sum_{n=0}^{N-1} x[n]W_N^{kn}$
	$x[n] = \text{IDFT}_N\{X(k)\} \overset{(3.4.3)}{=} \frac{1}{N}\sum_{k=0}^{N-1} X(k)e^{j2\pi kn/N} = \frac{1}{N}\sum_{k=0}^{N-1} X(k)W_N^{-kn}$
(1) Periodicity	$X(k) = X(k+mN)$ for any integer m
(2) Linearity	$ax[n] + by[n] \overset{\text{DFT}}{\leftrightarrow} aX(k) + bY(k)$
(3) Time reversal	$\tilde{x}[-n]r_N[n] \overset{\text{DFT}}{\leftrightarrow} \tilde{X}_N(-k)r_N[k]$ ($r_N[k] = u_s[k] - u_s[k-N]$) where $\tilde{x}_N[n]$ is the periodic repetition of $x[n]$ with period N
(4) Symmetry for real-valued sequences	Real-valued $x[n] = x_e[n] + x_o[n] \overset{\text{DFT}}{\leftrightarrow} X(k) = \tilde{X}_N^*(-k)r_N[k]$
	Real-valued and even $x_e[n] \overset{\text{DFT}}{\leftrightarrow} X_e(k) = \text{Re}\{X(k)\}$
	Real-valued and odd $x_o[n] \overset{\text{DFT}}{\leftrightarrow} X_o(k) = j\ \text{Im}\{X(k)\}$
(5) Conjugate	$x^*[n] \overset{\text{DFT}}{\leftrightarrow} \tilde{X}_N^*(-k)r_N[k]$
(6) Duality	$g[n] \overset{\text{DFT}}{\leftrightarrow} f(k) \Leftrightarrow f[n] \overset{\text{DFT}}{\leftrightarrow} N\tilde{g}_N(-k)r_N[k]$
(7) Time shifting (circular)	$\tilde{x}_N[n-M]r_N[n]$(one period of circular shift) $\overset{\text{DFT}}{\leftrightarrow} W_N^{Mk}X(k)$
(8) Frequency shifting (circular) (Complex translation)	$W_N{}^{-Ln}x[n] \overset{\text{DFT}}{\leftrightarrow} \tilde{X}_N(k-L)r_N[k]$(one period of circular shift)
(9) Real convolution	$\left(\sum_{i=0}^{N-1} \tilde{x}(i)\tilde{y}(n-i)\right) r_N[n] \overset{\text{DFT}}{\leftrightarrow} X(k)Y(k)$
(10) Complex (circular) convolution	$x[n]y[n] \overset{\text{DFT}}{\leftrightarrow} \frac{1}{2\pi}\left(\sum_{i=0}^{N-1} \tilde{X}_N(i)\tilde{Y}_N(k-i)\right) r_N[k]$
(11) Scaling	$x_{(K)}[n] = \begin{cases} x[n/K] & \text{for } n = rK \\ 0 & \text{elsewhere} \end{cases} \overset{\text{DFT}}{\leftrightarrow} \tilde{X}_N(Kk)r_N[k]$
(12) Parseval's relation	$\sum_{n=<N>} \|x[n]\|^2 = \frac{1}{N}\sum_{k=<N>} \|X(k)\|^2$ where $<N> = \{0, 1, \cdots, N-1\}$

Table B.6 DFS/DFT and DTFT for basic sequences

Sequence $x[n]$	DFS $\tilde{X}(k)$/DFT $X(k)$ (for periodic/aperiodic and finite-duration sequences)	DTFT $X(\Omega)$
(1) $\frac{1}{P}\sum_{k=<N>} X_k e^{j2\pi kn/P},\ P = NT$	$\tilde{X}(k) = \frac{1}{T}X_k$	$\frac{2\pi}{P}\sum_{k=-\infty}^{\infty} X_k\delta(\Omega - k\frac{2\pi}{N})$
(2) $e^{jK\Omega_0 n},\ \Omega_0 = \frac{2\pi}{N}$	$\tilde{X}(k) = \begin{cases} N \text{ for } k = K, K\pm N, K\pm 2N, \cdots \\ 0 \text{ elsewhere}\end{cases}$	$\overset{(3.1.5)}{\rightarrow} 2\pi\sum_{k=-\infty}^{\infty}\delta(\Omega - K\Omega_0 - 2\pi k)$
(3) $\cos K\Omega_0 n,\ \Omega_0 = \dfrac{2\pi}{N}$	$\tilde{X}(k) = \begin{cases} N/2 \text{ for } k = \pm K, \pm K\pm N, \cdots \\ 0 \quad \text{elsewhere}\end{cases}$	$\overset{(E3.7.2)}{\rightarrow} \pi\sum_{k=-\infty}^{\infty} \begin{matrix}\delta(\Omega + K\Omega_0 - 2\pi k) \\ +\delta(\Omega - K\Omega_0 - 2\pi k)\end{matrix}$
(4) $\sin K\Omega_0 n,\ \Omega_0 = \dfrac{2\pi}{N}$	$\tilde{X}(k) = \begin{cases} -jN/2 \text{ for } k = K, K\pm N, \cdots \\ jN/2 \quad \text{for } k = -K, -K\pm N, \cdots \\ 0 \quad \text{elsewhere}\end{cases}$	$\overset{(E3.7.1)}{\rightarrow} j\pi\sum_{k=-\infty}^{\infty} \begin{matrix}\delta(\Omega + K\Omega_0 - 2\pi k) \\ -\delta(\Omega - K\Omega_0 - 2\pi k)\end{matrix}$
(5) Constant sequence $x[n] = 1$	$\tilde{X}(k) = \begin{cases} N \text{ for } k = 0, \pm N, \pm 2N, \cdots \\ 0 \text{ elsewhere}\end{cases}$	$\overset{(E3.6.1)}{\rightarrow} 2\pi\sum_{k=-\infty}^{\infty}\delta(\Omega - 2\pi k)$
(6) $\sum_{i=-\infty}^{\infty}\delta[n - iN]$	$\tilde{X}(k) = 1\ \forall k$	$\frac{2\pi}{N}\sum_{k=-\infty}^{\infty}\delta(\Omega - k\Omega_0),\ \Omega_0 = \frac{2\pi}{N}$
(7) Rectangular wave $\sum_{i=-\infty}^{\infty} r_{2M+1}[n - iM],\ 2M+1 \le N$	$\tilde{X}(k) = \dfrac{\sin(k\Omega_0(2M+1)/2)}{\sin(k\Omega_0/2)},\ \Omega_0 = \dfrac{2\pi}{N}$	$\frac{2\pi}{N}\sum_{k=-\infty}^{\infty}\tilde{X}(k)\delta(\Omega - k\Omega_0)$
(8) Rectangular pulse $r_{2M+1}[n]$ $= u_s[n+M] - u_s[n-M-1]$	$X(k) = \dfrac{\sin(k\Omega_0(2M+1)/2)}{\sin(k\Omega_0/2)}$ with $\Omega_0 = \dfrac{2\pi}{N}$	$\overset{(E3.1.2)}{\rightarrow} \dfrac{\sin(\Omega(2M+1)/2)}{\sin(\Omega/2)}$

Table B.6 (continued)

Sequence $x[n]$	DFS $\tilde{X}(k)$/DFT $X(k)$ (for periodic/aperiodic and finite-duration sequences)	DTFT $X(\Omega)$
(9) $\frac{\sin(Bn)}{\pi n} = \frac{B}{\pi}\text{sinc}\left(\frac{Bn}{\pi}\right)$	N/A for an infinite-duration aperiodic sequence	$\overset{(E3.5.1)}{\rightarrow} \begin{cases} 1 \text{ for } \lvert\Omega - 2m\pi\rvert \le B \le \pi \\ 0 \text{ elsewhere} \end{cases}$
(10) $\delta[n]$	$X(k) = 1 \;\forall\; 0 \le k \le N-1$	$\overset{(E3.4.1)}{\rightarrow} 1$
(11) $u_s[n]$	N/A for an infinite-duration aperiodic sequence	$\overset{(E3.8.6)}{\rightarrow} \frac{1}{1-e^{-j\Omega n}} + \pi \sum_{i=-\infty}^{\infty} \delta(\Omega - 2\pi i)$
(12) $a^n u_s[n]$ with $\lvert a\rvert < 1$	N/A for an infinite-duration aperiodic sequence	$\overset{(E3.2.1)}{\rightarrow} \frac{1}{1-ae^{-j\Omega}}$
(13) $(n+1)a^n u_s[n],\ \lvert a\rvert < 1$	N/A for an infinite-duration aperiodic sequence	$\frac{1}{(1-ae^{-j\Omega})^2}$
(14) $a^{\lvert n\rvert}$ with $\lvert a\rvert < 1$	N/A for an infinite-duration aperiodic sequence	$\overset{(E3.3.1)}{\rightarrow} \frac{1-a^2}{1-2a\cos\Omega + a^2}$

Table B.7 Definitions and properties of the Laplace transform and z-transform

Laplace transform		z-transform	
$\mathbb{X}(s) = L\{x(t)\} = \int_{-\infty}^{\infty} x(t)e^{-st}dt$	Bilateral(two-sided)	$\mathbb{X}[z] = Z\{x[n]\} = \sum_{n=-\infty}^{\infty} x[n]z^{-n}$	
$X(s) = \mathcal{L}\{x(t)\} = \int_{0}^{\infty} x(t)e^{-st}dt$	Unilateral(one-sided)	$X[z] = \mathcal{Z}\{x[n]\} = \sum_{n=0}^{\infty} x[n]z^{-n}$	
(1) Linearity			
$aX(s)+bY(s)$	$ax(t)+by(t)$	$ax[n]+by[n]$	$aX[z]+bY[z]$
(2) Time shifting (Real translation)			
$e^{-st_1}\mathbb{X}(s)$ $e^{-st_1}\left\{X(s)+\int_{-t_1}^{0} x(\tau)e^{-s\tau}d\tau\right\}$ $e^{st_1}\left\{X(s)-\int_{0}^{t_1} x(\tau)e^{-s\tau}d\tau\right\}$	$x(t-t_1)$ $x(t-t_1),\ t_1>0$ $x(t+t_1),\ t_1>0$	$x[n-n_1]$ $x[n-n_1],\ n_1>0$ $x[n+n_1],\ n_1>0$	$z^{-n_1}\mathbb{X}[z]$ $z^{-n_1}\left\{X[z]+\sum_{m=-n_1}^{-1} x[m]z^{-m}\right\}$ $z^{n_1}\left\{X[z]-\sum_{m=0}^{n_1-1} x[m]z^{-m}\right\}$
(3) Frequency shifting (Complex translation)			
$X(s-s_1)$	$e^{s_1t}x(t)$	$z_1^n x[n]$	$X[z/z_1]$
(4) Real convolution			
$G(s)X(s)$	$g(t)*x(t)$	$g[n]*x[n]$	$G[z]X[z]$
(5) Time derivative (Differentiation property)		Differencing	
<Bilateral> $s\mathbb{X}(s)$ <Unilateral> $sX(s)-x(0)$	$\frac{d}{dt}x(t)$	$x[n]-x[n-1]$	<Bilateral> $(1-z^{-1})\mathbb{X}[z]$ <Unilateral> $(1-z^{-1})X[z]-x[-1]$
(6) Time integral (Integration property)		Summation	
<Bilateral> $\frac{1}{s}\mathbb{X}(s)$ <Unilateral> $\frac{1}{s}\left(X(s)+\int_{-\infty}^{t} x(\tau)\,d\tau\right)$	$\int_{-\infty}^{t} x(\tau)d\tau$	$\sum_{m=-\infty}^{n} x[m]$	<Bilateral> $\frac{1}{1-z^{-1}}\mathbb{X}[z]$ <Unilateral> $\frac{1}{1-z^{-1}}(X[z]+\sum_{m=-\infty}^{-1} x[m])$
(7) Complex derivative			
$\frac{d}{ds}X(s)$	$-t\,x(t)$	$-n\,x[n]$	$z\frac{d}{dz}X[z]$
(8) Complex convolution			
$\frac{1}{2\pi j}\int_{\sigma_0-\infty}^{\sigma_0+\infty} X(v)Y(s-v)dv$	$x(t)y(t)$	$x[n]y[n]$	$\frac{1}{2\pi j}\oint_c X[z/v]Y[v]dv$
(9) Time scaling			
$aX(as)$	$x(t/a)$	$x[n/M]$ (M : a positive integer)	$X[z^M]$
(10) Time reversal			
$\mathbb{X}(-s)$	$x(-t)$	$x[-n]$	$\mathbb{X}[z^{-1}]$
(11) Initial value theorem			
$x(0) = \lim_{s\to\infty} sX(s)$		$x[0] = \lim_{z\to\infty} X[z]$	
(12) Final value theorem			
$x(\infty) = \lim_{s\to 0} sX(s)$		$x[\infty] = \lim_{z\to 1}(1-z^{-1})X[z]$	

Table B.8 Laplace transforms and z-transforms of basic functions/sequences

$X(s) = \mathcal{L}\{x(t)\}$	$x(t)$		$x[n]$	$X[z] = \mathcal{Z}\{x[n]\}$
1	$\delta(t)$	(1)	$\frac{1}{T}\delta[n]$	$\frac{1}{T}$
$e^{-t_1 s}$	$\delta(t - t_1)$	(2)	$\frac{1}{T}\delta[n - n_1]$	$\frac{1}{T}z^{-n_1}$
$\frac{1}{s}$	$u_s(t)$	(3)	$u_s[n]$	$\frac{z}{z-1} = \frac{1}{1-z^{-1}}$
$\frac{1}{s^2}$	$t\, u_s(t)$	(4)	$nT\, u_s[n]$	$\frac{T\, z}{(z-1)^2}$
$\frac{M!}{s^{M+1}}$	$t^M\, u_s(t)$	(5)	$(nT)^M\, u_s[n]$	$\lim_{a\to 0}(-1)^M \frac{\partial^M}{\partial a^M}\left(\frac{z}{z-e^{-aT}}\right)$
$\frac{1}{s+a}$	$e^{-at}u_s(t)$	(6)	$e^{-aT\,n}\, u_s[n] = b^n\, u_s[n]$	$\frac{z}{z-e^{-aT}} = \frac{z}{z-b}(b = e^{-aT})$
$\frac{1}{(s+a)^2}$	$t\, e^{-at}\, u_s(t)$	(7)	$nT\, e^{-aT\,n}\, u_s[n]$	$\frac{T\, ze^{-aT}}{(z-e^{-aT})^2}$
$\frac{M!}{(s+a)^{M+1}}$	$t^M\, e^{-at}\, u_s(t)$	(8)	$(nT)^M e^{-aTn}\, u_s[n]$	$(-1)^M \frac{\partial^M}{\partial a^M}\left(\frac{z}{z-e^{-aT}}\right)$
$\frac{\omega}{s^2+\omega^2}$	$\sin\omega t\, u_s(t)$	(9)	$\sin(\omega T n)u_s[n]$	$\frac{z\sin\omega T}{z^2 - 2z\cos\omega T + 1}$
$\frac{\omega}{(s+\sigma)^2+\omega^2}$	$e^{-\sigma t}\sin\omega t\, u_s(t)$	(10)	$e^{-\sigma T n}\sin(\omega T n)u_s[n]$	$\frac{ze^{-\sigma T}\sin\omega T}{(z - e^{-\sigma T}\cos\omega T)^2 + e^{-2\sigma T}\sin^2\omega T}$
$\frac{s}{s^2+\omega^2}$	$\cos\omega t\, u_s(t)$	(11)	$\cos(\omega T n)u_s[n]$	$\frac{z(z-\cos\omega T)}{z^2 - 2z\cos\omega T + 1}$
$\frac{s+\sigma}{(s+\sigma)^2+\omega^2}$	$e^{-\sigma t}\cos\omega t\, u_s(t)$	(12)	$e^{-\sigma T n}\cos(\omega T n)u_s[n]$	$\frac{z(z - e^{-\sigma T}\cos\omega T)}{(z - e^{-\sigma T}\cos\omega T)^2 + e^{-2\sigma T}\sin^2\omega T}$

Table B.9 Bilateral z-transform of basic sequences

$x[n]$	$\mathbb{X}[z]=\mathbb{Z}\{x[n]\}=\sum_{n=-\infty}^{\infty}x[n]z^{-n}$	with ROC
(1) $\delta[n]$	1	All z
(2) $\delta[n-n_1]$	z^{-n_1} All z except 0 (for $m>0$) or ∞ (for $m<0$)	
(3) $u_s[n]$	$\frac{z}{z-1}=\frac{1}{1-z^{-1}}$	$\lvert z\rvert>1$
(4) $-u_s[-n-1]$	$\frac{z}{z-1}=\frac{1}{1-z^{-1}}$	$\lvert z\rvert<1$
(5) $e^{-\sigma Tn}u_s[n]=a^n u_s[n]$	$\frac{z}{z-a}=\frac{1}{1-az^{-1}}(a=e^{-\sigma T})$	$\lvert z\rvert>\lvert a\rvert$
(6) $-b^n u_s[-n-1]$	$\frac{z}{z-b}=\frac{1}{1-bz^{-1}}$	$\lvert z\rvert<\lvert b\rvert$
(7) $e^{-\sigma T\lvert n\rvert}=a^{\lvert n\rvert}$	$\frac{z}{z-a}-\frac{z}{z-1/a}(a=e^{-\sigma T})$	$\lvert a\rvert<\lvert z\rvert<\left\lvert\frac{1}{a}\right\rvert$
(8) $nu_s[n]$	$\frac{z}{(z-1)^2}$	$\lvert z\rvert>1$
(9) $-nu_s[-n-1]$	$\frac{z}{(z-1)^2}$	$\lvert z\rvert<1$
(10) $na^n u_s[n]$	$\frac{az}{(z-a)^2}$	$\lvert z\rvert>\lvert a\rvert$
(11) $-nb^n u_s[-n-1]$	$\frac{bz}{(z-b)^2}$	$\lvert z\rvert<\lvert b\rvert$
(12) $n^2\,a^n\,u_s[n]$	$\frac{az(z+a)}{(z-a)^3}$	$\lvert z\rvert>\lvert a\rvert$
(13) $-n^2b^n u_s[-n-1]$	$\frac{bz(z+b)}{(z-b)^3}$	$\lvert z\rvert<\lvert b\rvert$
(14) $\frac{n(n-1)\cdots(n-K+2)}{(K-1)!}a^{n-(K-1)}u_s[n]$	$\frac{z}{(z-a)^K}$	$\lvert z\rvert>\lvert a\rvert$
(15) $\sin(\Omega_1 n)u_s[n]$	$\frac{z\sin\Omega_1}{(z-\cos\Omega_1)^2+\sin^2\Omega_1}=\frac{z\sin\Omega_1}{z^2-2z\cos\Omega_1+1}$	$\lvert z\rvert>1$
(16) $\cos(\Omega_1 n)u_s[n]$	$\frac{z(z-\cos\Omega_1)}{(z-\cos\Omega_1)^2+\sin^2\Omega_1}=\frac{z(z-\cos\Omega_1)}{z^2-2z\cos\Omega_1+1}$	$\lvert z\rvert>1$
(17) $r^n\,\sin(\Omega_1 n)u_s[n]$	$\frac{zr\sin\Omega_1}{(z-r\cos\Omega_1)^2+r^2\sin^2\Omega_1}=\frac{zr\sin\Omega_1}{z^2-2zr\cos\Omega_1+r^2}$	$\lvert z\rvert>\lvert r\rvert$
(18) $r^n\,\cos(\Omega_1 n)u_s[n]$	$\frac{z(z-r\cos\Omega_1)}{(z-r\cos\Omega_1)^2+r^2\sin^2\Omega_1}=\frac{z(z-r\cos\Omega_1)}{z^2-2zr\cos\Omega_1+r^2}$	$\lvert z\rvert>\lvert r\rvert$

Appendix C
Operations on Complex Numbers, Vectors, and Matrices

C.1 Complex Addition

$$(a_1 + jb_1) + (a_2 + jb_2) = (a_1 + a_2) + j(b_1 + b_2) \tag{C.1}$$

C.2 Complex Multiplication

Rectangular form: $(a_1 + jb_1) \times (a_2 + jb_2) = (a_1a_2 - b_1b_2) + j(a_1b_2 + b_1a_2)$ (C.2a)

Polar form: $r_1\angle\theta_1 \times r_2\angle\theta_2 = r_1e^{j\theta_1}r_2e^{j\theta_2} = r_1r_2e^{j(\theta_1+\theta_2)} = r_1r_2\angle(\theta_1 + \theta_2)$ (C.2b)

C.3 Complex Division

$$\text{Rectangular form: } \frac{a_2 + jb_2}{a_1 + jb_1} = \frac{a_2 + jb_2}{a_1 + jb_1} \times \frac{a_1 - jb_1}{a_1 - jb_1} = \frac{a_1a_2 + b_1b_2}{a_1^2 + b_1^2} + j\frac{a_1b_2 - a_2b_1}{a_1^2 + b_1^2} \tag{C.3a}$$

$$\text{Polar form: } \frac{r_2\angle\theta_2}{r_1\angle\theta_1} = \frac{r_2e^{j\theta_2}}{r_1e^{j\theta_1}} = \frac{r_2}{r_1}e^{j(\theta_2-\theta_1)} = \frac{r_2}{r_1}\angle(\theta_2 - \theta_1) \tag{C.3b}$$

C.4 Conversion Between Rectangular Form and Polar/Exponential Form

$$a_r + ja_i = r\angle\theta = re^{j\theta} \text{ with } r = |a_r + ja_i| = \sqrt{a_r^2 + a_i^2} \text{ and } \theta = \tan^{-1}\frac{a_i}{a_r} \tag{C.4}$$

Here, r and θ are referred to as the *absolute value* and *argument* or *phase angle* of the complex number $a_r + ja_i$, respectively and j is the unit imaginary number $\sqrt{-1}$.

C.5 Operations on Complex Numbers Using MATLAB

If we do not use `i` and `j` for any other purpose, they represent the basic imaginary unit $\sqrt{-1}$ by default. Try typing the following statements into the MATLAB Command Window:

```
>>c1= 1+2i; c2= 3-4i;
>>c1*c2, c1/c2 % multiplication/division of complex numbers
>>r=abs(c2) % absolute value of the complex number c2
>>sqrt(real(c2)^2+imag(c2)^2) % equivalent to the absolute value
>>th=angle(c2) % phase angle of the complex number c2 in radians
>>atan2(imag(c2),real(c2)) % equivalent to the phase angle
>>imag(log(c2)) % equivalent to the phase angle
>>th*180/pi % radian-to-degree conversion
>>r*exp(j*th) % polar-to-rectangular conversion
  ans = 3.0000 - 4.0000i
>>C= [1+i  1-2i; -1+3i -1-4i] % a complex matrix
  C1 =  1.0000 + 1.0000i   1.0000 - 2.0000i
       -1.0000 + 3.0000i  -1.0000 - 4.0000i
>>C1= C' % conjugate transpose
  C1 =  1.0000 - 1.0000i  -1.0000 - 3.0000i
        1.0000 + 2.0000i  -1.0000 + 4.0000i
```

C.6 Matrix Addition and Subtraction[Y-1]

$$A + B = \begin{bmatrix} a_{11} & a_{12} & \cdot & a_{1N} \\ a_{21} & a_{22} & \cdot & a_{2N} \\ \cdot & \cdot & \cdot & \cdot \\ a_{M1} & a_{M2} & \cdot & a_{MN} \end{bmatrix} + \begin{bmatrix} b_{11} & b_{12} & \cdot & b_{1N} \\ b_{21} & b_{22} & \cdot & b_{2N} \\ \cdot & \cdot & \cdot & \cdot \\ b_{M1} & b_{M2} & \cdot & b_{MN} \end{bmatrix} = \begin{bmatrix} c_{11} & c_{12} & \cdot & c_{1N} \\ c_{21} & c_{22} & \cdot & c_{2N} \\ \cdot & \cdot & \cdot & \cdot \\ c_{M1} & c_{M2} & \cdot & c_{MN} \end{bmatrix} = C \tag{C.5a}$$

$$\text{with } a_{mn} + b_{mn} = c_{mn} \tag{C.5b}$$

C.7 Matrix Multiplication

$$AB = \begin{bmatrix} a_{11} & a_{12} & \cdot & a_{1K} \\ a_{21} & a_{22} & \cdot & a_{2K} \\ \cdot & \cdot & \cdot & \cdot \\ a_{M1} & a_{M2} & \cdot & a_{MK} \end{bmatrix} \begin{bmatrix} b_{11} & b_{12} & \cdot & b_{1N} \\ b_{21} & b_{22} & \cdot & b_{2N} \\ \cdot & \cdot & \cdot & \cdot \\ b_{K1} & b_{K2} & \cdot & b_{KN} \end{bmatrix} = \begin{bmatrix} c_{11} & c_{12} & \cdot & c_{1N} \\ c_{21} & c_{22} & \cdot & c_{2N} \\ \cdot & \cdot & \cdot & \cdot \\ c_{M1} & c_{M2} & \cdot & c_{MN} \end{bmatrix} = C \quad \text{(C.6a)}$$

$$\text{with } c_{mn} = \sum\nolimits_{k=1}^{K} a_{mk} b_{kn} \quad \text{(C.6b)}$$

(cf.) For this multiplication to be done, the number of columns of A must equal the number of rows of B.

(cf.) Note that the commutative law does not hold for the matrix multiplication, i.e., $AB \neq BA$.

C.8 Determinant

The *determinant* of a $K \times K$ (square) matrix $A = [a_{mn}]$ is defined by

$$\det(A) = |A| = \sum\nolimits_{k=0}^{K} a_{kn}(-1)^{k+n} M_{kn} \text{ or } \sum\nolimits_{k=0}^{K} a_{mk}(-1)^{m+k} M_{mk} \quad \text{(C.7a)}$$

$$\text{for any fixed } 1 \le n \le K \text{ or } 1 \le m \le K$$

where the minor M_{kn} is the determinant of the $(K-1) \times (K-1)$ (minor) matrix formed by removing the kth row and the nth column from A and $A_{kn} = (-1)^{k+n} M_{kn}$ is called the cofactor of a_{kn}.

Especially, the determinants of a 2×2 matrix $A_{2\times2}$ and a 3×3 matrix $A_{3\times3}$ are

$$\det(A_{2\times2}) = \begin{vmatrix} a_{11} & a_{12} \\ a_{21} & a_{22} \end{vmatrix} = \sum\nolimits_{k=1}^{2} a_{kn}(-1)^{k+n} M_{kn} = a_{11}a_{22} - a_{12}a_{21} \quad \text{(C.7b)}$$

$$\det(A_{3\times3}) = \begin{vmatrix} a_{11} & a_{12} & a_{13} \\ a_{21} & a_{22} & a_{23} \\ a_{31} & a_{32} & a_{33} \end{vmatrix} = a_{11} \begin{vmatrix} a_{22} & a_{23} \\ a_{32} & a_{33} \end{vmatrix} - a_{12} \begin{vmatrix} a_{21} & a_{23} \\ a_{31} & a_{33} \end{vmatrix} + a_{13} \begin{vmatrix} a_{21} & a_{22} \\ a_{31} & a_{32} \end{vmatrix}$$

$$= a_{11}(a_{22}a_{33} - a_{23}a_{32}) - a_{12}(a_{21}a_{33} - a_{23}a_{31}) + a_{13}(a_{21}a_{32} - a_{22}a_{31}) \quad \text{(C.7c)}$$

Note the following properties.

- If the determinant of a matrix is zero, the matrix is singular.
- The determinant of a matrix equals the product of the eigenvalues of a matrix.

- If A is upper/lower triangular having only zeros below/above the diagonal in each column, its determinant is the product of the diagonal elements.
- $\det(A^T) = \det(A)$; $\det(AB) = \det(A)\det(B)$; $\det(A^{-1}) = 1/\det(A)$

C.9 Eigenvalues and Eigenvectors of a Matrix[1]

The *eigenvalue* or characteristic value and its corresponding *eigenvector* or characteristic vector of an $N \times N$ matrix A are defined to be a scalar λ and a nonzero vector $\mathbf{v}$ satisfying

$$A\,\mathbf{v} = \lambda\,\mathbf{v} \Leftrightarrow (A - \lambda I)\mathbf{v} = \mathbf{0} \quad (\mathbf{v} \neq \mathbf{0}) \tag{C.8}$$

where $(\lambda,\ \mathbf{v})$ is called an eigenpair and there are N eigenpairs for an $N \times N$ matrix A.

The eigenvalues of a matrix can be computed as the roots of the characteristic equation

$$|A - \lambda I| = 0 \tag{C.9}$$

and the eigenvector corresponding to an eigenvalue λ_i can be obtained by substituting λ_i into Eq. (C.8) and solve it for $\mathbf{v}$.

Note the following properties.

- If A is symmetric, all the eigenvalues are real-valued.
- If A is symmetric and positive definite, all the eigenvalues are real and positive.
- If $\mathbf{v}$ is an eigenvector of A, so is $c\mathbf{v}$ for any nonzero scalar c.

C.10 Inverse Matrix

The *inverse* matrix of a $K \times K$ (square) matrix $A = [a_{mn}]$ is denoted by A^{-1} and defined to be a matrix which is pre-multiplied/post-multiplied by A to form an identity matrix, i.e., satisfies

$$A \times A^{-1} = A^{-1} \times A = I \tag{C.10}$$

An element of the inverse matrix $A^{-1} = [\alpha_{mn}]$ can be computed as

$$\alpha_{mn} = \frac{1}{\det(A)} A_{mn} = \frac{1}{|A|}(-1)^{m+n} M_{mn} \tag{C.11}$$

where M_{kn} is the minor of a_{kn} and $A_{kn} = (-1)^{k+n} M_{kn}$ is the cofactor of a_{kn}.

---

[1] See the website @http://www.sosmath.com/index.html or http://www.psc.edu/~burkardt/papers/linear_glossary.html.

Note that a square matrix A is invertible/nonsingular if and only if

- no eigenvalue of A is zero, or equivalently
- the rows (and the columns) of A are linearly independent, or equivalently
- the determinant of A is nonzero;

C.11 Symmetric/Hermitian Matrix

A square matrix A is said to be *symmetric*, if it is equal to its transpose, i.e.,

$$A^T \equiv A \tag{C.12}$$

A complex-valued matrix is said to be *Hermitian* if it is equal to its complex conjugate transpose, i.e.,

$$A \equiv A^{*T} \text{ where } ^* \text{ means the conjugate.} \tag{C.13}$$

Note the following properties of a symmetric/Hermitian matrix.

- All the eigenvalues are real.
- If all the eigenvalues are distinct, the eigenvectors can form an orthogonal/unitary matrix U.

C.12 Orthogonal/Unitary Matrix

A nonsingular (square) matrix A is said to be *orthogonal* if its transpose is equal to its inverse, i.e.,

$$A^T A \equiv I; \quad A^T \equiv A^{-1} \tag{C.14}$$

A complex-valued (square) matrix is said to be *unitary* if its conjugate transpose is equal to its inverse, i.e.,

$$A^{*T} A \equiv I; \quad A^{*T} \equiv A^{-1} \tag{C.15}$$

Note the following properties of an orthogonal/unitary matrix.

- The magnitude (absolute value) of every eigenvalue is one.
- The product of two orthogonal matrices is also orthogonal; $(AB)^{*T}\ (AB) = B^{*T}\ (A^{*T}\ A)B \equiv I$.

C.13 Permutation Matrix

A matrix P having only one nonzero element of value 1 in each row and column is called a *permutation* matrix and has the following properties.

- Pre-multiplication/post-multiplication of a matrix A by a permutation matrix P, i.e., PA or AP yields the row/column change of the matrix A, respectively.
- A permutation matrix A is orthogonal, i.e., $A^T A \equiv I$.

C.14 Rank

The *rank* of an $M \times N$ matrix is the number of linearly independent rows/columns and if it equals min (M, N), then the matrix is said to be of *maximal* or *full* rank; otherwise, the matrix is said to be *rank-deficient* or to have *rank-deficiency*.

C.15 Row Space and Null Space

The *row space* of an $M \times N$ matrix A, denoted by $\mathcal{R}(A)$, is the space spanned by the row vectors, i.e., the set of all possible linear combinations of row vectors of A that can be expressed by $A^T\boldsymbol{\alpha}$ with an M-dimensional column vector $\boldsymbol{\alpha}$. On the other hand, the *null space* of the matrix A, denoted by $\mathcal{N}(A)$, is the space orthogonal (perpendicular) to the row space, i.e., the set of all possible linear combinations of the N-dimensional vectors satisfying $A\mathbf{x} = \mathbf{0}$.

C.16 Row Echelon Form

A matrix is said to be of *row echelon form* [W-6] if

- each nonzero row having at least one nonzero element has a 1 as its first nonzero element, and
- the leading 1 in a row is in a column to the right of the leading 1 in the upper row.
- All-zero rows are below the rows that have at least one nonzero element.

A matrix is said to be of *reduced row echelon form* if it satisfies the above conditions and additionally, each column containing a leading 1 has no other nonzero elements.

Any matrix, singular or rectangular, can be transformed into this form through the Gaussian elimination procedure, i.e., a series of elementary row operations, or equivalently, by using the MATLAB built-in routine "`rref()`". For example, we have

$$A = \begin{bmatrix} 0 & 0 & 1 & 3 \\ 2 & 4 & 0 & -8 \\ 1 & 2 & 1 & -1 \end{bmatrix} \underset{\text{change}}{\overset{\text{row}}{\rightarrow}} \begin{bmatrix} 2 & 4 & 0 & -8 \\ 0 & 0 & 1 & 3 \\ 1 & 2 & 1 & -1 \end{bmatrix} \underset{\substack{\text{row} \\ \text{subtraction}}}{\overset{\substack{\text{row} \\ \text{division}}}{\rightarrow}} \begin{bmatrix} 1 & 2 & 0 & -4 \\ 0 & 0 & 1 & 3 \\ 0 & 0 & 1 & 3 \end{bmatrix} \underset{\text{subtraction}}{\overset{\text{row}}{\rightarrow}} \begin{bmatrix} 1 & 2 & 0 & -4 \\ 0 & 0 & 1 & 3 \\ 0 & 0 & 0 & 0 \end{bmatrix} = rref(A)$$

Once this form is obtained, it is easy to compute the rank, the determinant, and the inverse of the matrix, if only the matrix is invertible.

C.17 Positive Definiteness

A square matrix A is said to be *positive definite* if

$$\mathbf{x}^{*T} A\,\mathbf{x} > 0 \text{ for any nonzero vector } \mathbf{x} \tag{C.16}$$

A square matrix A is said to be *positive semi-definite* if

$$\mathbf{x}^{*T} A\,\mathbf{x} \geq 0 \text{ for any nonzero vector } \mathbf{x} \tag{C.17}$$

Note the following properties of a positive definite matrix A.

- A is nonsingular and all of its eigenvalues are positive, and
- the inverse of A is also positive definite.

There are similar definitions for *negative definiteness* and *negative semi-definiteness*.

Note the following property, which can be used to determine if a matrix is positive (semi-) definite or not. A square matrix is positive definite if and only if

(i) every diagonal element is positive and
(ii) every leading principal minor matrix has positive determinant.

On the other hand, a square matrix is positive semi-definite if and only if

(i) every diagonal element is nonnegative and
(ii) every principal minor matrix has nonnegative determinant.

Note also that the principal minor matrices are the sub-matrices taking the diagonal elements from the diagonal of the matrix A and, say for a 3×3 matrix, the principal minor matrices are

$$a_{11},\ a_{22},\ a_{33},\ \begin{bmatrix} a_{11} & a_{12} \\ a_{21} & a_{22} \end{bmatrix},\ \begin{bmatrix} a_{22} & a_{23} \\ a_{32} & a_{33} \end{bmatrix},\ \begin{bmatrix} a_{11} & a_{13} \\ a_{31} & a_{33} \end{bmatrix},\ \begin{bmatrix} a_{11} & a_{12} & a_{13} \\ a_{21} & a_{22} & a_{23} \\ a_{31} & a_{32} & a_{33} \end{bmatrix}$$

among which the leading ones are

$$a_{11},\ \begin{bmatrix} a_{11} & a_{12} \\ a_{21} & a_{22} \end{bmatrix},\ \begin{bmatrix} a_{11} & a_{12} & a_{13} \\ a_{21} & a_{22} & a_{23} \\ a_{31} & a_{32} & a_{33} \end{bmatrix}$$

C.18 Scalar(Dot) Product and Vector(Cross) Product

A *scalar product* of two N -dimensional vectors $\mathbf{x}$ and $\mathbf{y}$ is denoted by $\mathbf{x} \cdot \mathbf{y}$ and is defined by

$$\mathbf{x} \cdot \mathbf{y} = \sum_{n=1}^{N} x_n y_n = \mathbf{x}^T \mathbf{y} \tag{C.18}$$

An *outer product* of two 3-dimensional column vectors $\mathbf{x} = [x_1\ x_2\ x_3]^T$ and $\mathbf{y} = [y_1\ y_2\ y_3]^T$ is denoted by $\mathbf{x} \times \mathbf{y}$ and is defined by

$$\mathbf{x} \times \mathbf{y} = \begin{bmatrix} x_2 y_3 - x_3 y_2 \\ x_3 y_1 - x_1 y_3 \\ x_1 y_2 - x_2 y_1 \end{bmatrix} \tag{C.19}$$

C.19 Matrix Inversion Lemma

[Matrix Inversion Lemma]
Let A, C and $[C^{-1} + DA^{-1}B]$ be well-defined with non-singularity as well as compatible dimensions. Then we have

$$[A + BCD]^{-1} = A^{-1} - A^{-1}B[C^{-1} + DA^{-1}B]^{-1}DA^{-1} \tag{C.20}$$

Proof We will show that post-multiplying Eq. (C.20) by $[A + BCD]$ yields an identity matrix.

$$\begin{aligned}
&[A^{-1} - A^{-1}B[C^{-1} + DA^{-1}B]^{-1}DA^{-1}][A + BCD] \\
&= I + A^{-1}BCD - A^{-1}B[C^{-1} + DA^{-1}B]^{-1}D - A^{-1}B[C^{-1} + DA^{-1}B]^{-1}DA^{-1}BCD \\
&= I + A^{-1}BCD - A^{-1}B[C^{-1} + DA^{-1}B]^{-1}C^{-1}CD - A^{-1}B[C^{-1} + DA^{-1}B]^{-1}DA^{-1}BCD \\
&= I + A^{-1}BCD - A^{-1}B[C^{-1} + DA^{-1}B]^{-1}[C^{-1} + DA^{-1}B]CD \\
&= I + A^{-1}BCD - A^{-1}BCD \equiv I
\end{aligned}$$

C.20 Differentiation w.r.t. a Vector

The 1st derivative of a scalar-valued function $f(\mathbf{x})$ w.r.t. a vector $\mathbf{x} = [x_1\ x_2]^T$ is called the gradient of $f(\mathbf{x})$ and defined as

$$\nabla f(\mathbf{x}) = \frac{d}{d\mathbf{x}} f(\mathbf{x}) = \begin{bmatrix} \partial f / \partial x_1 \\ \partial f / \partial x_2 \end{bmatrix} \tag{C.21}$$

Based on this definition, we can write the following equation.

$$\frac{\partial}{\partial \mathbf{x}}\mathbf{x}^T\mathbf{y} = \frac{\partial}{\partial \mathbf{x}}\mathbf{y}^T\mathbf{x} = \frac{\partial}{\partial \mathbf{x}}(x_1y_1 + x_2y_2) = \begin{bmatrix} y_1 \\ y_2 \end{bmatrix} = \mathbf{y} \tag{C.22}$$

$$\frac{\partial}{\partial \mathbf{x}}\mathbf{x}^T\mathbf{x} = \frac{\partial}{\partial \mathbf{x}}(x_1^2 + x_2^2) = 2\begin{bmatrix} x_1 \\ x_2 \end{bmatrix} = 2\mathbf{x} \tag{C.23}$$

Also with an $M \times N$ matrix A, we have

$$\frac{\partial}{\partial \mathbf{x}}\mathbf{x}^T A\mathbf{y} = \frac{\partial}{\partial \mathbf{x}}\mathbf{y}^T A^T\mathbf{x} = A\mathbf{y} \tag{C.24a}$$

$$\frac{\partial}{\partial \mathbf{x}}\mathbf{y}^T A\mathbf{x} = \frac{\partial}{\partial \mathbf{x}}\mathbf{x}^T A^T\mathbf{y} = A^T\mathbf{y} \tag{C.24b}$$

where

$$\mathbf{x}^T A\mathbf{y} - \sum_{m=1}^{M}\sum_{n=1}^{N} a_{mn}x_m y_n \tag{C.25}$$

Especially for a square, symmetric matrix A with $M = N$, we have

$$\frac{\partial}{\partial \mathbf{x}}\mathbf{x}^T A\mathbf{x} = (A + A^T)\mathbf{x} \overset{\text{if } A \text{ is symmetric}}{\rightarrow} 2A\mathbf{x} \tag{C.26}$$

The 2nd derivative of a scalar function $f(\mathbf{x})$ w.r.t. a vector $\mathbf{x} = [x_1\ x_2]^T$ is called the Hessian of $f(\mathbf{x})$ and is defined as

$$H(\mathbf{x}) = \nabla^2 f(\mathbf{x}) = \frac{d^2}{d\mathbf{x}^2}f(\mathbf{x}) = \begin{bmatrix} \partial^2 f/\partial x_1^2 & \partial^2 f/\partial x_1\partial x_2 \\ \partial^2 f/\partial x_2\partial x_1 & \partial^2 f/\partial x_2^2 \end{bmatrix} \tag{C.27}$$

Based on this definition, we can write the following equation:

$$\frac{d^2}{d\mathbf{x}^2}\mathbf{x}^T A\mathbf{x} = A + A^T \overset{\text{if } A \text{ is symmetric}}{\rightarrow} 2A \tag{C.28}$$

On the other hand, the 1st derivative of a vector-valued function $\mathbf{f}(\mathbf{x})$ w.r.t. a vector $\mathbf{x} = [x_1\ x_2]^T$ is called the Jacobian of $f(\mathbf{x})$ and is defined as

$$J(\mathbf{x}) = \frac{d}{d\mathbf{x}}\mathbf{f}(\mathbf{x}) = \begin{bmatrix} \partial f_1/\partial x_1 & \partial f_1/\partial x_2 \\ \partial f_2/\partial x_1 & \partial f_2/\partial x_2 \end{bmatrix} \tag{C.29}$$

Appendix D
Useful Formulas

$\sin(A \pm 90°) = \pm \cos A$	(D.1)	$\cos(A \pm 90°) = \mp \sin A$	(D.2)
$\sin(A \pm 180°) = -\sin A$	(D.3)	$\cos(A \pm 180°) = -\cos A$	(D.4)
$\sin(A \pm B) = \sin A \cos B \pm \cos A \sin B$	(D.5)	$\tan(A \pm B) = \dfrac{\tan A \pm \tan B}{1 \mp \tan A \tan B}$	(D.7)
$\cos(A \pm B) = \cos A \cos B \mp \sin A \sin B$	(D.6)		
$\sin A \sin B = \frac{1}{2}\{\cos(A-B) - \cos(A+B)\}$	(D.8)	$\sin A \cos B = \frac{1}{2}\{\sin(A+B) + \sin(A-B)\}$	(D.9)
$\cos A \sin B = \frac{1}{2}\{\sin(A+B) - \sin(A-B)\}$	(D.10)	$\cos A \cos B = \frac{1}{2}\{\cos(A+B) + \cos(A-B)\}$	(D.11)
$\sin A + \sin B = 2\sin\left(\frac{A+B}{2}\right)\cos\left(\frac{A-B}{2}\right)$	(D.12)	$\cos A + \cos B = 2\cos\left(\frac{A+B}{2}\right)\cos\left(\frac{A-B}{2}\right)$	(D.13)
$\sin^2 A = \frac{1}{2}(1 - \cos 2A)$	(D.14)	$\cos^2 A = \frac{1}{2}(1 + \cos 2A)$	(D.15)
$\sin 2A = 2\sin A \cos A$	(D.16)	$\cos 2A = \cos^2 A - \sin^2 A = 1 - 2\sin^2 A = 2\cos^2 A - 1$	(D.17)
$a\cos A - b\sin A = \sqrt{a^2+b^2}\cos(A+\theta),\ \theta = \tan^{-1}(b/a)$			(D.18)
$a\sin A + b\cos A = \sqrt{a^2+b^2}\sin(A+\theta),\ \theta = \tan^{-1}(b/a)$			(D.19)
Euler identity: $e^{\pm j\theta} = \cos\theta \pm j\sin\theta$	(D.20)	$e^{j\theta} + e^{-j\theta} = 2\cos\theta$	(D.21)
		$e^{j\theta} - e^{-j\theta} = j2\sin\theta$	(D.22)
Sum of (infinite) geometric series: $\sum_{n=0}^{N} r^n = \dfrac{1-r^{N+1}}{1-r} \overset{N\to\infty}{\to} \sum_{n=0}^{\infty} r^n = \dfrac{1}{1-r},\ \lvert r \rvert < 1$			(D.23)
Binomial expansion: $(a+b)^N = \sum_{n=0}^{N} {}_N C_n a^{N-n} b^n$ with ${}_N C_n = {}_N C_{N-n} = \dfrac{{}_N P_n}{n!} = \dfrac{N!}{(N-n)!n!}$			(D.24)
$e^x = \sum_{n=0}^{\infty} \frac{1}{n!}x^n = 1 + \frac{1}{1!}x + \frac{1}{2!}x^2 + \frac{1}{3!}x^3 + \cdots \overset{x\to 0}{\to} e^x = 1 + x$			(D.25)
$\frac{d}{dt}t^n = nt^{n-1}$	(D.26)	$\int t^n\,dt = \frac{1}{n+1}t^{n+1}$ for $n \neq -1$	(D.32)
$\frac{d}{dt}e^{at} = ae^{at}$	(D.27)	$\int e^{at}\,dt = \frac{1}{a}e^{at}$	(D.33)
$\frac{d}{dt}\cos\omega t = -\omega\sin\omega t$	(D.28)	$\int \cos\omega t\,dt = \frac{1}{\omega}\sin\omega t$	(D.34)
$\frac{d}{dt}\sin\omega t = \omega\cos\omega t$	(D.29)	$\int \sin\omega t\,dt = -\frac{1}{\omega}\cos\omega t$	(D.35)
$\frac{d}{dt}(uv) = u\frac{d}{dt}v + v\frac{d}{dt}u$	(D.30)	$\int u\frac{dv}{dt}dt = uv - \int v\frac{du}{dt}dt$ (Partial integral)	(D.36)
$\frac{d}{dt}\left(\frac{v}{u}\right) = \dfrac{u(dv/dt) - v(du/dt)}{u^2}$	(D.31)		
$x(t) * \delta(t-t_1) \overset{(1.1.22)}{=} x(t-t_1)$	(D.37)	$\int_{-\infty}^{+\infty} x(t)\delta(t-t_1)\,dt \overset{(1.1.25)}{=} x(t_1)$	(D.38)
$\int_{-\infty}^{\infty} e^{j\omega t}\,dt \overset{(E2.7.1)}{=} 2\pi\delta(\omega)$	(D.39)	$\sum_{n=-\infty}^{\infty} e^{j\Omega n} \overset{(E3.6.1)}{=} 2\pi\sum_{i=-\infty}^{\infty}\delta(\Omega - 2\pi i)$	(D.40)

Appendix E
MATLAB

Once you installed MATLAB®, you can click the icon like the one at the right side to run MATLAB. Then you will see the MATLAB Command Window on your monitor as depicted in Fig. E.1, where a cursor appears (most likely blinking) to the right of the prompt like ">>" waiting for you to type in commands/statements. If you are running MATLAB of version 6.x or above, the main window has not only the command window, but also the workspace and command history boxes on the left-up/down side of the command window, in which you can see the contents of MATLAB memory and the commands you have typed into the Command window up to the present time, respectively (see Fig. E.1). You might clear the boxes by clicking the corresponding submenu under the "Edit" menu and even remove/restore them by clicking the corresponding submenu under the "Desktop/Desktop_Layout" menu.

How do we work with the MATLAB Command Window?

- By clicking "File" on the top menu and then "New"/"Open" in the File pull-down menu, you can create/edit any file with the MATLAB editor.

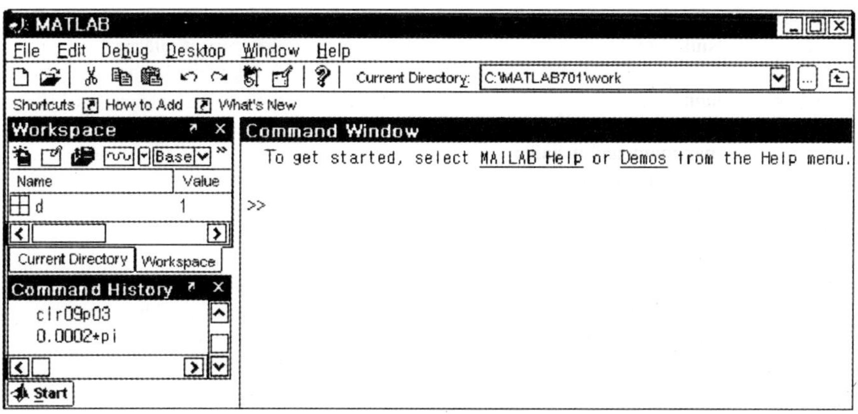

Fig. E.1 The MATLAB command window with the workspace and command history boxes

- By clicking "File" on the top menu and then "Set_Path" in the File pull-down menu, you can make the MATLAB search path list include or exclude the paths containing the files you want to or not to be run.
- If you are a beginner in MATLAB, then it may be worthwhile to click "Help" on the top menu, "Demos" in the Help pull-down menu, (double-)click any topic that you want to learn about, and watch the visual explanation about it.
- By typing any MATLAB commands/statements in the MATLAB Command Window, you can use various powerful mathematic/graphic functions of MATLAB.
- If you have an M-file which contains a series of commands/statements composed for performing a target procedure, you can type in the file name (without the extension ".m") to run it.

Note the following:

1) the index of an array in MATLAB starts from 1, not 0, and
2) a dot(.) must be put before an operator to make a term-wise (element-by-element) operation.

Table E.1 Functions and Variables in MATLAB

function	Remark	function	Remark
`cos(x)`		`exp(x)`	exponential function
`sin(x)`		`log(x)`	natural logarithm
`tan(x)`		`log10(x)`	common logarithm
`acos(x)`	$\cos^{-1}(x)$	`abs(x)`	absolute value
`asin(x)`	$\sin^{-1}(x)$	`angle(x)`	phase of a complex number[rad]
`atan(x)`	$-\pi/2 \le \tan^{-1}(x) \le \pi/2$	`sqrt(x)`	square root
`atan2(y,x)`	$-\pi \le \tan^{-1}(y,x) \le \pi$	`real(x)`	real part
`cosh(x)`	$(e^x + e^{-x})/2$	`imag(x)`	imaginary part
`sinh(x)`	$(e^x - e^{-x})/2$	`conj(x)`	complex conjugate
`tanh(x)`	$(e^x - e^{-x})/(e^x + e^{-x})$	`round(x)`	the nearest integer (round-off)
`acosh(x)`	$\cosh^{-1}(x)$	`fix(x)`	the nearest integer towards 0
`asinh(x)`	$\sinh^{-1}(x)$	`floor(x)`	the greatest integer $\le x$ rounding down
`atanh(x)`	$\tanh^{-1}(x)$	`ceil(x)`	the smallest integer $\ge x$ rounding up
`max`	maximum and its index	`sign(x)`	1(positive)/0/-1(negative)
`min`	minimum and its index	`mod(y,x)`	remainder of y/x
`sum`	sum	`rem(y,x)`	remainder of y/x
`prod`	product	`eval(f)`	evaluate an expression
`norm`	norm	`feval(f,a)`	function evaluation
`sort`	sort in the ascending order	`polyval`	value of a polynomial function
`clock`	present time	`poly`	polynomial with given roots
`find`	index of element(s)	`roots`	roots of polynomial
`tic`	start a stopwatch timer	`toc`	read the stopwatch timer (elapsed time from tic)
`date`	present date		

Table E.1 (continued)

function	Remark	function	Remark
reserved variables with special meaning			
`i, j`	$\sqrt{-1}$	`pi`	π
`eps`	machine epsilon (smallest positive number)	`Inf, inf`	largest number (∞)
realmax, realmin	largest/smallest positive number	`NaN`	Not_a_Number(undetermined)
`end`	the end of for-loop or if, while, case statement or an array index	`break`	exit while/for loop
`nargin`	# of input arguments	`nargout`	# of output arguments
`varargin`	variable input argument list	`varargout`	variable output argument list

Table E.2 Graphic line specifications used in the `plot()` command

Line type	Point type (Marker symbol)			Color	
- solid line	.(dot)	+(plus)	*(asterisk)	r: red	m: magenta
: dotted line	$\wedge$: Δ	>: >	°(circle)	g: green	y: yellow
-- dashed line	p: $\star$	v: ∇	x: x-mark	b: blue	c: cyan(sky blue)
-. dash-dot	d: $\Diamond$	<: <	s: square($\square$)	k: black	

Some of mathematical functions and special reserved constants/variables defined in MATLAB are listed in Table E.1. Table E.2 shows the graphic line specifications used in the plot() command.

E.1 Convolution and Deconvolution

```
C=conv(A,B)
```

- For two sequences $A = [a_1\ a_2\ \cdots\ a_N]$ and $B = [b_1\ b_2\ \cdots\ b_M]$, this routine computes their (linear) convolution $C = [c_1\ c_2\ \cdots\ c_{N+M-1}]$ as

$$c_{n+1} = \sum_{m=\max(0,n-N+1)}^{\min(n,M-1)} a_{n-m+1} b_{m+1} \tag{E.1}$$

- It can be used for multiplying two polynomials whose coefficients are A and B, respectively:

$$(a_1 x^{N-1} + \cdots + a_{N-1}x + a_N)(b_1 x^{M-1} + \cdots + b_{M-1}x + b_M)$$
$$= c_1 x^{N+M-2} + \cdots + c_{N+M-2}x + c_{N+M-1}$$

- `C=xcorr2(A,B)` performs two-dimensional convolution of matrices A and B.

```
function z=conv_circular(x,y,N)
% Circular convolution z(n)= (1/N) sum_m=0^N-1 x(m)*y(n-m)
if nargin<3, N=min(length(x),length(y)); end
x=x(1:N); y=y(1:N); y_crc= fliplr(y);
for n=1:N
   y_crc = [y_crc(N) y_crc(1:N-1)]; z(n) = x*y_crc.'/N;
end
```

`[Q,R]=deconv(C,A)`

- This routine deconvolves a vector A out of C by long division to find Q and R such that

$$\mathrm{C} = \mathrm{conv}(\mathrm{A}, \mathrm{Q}) + \mathrm{R}$$

- It can be used for polynomial division $C(x)/A(x) = Q(x) + R(x)/A(x)$.

E.2 Correlation

`[phi,lags]=xcorr(x,y,opt)`

- With no optional parameter opt, this routine computes the (linear) correlation as

$$\begin{aligned}\phi_{xy}[m] &= \sum\nolimits_{n=-\infty}^{\infty} x[n+m]y^*[n] = \sum\nolimits_{n=-\infty}^{\infty} x[n]y^*[n-m] \\ &= \sum\nolimits_{n=-\infty}^{\infty} x[-n+m]y^*[-n] \end{aligned} \tag{E.2}$$

together with the corresponding time lag vector `lags`.
- With optional parameter opt = “coeff”/“biased”/“unbiased”, it outputs a normalized version as Eqs. (1.4.16a)/(1.4.16b)/(1.4.16c), respectively.

```
function [phi,lags]=xcorr_my(x,y,opt)
% computes the crosscorrelation of two vectors x and y as phi(1:Nx+Ny-1)
if nargin<3, opt='reg'; if nargin<2, y=x; end, end
x=x(:).'; y=y(:).'; % make them row vectors
Nx=length(x); Ny=length(y); N=max(Nx,Ny);
for n=1:Nx-1
  N1=min(Nx-n,Ny); m=1:N1; phi(n+Ny)= x(m+n)*y(m)';
  if opt(1:3)=='unb', phi(n+Ny)= phi(n+Ny)/N1; end  % regular or unbiased
end
for n=1-Ny:0
  N2=min(Nx,Ny+n); m=1:N2; phi(n+Ny)= x(m)*y(m-n)';
  if opt(1:3)=='unb', phi(n+Ny)= phi(n+Ny)/N2; end  % regular or unbiased
end
if opt(1)=='b', phi= phi/N; % biased version (1.4.16b)
 elseif opt(1)=='c', phi=phi/sqrt((x*x')*(y*y')); % coefficient version
end
lags=[-(Ny-1):(Nx-1)];
```

```
function phi=xcorr_circular(x,y,N)
% Circular (cyclic) correlation of x and y with period N
if nargin<2, y=x; end
Nx = length(x); Ny = length(y);
if nargin<3, N=max(Nx,Ny); end
if Nx<N, x = [x zeros(1,N-Nx)]; elseif N<Nx, x = x(1:N); end
if Ny<N, y = [y zeros(1,N-Ny)]; elseif N<Ny, y = y(1:N); end
for n=1:N, phi(n) = x*y'/N; y = [y(N) y(1:N-1)]; end
```

`PHI=xcorr2(X,Y)`

– Two-dimensional cross-correlation of matrices X and Y

E.3 CTFS (Continuous-Time Fourier Series)

```
function [c,kk]=CTFS_exponential(x,P,N)
% Find the complex exponential Fourier coefficients c(k) for k=-N:N
% x: A periodic function with period P
% P: Period, N: Maximum frequency index to specify the frequency range
w0=2*pi/P; % the fundamental frequency [rad/s]
xexp_jkw0t_= [x '(t).*exp(-j*k*w0*t)'];
xexp_jkw0t= inline(xexp_jkw0t_,'t','k','w0');
kk=-N:N; tol=1e-6; % the frequency range tolerance on numerical error
for k=kk
   c(k+N+1)= quadl(xexp_jkw0t,-P/2,P/2,tol,[],k,w0); % Eq. (2.1.5b)
end
```

E.4 DTFT (Discrete-Time Fourier Transform)

```
function [X,ph]=DTFT(x,W,n0)
% DTFT of x[n] for frequency W regarding the 1st sample as the n0-th one.
Nt =length(x); n=0:Nt-1;
if nargin<3, n0 = -floor(Nt/2); end
X= x*exp(-j*(n+n0)'*W); % Eq.(3.1.1)
if nargout==2, ph=angle(X); X=abs(X); end
```

E.5 DFS/DFT (Discrete Fourier Series/Transform)

```
function [X,ph]=DFS(x,N,n0)
% N-point DFS/DFT of x[n] regarding the 1st sample as the n0-th one.
if nargin<3, n0 = 0; end
n=n0+[0:length(x)-1]; k=[0:N-1];
X= x*exp(-j*2*pi*n'*k/N); % Eq.(3.4.7)
If nargout==2, ph=angle(X); X=abs(X); end
```

E.6 FFT (Fast Fourier Transform)

`X=fft(x,N)` or `fft(x)`

- This routine computes the N-point DFT X(k) of a given sequence x[n], where the sequence will be zero-padded or truncated so that its resulting length will be N. If the 2nd input argument is not given, the DFT size will be set to the length of the sequence. In case the 1st input argument x is a matrix, X will also be a matrix, each column of which is the FFT of the corresponding column of x.
- Note that, to exploit the computational efficiency of FFT, the DFT size N should be chosen as a power of 2 greater than or equal to the length of the sequence.

`x=ifft(X,N)` or `ifft(X)`

- This routine computes the N-point IDFT x[n] of a given sequence X(k), where the sequence will be zero-padded or truncated so that its resulting length will be N. If the 2nd input argument is not given, the IDFT size will be set to the length of the sequence. In case the 1st input argument x is a matrix, x will also be a matrix, each column of which is the IDFT of the corresponding column of X.
- Note that, to exploit the computational efficiency of FFT, the IDFT size N should be chosen as a power of 2 greater than or equal to the length of the sequence.

`X=fft2(x,M,N)` or `fft2(x)`

- This routine computes the two-dimensional DFT X(k,l) of a given matrix x[m,n], where each column/row of x will be zero-padded or truncated so that the resulting matrix size will be M × N. If the 2nd and 3rd input arguments are not given, the DFT sizes will be set to the row/column size. In fact, "`fft2(x,M,N)`" is equivalent to "`fft(fft(x,M).',N).'`"

`x=ifft2(X,M,N)` or `ifft2(X)`

- This routine computes the two-dimensional IDFT x[m,n] of a given matrix X(k,l), where each column/row of X will be zero-padded or truncated so that the resulting matrix size will be M × N. If the 2nd and 3rd input arguments are not given, the IDFT sizes will be set to the row/column size. In fact, "`ifft2(x,M,N)`" is equivalent to "`ifft(ifft(x,M).',N).'`"

`y=fftshift(x)`

- This routine swaps the first and second halves of the input vector x if x is a vector and swaps quadrants one and three with quadrants two and four if x is a matrix. It is used to rearrange the FFT sequence so that the zero (DC) frequency component is located at the center with the negative/positive frequency ones on the left/right side.

E.7 Windowing

```
function xw=windowing(x,w,opt)
N= length(x);
if isnumeric(w)
  xw= x; Nw2=floor((N-w)/2); xw(1:Nw2)=0; xw(Nw2+w+1:end)=0;
 else
  switch lower(w(1:2))
    case {'bt','tt'}, w= bartlett(N); % window(bartlett,N)
    case 'bk', w= blackman(N); % window(@blackman,N)
    case 'cb', if nargin<3, r=100; else r=opt; end
               w= chebwin(N,r); % window(@chebwin,N,r)
    case 'gs', if nargin<3, alpha=2.5; else alpha=opt; end
               w= gausswin(N,alpha); % window(@gausswin,N,alpha)
    case 'hm', w= hamming(N); % window(@hamming,N)
    case 'hn', w= hanning(N); % window(@hanning,N)
    case 'ks', if nargin<3, beta=0.5; else beta=opt; end
               w= kaiser(N,beta); % window(@kaiser,N,beta)
    otherwise  w= x;
  end
  if size(x,1)==1,  w=w.';  end
  xw = x.*w;
end
```

(cf.) You can just type "window" or "wintool" into the MATLAB Command window to open the Window Design & Analysis Tool. Type "doc window/ signal"/"doc window" to see other usages of the signal processing toolbox or filter design toolbox function "window".

E.8 Spectrogram (FFT with Sliding Window)

`[X,F,T]=specgram(x,N,Fs,WND,Noverlap)` in MATLAB 6.x versions

- This STFT (Short-Time Fourier Transform) routine computes the spectrogram, i.e., N-point DFT X(k,m) of mth segment that is windowed by WND and overlaps with neighboring segment by Noverlap (whose default value is length(WND)/2), where the length of WND must be greater than Noverlap, but not greater than N.

- Input
  - WND: Window sequence given as a column vector of length Nw such that

    Noverlap < Nw <= N

    If a scalar is given as WND, a Hamming window of length N is used.
  - N: DFT size whose default value is min(256,length(x)).
  - Fs: Sampling frequency which does not affect the spectrogram but is used for scaling plots and its default value is 2 Hz.
  - You can use the default value for any parameter by leaving it off or using [].
- Output
  - X: STFT X(k,m) of x with time and frequency increasing linearly across its columns, from left to right and down its rows, starting at 0, respectively. The number of coumns of X will be Nc = fix((length(x)-Noverlap)/(length(WND)-Noverlap)). The number of rows of X will be Nr = ceil((N + 1)/2) or N depending on whether x is real or complex-valued.
  - F and T: Nr normalized frequencies (Ω/π) in Hz and Nc times in column vectors
- With no output arguments, the routine plots the absolute value of the spectrogram in the current figure, using imagesc(T,F,20*log10(abs(X))), where T is created with its starting point shifted by (Nw-Noverlap)/2. Then "axis xy", and "colormap(jet)" are used so that the low frequency content of the first portion of the signal is displayed in the lower left corner of the axes.

```
>>imagesc(T,F,20*log10(abs(X)+eps)); axis xy; colormap(jet)
```

(Example) Compute the spectrogram of a quadratic chirp.

```
Fs=1e3;  Ts=1/Fs;  t=0:Ts:2;  % 0~2 secs @ 1kHz sample rate
x= chirp(t,0,1,150,'q'); % quadratic chirp start @ 0 & cross 150Hz at t=1
N=64; Nhw=N;  % The DFT size and the length of the Hanning window
Noverlap=32; [X,F,T] = specgram(x,N,Fs,Nhw,Noverlap);
specgram(x,N,Fs,Nhw,Noverlap); % Display the spectrogram
title('Quadractic Chip: start at 0Hz and cross 150Hz at t=1sec')
```

`[X,F,T,P]=spectrogram(x,WND,Noverlap,N,Fs)` in MATLAB 7.x versions

- This STFT routine computes the spectrogram, i.e., N-point DFT X(k,m) of mth segment that is windowed by WND and overlaps with neighboring segment by Noverlap.
- By default, x is divided into eight segments. If x cannot be divided exactly into eight segments, it is truncated. Also by default, the DFT size N is set

to the maximum of 256 and the next power of 2 greater than the length of each segment of x. WND is a Hamming window of length N. Noverlap is the value producing 50% overlap between segments. Fs is the sampling frequency, which defaults to normalized frequency 1 Hz.

- Each column of X contains an estimate of the short-term, time-localized frequency content of x. Time increases across the columns of S and frequency increases down the rows.
- The number of rows of X will be Nr = ceil((N + 1)/2) or N depending on whether x is real-valued or complex-valued.
- The number of coumns of X will be Nc = fix((length(x)-Noverlap)/(length(WND)-Noverlap)).
- If WND is given as an integer, x is divided into segments of length WND and each segment is windowed by a Hamming window. If WND is a vector, x is divided into segments equal to the length of WND and then each segment is windowed by the window vector WND.
- Noverlap must be an integer smaller than WND (an integer) or its length if WND is a vector.
- If the 4th input argument is given as not a scalar meaning the DFT size, but a vector of frequency[Hz], spectrogram(x,WND,Noverlap,F) computes the spectrogram at the frequencies in F using the Goertzel algorithm.
- The 2nd output F is the frequency vector of length N and the 3rd output T the time vector of length Nc at which the spectrogram is computed, where each value of T corresponds to the center of each segment.
- The 4th output P is a matrix of the same size as x containing the PSD of each segment. For real x, each column of P has the one-sided periodogram estimate of the PSD of each segment. If x is complex or a frequency vector is given as the 4th input, P has the two-sided PSD.

E.9 Power Spectrum

```
[Pxy,F]=cpsd(x,y,WND,Noverlap,Nfft,Fs)
```

- This routine computes the one/two-sided CPSD (Cross Power Spectral Density) estimate Pxy of two real/complex-valued signal vectors x and y using Welch's averaged periodogram method. For real signals, Pxy has length (Nfft/2 + 1) if Nfft is even, and (Nfft + 1)/2 if Nfft is odd. For complex signals, Pxy always has length Nfft.
- By default, x is divided into 8 sections (with 50% overlap), each of which is windowed with a Hamming window and eight modified periodograms are computed and averaged.

- If WND is a vector, the signal vectors are divided into overlapping sections of length equal to the length of WND, and then each section is windowed with WND. If WND is an integer, x and y are divided into sections of length WND, and a Hamming window of equal length is used. If the length of x and y is such that it cannot be divided exactly into integer number of sections overlapped by Novelap samples (or 50% by default), they will be truncated accordingly. Noverlap must be an integer smaller than WND (given as an integer) or the length of WND (given as a vector).
- Nfft specifies the DFT size used to calculate the CPSD estimate and its default value is the maximum of 256 or the next power of 2 greater than the length of each section of x (and y).
- Note that if Nfft is greater than the segment length, the data is zero-padded. If Nfft is less than the segment, the segment is "wrapped" to make the length equal to Nfft. This produces the correct FFT when Nfft < L, L being signal or segment length.
- If the sampling frequency Fs is not given, F will be the vector of digital frequencies in rad/sample at which the PSD is estimated. For real signals, F will span the interval [0,pi] when Nfft is even and [0,pi) when Nfft is odd. For complex signals, F spans [0, 2*pi). If Fs[Hz] is given as the real frequency, F will span the interval [0,Fs/2] when Nfft is even and [0,Fs/2) when Nfft is odd. For complex signals, F spans the interval [0,Fs). If Fs is given as [] (an empty set), it will be set to 1 Hz so that F will be a vector of normalized frequencies.
- The string "twosided" or "onesided" may be placed in any position in the input argument list after Noverlap to specify the frequency range over which the CPSD estimated is found.
- cpsd(..) with no output arguments plots the CPSD estimate in dB per unit frequency.
- cpsd(x,x,. . .) can be used to find the APSD (auto power spectral density) of a signal x, which shows the distribution of signal power per unit frequency.

(cf.) An interactive GUI (graphic user interface) signal processing tool, named "SPTool", is available.

E.10 Impulse and Step Responses

```
[y,t,x]=impulse(sys,t_f) or [y,x,t]=impulse(B,A,t_f)
```

- This routine returns the impulse response `y(t)` (together with the time vector `t` starting from 0 to `t_f` and the state history x for a state-space model) of an LTI system with transfer function B(s)/A(s) or a *system model* sys that might have been created with either tf (transfer function), zpk (zero-pole-gain), or ss (state space) as below:

```
>>sys=tf(B,A) % continuous-time system B(s)/A(s)
>>sys=tf(B,A,Ts) %discrete-time system B[z]/A[z]
                  with sampling interval Ts
```

- For continuous-time systems, the final time or time vector t_f can be given in the form of t_0:dt:t_f where dt is the sample time of a discrete approximation to the continuous system. For discrete-time systems, it should be of the form t_0:Ts:t_f where Ts is the sampling period. Note that the impulse input is always assumed to arise at t = 0 (regardless of t_0).
- If t_f is not given, the simulation time range will somehow be chosen appropriately.
- `[y,x,t]=impulse(B,A,t_f)` returns an empty set [] for the 2nd output argument x.
- With no output arguments, the routine plots the impulse response with `plot()` for continuous-time systems and with `stair()` for discrete-time systems.

`[g,t]=impz(B,A,N,Fs)`

- This computes two N × 1 column vectors, g and t, where g consists of N samples of the impulse response of the digital filter B[z]/A[z] and t is the corresponding time vector consisting of N sample times spaced Ts = 1/Fs[s] apart. Note that Fs is 1 by default.
- The impulse response can also be obtained from `filter(B,A,[1 zeros(1,N-1)])`.
- With no output arguments, the routine plots the impulse response using `stem(t,g)`.

`[y,x]=dimpulse(A,B,C,D,IU,N)` or `y=dimpulse(B,A,N)`

- This returns N samples of the impulse response in y (together with the state history x for a state-space model) of a discrete-time LTI system described by state-space model (A,B,C,D) or transfer function B[z]/A[z] to the unit impulse input (applied to the IUth input terminal), where the state-space model is given as

$$\begin{aligned} x[n+1] &= Ax[n] + Bu[n] \\ y[n] &= Cx[n] + Du[n] \end{aligned}$$

`[y,t,x]=step(sys,t_f)` or `[y,x,t]=step(B,A,t_f)`

- This routine returns the step response `y(t)` (together with the time vector `t` starting from 0 to the final time `t_f` and the state history x for a state-space model) of the LTI model `sys` that might have been created with either `tf` (transfer function), `zpk` (zero-pole-gain), or `ss` (state space).
- For a continuous-time system, `t_f` can be given as a time vector in the form of `t_0:dt:t_f` where `dt` is the sample time of a discrete approximation to the continuous system. For a discrete-time system, it can be given in the form of `t_0:Ts:t_f` where `Ts` is the sampling interval. Note that the unit step input is always assumed to arise at $t = 0$ (regardless of `t_0`).
- If the final time or time vector `t_f` is not given, the simulation time range and number of points will somehow be chosen appropriately.
- `[y,x,t]=step(B,A,t_f)` returns an empty set [] for the 2nd output argument x.
- With no output arguments, the routine plots the step response with `plot()` for continuous-time systems and with `stair()` for discrete-time systems.

`[y,x]=dstep(A,B,C,D,IU,N)` or `y=dstep(B,A,N)`

- This returns N samples of the step response in y (together with the state history x for a state-space model) of a discrete-time LTI system described by state-space model (A,B,C,D) or transfer function B[z]/A[z] to the unit step input (applied to the IUth input terminal).

`[y,t,x]=lsim(sys,u,t,x0)`

- This routine returns the time response (together with the time vector `t` and the state history x) of the LTI system sys (with the initial state x0 at time t(1) for state-space models only) to the input signal u(t) where the input matrix u has as many columns as the input coefficient matrix B and its ith row specifies u(t(i)) where sys=ss(A,B,C,D).
- For discrete-time systems, u should be sampled at the same rate as sys and in this case, the input argument t is redundant and can be omitted or set to the empty matrix.
- For continuous-time systems, choose the sampling period $t(n + 1) - t(n)$ small enough to describe the input u accurately.

`[y,x]=dlsim(A,B,C,D,u,x0)` or `y=dlsim(B,A,u)`

- This routine returns the time response (together with the state history x) of a discrete-time LTI system described by state space model (A,B,C,D) or transfer function B[z]/A[z] (with the initial state x0 for state-space models only) to the input signal u[n] where the input matrix u has as many columns as the input coefficient matrix B.

E.11 Frequency Response

`G=freqs(B,A,w)` (`w` : analog frequency vector[rad/s])

– This routine returns the frequency response of a continuous-time system by substituting $s = j\omega$ ($\omega =$ `w`) into its system function $G(s) = B(s)/A(s) = (b_1 s^{NB-1} + \cdots + b_{NB})/(a_1 s^{NA-1} + \cdots + a_{NA})$:

$$G(\omega) = G(s)|_{s=j\omega} = \left. \frac{b_1 s^{NB-1} + \cdots + b_{NB-1}s + b_{NB}}{a_1 s^{NA-1} + \cdots + a_{NA-1}s + a_{NA}} \right|_{s=j\omega} \tag{E.3a}$$

`G=freqz(B,A,W)` (`W` : digital frequency vector[rad/sample])

– This routine returns the frequency response of a discrete-time system by substituting $z = e^{j\Omega}$ ($\Omega =$ `W`) into its system function $G[z] = B[z]/A[z] = (b_1 + \cdots + b_{NB} z^{-(NB-1)})/(a_1 + \cdots + a_{NA} z^{-(NA-1)})$:

$$G(\Omega) = G[z]|_{z=e^{j\Omega}} = \left. \frac{b_1 + b_2 z^{-1} + \cdots + b_{NB} z^{-(NB-1)}}{a_1 + a_2 z^{-1} + \cdots + a_{NA} z^{-(NA-1)}} \right|_{z=e^{j\Omega}} \tag{E.3b}$$

`[G,W]=freqz(B,A,N)`

(with `N`: number of digital frequency points around the upper half of the unit circle)

– This routine returns the frequency response of a discrete-time system by substituting $z = e^{j\Omega_k}(\Omega_k = 2k\pi/N$ with $k = 0 : N-1)$ into its system function $G[z] = B[z]/A[z]$ where the 2nd output argument is a column vector consisting of $\{\Omega_k = k\pi/N,\ k = 0 : N-1\}$.

`[G,W]=freqz(B,A,N,'whole')`

(with `N`: number of digital frequency points around the whole unit circle

– This routine returns the frequency response of a discrete-time system by substituting $z = e^{j\Omega_k}$ ($\Omega_k = 2k\pi/N$ with $k = 0 : N-1$) into its system function $G[z] = B[z]/A[z]$.

`G=freqz(B,A,f,fs)` (with f: digital frequency vector in Hz, `fs`: sampling frequency in Hz)

– This routine returns the frequency response of a discrete-time system by substituting $z = e^{j\Omega}$ ($\Omega = 2\pi f/f_s$) into its system function $G[z] = B[z]/A[z]$.

E.12 Filtering

```
[y,wf]=filter(B,A,x,w0) or
[y,wf]=filter(Gd_structure,x,w0)
```

- This routine computes the output y (together with the final condition wf) of the digital filter $G[z] = B[z]/A[z] = (b_1 + \cdots + b_{NB}z^{-(NB-1)})/(a_1 + \cdots + a_{NA}z^{-(NA-1)})$ (described by numerator B and denominator A or a filter structure constructed using `dfilt()`) to the input x (with the initial condition w0).
- Let every filter coefficient be divided by a_1 for normalization so that we have the system function of the digital filter as

$$G[z] = \frac{b_1 + b_2z^{-1} + \cdots + b_{NB}z^{-(NB-1)}}{a_1 + a_2z^{-1} + \cdots + a_{NA}z^{-(NA-1)}} \overset{b_i/a_1 \to b_i}{\underset{a_i/a_1 \to a_i}{=}} \frac{b_1 + b_2z^{-1} + \cdots + b_{NB}z^{-(NB-1)}}{1 + a_2z^{-1} + \cdots + a_{NA}z^{-(NA-1)}} \tag{E.4}$$

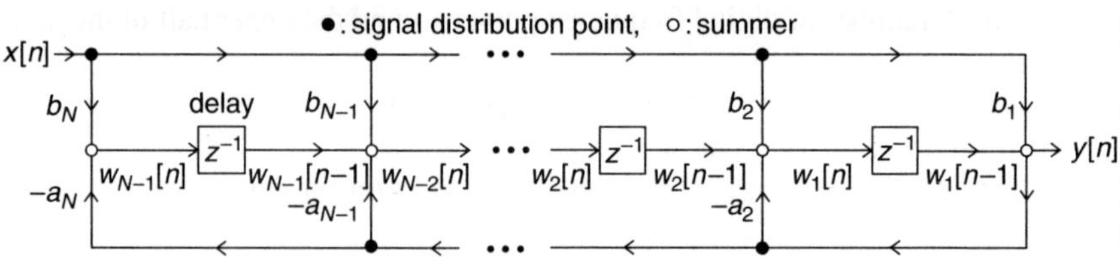

Fig. E.2 Implementation of an FIR filter in the transposed direct form II structure

The difference equation describing the input-output relationship of the filter

$$y[n] = -a_2y[n-1] - \cdots - a_{NA}y[n-NA+1] + b_1x[n] + \cdots + b_{NB}x[n-NB+1] \tag{E.5}$$

is implemented in the transposed direct form II structure (Fig. E.2) as

$$\begin{aligned}
y[n] &= w_1[n-1] + b_1x[n] \\
w_1[n] &= w_2[n-1] - a_2y[n] + b_2x[n] \\
&\cdots\cdots\cdots\cdots\cdots\cdots\cdots\cdots\cdots \\
w_{N-2}[n] &= w_{N-1}[n-1] - a_{N-1}y[n] + b_{N-1}x[n] \\
w_{N-1}[n] &= -a_Ny[n] + b_Nx[n]
\end{aligned}$$

where $N = \max(NA, NB)$ and the initial condition vector w0 of length $N-1$ can be determined from the past output/input by using `w0=filtic(B,A,yp,xp)`.

– When the input x is given as an $L \times M$ matrix, `filter()` operates on each column of the input matrix. If the initial state wi is nonzero, it must be given as an $(N-1) \times M$ matrix where $N = \max(NA, NB)$. The following program illustrates how to apply it to obtain the output(s) of a filter with some initial states sample by sample:

```
Ts=0.001; tt=[0:Ts:0.2]; x= [sin(2*pi*20*tt); cos(2*pi*50*tt)]; x=x.';
B=[1 1]; A=[1 1.5 0.5]; %numerator/denominator of filter transfer function
NB=length(B); NA=length(A); N=max(NA,NB)-1;
[Nx,M]=size(x); wi=zeros(NB-1,M); % input dimension and initial state
for n=1:Nx
   if n==1, [yn,wi]= filter(B,A,[zeros(1,M); x(n,:)],wi);
    else    [yn,wi]= filter(B,A,[x([n-1 n],:)],wi);
   end
   y(n,:) = yn(end,:);
end
```

```
function  [y,y0]=filterb(B, A, x, y0)
% Input:  x= [x(-NB+1),..,x(0),x(1),..,x(n)]: the past/future input
%         y0= [y(-NA+1),..,y(-1)]: the past output
% Output: y= [y(0),y(1),..,y(n)]: the output
%         y0= [y(n-NA+1),..,y(n)]: the last output
%          to be used for successive processing of the same filter
if  nargin<4,  y0=[];  end
NA= length(A);
A1=A;
for i=1:NA
  if A(i)==0, A1=A(i+1:NA); else break;  end
end
A=A1; NA= length(A); NB= length(B);  N= NA-1; % the order of filter
Ny0= length(y0);
if length(x)<NB,  x=[x zeros(1,NB-length(x))];  end
if Ny0<=N %the initial values of the output
  y0= [zeros(1,N-Ny0) y0];
 elseif Ny0>N
  y0= y0(Ny0-N+1:Ny0);
end
A1= A(NA:-1:2);
if A(1)~=1,  B= B/A(1); A1= A1/A(1);  end
for n= 1: length(x)-NB+1
   y(n)= B*x(n+NB-1:-1:n)';
   if NA>1,  y(n)= y(n)-A1*y0';  end
   y0= [y0(2:N) y(n)];
end
```

`Y=filter2(B,X)`

– This routine computes the output Y of the 2-D (two-dimensional) FIR filter described by matrix B where most of its work is done by using conv2.

E.13 Filter Design

E.13.1 Analog Filter Design

`[N,wc]=buttord(wp,ws,Rp,As,'s')`

- This routine selects the lowest order N and cutoff frequency wc of analog Butterworth filter that has the passband ripple<= Rp[dB] and stopband attenuation>= As[dB] for the passband edge frequency wp[rad/s] and stopband edge frequency ws[rad/s]. (Sect. 7.1)
- The cutoff frequency is the frequency at which the magnitude response is $1/\sqrt{2}$.
- Note that for BPF/BSF, the passband edge frequency wp and stopband edge frequency ws should be given as two-element vectors like [wp1 wp2] and [ws1 ws2].

`[B,A]=butter(N,wc,'s')`

- This routine designs an analog Butterworth filter, returning the numerator/denominator of system function of designed filter.
- `[B,A]=butter(N,wc,'s')` for an analog LPF of order N with cutoff frequency wc[rad/s]
- `butter(N,[wc1 wc2],'s')` for an analog BPF of order 2N with passband wc1 < w < wc2 [rad/s]
- `butter(N,[wc1 wc2],'stop','s')` for an analog BSF of order 2N with stopband wc1 < w < wc2 [rad/s]
- `butter(N,wc,'high','s')` for an analog HPF of order N with cutoff frequency wc[rad/s]
- Note that N and wc can be obtained from `[N,wc]=buttord(wp,ws,Rp,As,'s')`.

`[B,A]=cheby1(N,Rp,wpc,'s')`

- This routine designs an analog Chebyshev type I filter with passband ripple Rp[dB] and critical passband edge frequency wpc (Use Rp = 0.5 as a starting point, if not sure).
- Note that N and wpc can be obtained from `[N,wpc]=cheby1ord(wp,ws,Rp,As,'s')`.

`[B,A]=cheby2(N,As,wsc,'s')`

- This routine designs an analog Chebyshev type II filter with stopband attenuation As[dB] down and critical stopband edge frequency wsc (Use As = 20 as a starting point, if not sure).
- Note that N and wsc can be obtained from `[N,wsc]=cheby2ord(wp,ws,Rp,As,'s')`.

`[B,A]=ellip(N,Rp,As,wpc,'s')`

- This routine designs an analog Elliptic filter with passband ripple Rp, stopband attenuation As, and critical passband edge frequency wpc (Use Rp = 0.5[dB] & As = 20[dB], if unsure).
- Note that N and wpc can be obtained from `ellipord(wp,ws,Rp,As,'s')`.

E.13.2 Digital Filter Design – IIR (Infinite-duration Impulse Response) Filter

`[N,fc]=buttord(fp,fs,Rp,As)`

- This routine selects the lowest order N and normalized cutoff frequency fc of digital Butterworth filter that has passband ripple<= Rp[dB] and stopband attenuation>= As[dB] for passband edge frequency fp and stopband edge frequency fs.
- The cutoff frequency is the frequency at which the magnitude response is $1/\sqrt{2}$.
- As far as digital filter is concerned, every digital frequency $\Omega_i = \omega_i T = 2\pi f_i T$ [rad/sample] should be divided by π so that it can be normalized into the range [0, 1] where 1.0 corresponds to the digital frequency π [rad/sample].
- Note that for BPF/BSF, the passband edge frequency fp and stopband edge frequency fs should be given as two-dimensional vectors like [fp1 fp2] and [fs1 fs2].

`[B,A]=butter(N,fc)`

- This routine designs the system function of digital Butterworth filter.
- `[B,A]=butter(N,fc)`: digital LPF of order N with cutoff frequency fc
- `butter(N,[fc1 fc2])`: digital BPF of order 2N with passband fc1 < f < fc2
- `butter(N,[fc1 fc2],'stop')`: digital BSF of order 2N with stopband fc1 < f < fc2
- `butter(N,fc,'high')`: digital HPF of order N with cutoff frequency fc
- Note that N and fc can be obtained from `[N,fc]=buttord(fp,fs,Rp,As)`.

`[B,A]=cheby1(N,Rp,fpc,opt)`

- This routine designs a digital Chebyshev type I filter with passband ripple Rp[dB] and critical passband edge frequency fpc (Use Rp = 0.5 as a starting point, if not sure).
- Note that N and fpc can be obtained from `[N,fpc]=cheby1ord(fp,fs,Rp,As)`.

`[B,A]=cheby2(N,As,fsc,opt)`

- This routine designs a digital Chebyshev type II filter with stopband attenuation As[dB] down and critical stopband edge frequency fsc (Use As = 20 as a starting point, if not sure).
- Note that N and fsc can be obtained from `[N,fsc]=cheby2ord(fp,fs,Rp,As)`.

`[B,A]=ellip(N,Rp,As,fpc,opt)`

- This routine designs a digital elliptic filter with passband ripple Rp, stopband attenuation As, and critical passband edge frequency fpc (Use Rp = 0.5[dB] and As = 20[dB], if unsure).
- Note that N and fpc can be obtained from `ellipord(fp,fs,Rp,As)`.

E.13.3 Digital Filter Design – FIR (Finite-duration Impulse Response) Filter

`B=fir1(N,fc,opt)`

- This designs a linear-phase FIR filter by using the windowing method. (Sect. 7.2.2.1)
- The normalized cutoff frequency fc must be between 0 and 1, with 1.0 corresponding to the digital frequency π [rad/sample] or half the sampling rate.
- `B=fir1(N,fc)` designs an Nth order lowpass FIR filter with linear phase and returns the (real-valued) filter coefficients in a row vector B of length $N+1$ where the normalized gain of the filter at fc is -6 dB.
- `B=fir1(N,fc,'high')` designs an Nth order highpass FIR filter.
- If fc is a two-element vector such as fc = [fc1 fc2], `fir1(N,fc)` designs an Nth order BPF with passband fc1 < f < fc2, while `fir1(N,fc,'stop')` designs a bandstop filter.
- If fc is a multi-element vector such as fc = [fc1 fc2 fc3 fc4 … fcM], it designs an Nth order multi-band filter with bands 0 < f < fc1, fc1 < f < fc2, …, fcM < f < 1. Note that, depending on the value of a trailing input argument opt = "DC-1" or "DC-0", it makes the first

band a passband or stopband, that is, `B=fir1(N,fc,'DC-1')` and `B=fir1(N,fc,'DC-0')` make the first band a passband and a stopband, respectively.

- `B=fir1(N,fc,WND)` designs an Nth order FIR filter using the N + 1 length vector WND to window the impulse response. If empty or omitted (by default), fir1 uses a Hamming window of length N + 1. Other windows such as Boxcar, Hann, Bartlett, Blackman, Kaiser, and Chebwin can be specified with an optional trailing argument. For example, `fir1(N,fc,chebwin(N+1,R))` uses a Chebyshev window.
- For filters with a non-zero gain at fs/2 (half the sampling rate corresponding to the digital frequency π), e.g., HPF and BSF, N must be even. If N is an odd number for HPF or BSF, it will be incremented by one to make the filter order even and the window length will be N + 2.
- By default, the filter is scaled so that the center of the first passband has a unit magnitude after windowing. To prevent this scaling, use a trailing 'noscale' argument as illustrated by
  `fir1(N,fc,'high','noscale')` and
  `fir1(N,fc,WND,'noscale')`.

`B=fir2(N,F,A)`

- This designs a FIR filter by using the frequency sampling method. (See Example 7.4)
- `B=fir2(N,F,A)` designs an Nth order FIR filter with the frequency response specified by frequency vector F and amplitude vector A, and returns the filter coefficients in a row vector B of length N + 1.
- Vectors F and A specify the frequency and magnitude breakpoints such that plot(F,A) would show a plot of the desired frequency response, where the frequencies in F must be between $0.0 < F < 1.0$ (with 1.0 corresponding to half the sample frequency) and must also be in increasing order starting from 0.0 and ending with 1.0.
- The filter B is real, and has linear phase, consisting of symmetric coefficients obeying $B(k) = B(N + 2 - k)$ for $k = 1, 2, \ldots, N + 1$.
- By default, fir2 windows the impulse response with a Hamming window. Other windows can be specified with an optional trailing argument.
- For filters with a non-zero gain at fs/2 (half the sampling rate corresponding to the digital frequency π), e.g., HPF and BSF, N must be even. If N is an odd number for HPF or BSF, it will be incremented by one to make the filter order even and the window length will be N + 2.

`B=firpm (N, F, A)`

- This designs a Parks-McClellan optimal equiripple FIR filter with desired frequency response.

- `B=firpm (N, F, A)` returns a length $N+1$ linear phase (real, symmetric coefficients) FIR filter which has the best approximation to the desired frequency response described by F and A in the minimax sense that the maximum error is minimized.
- F is a vector of paired frequency band edges in ascending order between 0 and 1 (with 1.0 corresponding to half the sample frequency). A is a real vector (of the same size as F) specifying the desired amplitude of the frequency response, which is the line connecting the points (F(k),A(k)) and (F(k+1), A(k+1)) for odd k where the bands between F(k+1) and F(k+2) for odd k is treated as "transition bands" or "don't care" regions.
- For filters with a non-zero gain at fs/2 (half the sampling rate corresponding to the digital frequency π), e.g., HPF and BSF, N must be even. If N is an odd number for HPF or BSF, it will be incremented by one to make the filter order even. Alternatively, you can use a trailing 'h' flag to design a type 4 linear phase filter (Table 7.2) and avoid incrementing N.
- `B=firpm (N, F, A, W)` uses the weights in W to weight the error where W has one entry per band (so it is half the length of F and A) which tells fripm() how much emphasis to put on minimizing the error in each band relative to the other bands.
- `B=firpm (N, F, A, 'h')` and `B=firpm (N, F, A, W, 'Hilbert')` design filters having odd symmetry, that is, B(k) = −B(N + 2 − k) for k = 1, ..., N + 1. A special case is a Hilbert transformer having an approximate amplitude of 1 across the entire band, which will be designed by the following statement: `B=firpm (30, [.1 .9], [1 1], 'Hilbert')`.
- `B=firpm (N, F, A, 'd')` and `B=firpm (N, F, A, W, 'differentiator')` also design filters having odd symmetry, but with a special weighting scheme for non-zero amplitude bands. The weight is assumed to be equal to the inverse of frequency times the weight W. Thus the filter has a much better fit at low frequency than at high frequency. This designs FIR differentiators.
- `firpm()` normally designs symmetric (even) FIR filters while firpm(...,'h') and firpm(...,'d') design antisymmetric (odd) filters.

`B=firls (N, F, A)`

- This designs a linear-phase FIR filter using a least-squares error (LSE) method.
- Everything is the same with `firpm()`.

```
B=fircls (N, F, A, ub, lb)
```

- This designs a length N + 1 linear-phase FIR multi-band filter (with the frequency response amplitude A upperbounded/lowerbounded by ub/lb) using a constrained LSE.
- F is a vector of normalized transition frequencies in ascending order from 0 to 1 (corresponding to half the sampling rate), whose length is one plus the number of bands, i.e., length(A) + 1.
- A, ub, and lb are vectors of length(F)-1 piecewise constant desired frequency response amplitudes and their upper/lower bounds for each band.

(Ex) `B=fircls (10, [0 0.4 0.8 1], [0 1 0], [0.02 1.02 0.01], [-0.02 0.98 -0.01])`

```
B=fircls1 (N, fc, rp, rs, opt)
```

- This designs a length N + 1 linear-phase FIR LPF/HPF (with the cut-off frequency fc) using a constrained LSE method where rp/rs specify the passband/stopband ripple, respectively.
- With opt=`'high'`, it designs a HPF.

(Ex) `B=fircls1 (10, 0.7, 0.02, 0.01, 'high')`

(cf.) The GUI signal processing tool "sptool" manages a suite of four other GUIs: signal browser, filter designer, fvtool (for filter visualization), and spectrum viewer where "fdatool" is for filter design and analysis.

E.14 Filter Discretization

```
[Bz,Az]=impinvar (Bs, As, fs, tol)
```

- This uses the impulse-invariant method (Sect. 6.2.1) to convert a continuous-time system $G(s) = B(s)/A(s)$ into an equivalent discrete-time system $G[z] = B[z]/A[z]$ such that the impulse response is invariant except for being scaled by Ts = 1/fs.
- If you leave out the argument fs or specify it as the empty vector [], its default value 1 Hz will be taken.
- The fourth input argument tol specifies the tolerance to determine whether poles are repeated. A larger tolerance increases the likelihood that `impinvar()` interprets closely located poles as multiplicities (repeated ones). The default is 0.001, or 0.1% of a pole's magnitude. Note that the accuracy of the pole values is still limited to the accuracy obtainable by the `roots()` function.

```
[Bz,Az]=bilinear (Bs, As, fs, fp)
```

- This uses the *bilinear transformation* (BLT) method (Sect. 6.3.5) to convert a continuous-time system $G(s) = B(s)/A(s)$ into its *discrete equivalent* $G[z] = B[z]/A[z]$ such that the entire LHP on the s-plane maps into the unit circle on the z-plane in one-to-one correspondence.
- The numerator/denominator polynomial vector Bs and As of an analog filter are written in descending powers of s and the numerator/denominator polynomial vector Bz and Az of the discrete-time equivalent are written in descending powers of z (ascending powers of z^{-1}).
- The optional parameter fp[Hz] specifies a *prewarping match frequency* for which the frequency responses before and after BLT mapping match exactly.

```
sysd=c2d (sysc, Ts, method)
```

- This converts a continuous-time system (transfer) function sysc (that might have been created with either `tf` (transfer function), `zpk` (zero-pole-gain), or `ss` (state space)) into an equivalent discrete-time system function sysd with the sampling interval Ts[s] by using the discretization method among the following:

'zoh'	: Zero-order hold on the inputs (by default) - Sect. 6.2.2
'foh'	: Linear interpolation of inputs (triangle approximation)
'imp'	: Impulse-invariant discretization - Sect. 6.2.1
'tustin'	: Bilinear (Tustin) approximation - Sect. 6.3.5
'prewarp'	: Tustin approximation with frequency prewarping where the critical frequency Wc (in rad/s) is specified as fourth input, e.g., as sysd = c2d(sysc,Ts,'prewarp',Wc)
'matched'	: Matched pole-zero method (for SISO systems only) – Sect. 6.3.6

```
[Ad,Bd]=c2d (Ac, Bc, Ts)
```

- This converts a continuous-time state equation x'(t) = Ac*x(t) + Bc*u(t) into the zero-order-hold (z.o.h.) equivalent discrete-time state equation x[n+1] = Ad*x[n] + Bd*u[n] with the sampling interval Ts[s]. (Sect. 8.4.1)

```
sysc=d2c (sysd, method)
```

- This converts a discrete-time system function into an equivalent continuous-time system function.

E.15 Construction of Filters in Various Structures Using dfilt()

```
[SOS,Kc]=tf2sos(B,A); % Transfer function B(s)/A(s) to Cascade (Chap. 7)
Gd=dfilt.df1sos(SOS,Kc); % Cascade to Direct I form
Gd=dfilt.df1tsos(SOS,Kc); % Cascade to Direct I transposed form
Gd=dfilt.df2sos(SOS,Kc); % Cascade to Direct II form
Gd=dfilt.df2tsos(SOS,Kc); % Cascade to Direct II transposed form
Gd=dfilt.dffir(B); % FIR transfer function to Direct form
Gd=dfilt.dfsymfir(B); % FIR transfer function to Direct symmetric form
Gd=dfilt.dfasymfir(B); % FIR transfer function to Direct asymmetric form
Gd=dfilt.latticearma(r,p); % Lattice/Ladder coefficients to lattice ARMA
Gd=dfilt.latticeallpass(r); % Lattice coefficients to lattice Allpass
Gd=dfilt.latticear(r); % Lattice coefficients to lattice Allpole
Gd=dfilt.latticemamax(r); % Lattice coefficients to lattice MA max-phase
Gd=dfilt.latticemamin(r); % Lattice coefficients to lattice MA min-phase
[A,B,C,D]=tf2ss(B,A); G_ss=dfilt.statespace(A,B,C,D); % State space
G_cas= dfilt.cascade(Gd1,Gd2); % Cascade structure
G_par= dfilt.parallel(Gd1,Gd2); % Parallel structure
```

```
%sigApE_01.m: % To practice using dfilt()
Fs=5e4; T=1/Fs; % Sampling frequency and sampling period
ws1=2*pi*6e3; wp1=2*pi*1e4; wp2=2*pi*12e3; ws2=2*pi*15e3; Rp=2; As=25;
fp=[wp1 wp2]*T/pi; fs=[ws1 ws2]*T/pi; %Normalize edge freq into [0,1]
[N,fc]=cheb1ord(fp,fs,Rp,As) % Order & critical passband edge freq
[B,A]= cheby1(N,Rp,fc) % numerator/denominator of Chebyshev I BPF
fn=[0:511]/512; W=pi*fn;
 plot(fn,20*log10(abs(freqz(B,A,W))+eps)) % Frequency response
[SOS,Kc]= tf2sos(B,A)  % Cascade form realization
[BBc,AAc]= tf2cas(B,A) % Alternative
[BBp,AAp,Kp]= tf2par_z(B,A) % Parallel form realization: dir2par(B,A)
[r,p]= tf2latc(B,A) % Lattice/Ladder coefficients of lattice filter
G_df1sos= dfilt.df1sos(SOS,Kc); % Direct I form (Fig. 7.23(a))
 pause, plot(fn,20*log10(abs(freqz(G_df1sos,W))+eps),'r')
G_df1tsos=dfilt.df1tsos(SOS,Kc); % Direct I transposed form (Fig.7.23(b))
 pause, plot(fn,20*log10(abs(freqz(G_df1tsos,W))+eps))
G_df2sos= dfilt.df2sos(SOS,Kc); % Direct II form (Fig. 7.23(c))
 pause, plot(fn,20*log10(abs(freqz(G_df2sos,W))+eps),'r')
G_df2tsos=dfilt.df2tsos(SOS,Kc); %Direct II transposed form (Fig.7.23(d))
 pause, plot(fn,20*log10(abs(freqz(G_df2tsos,W))+eps))
G_latticeARMA= dfilt.latticearma(r,p); % Lattice ARMA (Fig.7.23(e))
 pause, plot(fn,20*log10(abs(freqz(G_latticeARMA,W))+eps),'r')
[A,B,C,D]=tf2ss(B,A); G_ss=dfilt.statespace(A,B,C,D); % State space
 pause, plot(fn,20*log10(abs(freqz(G_ss,W))+eps),'m')
G1=dfilt.df2tsos(BBc(1,:),AAc(1,:)); G2=dfilt.df2tsos(BBc(2,:),AAc(2,:));
G3=dfilt.df2tsos(BBc(3,:),AAc(3,:))
G_cascade= dfilt.cascade(G1,G2,G3); % Cascade form
plot(fn,20*log10(abs(freqz(G_cascade,W))+eps)), hold on
G1=dfilt.df2tsos(BBp(1,:),AAp(1,:)); G2=dfilt.df2tsos(BBp(2,:),AAp(2,:))
G3=dfilt.df2tsos(Kp,1);
G_parallel= dfilt.parallel(G1,G2,G3); % Parallel form
pause, plot(fn,20*log10(abs(freqz(G_parallel,W))+eps),'r')
G_latticeAR_allpass=dfilt.latticeallpass(r); %Lattice Allpass Fig.7.23(f)
G_latticeAR_allpole= dfilt.latticear(r); % Lattice Allpole Fig.7.23(f)
G_dffir= dfilt.dffir(B);
G_dfsymfir= dfilt.dfsymfir(B);   G_dfasymfir= dfilt.dfasymfir(B);
G_latticeMA_maxphase=dfilt.latticemamax(r); % MA max phase Fig.7.23(g)
G_latticeMA_minphase= dfilt.latticemamin(r); % MA min phase Fig.7.23(g)
```

```
function [BB,AA,K]=tf2par_s(B,A)
% Copyleft: Won Y. Yang, wyyang53@hanmail.net, CAU for academic use only
EPS= 1e-8;
B= B/A(1); A= A/A(1);
I= find(abs(B)>EPS); K= B(I(1)); B= B(I(1):end);
p= roots(A); p= cplxpair(p,EPS); Np= length(p);
NB= length(B); N= length(A); M= floor(Np/2);
for m=1:M
   m2= m*2; AA(m,:) = [1 -p(m2-1)-p(m2) p(m2-1)*p(m2)];
end
if Np>2*M
  AA(M+1,:)= [0 1 -p(Np)]; % For a single pole
end
M1= M+(Np>2*M); b= [zeros(1,Np-NB) B]; KM1= K/M1;
% In case B(s) and A(s) has the same degree, we let all the coefficients
%  of the 2^{nd}-order term in the numerator of each SOS be Bi1=1/M1:
if NB==N, b= b(2:end); end
for m=1:M1
   polynomial = 1; m2=2*m;
   for n=1:M1
      if n~=m, polynomial = conv(polynomial,AA(n,:)); end
   end
   if m<=M
     if M1>M, polynomial = polynomial(2:end); end
     if NB==N, b = b - [polynomial(2:end)*KM1 0 0]; end
     Ac(m2-1,:) = [polynomial 0];
     Ac(m2,:) = [0 polynomial];
    else
     if NB==N, b = b - [polynomial(2:end)*KM1 0]; end
     Ac(m2-1,:) = polynomial;
   end
end
Bc = b/Ac; Bc(find(abs(Bc)<EPS)) = 0;
for m=1:M1
   m2= 2*m;
   if m<=M
     BB(m,:) = [0 Bc(m2-1:m2)]; if NB==N, BB(m,1) = KM1; end
    else
     BB(m,:) = [0 0 Bc(end)]; if NB==N, BB(m,2) = KM1; end
   end
end
```

```
function [BB,AA,K]=tf2par_z(B,A)
% Copyleft: Won Y. Yang, wyyang53@hanmail.net, CAU for academic use only
if nargin<3, IR =0; end  % For default, inverse z-transform style
EPS= 1e-8; %1e-6;
B= B/A(1); A= A/A(1);
I= find(abs(B)>EPS); B= B(I(1):end);
if IR==0, [z,p,K]= residuez(B,A);  else  [z,p,K]= residue(B,A);  end
m=1;  Np=length(p); N=ceil(Np/2);
for i=1:N
   if abs(imag(p(m)))<EPS % Real pole
     if m+1<=Np & abs(imag(p(m+1)))<EPS % Subsequent real pole
       if abs(p(m)-p(m+1))<EPS % Double pole
         BB(i,:)= [z(m)+z(m+1) -z(m)*p(m) 0];
         AA(i,:)= [1 -2*p(m) p(m)^2];  m=m+2;
        elseif m+2<=Np&abs(p(m+1)-p(m+2))<EPS %Next two poles are double
```

```
        BB(i,:)=[0 z(m) 0]; AA(i,:)=[0 1 -p(m)]; m=m+1; % Single pole
       else
        BB(i,:)= [real([z(m)+z(m+1) -z(m)*p(m+1)-z(m+1)*p(m)]) 0];
        AA(i,:)= [1 real([-p(m)-p(m+1) p(m)*p(m+1)])];  m=m+2;
      end
     else
      BB(i,:)=[0  z(m) 0]; AA(i,:)=[0  1 -p(m)]; m=m+1; % Single pole
     end
    else % Two distinct real poles or Complex pole
     BB(i,:)= [real([z(m)+z(m+1) -z(m)*p(m+1)-z(m+1)*p(m)]) 0];
     AA(i,:)= [1 real([-p(m)-p(m+1) p(m)*p(m+1)])];  m=m+2;
   end
end
if IR~=0, BB(:,2:3)= BB(:,1:2); end
```

```
function y=filter_cas(B,A,x)
y=x;
[Nsection,NB]=size(B);
for k=1:Nsection
   Bk= B(k,:);  Ak= A(k,:);
   if abs(B(k,1))+abs(A(k,1))<1e-10
     Bk= B(k,2:3); Ak= A(k,2:3);
   end
   %if B(k,3)==0&A(k,3)==0, Bk=B(k,1:2); Ak=A(k,1:2); end
   %if B(k,2)==0&A(k,2)==0, Bk=B(k,1); Ak=A(k,1); end
   %if Bk(1)==0; Bk=Bk(2:length(Bk)); end
   y=filterb(Bk,Ak,[zeros(1,length(Bk)-1) y]);
end
```

```
function y=filter_par(B,A,x)
[Nsection,NB]=size(B);
y= zeros(1,length(x));
for k=1:Nsection
   Bk= B(k,:); Ak= A(k,:);
   while length(Bk)>1&abs(Bk(1))<eps, Bk=Bk(2:end); end
   while length(Ak)>1&abs(Ak(1))<eps, Ak=Ak(2:end); end
   if sum(abs(Bk))>eps
     y=y+filterb(Bk,Ak,[zeros(1,length(Bk)-1) x]);
   end
end
```

```
function  [r,p]=tf2latc_my(B,A)
if nargin>1&length(A)>1 %Recursive Lattice Filter
  % IIR System Function to Lattice Filter
  % ****************************************************
  %         B(1)+B(2)*z^-1 +B(3)*z^-2 +.....+B(NB)*z^(-NB+1)
  %  G[z] = ---------------------------------------------
  %         A(1)+A(2)*z^-1 +A(3)*z^-2 +......+A(NA)*z^(-NA+1)
  % ****************************************************
  N= length(A);
  AA= A;
  for k=1:N-1
    if abs(AA(k))<.0000001,  A= AA(k+1: N);
     else  break;
    end
  end
  N= length(A);
```

```
  if N<=1, error('LATTICED: length of polynomial is too short!'); end
  BB= B;
  NB= length(B);
  for k=1:NB-1
    if abs(BB(k))<.0000001,  B= BB(k+1: NB);
     else  break;
    end
  end
  if length(B) ~= N
    error('tf2latc_my: lengths of polynomials B and A  do not agree!');
  end
  S= B/A(1); V= A/A(1);
  for  i=N:-1:2
    p(i)= S(i);     ri1= V(i);
    W(1:i)= V(i:-1:1);
    if  abs(ri1)>=.99999
      error('tf2latc_my: ri1= V(i) is too large to maintain stability!');
    end
    V(1:i)= (V(1:i)-ri1*W(1:i))/(1-ri1*ri1);
    r(i-1)= ri1;
    S(1:i)= S(1:i) -p(i)*W(1:i);
  end
  p(1)= S(1);
  if nargout==0
    fprintf('\n\t  Recursive Lattice Filter Coefficients\n');
    for i=1:length(r),  fprintf('   r(%1d)=%7.4f', i, r(i));  end
    fprintf('\n');
    for i=1:length(p),  fprintf('   p(%1d)=%7.4f', i, p(i));  end
    fprintf('\n');
  end
else %Nonrecursive Lattice Filter
  % FIR System Function --> Nonrecursive Lattice-II Filter
  % ************************************************
  %  G[z]= B(1)+B(2)*z^-1 +B(3)*z^-2 +.....+B(NB)*z^(-NB+1)
  % ************************************************
  N= length(B);
  BB= B;
  for k=1:N-1
    if abs(BB(k))<.0000001,  B= BB(k+1: N);
     else  break;
    end
  end
  N= length(B);
  if N<=1, error('tf2latc_my: length of polynomial is too short!'); end
  V= B/B(1);
  for  i=N:-1:2
    ri1= V(i);
    W(1:i)= V(i:-1:1);
    if abs(abs(ri1)-1)<.001 %Nonrecursive Lattice cannot be unstable
      ri1 =ri1/abs(ri1)*.99;
    end
    V(1:i)= (V(1:i)-ri1*W(1:i))/(1-ri1*ri1);
    r(i-1)= ri1;
  end
  if nargout==0
    fprintf('\n\t  Nonrecursive Lattice Filter Coefficients\n');
    for i=1:length(r),  fprintf('   r(%1d)=%7.4f', i, r(i));  end
```

```
    fprintf('\n');
  end
end
```

```
function [y,w]=filter_latc_r(r,p,x,w)
%hopefully equivalent to latcfilt() inside MATLAB
%w contains the past history of internal state ..,w(n-1), w(n)
%1-step Lattice filtering the input x to yield the output y and update w
N= length(r); % the order of the lattice filter
if  length(p)~= N+1
   error('LATTICEF: length(p) must equal  length(r)+1!');
end
if  nargin<4,  w=[];  end
if  length(w)<N
   w=[zeros(1,N-length(w)) w];
end
for n=1:length(x)
  vi=x(n); %Current Input
  for i=N:-1:1
     vi= vi -r(i)*w(i);  w(i+1)= w(i) +r(i)*vi;
  end
  w(1)= vi;
  y(n)= p(:).'*w(:);
end
```

```
function [y,w]=filter_latc_nr(r,x,w)
% hopefully equivalent to latcfilt() inside MATLAB
% w contains the past history of internal state...,w(n-1), w(n)
%1-step Lattice filtering the input x to yield the output y and update w
N= length(r); % the order of the lattice filter
if  nargin<3,  w=[];  end
if  length(w)<N
   w=[zeros(1,N-length(w)) w];
end
for n=1:length(x)
  vi= x(n)+r*w(1:N)';
  y(n)= vi;
  for i=N:-1:1
     vi= vi -r(i)*w(i);  w(i+1)= w(i) +r(i)*vi;
  end
  w(1)= x(n);
end
```

E.16 System Identification from Impulse/Frequency Response

```
[B,A]=prony (g, NB, NA)
```

- This identifies the system function

$$G[z] = \frac{B[z]}{A[z]} = \frac{b_1 + b_2 z^{-1} + \cdots + b_{NB+1} z^{-NB}}{a_1 + a_2 z^{-1} + \cdots + a_{NA+1} z^{-NA}} \tag{E.6}$$

of a discrete-time system from its impulse response $g[n]$.

- The outputs B and A are numerator and denominator polynomial coefficient vectors of length NB + 1 and NA + 1, respectively.

`[B,A]=stmcb (g, NB, NA, Niter, Ai)` or `stmcb (y, x, NB, NA, Niter, Ai)`

- This identifies the system function $G[z] = B[z]/A[z]$ of a discrete-time system from its impulse response $g[n]$ or its output $y[n]$ and input $x[n]$ optionally with Niter (5 by default) iterations and initial estimate of the denominator coefficients Ai where y and x are of the same length.

`G=tfe (x, y, N, fs, WND, Noverlap)`

- This estimates the transfer function as frequency response $G[\Omega] = Y[\Omega]/X[\Omega]$ of a discrete-time system from its input $x[n]$ and output $y[n]$ optionally with N-point DFT (N = 256 by default), sampling frequency fs (2Hz by default), windowing sequence WND, and with each section overlapped by Noverlap samples.

`[B,A]=invfreqs (G, w, NB, NA, wt, Niter, tol)`

- This identifies the system function

$$G(s) = \frac{B(s)}{A(s)} = \frac{b_1 s^{NB} + b_2 s^{NB-1} + \cdots + b_{NB+1}}{a_1 s^{NA} + a_2 s^{NA-1} + \cdots + a_{NA+1}} \tag{E.7}$$

of a continuous-time system from its frequency response G(w)($G(\omega)$) specified for an analog frequency vector w(ω) optionally with a weighting vector wt (of the same length as w), within Niter iterations, and with fitting error tolerance tol.

`[B,A]=invfreqz (G, W, NB, NA, wt, Niter, tol)`

- This identifies the system function $G[z] = B[z]/A[z]$ of a discrete-time system from its frequency response G(W)($G(\Omega) = G[e^{j\Omega}]$) specified for a digital frequency vector W(Ω)($0 \le \Omega \le \pi$) optionally with a weighting vector wt (of the same length as W), within Niter iterations, and with fitting error tolerance tol.

E.17 Partial Fraction Expansion and (Inverse) Laplace/z-Transform

`[r,p,k]=residue (B,A)` or `[B,A]=residue (r, p, k)`

– This converts a rational function $B(s)/A(s)$, i.e., a ratio of numerator/denominator polynomials to partial fraction expansion form (or pole-residue representation), and back again.

$$\frac{B(s)}{A(s)} = \frac{b_1 s^{NB-1} + b_2 s^{NB-2} + \cdots + b_{NB}}{a_1 s^{N-1} + a_2 s^{N-2} + \cdots + a_N}$$
$$\rightleftharpoons \left(\sum_{n=1}^{N-1-L} \frac{r_n}{s - p_n} \right) + \frac{r_{N-L}}{s-p} + \cdots + \frac{r_{N-1}}{(s-p)^L} + k(s) \quad \text{(E.8)}$$

where the numerator vector r and pole vector p are column vectors and the quotient polynomial coefficient vector k is a row vector.

`[r,p,k]=residuez (B,A)` or `[B,A]=residuez (r,p,k)`

– This converts a rational function $B[z]/A[z]$, i.e., a ratio of numerator/denominator polynomials to partial fraction expansion form (or pole-residue representation), and back again.

$$\frac{B[z]}{A[z]} = \frac{b_1 + b_2 z^{-1} + \cdots + b_{NB} z^{-(NB-1)}}{a_1 + a_2 z^{-1} + \cdots + a_N z^{-(N-1)}}$$
$$\rightleftharpoons \left(\sum_{n=1}^{N-1-L} \frac{r_n z}{z - p} \right) + \frac{r_{N-L} z}{z-p} + \cdots + \frac{r_{N-1} z^L}{(z-p)^L} + k(z) \quad \text{(E.9)}$$

where the numerator vector r and pole vector p are column vectors and the quotient polynomial coefficient vector k is a row vector.

`Xs=laplace (xt)` and `xt=ilaplace (Xs)`

– These MATLAB commands compute the symbolic expressions for the Laplace transform $X(s) = \mathcal{L}\{x(t)\}$ of a symbolic function in t and the inverse Laplace transform $x(t) = \mathcal{L}^{-1}\{X(s)\}$ of a symbolic function in s, respectively.

`Xz=ztrans (xn)` and `xn=iztrans (Xz)`

– These MATLAB commands compute the symbolic expressions for the z -transform $X[z] = \mathcal{Z}\{x[n]\}$ of a symbolic function in n and the inverse z -transform $x[n] = \mathcal{Z}^{-1}\{X[z]\}$ of a symbolic function in z, respectively.

```
function x=ilaplace_my(B,A)
% To find the inverse Laplace transform of B(s)/A(s) using residue()
% Copyleft: Won Y. Yang, wyyang53@hanmail.net, CAU for academic use only
if ~isnumeric(B)
   [B,A]=numden(simple(B)); B=sym2poly(B); A=sym2poly(A);
end
[r,p,k]= residue(B,A); EPS = 1e-4;
N= length(r); x=[]; n=1;
while n<=N
   if n>1,  x = [x ' + '];  end
   if n<N & abs(imag(p(n)))>EPS & abs(sum(imag(p([n n+1]))))<EPS
     sigma=real(p(n)); w=imag(p(n)); Kc=2*real(r(n)); Ks=-2*imag(r(n));
     sigma_=num2str(sigma); w_=num2str(w); Kc_=num2str(Kc); Ks_=num2str(Ks);
     if abs(sigma)>EPS
       x = [x 'exp(' sigma_ '*t).*'];
       if abs(Kc)>EPS&abs(Ks)>EPS
         x = [x '(' Kc_ '*cos(' w_ '*t) + ' Ks_ '*sin(' w_ '*t))'];
        elseif abs(Kc)>EPS, x = [x  Kc_ '*cos(' w_ '*t)'];
        else   x = [x  Ks_ '*sin(' w_ '*t)'];
       end
     end
     n = n+2;
    elseif n<=N & abs(imag(r(n)))<EPS
     if abs(p(n))>EPS,   x = [x  num2str(r(n)) '*exp(' num2str(p(n)) '*t)'];
      else    x = [x  num2str(r(n))];
     end
     n = n+1;
   end
end
if ~isempty(k), x = [x ' + ' num2str(k(end)) '*dirac(t)']; end
```

E.18 Decimation, Interpolation, and Resampling

`y=decimate (x, M, N, 'fir')`

- This process filters the input sequence x with an LPF and then resamples the resulting smoothed signal at a lower rate, which is the opposite of interpolation.
- `y=decimate (x, M)` reduces the sample rate of x by a factor M to produce a decimated sequence y that is M times shorter than x. By default, it employs an eighth-order lowpass Chebyshev type I filter. Note that x must be longer than 3 times the filter order.
- `y=decimate (x, M, N)` uses an Nth order Chebyshev filter.
- `y=decimate (x, M, 'fir')` uses a 30-point FIR filter.
- `y=decimate (x, M, N, 'fir')` uses an N-point FIR filter.

`y=interp (x, M, L, fc)`

- This inserts zeros into the input sequence x and then apply an LPF.
- `y=interp (x,M)` increases the sampling rate of x by a factor of M to produce an interpolated sequence y that is M times longer than x.
- `y=interp(x,M,L,fc)` specifies L (= 4 by default) for filter length and fc (= 0.5 by default) for normalized cut-off frequency.
- `[y,B]=interp(x,M,L,fc)` returns a vector B of length 2*L*M + 1 containing the filter coefficients used for the interpolation.

`y=upfirdn(x,B,Mi,Md)`

- This can be used to change the sampling rate by a rational factor.
- It upsamples x by Mi (i.e., inserts (Mi-1) zeros between samples), applies the FIR filter B, and downsamples the output by Md (i.e., throws away (Md-1) samples between samples).
- The default values of Mi and Md are 1.

`y=resample(x,Mi,Md,B)`

- This can be used to change the sampling rate by a rational factor.
- `y=resample(x,Mi,Md)` resamples x at Mi/Md times the original sampling rate, using a polyphase filter implementation where Mi and Md must be positive integers. The length of y will be ceil(length(x)*Mi/Md). If x is a matrix, `resample()` works down the columns of x.
- It applies an anti-aliasing (lowpass) FIR filter (designed using firls() with a Kaiser window) to x during the resampling process.
- `y=resample(x,Mi,Md,B)` filters x using a FIR filter with coefficients B.
- `y=resample(x,Mi,Md,N)` uses a weighted sum of 2*N* max (1, Md/Mi) samples of x to compute each sample of y. Since the length of the FIR filter used by resample() is proportional to N(= 10 by default), you can increase N to get better accuracy at the expense of a longer computation time. If you let N = 0, a nearest neighbor interpolation is performed so that the output y(n) is x(round((n − 1)*Md/Mi) + 1) where y(n) = 0 if round((n − 1)*Md/Mi) + 1 > length(x)).
- `y=resample(x,Mi,Md,N,beta)` uses beta as the design parameter for the Kaiser window used to design the filter where its default value is 5.
- `[y,B]=resample(x,Mi,Md)` returns the filter coefficient vector B applied to x during the resampling process (after upsampling).
- If x is a matrix, the routine resamples the columns of x.

`y=detrend(x)`

- `y=detrend(x)` removes the best straight-line fit from vector x.
- `y=detrend(x,0)` removes the mean value from vector x.

E.19 Waveform Generation

`chirp(t,f0,t1,f1,'type')`	swept-frequency cosine generator
`pulstran(t,D,pulse)`	pulse train generator
`rectpuls(t,D)`	sampled aperiodic rectangular pulse generator
`square(t,D), sawtooth(t,D)`	square, sawtooth/triangular wave generator
`tripuls(t,D,skew)`	sampled aperiodic triangular pulse generator

E.20 Input/Output through File

```
% input_output_data.m
clear
x=[1 2]; y=[3 4 5];
save sig x y % save x and y in a MATLAB data file 'sig.mat'
clear('y')
display('After y has been cleared, does y exist?')
if (exist('y')~=0), disp('Yes'), y
 else  disp('No')
end
load sig y % read y from the MATLAB data file 'sig.mat'
disp('After y has been loaded the file sig.mat, does y exist?')
if isempty('y'), disp('No'), else disp('Yes'), y,  end
fprintf('x(2)=%5.2f \n', x(2))
save y.dat y /ascii % save y into the ASCII data file 'y.dat'
% The name of the ASCII data file must be the same as the variable name.
load y.dat % read y from the ASCII data file 'y.dat'
str='prod(y)'; % ready to compute the produce of the elements of y
eval(str) % evaluate the string expression 'prod(y)'
```

Appendix F
Simulink®

According to the MATLAB documentation [W-5], Simulink® is software for modeling, simulating, and analyzing dynamic systems. It supports linear and nonlinear systems, modeled in continuous time, sampled time, or a hybrid of the two. Systems can be multirate, i.e., have different parts that are sampled or updated at different rates.

Simulink® provides a graphical user interface (GUI) for building models as block diagrams, using click-and-drag mouse operations. Simulink includes a comprehensive block library of sinks, sources, linear and nonlinear components, and connectors. Models are hierarchical, so you can build models using both top-down and bottom-up approaches. You can view the system at a high level, then double-click blocks to go down through the levels to see increasing levels of model detail. This approach provides insight into how a model is organized and how its parts interact.

After you define a model, you can simulate it either from the Simulink menus or by entering commands in the MATLAB® Command Window. The menus are convenient for interactive work, while the command line is useful for running a batch of simulations with a parameter swept across a range of values. Using scopes and other display blocks, you can see the simulation results while the simulation runs. The simulation results can be put in the MATLAB workspace for postprocessing and visualization. MATLAB® and Simulink® are integrated so that you can simulate, analyze, and revise your models in either environment at any point.

To start Simulink, you must first start MATLAB. Consult your MATLAB documentation for more information. You can then start Simulink in two ways:

- Click the Simulink icon on the MATLAB toolbar.
- Type "simulink" (without the quotation marks) at the MATLAB prompt (like ">>") in the MATLAB Command Window or use the "sim()" command inside a MATLAB program.

On Microsoft Windows platforms, starting Simulink displays the Simulink Library Browser as depicted in Fig. F.1. The Library Browser displays a tree-structured view of the Simulink block libraries installed on your system, from which you can

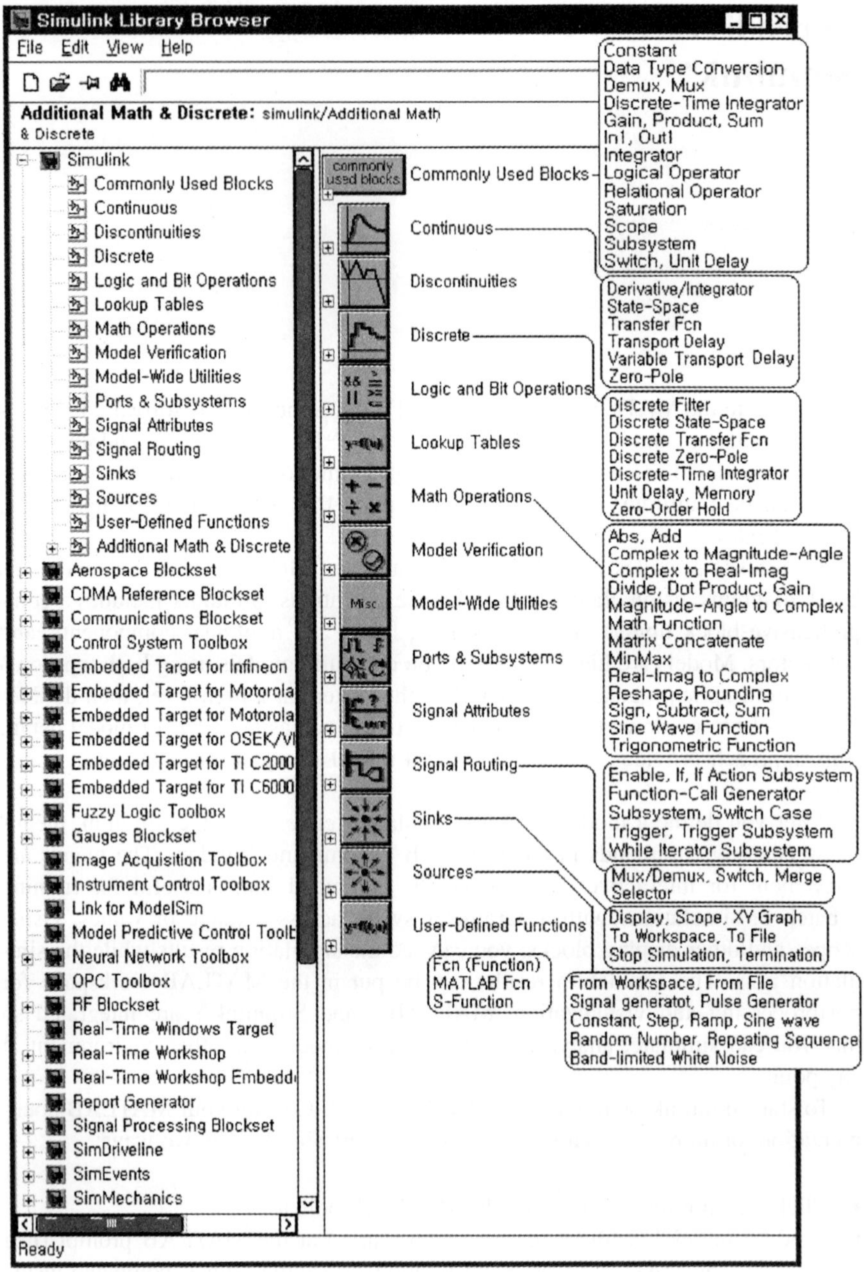

Fig. F.1 Simulink Library Browser window

copy and move (click/drag) the blocks into a model window and build models. The procedure of creating or editing a Simulink model is as follows:

- A new empty model file (window) can be created by selecting File/New menu or clicking "Create a new model" button on the library browser's toolbar. An existing one can be opened by selecting File/Open menu or clicking the Open button on the library browser's toolbar and then choosing/entering the file name for the model to edit.
- Copy and move (click/drag) the blocks you want into the model window.
- Connect the blocks by clicking at an (input/output) point of a block and drag to an (output/input) point of another block.
- To draw a branch line, position the pointer at a point on the existing line (wire), press and hold down the CTRL key, press the mouse button, and then drag the pointer to another point.

To simulate the created/edited Simulink model, select the Simulation/ Configuration_Parameters menu (in the model window) to open the Configuration Parameters window in which you can set the simulation parameters including the start/final(stop) times, and then press the CTRL + T key (on the keyboard) or select the Simulation/Start menu on the toolbar of the model window. To see the simulation results, double-click the Scope block.

Figures F.2 and F.3 show the Signal Processing and Communication Blocksets, which enables you to design and prototype various signal processing and communication systems using key signal processing algorithms and components in the Simulink® block format.

(cf.) If you see an error message that a Simulink model file cannot be saved because of character encoding problem, keep the erroneous part of the model file in mind, click OK to close the error message dialog block, use a text editor to modify the erroneous part in the designated file named "***.err", and then save it with the extension name "mdl" instead of "err" so that it can be opened in another Simulink session.

Signal Processing Blockset

- **Estimation**
  - Linear Prediction: Autocorrelation LPC, Levinson-Durbin, LPC to LSF/LSP Conversion
    LSF/LSP to LPC Conversion, LPC to/from RC, Autocorrelation, Cepstral Coefficients
  - Parametric Estimation: Burg AR Estimator, Covariance AR Estimator, Yule-Walker AR Estimator
  - Power Spectrum Estimation: Burg Method, Covariance Method; Magnitude FFT, Periodogram,
    Yule-Walker Method
- **Filtering**
  - Adaptive Filters: Block LMS Filter, Kalman Adaptive Filter, LMS Filter, RLS Filter
  - Filtering Designs: Analog Filter Design, Digital Filter, Digital Filter Design, Filter Realization Wizard
  - Multirate Filters: CIC Decimation/Interpolation, Dyadic Analysis/Synthesis Filter Bank
    FIR Decimation/Interpolation, FIR Rate Conversion
    Two-Channel Analysis/Synthesis Subband Filter
- **Math Functions**
  - Math Operations: Complex Exponential, Cumulative Product/Sum
    Difference, Normalization, dB Conversion, dB Gain
  - Matrices and LInear Algebra:
    - Linear System Solvers: Backward/Forward Substitution, Levinson-Durbin
      Cholesky/LDL/LU/QR/SVD Solver
    - Matrix Factorizations: Cholesky/LDL/LU/QR/SVD Factorization
    - Matrix Inverses: Cholesky/LDL/LU/Pseudo Inverse
    - Matrix Operations: Identity Matrix, Constant Diagonal Matrix, Create Diagonal Matrix
      Extract Diagonal/Triangular, Overwrite Values, Reciprocal Condition
      Matrix Concatenation/Product/Scaling/Square/Sum/Multiply/Transpose
      Matrix 1-Norm, Permute Matrix, Submatrix, Toeplitz
  - Polynomial Functions: Least Squares Polynomial Fit, Polynomial Evaluation
    Polynomial Stability Test
- **Platform Specific I/Q**
  - Windows (WIN32): From Wave Device/File, To Wave Device/File
- **Quantizers** : Quantizer, Uniform Encoder/Decoder
- **Signal Management**
  - Buffers: Buffer, Delay Line, Triggered Delay Line, Queue, Stack, Unbuffer
  - Indexing: Selector, Multiport Selector, Variable Selector, Flip, Submatrix, Overwrite Values
  - Signal Attributes: Check Signal Attributes, Convert 1-D to 2-D, Convert 2-D to 1-D
    Date Type Conversion, Frame Status Conversion, Inherity Complexity
  - Switches and Counters: Counter, Edge Detector, Event-Count Comparator
    Multiphase Clock, N-Sample Enable, N-Sample Switch
- **Signal Operation** Convolution, Downsample, Upsample, Interpolation, Unwrap, Window Function
  Pad, Zero-Pad, Repeat, S/H, Integer Delay, Variable Integer/Fractional Delay
- **Statistics** : Autocorrelation, Correlation, Detrend, Histogram, Maximum, Minimum, Mean, Median,
  RMS, Standard Deviation, Variance, Sort
- **Transforms** : Analytic Signal, Real/Complex Cepstrum
  DCT, DWT, FFT, IDCT, IDWT, IFFT, Magnitude FFT
- **DSP Sinks** : Display, Matrix Viewer, Spectrum Scope, Time Scope, Vector Scope
  Signal To Workspace, Triggered To Workspace
- **DSP Sources** : Chirp, Constant Ramp, Discrete Impulse, DSP Constant, Sine Wave, Random Source
  Identity Matrix, Constant Diagonal Matrix, Multiphase Clock, N-Sample Enable
  Signal From Workspace, Triggered Signal From Workspace

Fig. F.2 Signal Processing Blockset available in the Simulink Library Browser window

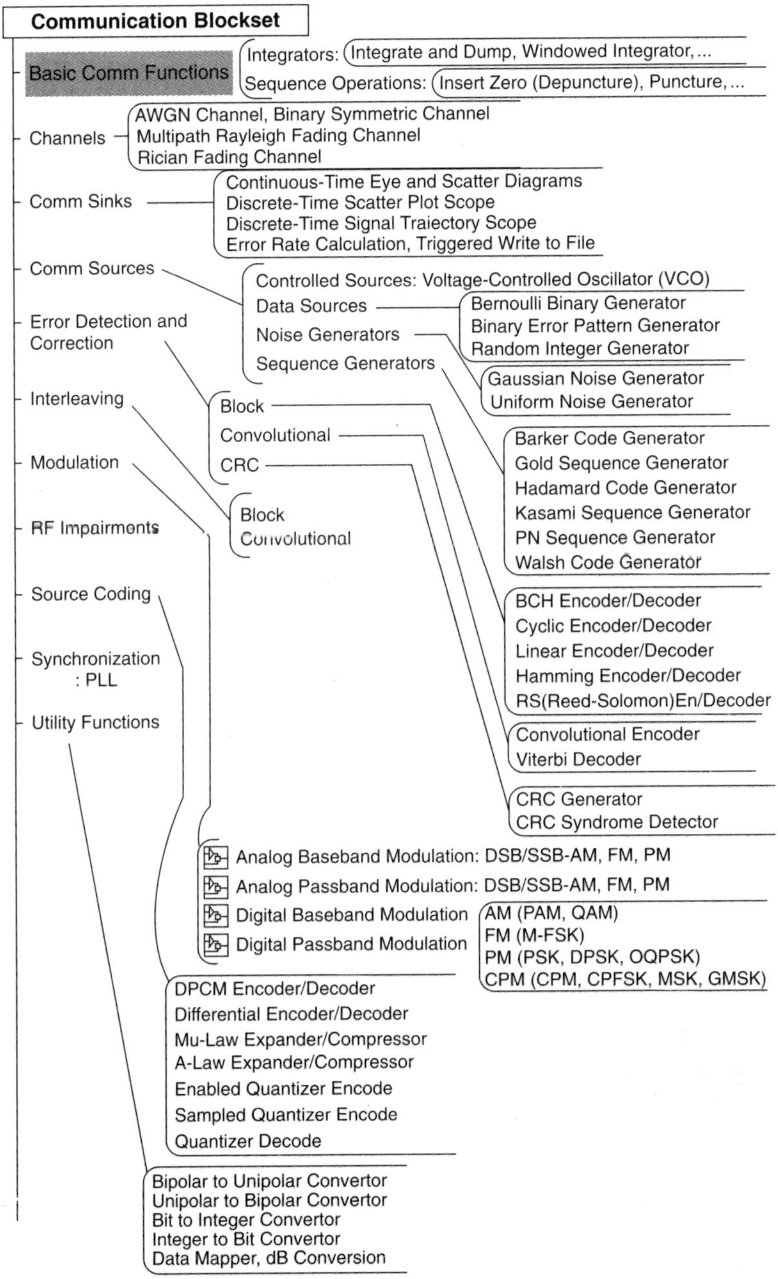

Fig. F.3 Communication Blockset available in the Simulink Library Browser window

References

[D-1] Denbigh, P., *System Analysis and Signal Processing: with emphasis on the use of MATLAB*, Prentice Hall, Inc., Englewood Cliff, N.J., 1998.

[F-1] Franklin, G. F., J. D. Powell, and M. L. Workman, *Digital Control of Dynamic Systems*, 2nd ed., Addison-Welsey Publishing Company, New York, 1990.

[G-1] Gopal, M., *Digital Control Engineering*, John Wiley & Sons, Singapore, 1988.

[G-2] Goertzel, G., "An Algorithm for the Evaluation of Finite Trigonometric Series," Amer. Math. Monthly, Vol. 65, Jan. 1958, pp. 34–35.

[J-1] Jaeger, R. C., "Tutorial: Analog data acquisition technology, part I; digital-to-analog conversion", IEEE MICRO, Vol. 2, No 3, pp. 20–37, 1982a.

[J-2] Jaeger, R. C., "Tutorial: Analog data acquisition technology, part II; analog-to-digital conversion", IEEE MICRO, Vol. 2, No 3, pp. 46–56, 1982b.

[K-1] Kreyszig, E., *Advanced Engineering Mathematics*, John Wiley & Sons, Inc., New York, 1983.

[K-2] Kuc, R., *Introduction to Digital Signal Processing*, McGraw-Hill Book Company, New York, 1988.

[L-1] Ludeman, L. C., *Fundamentals of Digital Signal Processing*, John Wiley & Sons, Inc., New York, 1987.

[O-1] Oppenheim, A. V., A. S. Willsky, and I. T. Young, *Signals and Systems*, Prentice Hall, Inc., Englewood Cliff, NJ, 1983.

[O-2] Oppenheim, A. V. and R. W. Schafer, *Digital Signal Processing*, Prentice Hall, Inc., Englewood Cliff, NJ, 1975.

[P-1] Phillips, C. L. and H. T. Nagle, *Digital Control System Analysis and Design*, 2nd ed., Prentice Hall, Inc., Englewood Cliff, NJ, 1989.

[S-1] Soliman, S. S. and M. D. Srinath, *Continuous and Discrete Signals and Systems*, Prentice Hall, Inc., Englewood Cliff NJ, 1999.

[S-2] Stearns, S. D. and D. R. Hush, *Digital Signal Analysis*, 2nd ed., Prentice Hall, Inc., Englewood Cliff, NJ, 1990.

[W-1] Web site <http://en.wikipedia.org/wiki/Main_Page> (Wikipedia: The Free Encyclopedia)

[W-2] Website <http://www.allaboutcircuits.com/vol_4/chpt_13/> (analog-to-digital converter)

[W-3] Website <http://www.mathworks.com/>

[W-4] Website <http://www.mathworks.com/access/helpdesk/help/pdf_doc/signal/signal_tb.pdf>

[W-5] Website <http://www.mathworks.com/access/helpdesk/help/toolbox/simulink/>

[W-6] Website <http://mathworld.wolfram.com/>

[Y-1] Yang, W. Y., W. Cao, T.-S. Chung, and J. Morris, *Applied Numerical Methods Using MATLAB*, John Wiley & Sons, Inc., Hoboken NJ, 2005.

[Y-2] Yang, W. Y. and S. C. Lee, *Circuit Systems with MATLAB and PSpice*, John Wiley & Sons, Inc., Hoboken, NJ, 2007.

Index

Index for MATLAB routines

Index for Remarks